T0143069

Lecture Notes
in Computational Science
and Engineering

116

Editors:

Timothy J. Barth
Michael Griebel
David E. Keyes
Risto M. Nieminen
Dirk Roose
Tamar Schlick

More information about this series at http://www.springer.com/series/3527

Chang-Ock Lee • Xiao-Chuan Cai •
David E. Keyes • Hyea Hyun Kim • Axel Klawonn •
Eun-Jae Park • Olof B. Widlund

Editors

Domain Decomposition Methods in Science and Engineering XXIII

 Springer

Editors

Chang-Ock Lee
Dept. of Mathematical Sciences
KAIST
Daejeon, Republic of Korea

Xiao-Chuan Cai
Dept. of Computer Science
University of Colorado
Boulder, CO, USA

David E. Keyes
Applied Math & Computational Science
KAUST
Thuwal, Saudi Arabia

Hyea Hyun Kim
Dept. of Applied Mathematics
Kyung Hee University
Yongin, Republic of Korea

Axel Klawonn
Mathematisches Institut
Universität zu Köln
Köln, Germany

Eun-Jae Park
Dept. of Computational Science &
  Engineering
Yonsei University
Seoul, Republic of Korea

Olof B. Widlund
Courant Institute of Mathematical Sciences
New York University
New York, NY, USA

ISSN 1439-7358   ISSN 2197-7100   (electronic)
Lecture Notes in Computational Science and Engineering
ISBN 978-3-319-84894-5   ISBN 978-3-319-52389-7   (eBook)
DOI 10.1007/978-3-319-52389-7

Mathematics Subject Classification (2010): 65F10, 65N30, 65N55

Printed on acid-free paper

This Springer imprint is published by Springer Nature
The registered company is Springer International Publishing AG
The registered company address is: Gewerbestrasse 11, 6330 Cham, Switzerland

Preface of DD23 Book of Proceedings

The proceedings of the 23rd International Conference on Domain Decomposition Methods contain developments up to 2015 in various aspects of domain decomposition methods bringing together mathematicians, computational scientists, and engineers who are working on numerical analysis, scientific computing, and computational science with industrial applications. The conference was held on Jeju Island, Korea, July 6–10, 2015.

Background of the Conference Series

The International Conference on Domain Decomposition Methods has been held in 14 countries throughout Asia, Europe, and North America beginning in Paris in 1987. Held annually for the first 14 meetings, it has been spaced out since DD15 at roughly 18-month intervals. A complete list of the past meetings appears below. The 23rd International Conference on Domain Decomposition Methods was the first one held in Korea, and it took place on the beautiful Jeju Island.

The main technical content of the DD conference series has always been mathematical, but the principal motivation was and is to make efficient use of distributed memory computers for complex applications arising in science and engineering. As we approach the dawn of exascale computing, where we will command 10^{18} floating-point operations per second, clearly efficient and mathematically well-founded methods for the solution of large-scale systems become more and more important—as does their sound realization in the framework of modern HPC architectures. In fact, the massive parallelism, which makes exascale computing possible, requires the development of new solution methods, which are capable of efficiently exploiting this large number of cores as the connected hierarchies for memory access. Ongoing developments such as parallelization in time asynchronous iterative methods or nonlinear domain decomposition methods show that this massive parallelism does not only demand for new solution and discretization methods but also allows to foster the development of new approaches.

The progress obtained in domain decomposition techniques during the last decades has led to a broadening of the conference program in terms of methods and applications. Multiphysics, nonlinear problems, and space-time decomposition methods are more prominent these days than they have been previously. Domain decomposition has always been an active and vivid field, and this conference series is representing well the highly active and fast advancing scientific community behind it. This is also due to the fact that there is basically no alternative to domain decomposition methods as a general approach for massively parallel simulations at a large scale. Thus, with growing scale and growing hardware capabilities, also the methods can—and have to—improve.

However, even if domain decomposition methods are motivated historically by the need for efficient simulation tools for large-scale applications, there are also many interesting aspects of domain decomposition, which are not necessarily moti- vated by the need for massive parallelism. Examples are the choice of transmission conditions between subdomains, new coupling strategies, or the principal handling of interface conditions in problem classes such as fluid-structure interaction or contact problems in elasticity.

While research in domain decomposition methods is presented at numerous venues, the International Conference on Domain Decomposition Methods is the only regularly occurring international forum dedicated to interdisciplinary tech- nical interactions between theoreticians and practitioners working in the develop- ment, analysis, software implementation, and application of domain decomposition methods.

The list of previous Domain Decomposition Conferences is the following:

1. Paris, France, January 7–9, 1987
2. Los Angeles, USA, January 14–16, 1988
3. Houston, USA, March 20–22, 1989
4. Moscow, USSR, May 21–25, 1990
5. Norfolk, USA, May 6–8, 1991
6. Como, Italy, June 15–19, 1992
7. University Park, Pennsylvania, USA, October 27–30, 1993
8. Beijing, China, May 16–19, 1995
9. Ullensvang, Norway, June 3–8, 1996
10. Boulder, USA, August 10–14, 1997
11. Greenwich, UK, July 20–24, 1998
12. Chiba, Japan, October 25–29, 1999
13. Lyon, France, October 9–12, 2000
14. Cocoyoc, Mexico, January 6–11, 2002
15. Berlin, Germany, July 21–25, 2003
16. New York, USA, January 12–15, 2005
17. St. Wolfgang-Strobl, Austria, July 3–7, 2006
18. Jerusalem, Israel, January 12–17, 2008
19. Zhangjiajie, China, August 17–22, 2009
20. San Diego, California, USA, February 7–11, 2011

21. Rennes, France, June 25–29, 2012
22. Lugano, Switzerland, September 16–20, 2013
23. Jeju Island, Korea, July 6–10, 2015

International Scientific Committee on Domain Decomposition Methods

- Petter Bjørstad, University of Bergen, Norway
- Susanne Brenner, Louisiana State University, USA
- Xiao-Chuan Cai, CU Boulder, USA
- Martin Gander, University of Geneva, Switzerland
- Laurence Halpern, University Paris 13, France
- David E. Keyes, KAUST, Saudi Arabia
- Hyea Hyun Kim, Kyung Hee University, Korea
- Axel Klawonn, Universität zu Köln, Germany
- Ralf Kornhuber, Freie Universität Berlin, Germany
- Ulrich Langer, University of Linz, Austria
- Alfio Quarteroni, EPFL, Switzerland
- Olof B. Widlund, Courant Institute, USA
- Jinchao Xu, Penn State, USA
- Jun Zou, Chinese University of Hong Kong, Hong Kong

About the 23rd Conference

The 23rd International Conference on Domain Decomposition Methods had 108
participants from over 22 countries. It was the first one to be held in Korea.

As in previous meetings, DD23 featured a well-balanced mixture of established
and new topics, such as space-time domain decomposition methods, isogeometric
analysis, exploitation of modern HPC architectures, optimal control and inverse
problems, and electromagnetic problems. From the conference program, it is evident
that the growing capabilities in terms of theory and available hardware allow for
increasingly complex nonlinear and multiscale simulations, confirming the huge
potential and flexibility of the domain decomposition idea. The conference, which
was organized over an entire week, featured presentations of three different types:
The conference contained:

- Eleven invited presentations, fostering also younger scientists and their scientific
  development, selected by the International Scientific Committee
- A poster session, which also gave rise to intense discussions with the mostly
  younger presenting scientists
- Nine minisymposia, arranged around a special topic
- Seven sessions of contributed talks

The present proceedings volume contains a selection of 42 papers, split into 8 plenary papers, 21 minisymposium papers, and 13 contributed papers and posters.

Sponsoring Organizations

- KAIST Mathematics Research Station
- National Institute for Mathematical Sciences
- The Korean Federation of Science and Technology Societies
- KISTI Supercomputing Center
- A3 Foresight Program
- NVIDIA
- Jeju Convention & Visitors Bureau

The organizing committee would like to thank the sponsors for the financial support.

Local Organizing/Program Committee Members

- Chang-Ock Lee (KAIST; Chair)
- Kum Won Cho (KISTI)
- Taeyoung Ha (NIMS)
- Hyeonseong Jin (Jeju National University)
- Hyea Hyun Kim (Kyung Hee University)
- Eun-Hee Park (Kangwon National University)
- Eun-Jae Park (Yonsei University)

Research Activity in Domain Decomposition According to DD23 and Its Proceedings

The conference and the proceedings contain three parts: the plenary presentations, the minisymposium presentation, and the contributed talks and posters.

Plenary Presentations

The plenary presentations of the conference have been dealing with established topics in domain decomposition as well as with new approaches:

- Global convergence rates of some multilevel methods for variational and quasi-variational inequalities, Lori Badea (Institute of Mathematics of the Romanian Academy, Romania)

- Robust solution strategies for fluid-structure interaction problems with applications, Yuri Bazilevs (University of California, San Diego, USA)
- BDDC algorithms for discontinuous Petrov-Galerkin methods, Clark Dohrmann (Sandia National Laboratories, USA)
- Schwarz methods for the time-parallel solution of parabolic control problems, Felix Kwok (Hong Kong Baptist University, Hong Kong)
- Computational science activities in Korea, Jysoo Lee (KISTI, Korea)
- Recent advances in robust coarse space construction, Frédéric Nataf (Université Paris 6, France)
- Domain decomposition preconditioners for isogeometric discretizations, Luca F. Pavarino (University of Milano, Italy)
- Development of nonlinear structural analysis using co-rotational finite elements with improved domain decomposition method, SangJoon Shin (Seoul National University, Korea)
- Adaptive coarse spaces and multiple search directions: tools for robust domain decomposition algorithms, Nicole Spillane (Universidad de Chile, Chile)
- Element-based algebraic coarse spaces with applications, Panayot Vassilevski (Lawrence Livermore National Laboratory, USA)
- Preconditioning for nonsymmetry and time dependence, Andrew Wathen (University of Oxford, United Kingdom)

Minisymposia

There are nine minisymposia organized within DD23:

1. Space-time domain decomposition methods (Ulrich Langer, Olaf Steinbach)
   The space-time discretization of transient partial differential equations by using general space-time finite and boundary elements in the space-time computational domain allows for an almost optimal, adaptive space-time resolution of wave fronts and moving geometries. The global solution of the resulting systems of algebraic equations can easily be done in parallel, but requires appropriate preconditioning techniques by means of multilevel and domain decomposition methods. This minisymposium presents recent results on general space-time discretizations and parallel solution strategies.
2. Domain decomposition with adaptive coarse spaces in finite element and isogeometric applications (Durkbin Cho, Luca F. Pavarino, Olof B. Widlund)
   The aim of the minisymposium is to bring together researchers in both fields of finite elements and isogeometric analysis (IGA) to discuss the latest research developments in domain decomposition methods with adaptive coarse spaces. While coarse spaces are essential for the design of scalable algorithms, they can become quite expensive for problems with a large number of subdomains, or very irregular coefficients/domains, or for IGA discretizations where the high irregularity of the NURBS basis functions yields large interface and coarse

problems. This minisymposium will focus on recently proposed novel adaptive coarse spaces, generalized eigenproblems, and primal constraints selection.

3. Domain decomposition and high-performance computing (Santiago Badia, Jakub Šístek, Kab Seok Kang)

   The next generation of supercomputers, able to reach 1 exaflop/s, is expected to reach billions of cores. The success of domain decomposition for large-scale scientific computing will be strongly related to the ability to efficiently exploit extreme core counts. This MS is mainly oriented to novel algorithmic and implementation strategies that will boost the scalability of domain decomposition methods and their application for large-scale problems. Since large-scale computing is demanded by the most complex applications, generally multiscale, multiphysics, nonlinear, and/or transient in nature, tailored algorithms for these types of applications will be particularly relevant.

4. Domain decomposition methods and parallel computing for optimal control and inverse problems (Huibin Chang, Xue-Cheng Tai, Jun Zou)

   This minisymposium will bring together active experts working on domain decomposition methods and parallel computing for large-scale ill-posed problems from image processing, optimal control, and inverse problems to discuss and exchange the latest developments in these areas.

5. Efficient solvers for electromagnetic problems (Victorita Dolean, Zhen Peng)

   In this minisymposium we explore domain decomposition-type solvers for electromagnetic wave propagation problems. These problems are very challenging (especially in time-harmonic regime where the problem is indefinite in nature and most of the iterative solvers will fail) . The mini-symposium will discuss different areas of recent progress as parallel domain decomposition libraries, sweeping preconditioners, iterative methods based on multi-trace formulations, or new results on optimized Schwarz methods.

6. Domain decomposition methods for multiscale PDEs (Eric Chung, Hyea Hyun Kim)

   It is well known that classical ways to construct coarse spaces are not robust and give large condition numbers depending on the heterogeneities and contrasts of the coefficients. Recently, there are increasing interests in constructing domain decomposition methods with enriched coarse spaces or adaptive coarse spaces. The purpose of this minisymposium is to bring together researchers in the area of domain decomposition methods for PDEs with highly oscillatory coefficients and provide a forum for them to present the latest findings.

7. Birthday minisymposium Ralf Kornhuber (60th Birthday) (Rolf Krause, Martin Gander)

   This MS will bring together talks which are related to the scientific work of Ralf Kornhuber. This includes fast numerical methods for variational inequalities, multigrid methods, numerical methods for phase field equations, and biomechanics.

8. Recent approaches to nonlinear doman decomposition methods (Axel Klawonn, Oliver Rheinbach)

For a few decades already, Newton-Krylov algorithms with suitable pre-conditioners such as domain decomposition (DD) or multigrid (MG) methods (Newton-Krylov-DD or Newton-Krylov-MG) have been the workhorse for the parallel solution of nonlinear implicit problems. The standard Newton-Krylov approaches are based on a global linearization and the efficient parallel solution of the resulting linear (tangent) systems in each linearization step ("first linearize, then decompose"). Increasing local computational work and reducing communication are key ingredient for the efficient use of future exascale machines. In Newton-Krylov-DD/Newton-Krylov-MG methods, these aspects can be mainly treated at the level of the solution of the linear systems by the preconditioned Krylov methods. Computational work can be localized, and communication can be reduced by a complete reordering of operations: the nonlinear problem is first decomposed and then linearized, leading to nonlinear domain decomposition methods. An early approach in this direction is the ASPIN (additive Schwarz preconditioned inexact Newton) method by Cai and Keyes. Recently, there has been work on nonlinear FETI-DP and BDDC methods by Klawonn, Lanser, and Rheinbach. In this minisymposium, recent approaches to nonlinear domain decomposition methods will be presented.

9. Tutorial for domain decomposition on heterogenous HPC (Junard Lee)

At this minisymposium, we will have a tutorial session. We will cover heterogeneous HPC architecture, CUDA programming language, Open ACC directives, and how to implement these technologies to accelerate PDE solvers specially domain decomposition method.

Contributed Presentations and Posters

The contributed talks have been distributed over seven different sessions:

1. Domain Decomposition Methods for Applications
2. Optimized Schwarz Methods
3. Fast Solvers for Nonlinear and Unsteady Problems
4. Domain Decomposition Methods with Lagrange Multipliers
5. Efficient Methods and Solvers for Applications
6. Multiphysics Problems
7. Coarse Space Selection Strategies

The proceedings part with poster presentations is also a real treasure trove for new ideas in domain decomposition methods.

Acknowledgements In closing, we would like to thank all the participants gathered on Jeju Island for their contributions to the scientific success of this conference. Moreover, it is our pleasure to express our sincere thanks to everybody who has supported this conference on the administrative side. This includes the chairs of the conference sessions, the volunteers from KAIST and Jeju

National University helping on the practical and technical issues, and last but not least the KSIAM staff who has provided invaluable support.

Daejeon, Republic of Korea   C.-O. Lee
Boulder, CO, USA   X.-C. Cai
Thuwal, Saudi Arabia   D.E. Keyes
Yongin, Republic of Korea   H.H. Kim
Köln, Germany   A. Klawonn
Seoul, Republic of Korea   E.-J. Park
New York, NY, USA   O.B. Widlund
November 24, 2016

Organization

Program Chairs

Chang-Ock Lee   KAIST

Program Committee

Xiao-Chuan Cai   University of Colorado at Boulder
David E. Keyes   KAUST
Hyea Hyun Kim   Kyung Hee University
Axel Klawonn   Universität zu Köln
Eun-Jae Park   Yonsei University
Olof B. Widlund   Courant Institute

Contents

Part II Talks in Minisymposia (MT)

Part I
Plenary Talks (PT)

Global Convergence Rates of Some Multilevel Methods for Variational and Quasi-Variational Inequalities

Lori Badea

1 Introduction

The first multilevel method for variational inequalities has been proposed in Mandel (1984a) for complementarity problems. An upper bound of the asymptotic convergence rate of this method is derived in Mandel (1984b). The method has been studied later in Kornhuber (1994) in two variants, standard monotone multigrid method and truncated monotone multigrid method. These methods have been extended to variational inequalities of the second kind in Kornhuber (1996, 2002). Also, versions of this method have been applied to Signorini's problem in elasticity in Kornhuber and Krause (2001). In Badea (2003, 2006) global convergence rates of some projected multilevel relaxation methods of multiplicative type are given. Also, a global convergence rate was derived in Badea (2008) for a two-level additive method. Two-level methods for variational inequalities of the second kind and for some quasi variational inequalities have been analyzed in Badea and Krause (2012). In Badea (2014), it was theoretically justified the global convergence rate of the standard monotone multigrid methods and, in Badea (2015), this result has been extended to the hybrid algorithms, where the type of the iterations on the levels is different from the type of the iterations over the levels. Finally, a multigrid method for inequalities containing a term given by a Lipschitz operator is analyzed in Badea (2016). Evidently, the above list of citations is not exhaustive and, for further information, we can see the review article (Gräser and Kornhuber, 2009).

L. Badea (✉)
Institute of Mathematics of the Romanian Academy, P.O. Box 1-764, RO-014700 Bucharest, Romania
e-mail: lori.badea@imar.ro

© Springer International Publishing AG 2017   3
C.-O. Lee et al. (eds.), *Domain Decomposition Methods in Science and Engineering XXIII*, Lecture Notes in Computational Science and Engineering 116, DOI 10.1007/978-3-319-52389-7_1

This is a review paper regarding the convergence rate of some multilevel methods for variational inequalities and also, for more complicated problems such as variational inequalities of the second kind, quasi-variational inequalities and inequalities with a term containing a Lipschitz operator. The methods are first introduced as some subspace correction algorithms in a reflexive Banach space and, under some assumptions, general convergence results (error estimations, included) are given. In the finite element spaces, we prove that these assumptions are satisfied and that the introduced algorithms are in fact one-, two-, multilevel or multigrid methods. The constants in the error estimations are explicitly written in functions of the overlapping and mesh parameters for the one- and two-level methods and in function of the number of levels for the multigrid methods.

In this paper, we denote by V a reflexive Banach space and $K \subset V$ is a non empty closed convex subset. Also, $F : K \to \mathbf{R}$ is a Gâteaux differentiable functional and we assume that there exist two real numbers $p, q > 1$ such that for any $M > 0$ there exist $\alpha_M, \beta_M > 0$ for which

$$\alpha_M ||v - u||^p \leq < F'(v) - F'(u), v - u >$$
$$\text{and } ||F'(v) - F'(u)||_{V'} \leq \beta_M ||v - u||^{q-1},$$

for any $u, v \in K, ||u||, ||v|| \leq M$. In view of these properties, we can prove that F is a convex functional and $1 < q \leq 2 \leq p$.

2   One- and Two-Level Methods

In this section we introduce one- and two-level methods of multiplicative type, first as a general subspace correction algorithm. Details concerning the proof of its global convergence can be found in Badea (2003). The one- and two-level methods are derived from this algorithm by the introduction of the finite element spaces and details are given in Badea (2006). Similar results can be proved for the additive variant of the methods [see Badea (2008)].

We consider the variational inequality

$$u \in K : \ < F'(u), v - u > \geq 0, \text{ for any } v \in K, \tag{1}$$

and if K is not bounded, we suppose that F is coercive, i.e. $F(v) \to \infty$ as $||v|| \to \infty$. Then, problem (1) has an unique solution. Let $V_1, \cdots, V_m$ be some closed subspaces of V for which we make the following.

Assumption 1 *There exists a constant $C_0 > 0$ such that for any $w, v \in K$ and $w_i \in V_i$ with $w + \sum_{j=1}^{i} w_j \in K$, $i = 1, \cdots, m$, there exist $v_i \in V_i$, $i = 1, \cdots, m$, satisfying*

$$w + \sum_{j=1}^{i-1} w_j + v_i \in K, \quad v - w = \sum_{i=1}^{m} v_i, \quad \sum_{i=1}^{m} ||v_i||^p \leq C_0^p \left(||v - w||^p + \sum_{i=1}^{m} ||w_i||^p \right).$$

For linear problems, the last condition has a more simple form and is named the stability condition of the space decomposition. To solve problem (1), we introduce the following subspace correction algorithm.

Algorithm 1 *We start the algorithm with an arbitrary $u^0 \in K$. At iteration $n + 1$, having $u^n \in K$, $n \geq 0$, we sequentially compute for $i = 1, \cdots, m$,*

$$w_i^{n+1} \in V_i, \quad u^{n+\frac{i-1}{m}} + w_i^{n+1} \in K : \langle F'(u^{n+\frac{i-1}{m}} + w_i^{n+1}), v_i - w_i^{n+1} \rangle \geq 0,$$

for any $v_i \in V_i$, $u^{n+\frac{i-1}{m}} + v_i \in K$, and then we update $u^{n+\frac{i}{m}} = u^{n+\frac{i-1}{m}} + w_i^{n+1}$.

The following result proves the global convergence of this algorithm [see Theorem 2 in Badea (2003)].

Theorem 1 *On the above conditions on the spaces and the functional F, if Assumption 1 holds, then there exists an $M > 0$ such that $||u^n|| \leq M$, for any $n \geq 0$, and we have the following error estimations:*

(i) *if $p = q = 2$ we have* $||u^n - u||^2 \leq \frac{2}{\alpha_M} \left(\frac{\tilde{C}_1}{\tilde{C}_1 + 1} \right)^n \left[F(u^0) - F(u) \right].$

(ii) *if $p > q$ we have* $||u - u^n||^p \leq \frac{p}{\alpha_M} \dfrac{F(u^0) - F(u)}{\left[1 + n\tilde{C}_2 (F(u^0) - F(u))^{\frac{q-1}{q-1}} \right]^{\frac{q-1}{p-q}}},$

where

$$\tilde{C}_1 = \beta_M (\tfrac{p}{\alpha_M})^{\frac{q}{p}} m^{2 - \frac{q}{p}} \left[(1 + 2C_0) \left(F(u^0) - F(u) \right)^{\frac{p-q}{p(p-1)}} + \left(\beta_M (\tfrac{p}{\alpha_M})^{\frac{q}{p}} m^{2 - \frac{q}{p}} \right)^{\frac{1}{p-1}} C_0^{\frac{p}{p-1}} / \eta^{\frac{1}{p-1}} \right] / (1 - \eta) \text{ and}$$

$$\tilde{C}_2 = \frac{p - q}{(p - 1) \left(F(u^0) - F(u) \right)^{\frac{p-q}{q-1}} + (q - 1) \hat{C}^{\frac{p-1}{q-1}}}.$$

The value of η in the expression of $\tilde{C}_1$ can be arbitrary in $(0, 1)$, but we can also chose a $\eta_0 \in (0, 1)$ such that $\tilde{C}_1(\eta_0) \leq \tilde{C}_1(\eta)$ for any $\eta \in (0, 1)$.

One-level methods are obtained from Algorithm 1 by using the finite element spaces. To this end, we consider a simplicial regular mesh partition $\mathcal{T}_h$ of mesh size h over $\Omega \subset \mathbf{R}^d$. Also, let $\Omega = \cup_{i=1}^m \Omega_i$ be a domain decomposition of Ω, the overlapping parameter being δ, and we assume that $\mathcal{T}_h$ supplies a mesh partition for each subdomain Ω_i, $i = 1, \ldots, m$. In Ω, we use the linear finite element space V_h whose functions vanish on the boundary of Ω and, for each $i = 1, \ldots, m$, we consider the linear finite element space $V_h^i \subset V_h$ whose functions vanish outside Ω_i. Spaces V_h and V_h^i, $i = 1, \ldots, m$, are considered as subspaces of $W^{1,\sigma}$, $1 \leq \sigma \leq \infty$, and let $K_h \subset V_h$ be a convex set satisfying.

Property 1 If $v, w \in K_h$, and if $\theta \in C^0(\bar{\Omega})$, $\theta|_\tau \in C^1(\tau)$ for any $\tau \in \mathcal{T}_h$, and $0 \leq \theta \leq 1$, then $L_h(\theta v + (1 - \theta)w) \in K_h$, where L_h is the P_1-Lagrangian interpolation.

We see that the convex sets of obstacle type satisfy this property, and we have (see Proposition 3.1 in Badea (2006) for the proof)

Proposition 1 *Assumption 1 holds for the linear finite element spaces, $V = V_h$ and $V_i = V_h^i$, $i = 1, \ldots, m$, and for any convex set $K = K_h \subset V_h$ having Property 1. The constant C_0 in Assumption 1 can be written as $C_0 = C(m + 1)(1 + \frac{m-1}{\delta})$, where C is independent of the mesh parameter and the domain decomposition.*

In the case of the two-level methods, we consider two regular simplicial mesh partitions $\mathcal{T}_h$ and $\mathcal{T}_H$ on $\Omega \subset \mathbf{R}^d$, $\mathcal{T}_h$ being a refinement of $\mathcal{T}_H$. Besides the finite element spaces V_h, V_h^i, $i = 1, \ldots, m$ and the convex set K_h, defined for the one-level methods, we introduce the linear finite element space V_H^0 corresponding to the H-level, whose functions vanish on the boundary of Ω. The two-level method is obtained from the general subspace correction Algorithm 1 for $V = V_h$, $K = K_h$, and the subspaces $V_0 = V_H^0$, $V_1 = V_h^1$, $V_2 = V_h^2$, $\ldots$, $V_m = V_h^m$. Also, these spaces are considered as subspaces of $W^{1,\sigma}$, $1 \leq \sigma \leq \infty$, and we have the following (see Proposition 4.1 in Badea (2006) for the proof)

Proposition 2 *Assumption 1 is satisfied for the linear finite element spaces $V = V_h$ and $V_0 = V_H^0$, $V_i = V_h^i$, $i = 1, \ldots, m$, and any convex set $K = K_h$ having Property 1. The constant C_0 can be taken of the form $C_0 = Cm \left(1 + (m - 1)\frac{H}{\delta}\right) C_{d,\sigma}(H, h)$, where C is independent of the mesh and domain decomposition parameters, and*

$$C_{d,\sigma}(H, h) = \begin{cases} 1 & \text{if } d = \sigma = 1 \text{ or } 1 \leq d < \sigma \leq \infty \\ \left(\ln \frac{H}{h} + 1\right)^{\frac{d-1}{d}} & \text{if } 1 < d = \sigma < \infty \\ \left(\frac{H}{h}\right)^{\frac{d-\sigma}{\sigma}} & \text{if } 1 \leq \sigma < d < \infty. \end{cases}$$

Some numerical results have been given in Badea (2009) to compare the convergence of the one-level and two-level methods. They concern the two-obstacle problem of a nonlinear elastic membrane,

$$u \in [a, b] : \int_\Omega |\nabla u|^{\sigma-2} \nabla u \nabla(v - u) \geq 0, \text{ for any } v \in [a, b] \tag{2}$$

where $\Omega \subset \mathbf{R}^2$, $K = [a,b]$, $a \leq b$, $a, b \in W_0^{1,\sigma}(\Omega)$, $1 < \sigma < \infty$. These numerical experiments have confirmed the previous theoretical results.

3 Multilevel and Multigrid Methods

Details concerning the results in this section can be found in Badea (2014, 2015). As in the case of the one- and two-level methods, we consider problem (1). Let V_j, $j = 1, \ldots, J$, be closed subspaces of $V = V_J$ which will be associated with the level discretizations, and V_{ji}, $i = 1, \ldots, I_j$, be closed subspaces of V_j which will be associated with the domain decompositions on the levels. We consider $K \subset V$ a non empty closed convex subset and write $I = \max_{j=J,\ldots,1} I_j$.

To get sharper error estimations in the case of the multigrid method, we consider some constants $0 < \beta_{jk} \leq 1$, $\beta_{jk} = \beta_{kj}$, j, $k = J, \ldots, 1$, for which $\langle F'(v + v_{ji}) - F'(v), v_{kl} \rangle \leq \beta_M \beta_{jk} \|v_{ji}\|^{q-1} \|v_{kl}\|$, for any $v \in V$, $v_{ji} \in V_{ji}$, $v_{kl} \in V_{kl}$ with $\|v\|$, $\|v + v_{ji}\|$, $\|v_{kl}\| \leq M$, $i = 1, \ldots, I_j$ and $l = 1, \ldots, I_l$. Also, we fix a constant $\frac{p}{p-q+1} \leq \sigma \leq p$ and assume that there exists a constant C_1 such that $\|\sum_{j=1}^{J} \sum_{i=1}^{I_j} w_{ji}\| \leq C_1 (\sum_{j=1}^{J} \sum_{i=1}^{I_j} \|w_{ji}\|^\sigma)^{\frac{1}{\sigma}}$, for any $w_{ji} \in V_{ji}$, $j = J, \ldots, 1$, $i = 1, \ldots, I_j$. Evidently, in general, we can take $\beta_{jk} = 1$, j, $k = J, \ldots, 1$ and $C_1 = (IJ)^{\frac{\sigma-1}{\sigma}}$. In the multigrid methods, the convex sets where we look for the corrections are iteratively constructed from a level to another during the iterations in function of the current approximation. In this general background we make the following.

Assumption 2 *For a given* $w \in K$, *we recursively introduce the level convex sets* $\mathcal{K}_j$, $j = J, J - 1, \ldots, 1$, *satisfying*

- *at level J: we assume that* $0 \in \mathcal{K}_J$, $\mathcal{K}_J \subset \{v_J \in V_J : w + v_J \in K\}$ *and consider a* $w_J \in \mathcal{K}_J$,
- *at a level* $J - 1 \geq j \geq 1$: *we assume that* $0 \in \mathcal{K}_j$, $\mathcal{K}_j \subset \{v_j \in V_j : w + w_J + \ldots + w_{j+1} + v_j \in K\}$ *and consider a* $w_j \in \mathcal{K}_j$.

Also, we make a similar assumption with that in the case of the -one and two-level methods,

Assumption 3 *There exists two constants* C_2, $C_3 > 0$ *such that for any* $w \in K$, $w_{ji} \in V_{ji}$, $w_{j1} + \ldots + w_{ji} \in \mathcal{K}_j$, $j = J, \ldots, 1$, $i = 1, \ldots, I_j$, *and* $u \in K$, *there exist* $u_{ji} \in V_{ji}$, $j = J, \ldots, 1$, $i = 1, \ldots, I_j$, *which satisfy*

$$u_{j1} \in \mathcal{K}_j \text{ and } w_{j1} + \ldots + w_{ji-1} + u_{ji} \in \mathcal{K}_j, \ i = 2, \ldots, I_j, \ j = J, \ldots, 1,$$

$$u - w = \sum_{j=1}^{J} \sum_{i=1}^{I_j} u_{ji}, \quad \sum_{j=1}^{J} \sum_{i=1}^{I_j} \|u_{ji}\|^\sigma \leq C_2^\sigma \|u - w\|^\sigma + C_3^\sigma \sum_{j=1}^{J} \sum_{i=1}^{I_j} \|w_{ji}\|^\sigma$$

The convex sets $\mathcal{K}_j, j = J, \dots, 1$, are constructed as in Assumption 2 with the above

w and $w_j = \sum_{i=1}^{I_j} w_{ji}, j = J, \dots, 1$.

The general subspace correction algorithm corresponding to the multigrid method is written as [see Algorithm 2.2 in Badea (2014) or Algorithm 1.1 in Badea (2015)],

Algorithm 2 *We start with an arbitrary $u^0 \in K$. At iteration $n+1$ we have $u^n \in K$, $n \geq 0$, and successively perform:*

- *at level J: as in Assumption 2, with $w = u^n$, we construct $\mathcal{K}_J$.*

  Then, we write $w_J^n = 0$, and, for $i = 1, \dots, I_J$, we successively calculate $w_{Ji}^{n+1} \in V_{Ji}, w_J^{n+\frac{i-1}{I_J}} + w_{Ji}^{n+1} \in \mathcal{K}_J$,

$$\langle F'(u^n + w_J^{n+\frac{i-1}{I_J}} + w_{Ji}^{n+1}), v_{Ji} - w_{Ji}^{n+1} \rangle \geq 0$$

for any $v_{Ji} \in V_{Ji}, w_J^{n+\frac{i-1}{I_J}} + v_{Ji} \in \mathcal{K}_J$, and write $w_J^{n+\frac{i}{I_J}} = w_J^{n+\frac{i-1}{I_J}} + w_{Ji}^{n+1}$.

- *at a level $J - 1 \geq j \geq 1$: as in Assumption 2, we construct $\mathcal{K}_j$ with $w = u^n$ and $w_J = w_J^{n+1}, \dots, w_{j+1} = w_{j+1}^{n+1}$.*

Then, we write $w_j^n = 0$, and for $i = 1, \dots, I_j$, we successively calculate $w_{ji}^{n+1} \in V_{ji}, w_j^{n+\frac{i-1}{I_j}} + w_{ji}^{n+1} \in \mathcal{K}_j$,

$$\langle F'(u^n + \sum_{k=j+1}^{J} w_k^{n+1} + w_j^{n+\frac{i-1}{I_j}} + w_{ji}^{n+1}), v_{ji} - w_{ji}^{n+1} \rangle \geq 0$$

for any $v_{ji} \in V_{ji}, w_j^{n+\frac{i-1}{I_j}} + v_{ji} \in \mathcal{K}_j$, and write $w_j^{n+\frac{i}{I_j}} = w_j^{n+\frac{i-1}{I_j}} + w_{ji}^{n+1}$.

- *we write $u^{n+1} = u^n + \sum_{j=1}^{J} w_j^{n+1}$.*

Convergence of this algorithm is given by [see Theorem 1.1 in Badea (2015)]

Theorem 2 *Under the above conditions on the spaces and the functional F, if Assumptions 2 and 3 hold, then there exists an $M > 0$ such that $||u^n|| \leq M$, for any $n \geq 0$, and we have the following error estimations:*

(i) *if $p = q = 2$ we have $||u^n - u||^2 \leq \frac{2}{\alpha_M}(\frac{\tilde{C}_1}{\tilde{C}_1+1})^n [F(u^0) - F(u)]$,*

(ii) *if $p > q$ we have $||u - u^n||^p \leq \frac{p}{\alpha_M} \frac{F(u^0)-F(u)}{[1+n\tilde{C}_2(F(u^0)-F(u))^{\frac{p-q}{q-1}}]^{\frac{q-1}{p-q}}}$,* *where*

$$\tilde{C}_1 = \frac{1}{C_2\varepsilon}\left[\frac{C_2}{\varepsilon} + 1 + C_1C_2 + C_3\right],$$

$$\tilde{C}_2 = \frac{p-q}{(p-1)(F(u^0)-F(u))^{\frac{p-q}{q-1}} + (q-1)\tilde{C}_3^{\frac{q-1}{q-1}}} \quad with$$

$$\tilde{C}_3 = \frac{\frac{\alpha_M}{p}}{C_2\varepsilon}\left[\frac{C_2}{\varepsilon^{\frac{1}{p-1}}(\frac{\alpha_M}{p})^{\frac{q-1}{p-1}}} + \frac{(1+C_1C_2+C_3)(IJ)^{\frac{p-\sigma}{p\sigma}}}{(\frac{\alpha_M}{p})^{\frac{q}{p}}}(F(u^0)-F(u))^{\frac{p-q}{p(p-1)}}\right]$$

$$\varepsilon = \frac{\alpha_M}{p}\frac{1}{2C_2\beta_M I^{\frac{\sigma-1}{\sigma}+\frac{p-q+1}{p}}J^{\frac{\sigma-1}{\sigma}-\frac{q-1}{p}}(\max_{k=1,\cdots,J}\sum_{j=1}^{J}\beta_{kj})}.$$

To get the multilevel method corresponding to Algorithm 2, we consider a family of regular meshes $\mathcal{T}_{h_j}$ of mesh sizes $h_j, j = 1,\ldots,J$, over the domain $\Omega \subset \mathbf{R}^d$ and assume that $\mathcal{T}_{h_{j+1}}$ is a refinement of $\mathcal{T}_{h_j}$. Let, at each level $j = 1,\ldots,J$, $\{\Omega_j^i\}_{1\le i\le I_j}$ be an overlapping decomposition of Ω, of overlapping size δ_j. We also assume that, for $1 \le i \le I_j$, the mesh partition $\mathcal{T}_{h_j}$ of Ω supplies a mesh partition for each Ω_j^i, $\mathrm{diam}(\Omega_{j+1}^i) \le Ch_j$ and $I_1 = 1$.

We introduce the linear finite element spaces, $V_{h_j} = \{v \in C(\bar{\Omega}_j) : v|_\tau \in P_1(\tau), \tau \in \mathcal{T}_{h_j}, v = 0 \text{ on } \partial\Omega_j\}, j = 1,\ldots,J$, corresponding to the level meshes, and $V_{h_j}^i = \{v \in V_{h_j} : v = 0 \text{ in } \Omega_j\backslash\Omega_j^i\}, i = 1,\ldots,I_j$, associated with the level decompositions. Spaces $V_{h_j} \, j = 1,\ldots,J-1$, will be considered as subspaces of $W^{1,\sigma}, 1 \le \sigma \le \infty$.

The multilevel and multigrid methods will be obtained from Algorithm 2 for a two sided obstacle problem (1), i.e. the convex set is of the form $K = \{v \in V_{h_J} : \varphi \le v \le \psi\}$, with $\varphi, \psi \in V_{h_J}, \varphi \le \psi$. Concerning the construction of the level convex sets, we have [Proposition 3.1 in Badea (2014)]

Proposition 3 *Assumption 2 holds for the convex sets $\mathcal{K}_j, j = J,\ldots,1$, defined as,*

- *for $w \in K$, at the level J, we take $\varphi_J = \varphi - w, \psi_J = \psi - w, \mathcal{K}_J = [\varphi_J, \psi_J]$, and consider an $w_J \in \mathcal{K}_J$,*
- *at a level $j = J-1,\ldots,1$, we define $\varphi_j = I_{h_j}(\varphi_{j+1} - w_{j+1}), \psi_j = I_{h_j}(\psi_{j+1} - w_{j+1}), \mathcal{K}_j = [\varphi_j, \psi_j]$, and consider an $w_j \in \mathcal{K}_j, I_{h_j} : V_{h_{j+1}} \to V_{h_j}, j = 1,\ldots,J-1$, being some nonlinear interpolation operators between two consecutive levels.*

Also, our second assumption holds [see Proposition 2 in Badea (2015)],

Proposition 4 *Assumption 3 holds for the convex sets* $\mathcal{K}_j, j = J, \ldots, 1,$ *defined in Proposition 3. The constants* C_2 *and* C_3 *are written as*

$$C_2 = CI^{\frac{\sigma+1}{\sigma}}(I+1)^{\frac{\sigma-1}{\sigma}}(J-1)^{\frac{\sigma-1}{\sigma}}[\textstyle\sum_{j=2}^{J} C_{d,\sigma}(h_{j-1}, h_J)^\sigma]^{\frac{1}{\sigma}}$$
$$C_3 = CI^2(I+1)^{\frac{\sigma-1}{\sigma}}(J-1)^{\frac{\sigma-1}{\sigma}}[\textstyle\sum_{j=2}^{J} C_{d,\sigma}(h_{j-1}, h_J)^\sigma]^{\frac{1}{\sigma}}$$

We proved that Assumptions 2 and 3 hold, and have explicitly written constants C_2 and C_3 in function of the mesh and overlapping parameters. We can then conclude from Theorem 2 that Algorithm 2 is globally convergent. Convergence rates given in Theorem 2 depend on the functional F, the maximum number of the subdomains on each level, I, and the number of levels J. Since the number of subdomains on levels can be associated with the number of colors needed to mark the subdomains such that the subdomains with the same color do not intersect with each other, we can conclude that the convergence rate essentially depends on the number of levels J.

In the general framework of multilevel methods we take $C_1 = CJ^{\frac{\sigma-1}{\sigma}} \max_{k=1,\cdots,J}$ $\sum_{j=1}^{J} \beta_{kj} = J$ and, as functions depending only of J, we have

$$C_2 = C(J-1)^{\frac{\sigma-1}{\sigma}} S_{d,\sigma}(J) \text{ and } C_3 = C(J-1)^{\frac{\sigma-1}{\sigma}} S_{d,\sigma}(J) \text{ where}$$

$$S_{d,\sigma}(J) = \left[\sum_{j=2}^{J} C_{d,\sigma}(h_{j-1}, h_J)^\sigma \right]^{\frac{1}{\sigma}} = \begin{cases} (J-1)^{\frac{1}{\sigma}} & \text{if } d = \sigma = 1 \\ & \text{or } 1 \le d < \sigma < \infty \\ CJ & \text{if } 1 < d = \sigma < \infty \\ C^J & \text{if } 1 \le \sigma < d < \infty. \end{cases}$$

In the above multilevel methods a mesh is the refinement of that one on the previous level, but the domain decompositions are almost independent from one level to another. We obtain similar multigrid methods by decomposing the domain by the supports of the nodal basis functions of each level. Consequently, the subspaces $V_{h_j}^i$, $i = 1, \ldots, I_j$, are one-dimensional spaces generated by the nodal basis functions associated with the nodes of $\mathcal{T}_{h_j}$, $j = J, \ldots, 1$. In the case of the multigrid methods, we can take $C_1 = C$ and $\max_{k=1,\cdots,J} \sum_{j=1}^{J} \beta_{kj} = C$. Now we can write the convergence rate of the multigrid method corresponding to Algorithm 2 in function of the number of levels J for a given particular problem. In Badea (2014), the convergence rate of the multigrid method for the example in (2) has been written.

Remark 1 (See also Badea (2014))

1. The above results referred to problems in $W^{1,\sigma}$ with Dirichlet boundary conditions, but they also hold for Neumann or mixed boundary conditions.
2. Similar convergence results can be obtained for problems in $(W^{1,\sigma})^d$.
3. The analysis and the estimations of the global convergence rate which are given above refers to two sided obstacle problems which arise from the minimization of functionals defined on $W^{1,\sigma}$, $1 < \sigma < \infty$.

4. We can compare the convergence rates we have obtained with similar ones in the literature in the case of H^1 ($p = q = 2$) and $d = 2$. In this case, we get that the global convergence rate of Algorithm 2 is $1 - \frac{1}{1+CJ^3}$. The same estimate, of $1 - \frac{1}{1+CJ^3}$, is obtained by R. Kornhuber for the asymptotic convergence rate of the standard monotone multigrid methods for the complementarity problems.

Algorithm 2 is of multiplicative type over the levels as well as on each level, i.e. the current correction is found in function of all corrections on both the previous levels and the current level. We can also imagine hybrid algorithms where the type of the iteration over the levels is different from the type of the iteration on the levels. This idea can be also found in Smith et al. (1996). In Badea (2015), such hybrid algorithms (multiplicative over the levels—additive on levels, additive over the levels—multiplicative on levels and additive over the levels as well as on levels) have been introduced and analyzed in a similar manner with that of Algorithm 2. The following remark contains some conclusions withdrawn in Badea (2015) concerning the convergence rate (expressed only in function of J) of these hybrid algorithms for problem (2).

Remark 2

1. Regardless of the iteration type on levels, algorithms having the same type of iterations over the levels have the same convergence rate, provided that additive iterations on levels are parallelized.
2. The algorithms which are of multiplicative type over the levels converge better, by a factor of between $1/J$ and 1 (depending on σ), than their additive similar variants.

4 One- and Two-Level Methods for Variational Inequalities of the Second Kind and Quasi-Variational Inequalities

The results in this section are detailed in Badea and Krause (2012) where one- and two-level methods have been introduced and analyzed for the second kind and quasi-variational inequalities. In the case of the variational inequalities of the second kind, let $\varphi : K \to \mathbf{R}$ be a convex, lower semicontinuous, not differentiable functional and, if K is not bounded, we assume that $F + \varphi$ is coercive, i.e. $F(v) + \varphi(v) \to \infty$, as $\|v\| \to \infty$, $v \in K$. We consider the variational of the second kind

$$u \in K : \langle F'(u), v - u \rangle + \varphi(v) - \varphi(u) \geq 0, \text{ for any } v \in K \qquad (3)$$

which, in view of the properties of F and φ, has a unique solution. An example of such a problem is given by the contact problems with Tresca friction. To solve problem (3), we introduce

Algorithm 3 *We start the algorithm with an arbitrary $u^0 \in K$. At iteration $n +$ 1, having $u^n \in K$, $n \geq 0$, we compute sequentially for $i = 1, \cdots, m$, the local corrections $w_i^{n+1} \in V_i$, $u^{n+\frac{i-1}{m}} + w_i^{n+1} \in K$ as the solution of the variational inequality*

$$\langle F'(u^{n+\frac{i-1}{m}} + w_i^{n+1}), v_i - w_i^{n+1}\rangle + \varphi(u^{n+\frac{i-1}{m}} + v_i) - \varphi(u^{n+\frac{i-1}{m}} + w_i^{n+1}) \geq 0,$$

for any $v_i \in V_i$, $u^{n+\frac{i-1}{m}} + v_i \in K$, and then we update $u^{n+\frac{i}{m}} = u^{n+\frac{i-1}{m}} + w_i^{n+1}$.

To prove the convergence of the algorithm, we introduce a technical assumption,

$$\sum_{i=1}^m [\varphi(w + \sum_{j=1}^{i-1} w_j + v_i) - \varphi(w + \sum_{j=1}^{i-1} w_j + w_i)] \leq \varphi(v) - \varphi(w + \sum_{i=1}^m w_i)$$

for $v, w \in K$, and $v_i, w_i \in V_i$, $i = 1, \ldots, m$, in Assumption 1. In general, φ has not such a property and to show that this assumption holds when the finite element spaces are used, we have to take a numerical approximation of φ. The convergence of Algorithm 3 is proved by the following

Theorem 3 *Under the above assumptions on V, F and φ, let u be the solution of the problem and u^n, $n \geq 0$, be its approximations obtained from Algorithm 3. If Assumption 1 holds, then there exists $M > 0$ such that such that $\|u^{n+\frac{i}{m}}\| \leq M$, $n \geq 0, 1 \leq i \leq m$, and we have the following error estimations:*

(i) $\|u^n - u\|^2 \leq \frac{p}{\alpha_M} \left(\frac{\tilde{C}_1}{\tilde{C}_1 + 1}\right)^n [F(u^0) + \varphi(u^0) - F(u) - \varphi(u)]$ *if $p = q = 2$,*

(ii) $\|u - u^n\|^p \leq \frac{p}{\alpha_M} \dfrac{F(u^0) + \varphi(u^0) - F(u) - \varphi(u)}{\left[1 + n\tilde{C}_2(F(u^0) + \varphi(u^0) - F(u) - \varphi(u))^{\frac{p-q}{q-1}}\right]^{\frac{q-1}{p-q}}}$ *if $p > q$,*

where

$$\tilde{C}_1 = \beta_M(1 + 2C_0)m^{2-\frac{q}{p}} \left(\frac{p}{\alpha_M}\right)^{\frac{q}{p}} \left(F(u^0) - F(u) + \varphi(u^0) - \varphi(u)\right)^{\frac{p-q}{p(p-1)}} +$$
$$\beta_M C_0 m^{\frac{p-q+1}{p}} \frac{1}{\varepsilon^{\frac{1}{p-1}}} \left(\frac{p}{\alpha_M}\right)^{\frac{q-1}{p-1}} \text{ with } \varepsilon = \alpha_M / \left(p\beta_M C_0 m^{\frac{p-q+1}{p}}\right),$$

$$\tilde{C}_2 = \frac{p - q}{(p-1)(F(u^0) + \varphi(u^0) - F(u) - \varphi(u))^{\frac{p-q}{q-1}} + (q - 1)C_1^{\frac{p-1}{q-1}}}$$

In the case of the quasivariational inequalities, we consider only the case of $p = q = 2$ and let $\varphi : K \times K \to \mathbf{R}$ be a functional such that, for any $u \in K$, $\varphi(u, \cdot) : K \to \mathbf{R}$ is convex, lower semicontinuous and, if K is not bounded, $F(\cdot) + \varphi(u, \cdot)$ is coercive, i.e. $F(v) + \varphi(u, v) \to \infty$ as $\|v\| \to \infty$, $v \in K$. We assume that for any $M > 0$ there exists a constant $c_M > 0$ such that

$$|\varphi(v_1, w_2) + \varphi(v_2, w_1) - \varphi(v_1, w_1) - \varphi(v_2, w_2)| \leq c_M \|v_1 - v_2\| \|w_1 - w_2\|$$

for any v_1, v_2, w_1 $w_2 \in K$, $||v_1||$, $||v_2||$, $||w_1||$ $||w_2|| \leq M$. If φ has the above property, the quasi-variational inequality

$$u \in K : \langle F'(u), v - u \rangle + \varphi(u, v) - \varphi(u, u) \geq 0, \text{ for any } v \in K$$

has a unique solution. An example of such a problem is given by the contact problems with non-local Coulomb friction. We can write three algorithms depending on the first argument of φ.

Algorithm 4 *We start the algorithm with an arbitrary $u^0 \in K$. At iteration $n +$ 1, having $u^n \in K$, $n \geq 0$, we compute sequentially for $i = 1, \cdots, m$, the local corrections $w_i^{n+1} \in V_i$, $u^{n+\frac{i-1}{m}} + w_i^{n+1} \in K$, satisfying*

$$\langle F'(u^{n+\frac{i-1}{m}} + w_i^{n+1}), v_i - w_i^{n+1} \rangle + \varphi(v_i^{n+1}, u^{n+\frac{i-1}{m}} + v_i)$$
$$-\varphi(v_i^{n+1}, u^{n+\frac{i-1}{m}} + w_i^{n+1}) \geq 0,$$

for any $v_i \in V_i$, $u^{n+\frac{i-1}{m}} + v_i \in K$, and then we update $u^{n+\frac{i}{m}} = u^{n+\frac{i-1}{m}} + w_i^{n+1}$.

Above, the first argument v_i^{n+1} of φ can be taken either $u^{n+\frac{i-1}{m}} + w_i^{n+1}$ or $u^{n+\frac{i-1}{m}}$ or even u^n. As we shall see in the next convergence theorem, the three variants of the algorithm are convergent. Similarly with the case of the inequalities of the second kind, we introduce the technical assumption

$$\sum_{i=1}^{m} [\varphi(u, w + \sum_{j=1}^{i-1} w_j + v_i) - \varphi(u, w + \sum_{j=1}^{i} w_j)] \leq \varphi(u, v) - \varphi(u, w + \sum_{i=1}^{m} w_i)$$

for any $u \in K$ and for $v, w \in K$ and $v_i, w_i \in V_i$, $u^{n+\frac{i-1}{m}} + v_i \in K$, $i = 1, \ldots, m$, in Assumption 1. Also, in the finite element spaces, φ of the continuous problem is numerically approximated in order to get the above assumption satisfied. Convergence of the three algorithms is proved by

Theorem 4 *Under the above assumptions on V, F and φ, let u be the solution of the problem and u^n, $n \geq 0$, be its approximations obtained from one of the variants of Algorithm 4. If Assumption 1 holds, and if $\frac{\alpha_M}{2} \geq mc_M + \sqrt{2m(25C_0 + 8)}\beta_M c_M$, for any $M > 0$, then there exists an $M > 0$ such that $\|u^{n+\frac{i}{m}}\| \leq M$, $n \geq 0, 1 \leq i \leq m$, and we have the following error estimation*

$$\|u^n - u\|^2 \leq \frac{2}{\alpha_M} \left(\frac{\tilde{C}_1}{\tilde{C}_1 + 1} \right)^n [F(u^0) + \varphi(u, u^0) - F(u) - \varphi(u, u)].$$

where

$$\tilde{C}_1 = \tilde{C}_2/\tilde{C}_3 \text{ with } \tilde{C}_2 = \beta_M m(1 + 2C_0 + \tfrac{C_0}{\varepsilon_1}) + c_M m(1 + 2C_0 + \tfrac{1+3C_0}{\varepsilon_2}),$$
$$\tilde{C}_3 = \tfrac{\alpha_M}{2} - c_M(1 + \varepsilon_3)m \text{ and } \varepsilon_1 = \varepsilon_2 = \tfrac{2c_M m}{\tfrac{\alpha_M}{2} - c_M m}, \quad \varepsilon_3 = \tfrac{\tfrac{\alpha_M}{2} - c_M m}{2c_M m}.$$

Remark 3

1. Extension of the previous methods (given for variational inequalities of the second kind and quasi-variational inequalities) to methods with more than two levels, having an optimal rate of convergence, is not very evident because of the technical conditions we have introduced, which are not satisfied when the domain decompositions on the coarse levels are considered.
2. By using Newton linearizations of φ, R. Kornhuber introduced multigrid methods for complementarity problems and estimated the asymptotic convergence rates.

5 Multigrid Methods for Inequalities with a Term Given by a Lipschitz Operator

In this section, we estimate the global convergence rate of a multigrid method for the particular case of quasi-variational inequalities when the inequality contains a term given by a Lipschitz operator. Details concerning the results of this section can be found in Badea (2016). As in the previous section, we consider the case when $p = q = 2$ and $\alpha_M = \alpha$, $\beta_M = \beta$, i.e. they not depend on M. Let $T : V \to V'$ be a Lipschitz continuous operator $||T(v) - T(u)||_{V'} \le \gamma ||v - u||$ for any $v, u \in V$, and we consider the problem

$$u \in K : \langle F'(u), v - u \rangle + \langle T(u), v - u \rangle \ge 0 \text{ for any } v \in K.$$

In the following algorithm, each iteration contains κ intermediate iterations in which the argument of T is kept unchanged.

Algorithm 5 *We start the algorithm with an arbitrary $u^0 \in K$. Assuming that at iteration $n + 1$ we have $u^n \in K$, $n \ge 0$, we write $\tilde{u}^n = u^n$ and carry out the following two steps:*

1. *We perform $\kappa \ge 1$ iterations of Algorithm 2 starting with $\tilde{u}^n$ and keeping the argument of T equal with u^n, i.e. we apply Algorithm 2 to the inequality*

$$\tilde{u} \in K : \langle F'(\tilde{u}), v - \tilde{u} \rangle + \langle T(u^n), v - \tilde{u} \rangle \ge 0 \text{ for any } v \in K$$

   After the κ iterations we get the approximation $\tilde{u}^{n+\kappa}$ of $\tilde{u}$.
2. *We write $u^{n+1} = \tilde{u}^{n+\kappa}$.*

Convergence condition of Theorem 4 depends on the number m of the subspaces in the one- or two-level methods. We will see in the next theorem that if the Lipschitz constant of the operator T is small enough, the convergence condition of the above algorithm is independent of the number of levels and the number of subdomains on the levels.

Theorem 5 *We assume that V, F and T satisfy the above conditions and that Assumptions 2–3 hold. Then, if $\gamma/\alpha < 1/2$ and κ satisfies $(\frac{\tilde{C}}{\tilde{C}+1})^\kappa < \frac{1-2\frac{\gamma}{\alpha}}{1+3\frac{\gamma}{\alpha}+4\frac{\gamma^2}{\alpha^2}+\frac{\gamma^3}{\alpha^3}}$, Algorithm 5 is convergent and we have the following error estimation*

$$\|u^n - u\|^2 \leq \frac{2}{\alpha}[2\frac{\gamma}{\alpha} + (\frac{\tilde{C}}{\tilde{C}+1})^\kappa(1 + 3\frac{\gamma}{\alpha} + 4\frac{\gamma^2}{\alpha^2} + \frac{\gamma^3}{\alpha^3})]^n$$
$$\cdot[F(u^0) + \langle T(u), u^0\rangle - F(u) - \langle T(u), u\rangle],$$

where $\tilde{C} = \frac{1}{C_2\varepsilon}\left[1 + C_2 + C_1C_2 + \frac{C_2}{\varepsilon}\right]$, $\varepsilon = \frac{\alpha}{2\beta I(\max_{k=1,\cdots,J}\sum_{j=1}^{J}\beta_{kj})C_2}$.

References

L. Badea, Convergence rate of a multiplicative Schwarz method for strongly nonlinear variational inequalities, in *Analysis and Optimization of Differential Systems*, ed. by V. Barbu et al. (Kluwer Academic Publishers, Dordrecht, 2003), pp. 31–42

L. Badea, Convergence rate of a Schwarz multilevel method for the constrained minimization of non-quadratic functionals. SIAM J. Numer. Anal. **44**(2), 449–477 (2006)

L. Badea, Additive Schwarz method for the constrained minimization of functionals in reflexive banach spaces, in *Domain Decomposition Methods in Science and Engineering XVII, LNSE 60 U*, ed. by Langer et al. (Springer, Berlin, Heidelberg, 2008), pp. 427–434

L. Badea, One- and two-level domain decomposition methods for nonlinear problems, in *Proceedings of the First International Conference on Parallel, Distributed and Grid Computing for Engineering*, ed. by B.H.V. Topping, P. Iványi (Civil-Comp Press, Stirlingshire, 2009), p. 6

L. Badea, Global convergence rate of a standard multigrid method for variational inequalities. IMA J. Numer. Anal. **34**(1), 197–216 (2014)

L. Badea, Convergence rate of some hybrid multigrid methods for variational inequalities. J. Numer. Math. **23**(3), 195–210 (2015)

L. Badea, Globally convergent multigrid method for variational inequalities with a nonlinear term, in *Domain Decomposition Methods in Science and Engineering XXII, LNCSE 104*, ed. by T. Dickopf et al. (Springer, Heidelberg, 2016), pp. 427–435

L. Badea, R. Krause, One- and two-level Schwarz methods for inequalities of the second kind and their application to frictional contact. Numer. Math. **120**(4), 573–599 (2012)

C. Gräser, R. Kornhuber, Multigrid methods for obstacle problems. J. Comput. Math. **27**(1), 1–44 (2009)

R. Kornhuber, Monotone multigrid methods for elliptic variational inequalities I. Numer. Math. **69**, 167–184 (1994)

R. Kornhuber, Monotone multigrid methods for elliptic variational inequalities II. Numer. Math. **72**, 481–499 (1996)

R. Kornhuber, On constrained Newton linearization and multigrid for variational inequalities. Numer. Math. **91**, 699–721 (2002)

R. Kornhuber, R. Krause, Adaptive multigrid methods for Signorini's problem in linear elasticity. Comp. Visual. Sci. **4**, 9–20 (2001)

J. Mandel, A multilevel iterative method for symmetric, positive definite linear complementarity problems. Appl. Math. Opt. **11**, 77–95 (1984a)

J. Mandel, Etude algébrique d'une méthode multigrille pour quelques problèmes de frontière libre. C. R. Acad. Sci. Ser. I **298**, 469–472 (1984b)

B.F. Smith, P.E. Bjørstad, W. Gropp, *Domain Decomposition. Parallel Multilevel Methods for Elliptic Partial Differential Equations* (Cambridge University Press, Cambridge, 1996)

Parallel Sum Primal Spaces for Isogeometric Deluxe BDDC Preconditioners

L. Beirão daVeiga, L.F. Pavarino, S. Scacchi, O.B. Widlund, and S. Zampini

1 Introduction

In this paper, we study the adaptive selection of primal constraints in BDDC deluxe preconditioners applied to isogeometric discretizations of scalar elliptic problems. The main objective of this work is to significantly reduce the coarse space dimensions of the BDDC isogeometric preconditioners developed in our previous works, Beirão da Veiga et al. (2013a, 2014b), while retaining their fast and scalable convergence rates.

Recent works on adaptive selection of primal constraints have focused on constraints associated with the interface between pairs of subdomains, i.e. edges in 2D and faces in 3D; see Dohrmann and Pechstein (2011), Mandel et al. (2012),

L. Beirão da Veiga
Dipartimento di Matematica ed Applicazioni, Via Cozzi 55, 20125 Milano, Italy
e-mail: lourenco.beirao@unimib.it

L.F. Pavarino (✉)
Dipartimento di Matematica, Università di Pavia, Via Ferrata 5, 27100 Pavia, Italy
e-mail: luca.pavarino@unipv.it

S. Scacchi
Dipartimento di Matematica, Università di Milano, Via Saldini 50, 20133 Milano, Italy
e-mail: simone.scacchi@unimi.it

O.B. Widlund
Courant Institute of Mathematical Sciences, 251 Mercer Street, New York, NY 10012, USA
e-mail: widlund@cims.nyu.edu

S. Zampini
Extreme Computing Research Center, King Abdullah University of Science and Technology
(KAUST), Thuwal, Saudi Arabia
e-mail: stefano.zampini@kaust.edu.sa

© Springer International Publishing AG 2017   17
C.-O. Lee et al. (eds.), *Domain Decomposition Methods in Science
and Engineering XXIII*, Lecture Notes in Computational Science
and Engineering 116, DOI 10.1007/978-3-319-52389-7_2

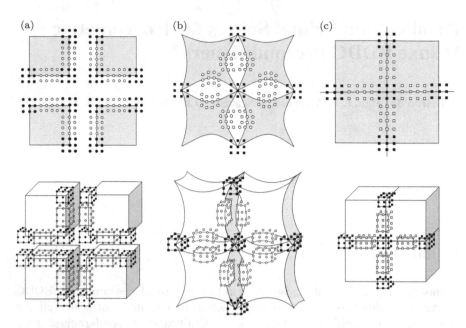

Fig. 1 Schematic illustration in index space of 2D (*top row*) and 3D (*bottom row*) "fat" interface equivalence classes for a configuration with four subdomains with $p = 3, \kappa = 2$: vertex variables are *black*, while edge variables are *white*; dual variables are denoted by *circles*, while primal variables by *square*. The figure shows the following configurations: (**a**) not assembled (all vertex and edge variables are dual); (**b**) partially assembled (all fat vertex variables are assembled); (**c**) fully assembled (all vertex and edge variables are primal)

Pechstein and Dohrmann (2013), Spillane et al. (2013), Klawonn et al. (2014a,b, 2015a,b, 2016), Kim and Chung (2015). The more complex case with constraints associated with three or more subdomains appears in isogeometric discretizations already for vertex constraints in 2D, where four subdomains are involved for each fat vertex (in 3D the subdomains involved for each vertex constraint becomes eight), see Fig. 1. Fewer works have considered these more general cases, see e.g. Mandel et al. (2012), Kim et al. (2015), Klawonn et al. (2015a), Calvo and Widlund (2016), and our previous work Beirão da Veiga et al. (2016), where we have constructed and compared four different strategies for the adaptive selection of primal constraints. Here we focus on a promising strategy based on generalized eigenvalue problems involving parallel sums of local Schur complement blocks. The resulting isogeometric BDDC algorithm is scalable, quasi-optimal and robust with respect to both increasing polynomial degree of the isogeometric basis functions employed and the presence of discontinuous elliptic coefficients across subdomain interfaces.

For earlier work on the iterative solution of isogeometric approximations, see Beirão da Veiga et al. (2013b), Collier et al. (2013), Gahalaut et al. (2013), Kleiss et al. (2012).

2 Model Elliptic Problem and Isogeometric Analysis (IGA)

Given a bounded and connected domain $\Omega \subset \mathbb{R}^d$, $d = 2, 3$, obtained by a CAD program, a right-hand side $f \in L^2(\Omega)$ and a scalar field ρ satisfying $0 < \rho_{min} \leq \rho(x) \leq \rho_{max}$, $\forall x \in \Omega$, we consider the model scalar elliptic problem

$$-\nabla \cdot (\rho \nabla u) = f \text{ in } \Omega, \quad u = 0 \text{ on } \partial\Omega, \tag{1}$$

and discretize it with IGA based on B-splines and NURBS basis functions; see, e.g., Hughes et al. (2005), Cottrell et al. (2009), Beirão da Veiga et al. (2014a). Given univariate B-spline basis functions $N_i^p(\xi)$ of degree p associated to the knot vector $\{\xi_1 = 0, \ldots, \xi_{n+p+1} = 1\}$ defined on the parametric interval $\widehat{I} := (0, 1)$, we define by a 2D tensor product (the 3D case is analogous) the 2D parametric space $\widehat{\Omega} := (0, 1) \times (0, 1)$, the $n \times m$ mesh of control points $\mathbf{C}_{i,j}$ associated with the knot vectors $\{\xi_1 = 0, \ldots, \xi_{n+p+1} = 1\}$ and $\{\eta_1 = 0, \ldots, \eta_{m+q+1} = 1\}$, the bivariate B-spline basis functions by $B_{i,j}^{p,q}(\xi, \eta) = N_i^p(\xi) M_j^q(\eta)$, and the bivariate B-spline discrete space as

$$\widehat{S}_h := \text{span}\{B_{i,j}^{p,q}(\xi, \eta), \ i = 1, \ldots, n, j = 1, \ldots, m\}. \tag{2}$$

Analogously, the NURBS space is the span of NURBS (Non-uniform rational Basis spline) basis functions defined in one dimension by

$$R_i^p(\xi) := \frac{N_i^p(\xi)\omega_i}{\sum_{k=1}^n N_k^p(\xi)\omega_k} = \frac{N_i^p(\xi)\omega_i}{w(\xi)}, \tag{3}$$

with the weight function $w(\xi) := \sum_{k=1}^n N_k^p(\xi)\omega_k \in \widehat{S}_h$, and in two dimensions by a tensor product

$$R_{i,j}^{p,q}(\xi, \eta) := \frac{B_{i,j}^{p,q}(\xi, \eta)\omega_{i,j}}{\sum_{k=1}^n \sum_{\ell=1}^m B_{k,\ell}^{p,q}(\xi, \eta)\omega_{k,\ell}} = \frac{B_{i,j}^{p,q}(\xi, \eta)\omega_{i,j}}{w(\xi, \eta)}, \tag{4}$$

where $w(\xi, \eta)$ is the weight function and $\omega_{k,\ell}$ are positive weights associated with a $n \times m$ net of control points. The discrete NURBS space on Ω is defined as the span of the *push-forward* of the NURBS basis functions (4), i.e.,

$$\mathcal{N}_h := \text{span}\{R_{i,j}^{p,q} \circ \mathbf{F}^{-1}, \text{ with } i = 1, \ldots, n; j = 1, \ldots, m\}, \tag{5}$$

with $\mathbf{F} : \widehat{\Omega} \to \Omega$, the geometrical map between parameter and physical spaces $\mathbf{F}(\xi, \eta) = \sum_{i=1}^n \sum_{j=1}^m R_{i,j}^{p,q}(\xi, \eta)\mathbf{C}_{i,j}$. The spline space in the parameter space is then defined as

$$\widehat{V}_h := [\widehat{S}_h \cap H_0^1(\widehat{\Omega})]^2 = [\text{span}\{B_{i,j}^{p,q}(\xi, \eta), \ i = 2, \ldots, n-1, j = 2, \ldots, m-1\}]^2,$$

and the NURBS space in physical space as

$$U_h := [\mathcal{N}_h \cap H_0^1(\Omega)]^2 = [\text{span}\{R_{i,j}^{p,q} \circ \mathbf{F}^{-1}, \text{ with } i = 2, \ldots, n-1, j = 2, \ldots, m-1\}]^2.$$

The IGA formulation of problem (1) then reads: Find $u_h \in U_h$ such that:

$$a(u_h, v_h) = <f, v_h > \qquad \forall v \in U_h, \tag{6}$$

with the bilinear form $a(u_h, v_h) = \int_\Omega \rho \nabla u_h \nabla v_h dx$ and the right-hand side $<f, v_h> = \int_\Omega f v_h dx$. The matrix form of (6) is the linear system

$$Au_h = f_h, \tag{7}$$

with a symmetric, positive definite stiffness matrix A.

3 Isogeometric BDDC Deluxe Preconditioners

Knots and Subdomain Decomposition. By partitioning the associated knot vector, we decompose the reference interval $\widehat{I}$ into quasi-uniform subintervals $\widehat{I}_k = (\xi_{i_k}, \xi_{i_{k+1}})$ of characteristic diameter H and we extend this decomposition to more dimensions by tensor products, e.g., in two dimension

$$\widehat{I}_k = (\xi_{i_k}, \xi_{i_{k+1}}), \quad \widehat{I}_l = (\eta_{j_l}, \eta_{j_{l+1}}), \quad \widehat{\Omega}_{kl} = \widehat{I}_k \times \widehat{I}_l, \quad 1 \le k \le N_1, \ 1 \le l \le N_2.$$

For simplicity, we reindex the subdomains using only one index to obtain the decomposition of our reference domain $\widehat{\Omega} = \bigcup_{k=1\ldots K} \widehat{\Omega}^{(k)}$, into $K = N_1 N_2$ subdomains. We assume that both the coarse subdomains mesh and the fine element mesh defined by the knot vectors mesh are *shape regular* and quasi-uniform.

The Schur Complement System. Denote by $\Gamma := \left(\bigcup_{k=1}^K \partial \widehat{\Omega}^{(k)}\right) \backslash \partial \widehat{\Omega}$ the subdomain interface and by $\Theta_\Gamma = \{(i,j) : \text{supp}(B_{i,j}^{p,q}) \cap \Gamma \ne \emptyset\}$ the set of indices associated with the "fat" interface, consisting of several layers of knots associated with the basis functions with support intersecting two or more subdomains, see, e.g., Fig. 1.

As in classical iterative substructuring, we reduce the original system (7) to one on the interface by static condensation, i.e., we eliminate the interior degrees of freedom (denoted by subscript I) associated with the basis functions with support in only one subdomain and interface degrees of freedom (denoted by subscript Γ), obtaining the Schur complement system

$$\widehat{S}_\Gamma w = \widehat{f}, \tag{8}$$

where using the same subscripts I and Γ on matrix and vector blocks, we have $\widehat{S}_\Gamma = A_{\Gamma\Gamma} - A_{\Gamma I}A_{II}^{-1}A_{\Gamma I}^T$, $\widehat{f} = f_\Gamma - A_{\Gamma I}A_{II}^{-1}f_I$. The Schur complement system (8) is solved by a Preconditioned Conjugate Gradient (PCG) iteration, where $\widehat{S}_\Gamma$ is never explicitly formed since the action of $\widehat{S}_\Gamma$ on a vector is computed by solving Dirichlet problems for individual subdomains and some sparse matrix-vector multiplications, which are also needed when working with the local Schur complements required by the application of the BDDC preconditioner defined below. The preconditioned Schur complement system solved by PCG is then

$$M_{\text{BDDC}}^{-1}\widehat{S}_\Gamma w = M_{\text{BDDC}}^{-1}\widehat{f}, \tag{9}$$

where M_{BDDC}^{-1} is the BDDC preconditioner, defined in (11) below.

The BDDC Preconditioner. We denote by $A^{(k)}$ the local stiffness matrix associated with the subdomain $\bar{\Omega}^{(k)}$. After partitioning the local degrees of freedom into those in the interior (I) and those on the interface (Γ), as before, we further partition the latter into dual (Δ) and primal (Π) degrees of freedom. The associated primal basis functions will be made continuous across the interface by subassembling them among their supporting elements. The dual basis functions can be discontinuous across the interface and will vanish at the primal degrees of freedom. Specific choices for the selection of primal degrees of freedom will be given below. According to this splitting, $A^{(k)}$ can then be written as

$$A^{(k)} = \begin{bmatrix} A_{II}^{(k)} & A_{\Gamma I}^{(k)^T} \\ A_{\Gamma I}^{(k)} & A_{\Gamma\Gamma}^{(k)} \end{bmatrix} = \begin{bmatrix} A_{II}^{(k)} & A_{\Delta I}^{(k)^T} & A_{\Pi I}^{(k)^T} \\ A_{\Delta I}^{(k)} & A_{\Delta\Delta}^{(k)} & A_{\Pi\Delta}^{(k)^T} \\ A_{\Pi I}^{(k)} & A_{\Pi\Delta}^{(k)} & A_{\Pi\Pi}^{(k)} \end{bmatrix}. \tag{10}$$

The BDDC preconditioner can be written as

$$M_{\text{BDDC}}^{-1} = \widetilde{R}_{D,\Gamma}^T \widetilde{S}_\Gamma^{-1} \widetilde{R}_{D,\Gamma}, \quad \text{where} \tag{11}$$

$$\widetilde{S}_\Gamma^{-1} = \widetilde{R}_{\Gamma\Delta}^T \left(\sum_{k=1}^K \begin{bmatrix} 0 & R_\Delta^{(k)^T} \end{bmatrix} \begin{bmatrix} A_{II}^{(k)} & A_{\Delta I}^{(k)^T} \\ A_{\Delta I}^{(k)} & A_{\Delta\Delta}^{(k)} \end{bmatrix}^{-1} \begin{bmatrix} 0 \\ R_\Delta^{(k)} \end{bmatrix} \right) \widetilde{R}_{\Gamma\Delta} + \Phi S_{\Pi\Pi}^{-1}\Phi^T.$$

Here $S_{\Pi\Pi}$ is the BDDC coarse matrix, Φ is a matrix mapping primal degrees of freedom to interface variables defined in (18) below, and $\widetilde{R}_{\Gamma\Delta}$ and $R_\Delta^{(k)}$ are appropriate restriction matrices; see, e.g., Li and Widlund (2006). The matrix $\widetilde{R}_{D,\Gamma}^T$ defines the BDDC scaling adopted, that here will be the deluxe scaling defined in (12), (13) below. We note that the choices of primal constraints and scaling are fundamental for the construction of efficient BDDC preconditioners.

In our previous works Beirão da Veiga et al. (2013a, 2014b), we proved, with an appropriate choice of primal constraints, that the condition number of the resulting

BDDC preconditioner satisfies a classical polylogarithmic bound

$$\text{cond}\left(M_{BDDC}^{-1}\widehat{S}_\Gamma\right) \leq C(1 + \log(H/h))^2,$$

with $C > 0$ independent of h, H and the jumps of the coefficient ρ across the interface Γ.

Deluxe Scaling (Dohrmann and Widlund (2013)). We split the interface Γ into equivalence classes, associated with subdomain vertices ($\mathcal{V}$), edges ($\mathcal{E}$), and in three-dimensions faces ($\mathcal{F}$), defined by the set of indices of the degrees of freedom belonging to the analogous subdomain boundaries. For simplicity, we define here the deluxe scaling for the class of $\mathcal{F}$ with only two elements, k, j, as for an edge in two dimensions or a face in three dimensions. Consider the local Schur complements $S^{(k)}$ and $S^{(j)}$ associated to subdomains $\Omega^{(k)}$ and $\Omega^{(j)}$, respectively. We define two principal minors, $S_{\mathcal{F}}^{(k)}$ and $S_{\mathcal{F}}^{(j)}$, obtained by removing all rows and columns which do not belong to the degrees of freedom which are common only to the fat boundaries of $\Omega^{(k)}$ and $\Omega^{(j)}$. The deluxe scaling across $\mathcal{F}$ is then defined by

$$D_{\mathcal{F}}^{(k)} := S_{\mathcal{F}}^{(k)}\left(S_{\mathcal{F}}^{(k)} + S_{\mathcal{F}}^{(j)}\right)^{-1}. \tag{12}$$

If these Schur complements have small dimensions, they can be computed explicitly, otherwise the action of $\left(S_{\mathcal{F}}^{(k)} + S_{\mathcal{F}}^{(j)}\right)^{-1}$ can be computed by solving a Dirichlet problem on the union of the relevant subdomains with a zero right hand side in the interiors of the subdomains. While these strategies are viable in two dimensions, in our three-dimensional tests we use the numerical factorization package MUMPS, see Amestoy et al. (2001), which computes explicitly the subdomain Schur complements (14) while factoring the subdomain problem (10).

We then define the block-diagonal scaling matrix

$$D^{(k)} = diag(D_{\mathcal{F}_{j_1}}^{(k)}, D_{\mathcal{F}_{j_2}}^{(k)}, \ldots, D_{\mathcal{F}_{j_k}}^{(k)}),$$

where $j_1, j_2, \ldots, j_k$ are the indices of all the $\Omega^{(j)}$, $j \neq k$, that share a face $\mathcal{F}$ with $\Omega^{(k)}$. We can now define the scaled local operators by $R_{D,\Gamma}^{(k)} := D^{(k)}R_\Gamma^{(k)}$ and the global scaled operator by

$$\widetilde{R}_{D,\Gamma} := \oplus_{k=1}^K R_{D,\Gamma}^{(k)}. \tag{13}$$

Generalized Eigenvalue Problems and Parallel Sums. Consider a fat edge $\mathcal{E}$ of a subdomain $\Omega^{(k)}$ and its complement $\mathcal{E}' := \Gamma_i \setminus \mathcal{E}$. We write the local Schur complement associated to $\Omega^{(k)}$ as

$$S^{(k)} = \begin{pmatrix} S_{\mathcal{E}'\mathcal{E}'}^{(k)} & S_{\mathcal{E}'\mathcal{E}}^{(k)^T} \\ S_{\mathcal{E}'\mathcal{E}}^{(k)} & S_{\mathcal{E}\mathcal{E}}^{(k)} \end{pmatrix},$$

and we define the Schur complement of a Schur complement by

$$\widetilde{S}_{\mathcal{E}\mathcal{E}}^{(k)} := S_{\mathcal{E}\mathcal{E}}^{(k)} - S_{\mathcal{E}'\mathcal{E}}^{(k)} S_{\mathcal{E}'\mathcal{E}'}^{(k)-1} S_{\mathcal{E}'\mathcal{E}}^{(k)^T}. \tag{14}$$

Analogous blocks $S_{\mathcal{V}\mathcal{V}}^{(k)}$, $\widetilde{S}_{\mathcal{V}\mathcal{V}}^{(k)}$ are defined for a fat vertex $\mathcal{V}$ of $\Omega^{(k)}$ and blocks $S_{\mathcal{F}\mathcal{F}}^{(k)}$, $\widetilde{S}_{\mathcal{F}\mathcal{F}}^{(k)}$ for a fat face $\mathcal{F}$ of $\Omega^{(k)}$. We note that these blocks are only positive semidefinite for subdomains in the interior of the domain Ω. In the definition of the parallel sum given in (15) below, we handle any such singular matrices by using generalized inverses or by adding to any singular $S^{(k)}$ the term ϵI, with $\epsilon > 0$ small compared with the eigenvalues of $S^{(k)}$.

Our adaptive selection of primal constraints will be based on generalized eigenvalue problems (GEP) based on the following definition of parallel sum [see Anderson and Duffin (1969), Tian (2002)] of r positive definite matrices $A^{(1)}, A^{(2)}, \cdots, A^{(r)}$ as

$$A^{(1)} : A^{(2)} \cdots : A^{(r)} := \left(A^{(1)^{-1}} + A^{(2)^{-1}} + \cdots + A^{(r)^{-1}} \right)^{-1}. \tag{15}$$

We define a first GEP $\mathcal{V}_{par}$ as follows: let $\mathcal{V}$ be a fat vertex in 2D shared by four subdomains $\Omega^{(i)}$, $\Omega^{(j)}$, $\Omega^{(k)}$, $\Omega^{(\ell)}$, and define the GEP

$$\left(\widetilde{S}_{\mathcal{V}\mathcal{V}}^{(i)} : \widetilde{S}_{\mathcal{V}\mathcal{V}}^{(j)} : \widetilde{S}_{\mathcal{V}\mathcal{V}}^{(k)} : \widetilde{S}_{\mathcal{V}\mathcal{V}}^{(\ell)} \right) \phi = \lambda \left(S_{\mathcal{V}\mathcal{V}}^{(i)} : S_{\mathcal{V}\mathcal{V}}^{(j)} : S_{\mathcal{V}\mathcal{V}}^{(k)} : S_{\mathcal{V}\mathcal{V}}^{(\ell)} \right) \phi. \tag{16}$$

We define another GEP $\mathcal{E}_{par}$ as follows: Let $\mathcal{E}$ be a fat edge in 2D shared by two subdomains $\Omega^{(i)}$, $\Omega^{(j)}$, and define the GEP

$$\left(\widetilde{S}_{\mathcal{E}\mathcal{E}}^{(i)} : \widetilde{S}_{\mathcal{E}\mathcal{E}}^{(j)} \right) \phi = \lambda \left(S_{\mathcal{E}\mathcal{E}}^{(i)} : S_{\mathcal{E}\mathcal{E}}^{(j)} \right) \phi. \tag{17}$$

The analogous GEP $\mathcal{V}_{par}$ for a fat vertex in 3D will involve parallel sums with eight terms, while four terms will be involved for a fat edge in 3D and two terms for a fat face in 3D (since we are considering IGA regular decompositions). Alternative choices of generalized eigenvalue problems based on both parallel and standard sums of matrices can be found in Beirão da Veiga et al. (2016).

Adaptive Choices of Reduced Sets of Primal Constraints. Inspired by the techniques of Dohrmann and Pechstein (2011), we propose an adaptive selection of primal constraints, driven by the desire to reduce the expensive fat vertex/edge/face primal constrains used in the standard or deluxe BDDC method. In order to construct the BDDC primal space, we select a threshold $0 < \theta < 1$, a set of GEPs associated to the equivalence classes considered (subdomain vertices and/or edges and/or faces) and for each equivalence class, we use the following two-step strategy:

(a) select the eigenvectors $\{v_1, v_2, \ldots, v_{N_c}\}$ of the generalized eigenproblem (16) that are associated to the eigenvalues $\{\lambda_1, \lambda_2, \ldots, \lambda_{N_c}\}$ smaller than θ;

(b) perform the following BDDC change of basis in order to introduce the selected eigenvectors as new primal constraints:

(b1) denoting by $\widetilde{S}_V \phi = \lambda S_V \phi$ the eigenproblem (16), compute the matrix

$$A_V = S_V[v_1 v_2, \ldots, v_{N_c}] \in \mathbb{R}^{n \times N_c},$$

with n the size of the v_i, $i = 1, \ldots, N_c$, and $N_c \leq n$ the number of primal constraints selected;

(b2) compute the SVD decomposition of A_V, i.e. the matrices U, S, V such that $A_V = USV^T$ and denote by C^T the first N_c columns of U;

(b3) compute the QR factorization $C^T = QR$, where $Q = [Q_{range} \ Q_{null}] \in \mathbb{R}^{n \times n}$, with $Q_{range} \in \mathbb{R}^{n \times N_c}$ and $Q_{null} \in \mathbb{R}^{n \times (n-N_c)}$ spanning the range and the kernel of C^T, respectively, and $R = \begin{bmatrix} \widetilde{R} \\ 0 \end{bmatrix} \in \mathbb{R}^{n \times N_c}$, with $\widetilde{R} \in \mathbb{R}^{N_c \times N_c}$ upper triangular;

(b4) construct the matrix Φ realizing the BDDC change of basis as

$$\Phi = [Q_{range}\widetilde{R}^{-T} \ Q_{null}]. \tag{18}$$

We denote the resulting primal spaces with the same name as the associated GEP they are based on. Among the possible combinations, we will consider the primal spaces $\mathcal{V}_{par}$ and $\mathcal{E}_{par}$ in 2D, while in 3D we will need the richer primal space $\mathcal{VEF}_{par}$ employing GEP $\mathcal{V}_{par}, \mathcal{E}_{par}, \mathcal{F}_{par}$.

4 Numerical Results

We now present the results of numerical experiments with the model problem (1) discretized on a 2D quarter-ring domain (see Fig. 2a) and on a 3D twisted domain (see Fig. 3a) using isogeometric NURBS spaces with mesh size h, polynomial degree p and regularity k. The domain is decomposed into K non-overlapping subdomains of characteristic size H, as described in Sect. 3. The Schur complement problems are solved by the PCG method with the isogeometric BDDC deluxe preconditioner described before, with a zero initial guess and a stopping criterion of a 10^{-6} reduction of the Euclidean norm of the PCG residual. In the tests, we study how the convergence rate of the BDDC preconditioner depends on h, K, p, k, and jumps in the coefficient of the elliptic problem. In all tests, the BDDC condition number is essentially the maximum eigenvalue of the preconditioned operator, since its minimum eigenvalue is always very close to 1. The 2D tests have been performed with a MATLAB code based on the GeoPDEs library, De Falco et al. (2011), while the 3D parallel tests have been performed using the PETSc library, Balay and et al.

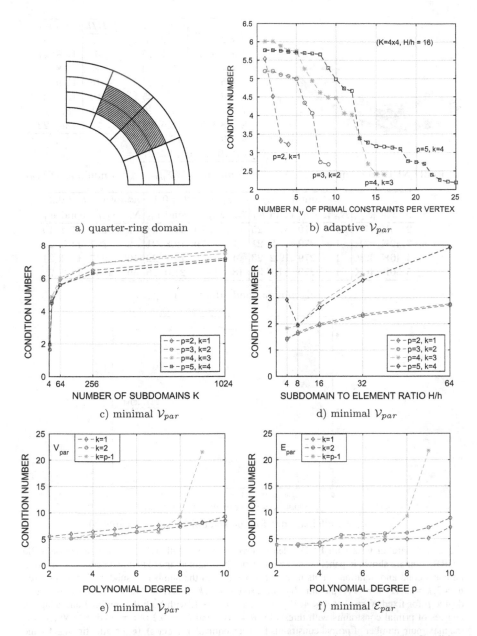

Fig. 2 2D tests with BDDC deluxe preconditioner. (**a**) Quarter-ring domain. (**b**) Condition numbers with adaptive coarse space V_{par} as a function of the number of vertex primal constraints for fixed $K = 4 \times 4, H/h = 16$, various degrees p and maximal regularity $k = p - 1$. The other panels (**c**)–(**f**) show the BDDC condition numbers with minimal ($N_C^V = 1$) primal space V_{par} as a function of: (**c**) the number of subdomains K for fixed $H/h = 8$; (**d**) the ratio H/h for fixed $K = 4 \times 4$; (**e**) the polynomial degree p for different regularity $k = 1, 2, p - 1$ and fixed $K = 4 \times 4, H/h = 16$. The last panel (**f**) is the analog of (**e**) but with minimal $\mathcal{E}_{par}$ coarse space

a) 3D NURBS domain

K	cond	n_{it}
$2 \times 2 \times 2$	2.2	8
$3 \times 3 \times 3$	10.1	16
$4 \times 4 \times 4$	13.4	22
$5 \times 5 \times 5$	15.4	24
$6 \times 6 \times 6$	16.8	25
$7 \times 7 \times 7$	17.8	26
$8 \times 8 \times 8$	18.5	26
$9 \times 9 \times 9$	19.8	27
$10 \times 10 \times 10$	19.6	27

b) minimal $\mathcal{VEF}_{par}$

H/h	cond	n_{it}
6	13.4	22
7	12.8	21
8	12.8	21
9	12.9	21
10	13.1	21
11	13.3	22
12	13.6	22

c) minimal $\mathcal{VEF}_{par}$

			minimal			adaptive ($\theta = 0.1$)			adaptive ($\theta = 0.2$)		
p	$|A|$	$|\widehat{S}_\Gamma|$	N_c	$|S_{\Pi\Pi}|$	cond n_{it}	N_c	$|S_{\Pi\Pi}|$	cond n_{it}	N_c	$|S_{\Pi\Pi}|$	cond n_{it}
2	39K	17K	1	279	31.9 25	1	279	31.8 24	2	291	17.4 19
3	42K	25K	1	279	12.8 21	1	279	12.8 21	2	287	11.5 20
4	46K	32K	1	279	19.2 23	4	350	14.7 22	16	967	14.2 21
5	50K	40K	1	279	44.1 32	18	1150	21.0 26	49	4354	15.3 22

d) minimal and adaptive $\mathcal{VEF}_{par}$

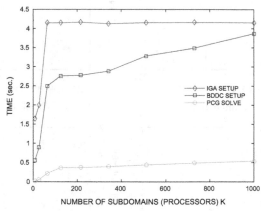

e) Parallel timings for the scalability test of Table b)

Fig. 3 3D parallel tests with BDDC deluxe preconditioner with $\mathcal{VEF}_{par}$ coarse space on a 3D NURBS domain shown in panel (**a**) and with each subdomain assigned to one processor. Condition numbers cond and iteration counts n_{it} as functions of: (**b**) the number of subdomains K for fixed $p = 3, k = 2, H/h = 6$; (**c**) the ratio H/h for fixed $p = 3, k = 2, K = 4 \times 4 \times 4$; (**d**) the polynomial degree p for fixed $K = 4 \times 4 \times 4, H/h = 8, k = p - 1$, with both the minimal and adaptive choices of primal constraints with thresholds $\theta = 0.1$ and $\theta = 0.2$ ($N_c = \max(N_c^V, N_c^E, N_c^F)$ is the maximum number of primal constraints for each equivalence class): (**e**) parallel timings for the scalability test of Table (**b**)

(2015), with the PCBDDC preconditioner [contributed to the PETSc library by S. Zampini, see Zampini (2016)], the PetIGA library, Dalcin et al. (2016), and run on the parallel machine Shaheen XC40 of KAUST.

2D Tests with $\mathcal{V}_{par}$ and $\mathcal{E}_{par}$. Figure 2 reports the results of several tests for various degrees p and maximal regularity $k = p - 1$ with the BDDC deluxe preconditioner with $\mathcal{V}_{par}$ coarse space on the quarter-ring domain shown in panel (a). Panel (b) shows that the condition number improves when the number of vertex primal constraints per vertex is increased from the minimal value $N_C^V = 1$ to the maximal value $N_C^V = (k + 1)^2$ (here $K = 4 \times 4, H/h = 16$ are fixed). For $p \geq 3$, the improvement is minimal when only a few vertex functions are added to the $\mathcal{V}_{par}$ primal space, but the improvement becomes substantial when about $p^2/3$ vertex functions are added.

Panel (c) show the scalability of the deluxe BDDC with minimal ($N_C^V = 1$) primal space $\mathcal{V}_{par}$ for increasing number of subdomains K (for fixed $H/h = 8$), while Panel (d) shows the quasi-optimality of deluxe BDDC with minimal $\mathcal{V}_{par}$ for increasing ratio H/h (for fixed $K = 4 \times 4$). Panels (e) and (f) show the robustness of both minimal $\mathcal{V}_{par}$ and minimal $\mathcal{E}_{par}$ with respect to the polynomial degree p, with $\mathcal{E}_{par}$ yielding slightly better results than $\mathcal{V}_{par}$. In both cases, robustness is lost in case of maximal regularity $k = p - 1$ and high degree $p \geq 8$, but it could be recovered by increasing the primal space, i.e. by considering $N_C^V \geq 1$.

3D Parallel Tests with $\mathcal{VEF}_{par}$. Figure 3 reports the condition numbers cond and iteration counts n_{it} for BDDC deluxe with $\mathcal{VEF}_{par}$ coarse space on a 3D NURBS domain shown in Panel (a). The tests have been run on the parallel machine Shaheen XC40 of KAUST, with a number of processors equal to the number of subdomains K. The minimal $\mathcal{V}_{par}$ and $\mathcal{VE}_{par}$ coarse spaces did not work well in 3D, yielding high condition numbers ($\geq 10^3$) already for low polynomial degree, so we report only the results with $\mathcal{VEF}_{par}$. Table (b) shows the scalability of $\mathcal{VEF}_{par}$ for an increasing number of subdomains K for fixed $p = 3, k = 2, H/h = 6$. The associated timings (for both the preconditioner setup and the PCG solve) are plotted in panel (e). Table (c) shows the quasi-optimality of $\mathcal{VEF}_{par}$ for an increasing ratio H/h, for fixed $p = 3, k = 2, K = 4 \times 4 \times 4$. Table (d) reports the results for an increasing polynomial degree p for fixed $K = 4 \times 4 \times 4, H/h = 8, k = p-1$, with both the minimal ($N_c = 1$) and adaptive choice ($N_c \geq 1$) of primal constraints, where $N_c = \max(N_c^V, N_c^E, N_c^F)$ is the maximum number of primal constraints over all equivalence classes (fat vertices, edges, faces). The table reports also on the dimension $|A|$ of the stiffness matrix, $|\widehat{S}_\Gamma|$ of the Schur complement, and $|S_{\Pi\Pi}|$ of the coarse space. As in the 2D tests, the minimal primal space loses robustness for increasing p (except the initial condition number drop from $p = 2$ to $p = 3$), but robustness can be recovered by adaptively increasing the number of primal constraints.

Acknowledgements For computer time, this research used the resources of the Supercomputing Laboratory at King Abdullah University of Science and Technology (KAUST) in Thuwal, Saudi Arabia. The Authors would like to thank L. Dalcin for the 3D NURBS geometry.

References

P.R. Amestoy, I.S. Duff, J.-Y. L'Excellent, J. Koster, A fully asynchronous multifrontal solver using distributed dynamic scheduling. SIAM J. Matrix Anal. Appl. **23**(1), 15–41 (2001)

W.N. Anderson Jr., R.J. Duffin, Series and parallel addition of matrices. J. Math. Anal. Appl. **26**, 576–594 (1969)

S. Balay et al., PETSc Web page (2015), http://www.mcs.anl.gov/petsc

L. Beirão da Veiga, D. Cho, L.F. Pavarino, S. Scacchi, BDDC preconditioners for Isogeometric Analysis. Math. Mod. Meth. Appl. Sci. **23**, 1099–1142 (2013a)

L. Beirão da Veiga, D. Cho, L.F. Pavarino, S. Scacchi, Isogeometric Schwarz preconditioners for linear elasticity systems. Comp. Meth. Appl. Mech. Eng. **253**, 439–454 (2013b)

L. Beirão da Veiga, A. Buffa, G. Sangalli, R. Vazquez, Mathematical analysis of variational isogeometric methods. ACTA Numer. **23**, 157–287 (2014a)

L. Beirão da Veiga, L.F. Pavarino, S. Scacchi, O.B. Widlund, S. Zampini, Isogeometric BDDC preconditioners with deluxe scaling. SIAM J. Sci. Comp. **36**, A1118–A1139 (2014b)

L. Beirão da Veiga, L.F. Pavarino, S. Scacchi, O.B. Widlund, S. Zampini, Adaptive selection of primal constraints for isogeometric BDDC deluxe preconditioners SIAM J. Sci. Comput. (2017), to appear

J.G. Calvo, O.B. Widlund, An adaptive choice of primal constraints for BDDC domain decomposition algorithms. Electron. Trans. Numer. Anal. **45**, 524–544 (2016)

N. Collier, L. Dalcin, D. Pardo, V.M. Calo, The cost of continuity: performance of iterative solvers on isogeometric finite elements. SIAM J. Sci. Comput. **35**(2), A767–A784 (2013)

J.A. Cottrell, T.J.R. Hughes, Y. Bazilevs, *Isogeometric Analysis. Towards integration of CAD and FEA* (Wiley, New York, 2009)

L. Dalcin, N. Collier, P. Vignal, A.M.A. Côrtes, V.M. Calo, PetIGA: a framework for high-performance isogeometric analysis. Comput. Meth. Appl. Mech. Eng. **308**, 151–181 (2016). ISSN: 0045-7825

C. De Falco, A. Reali, R. Vazquez, GeoPDEs: a research tool for Isogeometric Analysis of PDEs. Adv. Eng. Softw. **42**, 1020–1034 (2011)

C.R. Dohrmann, C. Pechstein, Constraint and weight selection algorithms for BDDC. Slides of a talk by Dohrmann at DD21 in Rennes, June 2011. http://www.numa.uni-linz.ac.at/~clemens/dohrmann-pechstein-dd21-talk.pdf.

C.R. Dohrmann, O.B. Widlund, Some recent tools and a BDDC algorithm for 3D problems in H(curl), in *Domain Decomp. Meth. Sci. Engrg. XX, San Diego, CA, 2011*. Springer LNCSE, vol. 91 (2013), pp. 15–26

K. Gahalaut, J. Kraus, S. Tomar, Multigrid methods for isogeometric discretization. Comp. Meth. Appl. Mech. Eng. **253**, 413–425 (2013)

T.J.R. Hughes, J.A. Cottrell, Y. Bazilevs, Isogeometric analysis: CAD, finite elements, NURBS, exact geometry, and mesh refinement. Comp. Meth. Appl. Mech. Eng. **194**, 4135–4195 (2005)

H.H. Kim, E.T. Chung, A BDDC algorithm with optimally enriched coarse space for two-dimensional elliptic problems with oscillatory and high contrast coefficients. Multiscale Model. Simul. **13**(2), 571–593 (2015)

H.H. Kim, E.T. Chung, J. Wang, BDDC and FETI-DP algorithms with adaptive coarse spaces for three-dimensional elliptic problems with oscillatory and high contrast coefficients (2015, submitted). arXiv:1606.07560

A. Klawonn, M. Lanser, P. Radtke, O. Rheinbach, On an adaptive coarse space and on nonlinear domain decomposition, in *Domain Decomp. Meth. Sci. Engrg. XXI, Rennes, 2012*. Springer LNCSE, vol. 98 (2014a)

A. Klawonn, P. Radtke, O. Rheinbach, FETI-DP with different scalings for adaptive coarse spaces. Proc. Appl. Math. Mech. **14**(1), 835–836 (2014b)

A. Klawonn, M. Kühn, O. Rheinbach, Adaptive coarse spaces for FETI-DP in three dimensions. Technical report 2015-11, Mathematik und Informatik, Bergakademie Freiberg, 2015a

A. Klawonn, P. Radtke, O. Rheinbach. FETI–DP methods with an adaptive coarse space. SIAM J. Numer. Anal. **53**(1), 297–320 (2015b)

A. Klawonn, P. Radtke, O. Rheinbach, A comparison of adaptive coarse spaces for iterative substructuring in two dimensions. Elec. Trans. Numer. Anal. **45**, 75–106 (2016)

S.K. Kleiss, C. Pechstein, B. Jüttler, S. Tomar, IETI - Isogeometric Tearing and Interconnecting. Comp. Meth. Appl. Mech. Eng. **247–248**, 201–215 (2012)

J. Li, O.B. Widlund, FETI-DP, BDDC, and block Cholesky methods. Int. J. Numer. Meth. Eng. **66**, 250–271 (2006)

J. Mandel, B. Sousedík, J. Šístek, Adaptive BDDC in three dimensions. Math. Comput. Simul. **82**(10), 1812–1831 (2012)

C. Pechstein, C.R. Dohrmann, Modern domain decomposition methods ? BDDC, deluxe scaling, and an algebraic approach. 2013. Seminar talk, Linz, December 2013. http://people.ricam. oeaw.ac.at/c.pechstein/pechstein-bddc2013.pdf.

N. Spillane, V. Dolean, P. Hauret, P. Nataf, J. Rixen, Solving generalized eigenvalue problems on the interface to build a robust two-level FETI method. C. R. Math. Acad. Sci. Paris **351**(5–6), 197–201 (2013)

Y. Tian, How to express a parallel sum of k matrices. J. Math. Anal. Appl. **266**(2), 333–341 (2002)

S. Zampini, PCBDDC: a class of robust dual-primal preconditioners in PETSc. SIAM J. Sci. Comput. **38**(5), S282–S306 (2016)

Development of Nonlinear Structural Analysis Using Co-rotational Finite Elements with Improved Domain Decomposition Method

Haeseong Cho, JunYoung Kwak, Hyunshig Joo, and SangJoon Shin

1 Introduction

Recent advances in computational science and technologies induce increasing size of the engineering problems, and impact the fields of computational fluids and structural dynamics as well as multi-physics problems, such as fluid-structure interactions. At the same time, structural components used in many engineering applications show geometrically nonlinear characteristics. Therefore, development of effective solution methodologies for large-size nonlinear structural problems is required seriously in the fields of the mechanical and aerospace engineering. Especially, general finite element methods require a large number of elements in order to predict precise stress or deformation, resulting in increased computational costs due to enlarged computational time and memory requirement. Therefore, careful selection of grid size and solution methodology becomes important.

One of the most successful approaches for large-size finite element analysis is the finite element tearing and interconnecting (FETI) method proposed by Farhat and Roux (1991). The basic idea of FETI is to decompose the computational

H. Cho • H. Joo
Department of Mechanical and Aerospace Engineering, Seoul National University, Seoul, South Korea
e-mail: nicejjo@snu.ac.kr; hyunshigjoo@snu.ac.kr

J.Y. Kwak
Rocket Engine Team, Korea Aerospace Research Institute, Daejeon, South Korea
e-mail: kjy84@kari.re.kr

S. Shin (✉)
Department of Mechanical and Aerospace Engineering, Institute of Advanced Aerospace Technology, Seoul National University, Seoul, South Korea
e-mail: ssjoon@snu.ac.kr

© Springer International Publishing AG 2017
C.-O. Lee et al. (eds.), *Domain Decomposition Methods in Science and Engineering XXIII*, Lecture Notes in Computational Science and Engineering 116, DOI 10.1007/978-3-319-52389-7_3

domain into non-overlapping sub-domains. Lagrange multipliers are used to enforce compatibility of the degrees of freedom along the interfaces between the sub-domains. The manner of handling such interfaces can distinguish the interface problem. Recently, the dual-primal FETI (FETI-DP) method (Farhat et al., 2000) was proposed; it is a dual sub-structuring method, which introduces Lagrange multipliers and a small number of coarse mesh nodes to enforce the continuity at sub-domain interfaces. The resulting dual problem is then solved by seeking a saddle-point of the relevant Lagrangian functional. The FETI-DP method is a standard preconditioned conjugate algorithm, which may use an arbitrary initial guess. Thus, the solution of the interface problem is obtained using an iterative process, which requires an adequate pre-conditioner. Therefore, to improve solution convergence, iterative solvers rely on various types of preconditioning techniques. By observing such limitation, the combination of domain decomposition methods with the direct solvers was significantly investigated, an approach that seems to have received little attention thus far (Guèye et al., 2011). Bauchau (2010) suggested the use of an augmented Lagrangian formulation (ALF) in conjunction with both global and local Lagrange multipliers. The use of augmented Lagrangian terms was considered to improve the conditioning of the flexibility matrix, thereby increasing the convergence performance of the iterative procedure used to solve the interface problem. As a preliminary step to the present effort, the authors proposed an improved domain decomposition approach, the FETI-Local, and the FETI algorithm was developed for multibody type structures (Kwak et al., 2014). Moreover, in order to improve the computational efficiency, a parallel version of the column solver was employed to deal with the interface problem (Kwak et al., 2015).

On the other hand, a co-rotational (CR) formulation has been developed and improved in accordance with an increased amount of interest during the last few decades to analyze the geometrical nonlinearity of structures (Felippa and Haugen, 2005). The main advantage of the CR framework is that it leads to an artificial separation between the material and any geometrical nonlinearity. This concept was originally developed by Rankin et al. during the formulating procedure of what is known as the element-independent co-rotational (EICR) description (Rankin and Brogan, 1986). In addition, Felippa et al. concluded that the CR formulation would be extremely useful for elements of a simple geometry; they were able to provide a reasonable solution to the localized failure problem as well (Felippa and Haugen, 2005). However, such nonlinear structural analysis would be confronted with the significant computational problem with increasing computational costs due to enlarged computational time and memory requirement, followed by prediction of precise stress and large deformation. Thus, an effective solution methodology for large-size nonlinear structural problem would be suggested through an extension of the CR framework into the FETI-Local method.

This manuscript is organized as follows. Formulation procedure of the FETI-Local method will be described. After that, derivation of the CR framework will be introduced. Then, unified computational algorithm of the FETI-Local and the CR framework will be described. Finally, computational cost and scalability results obtained by the proposed approach will be presented.

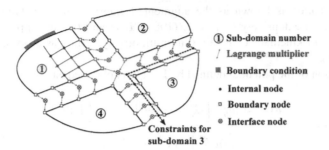

Fig. 1 Planar solid separated into four non-overlapping sub-domains by following the FETI-Local

2 Domain Decomposition Method: FETI-Local

Consider a planar solid depicted in Fig. 1. To develop a parallel solution algorithm for this problem, the solid is partitioned into N_s non-overlapping sub-domains. Each of these sub-domains could themselves be multibody systems comprising both elastic elements and nonlinear kinematic constraints. The FETI-Local uses local Lagrange multipliers to impose continuity of displacements at the nodes corresponding to adjacent sub-domains with those corresponding to the coarse mesh nodes. At corner nodes, i.e., at sub-domain cross-points, a single interface node is defined, and Lagrange multipliers are used to enforce equality of the displacements at the coarse mesh with those corresponding to all the adjacent nodes. Because four sub-domains are associated at this node, four boundary nodes would be created, one for each sub-domain. Note that for multiple connections, constraints and Lagrange multipliers remain localized, i.e., each associated with a single sub-domain. In finite element formulations, this approach has been used to enforce the continuity of displacement fields between adjacent incompatible elements (Tong and Pian, 1973). The same approach, called "localized version of the method of Lagrange multipliers," has been advocated by Park et al. (2000).

In the FETI-Local method, the kinematic continuity conditions between sub-domain interfaces is enforced via the localized Lagrange multiplier technique. Let $\underline{u}_b^{[j]}$ and $\underline{c}^{[j]}$ denote the arrays of dofs at a boundary node and at an interface node, respectively. Kinematic constraint j is written as $\underline{C}^{[j]} = \underline{u}_b^{[j]} - \underline{c}^{[j]} = \underline{0}$ and the associated potential is

$$V_c^{[j]} = s\underline{\lambda}^{[j]T}\underline{C}^{[j]} + \frac{p}{2}\underline{C}^{[j]T}\underline{C}^{[j]}, \tag{1}$$

where $\underline{\lambda}^{[j]}$ is the array of Lagrange multipliers used to enforce the constraint, and s the scaling factor for those multipliers. The second term of the potential is a penalty term and p is the penalty coefficient. The potential defined by Eq. (1) combines the localized Lagrange multiplier technique with the penalty method.

This combination is known as the augmented Lagrangian formulation and has been examined extensively Gill et al. (1984). It is an effective approach for the enforcement of kinematic constraints in multibody dynamics, as proposed by Bayo et al. (1991).

A variation of the potential defined by Eq. (1) is obtained easily.

$$
\begin{aligned}
\delta V_c^{[j]} &= \delta \underline{u}_b^{[j]T} \left[s\underline{\lambda}^{[j]} + p\underline{C}^{[j]} \right] + \delta \underline{\lambda}^{[j]T} \left[s\underline{C}^{[j]} \right] \\
&\quad + \delta \underline{c}^{[j]T} \left[-s\underline{\lambda}^{[j]} - p\underline{C}^{[j]} \right],
\end{aligned}
\tag{2}
$$

The Lagrange multipliers become localized in the formulation, i.e., Lagrange multipliers are associated with one sub-domain unequivocally. The potential of kinematic constraint involves two types of dofs, the sub-domain dofs, $\underline{u}_b^{[j]}$ and $\underline{\lambda}^{[j]}$, and the interface dofs, $\underline{c}^{[j]}$. The constraint forces and stiffness matrix are partitioned to reflect this fact

$$
\underline{f}^{[j]} = \begin{Bmatrix} \underline{f}_b^{[j]} \\ \underline{f}_c^{[j]} \end{Bmatrix}, \quad
\underline{\underline{k}}^{[j]} = \begin{bmatrix} \underline{\underline{k}}_{bb}^{[j]} & \underline{\underline{k}}_{bc}^{[j]} \\ \underline{\underline{k}}_{bc}^{[j]T} & \underline{\underline{k}}_{cc}^{[j]} \end{bmatrix}.
\tag{3}
$$

Subscripts $(\cdot)_b$ and $(\cdot)_c$ denote dofs associated with boundary and interface nodes, respectively. Partitioning the constraint forces can be defined as follows.

$$
\underline{f}_b^{[j]} = \begin{Bmatrix} s\underline{\lambda}^{[j]} + p\underline{C}^{[j]} \\ s\underline{C}^{[j]} \end{Bmatrix}, \quad
\underline{f}_c^{[j]} = -\left\{ s\underline{\lambda}^{[j]} + p\underline{C}^{[j]} \right\}.
\tag{4}
$$

A similar operation for the constraint stiffness matrix leads to

$$
\underline{\underline{k}}_{bb}^{[j]} = \begin{bmatrix} p\underline{\underline{I}} & s\underline{\underline{I}} \\ s\underline{\underline{I}} & \underline{\underline{0}} \end{bmatrix}, \quad
\underline{\underline{k}}_{cc}^{[j]} = \begin{bmatrix} p\underline{\underline{I}} \end{bmatrix}, \quad
\underline{\underline{k}}_{bc}^{[j]} = \begin{bmatrix} -p\underline{\underline{I}} \\ -s\underline{\underline{I}} \end{bmatrix}.
\tag{5}
$$

Each constraint element contributes constraint forces and stiffness matrices defined by Eqs. (4) and (5), respectively. Using the standard assembly procedure used in the finite element method, the force arrays and stiffness matrices generated by all the constraint elements associated with sub-domain i are assembled into the following sub-domain arrays and matrices

$$
\underline{\check{F}}_b^{(i)} = \sum_{j=1}^{N_b^{(i)}} \underline{\underline{B}}_b^{[j]T} \underline{f}_b^{[j]}, \quad
\underline{\underline{\check{K}}}_{bb}^{(i)} = \sum_{j=1}^{N_b^{(i)}} \underline{\underline{B}}_b^{[j]T} \underline{\underline{k}}_{bb}^{[j]} \underline{\underline{B}}_b^{[j]},
\tag{6}
$$

where $\underline{\underline{B}}_b^{[j]}$ is the Boolean matrices used for the assembly process, i.e., $\underline{u}_b^{[j]} = \underline{\underline{B}}_b^{[j]} \underline{\check{u}}^{(i)}$. Of course, the assembly procedure can be performed in parallel for all sub-domains. Similarly, the constraint elements contribute force arrays and stiffness matrices to the interface problem,

$$\underline{F}_c^{(i)} = \sum_{j=1}^{N_b^{(i)}} \underline{\underline{B}}_c^{[j]T} \underline{f}_c^{[j]}, \quad \underline{\underline{K}}_{cc}^{(i)} = \sum_{j=1}^{N_b^{(i)}} \underline{\underline{B}}_c^{[j]T} \underline{\underline{k}}_{cc}^{[j]} \underline{\underline{B}}_c^{[j]}, \tag{7}$$

where $\underline{\underline{B}}_c^{[j]}$ is the Boolean matrices used for the assembly process, i.e., $\underline{c}^{[j]} = \underline{\underline{B}}_c^{[j]} \underline{c}$. Finally, the constraint coupling stiffness is assembled to find

$$\underline{\underline{K}}_{bc}^{(i)} = \sum_{j=1}^{N_b^{(i)}} \underline{\underline{B}}_b^{[j]T} \underline{\underline{k}}_{bc}^{[j]} \underline{\underline{B}}_c^{[j]}. \tag{8}$$

By considering the potential energy of the system composed of the strain energy (A)/the work done by external force(Φ)/additional energy induced by Lagrange multipliers(V_c), $\Pi = A + \Phi + V_c$, and the principle of minimum total potential energy, the governing equations can be expressed as

$$\begin{bmatrix} \text{diag}(\underline{\underline{\check{K}}}^{(\alpha)} + \underline{\underline{\check{K}}}_{bb}^{(\alpha)}) & \underline{\underline{K}}_{bc} \\ \underline{\underline{K}}_{bc}^T & \underline{\underline{K}}_{cc} \end{bmatrix} \begin{Bmatrix} \underline{\check{u}} \\ \underline{c} \end{Bmatrix} = \begin{Bmatrix} \underline{\check{Q}} - \underline{\check{F}}_b \\ -\underline{F}_c \end{Bmatrix}, \tag{9}$$

where $\underline{\check{Q}}^T = [\underline{Q}^T, 0]$ and $\underline{\check{u}}$ is the displacement of the sub-domain. The sub-domain stiffness matrix $\underline{\underline{\check{K}}}^{(\alpha)}$ is now

$$\underline{\underline{\check{K}}}^{(\alpha)} = \begin{bmatrix} \underline{\underline{K}}^{(\alpha)} & \underline{0} \\ \underline{0} & \underline{0} \end{bmatrix}. \tag{10}$$

Arrays $\underline{\check{F}}_b$ and $\underline{F}_c$ are the assembly of their sub-domain counterparts, $\underline{\check{F}}_b^{(i)}$ and $\underline{F}_c^{(i)}$, respectively, $\underline{\underline{K}}_{cc} = \sum_{i=1}^{N_s} \underline{\underline{K}}_{cc}^{(i)}$ and

$$\underline{\underline{K}}_{bc}^T = \begin{bmatrix} \underline{\underline{K}}_{bc}^{(1)T}, \underline{\underline{K}}_{bc}^{(2)T}, \dots, \underline{\underline{K}}_{bc}^{(N_s)T} \end{bmatrix}. \tag{11}$$

The block-diagonal nature of the leading entry of the system matrix makes this approach amenable to parallel solution algorithms.

3 Co-rotational (CR) Finite Elements

Figure 2 shows the coordinates defined in the present CR framework and rotational transformations when obeying the elemental kinematics. Beginning with the fixed frame, a rotational operator, $\underline{\underline{R}}_o$, can be defined by tracking the elemental initial state.

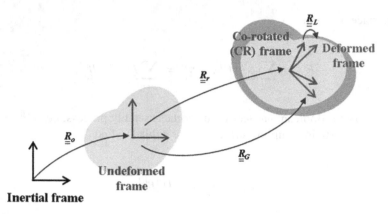

Fig. 2 Coordinate in the CR framework

The rotational operator, $\underset{=}{R}_G$, can be defined by elemental rotational displacement referring to an undeformed configuration. The complete behavior included in this case can be decomposed into rigid body rotation and elastic deformational rotation. According to such kinematics, the origin of each coordinate is taken at the centroid of the triangle.

In the CR formulation, the existing linearized formulation is selected for the local system matrices, i.e., the stiffness matrix and the internal load vector. These physical variables is re-expressed between the local and global quantities by the introduction of a transformation matrix. The virtual work with respect to the local and global systems can be obtained in terms of the local and global internal load vectors and displacements.

$$V = \delta q_G^T f_G = \delta q_L^T f_L = \delta q_G^T \underset{=}{B}^T f_L \tag{12}$$

Hence the global internal load vector is obtained with Eq. (12) by taking the transformation matrix, $\underset{=}{B}$, into account.

$$f_G = \underset{=}{B}^T f_L, \quad f_L = \left\{ f_L^i \right\}^T \quad i = 1, 2, \ldots, N_e, \tag{13a}$$

$$f_L^i = \left\{ n_1^i, n_2^i, m^i \right\}^T \quad i = 1, 2, \ldots, N_e. \tag{13b}$$

By the differentiation of Eq. (12) with respect to the displacements, the internal load vector can then be

$$\delta f_G = \underset{=}{K}_G \delta q_G \tag{14}$$

In addition, by Eqs. (12) and (14) the global stiffness matrix $\underline{\underline{K}}_G$ can be derived as shown below.

$$\underline{\underline{K}}_G = \underline{\underline{B}}^T \underline{\underline{K}}_L \underline{\underline{B}} + \underline{\underline{K}}_T, \quad \underline{\underline{K}}_T = \frac{\delta \underline{f}_G}{\delta \underline{q}_G} = \frac{\delta(\underline{\underline{B}}^T \underline{f}_L)}{\delta \underline{q}_G} \tag{15}$$

In the present transformation procedure regarding the load vector and stiffness matrix, the computed local elemental loads can naturally be related to the CR frame rather than to the final deformed frame. Thus, the local internal load can not be a self-equilibrating set of loads under the deformed frame. Introducing the projector matrix $\underline{\underline{P}}$, resolves this problem Rankin and Brogan (1986). The projector matrix $\underline{\underline{P}}$ can be considered as a type of 3×3 block matrix related to the elemental nodes, $\underline{\underline{P}}^{ij}$. The derivative form of $\underline{\underline{P}}$ is obtained as follows.

$$\underline{\underline{P}}_{ij} = \begin{bmatrix} \frac{\partial u_L^i}{\partial u_G^j} & \frac{\partial u_L^i}{\partial \theta_G^j} \\ \frac{\partial \theta_L^i}{\partial u_G^j} & \frac{\partial \theta_L^i}{\partial \theta_G^j} \end{bmatrix} \tag{16}$$

Using the differentiation of the local translational and rotational components, it can be

$$\underline{\underline{P}}_{ij} = \underline{\underline{I}}_3 \delta_{ij} - \underline{\Xi}^i \underline{\Gamma}^{jT} \tag{17}$$

where δ_{ij} is Kronecker's delta. Let $\underline{r}_o^i = \underline{r}_G^i + \underline{u}_L^i$ and then $\underline{\Xi}^i$, $\underline{\Gamma}^j$ can be

$$\underline{\Xi}^i = \left\{ -r_{o,2}^i, r_{o,1}^i, 1 \right\}^T \tag{18a}$$

$$\underline{\Gamma}^j = s_r^{-1} \left\{ -r_{G,2}^j, r_{G,1}^j, 0 \right\}^T \tag{18b}$$

After the projector matrix for the element is constructed, the transformation matrix between the local and global internal load vectors can be expressed in terms of the projector matrix.

$$\underline{f}_G = \underline{\underline{B}}^T \underline{f}_L = \underline{\underline{E}} \underline{\underline{P}}^T \underline{f}_L \tag{19}$$

Here, the matrix $\underline{\underline{E}} = \text{diag}(\underline{\underline{R}}_r, \underline{\underline{R}}_r, \underline{\underline{R}}_r)$. Taking the variation of $\underline{f}_G$, the resulting global stiffness matrix $\underline{\underline{K}}_G$ can be

$$\underline{\underline{K}}_G = \underline{\underline{E}} \underline{\underline{P}}^T \underline{\underline{K}}_L \underline{\underline{P}} \underline{\underline{E}}^T + \underline{\underline{E}} \left[-\underline{\Gamma} \underline{F}_1^T \underline{\underline{P}} - \underline{F}_2 \underline{\Gamma}^T \right] \underline{\underline{E}}^T \tag{20}$$

where the vectors $\underline{F}_1$ and $\underline{F}_2$ are expressed in terms of $\underline{F}_t = \underline{\underline{P}}^T \underline{f}_L$.

4   Unified Computational Algorithm

The FETI-Local proceeds in the three computational steps as follows. Step I sets up the structural interface problem, Step II evaluates the solution of the structural interface problem, and Step III recovers the solution in each sub-domain. In order to involve nonlinear structural analysis, iterative computational algorithm is developed. A load incremental Newton-Rhapson iterative scheme is employed. The unified computational algorithm is depicted in Fig. 3. The purpose of Step I is to set up the interface problem. For each sub-domain, this involves the evaluation and assembly of the stiffness matrix, the factorization of the stiffness matrix, and the assembly of the interface stiffness matrix. In Step II, the solution of the interface problem is computed first. In this step, the stiffness matrix corresponding to the interface nodes existing in the individual sub-domains needs to be distributed to each processor. Using the MPI_REDUCE routine, the matrix data are collected to a root process. In Step III, the final solution for each sub-domain is obtained by the linear solver. From Step II, array c, degrees of freedom at the interface nodes, is obtained. Thus, the displacement of each sub-domain is obtained easily. The MPI_BCAST routine sends the value of array to all the other processes first, and then, the solution of a linear equation for each sub-domain. In order to handle the sparsity of the system matrix generated in each computational step, i.e. Eq. (9), the sparse linear solver, PARDISO, is implemented. Such process is illustrated in Fig. 4.

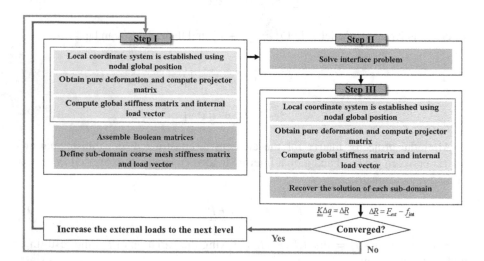

Fig. 3 Unified computational algorithm

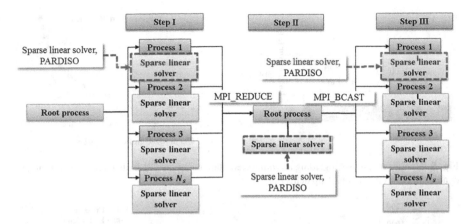

Fig. 4 Parallel implementation of the FETI-Local

5 Numerical Investigation Regarding Nonlinear Problems

Numerical assessment of the present FETI-Local method was performed by comparing with the standard FETI method by iterative solvers in the previous studies conducted by the present authors (Kwak et al., 2014, 2015). The present approach developed herein is applied to the solution of a static, two-dimensional nonlinear problems. The parallel computations were executed in the TACHYON system (Anonymous, 2017), which is one of the supercomputers operated by Korea Institute of Science and Technology Information. Section 5.1 will discuss the results for the two-dimensional configuration: the computational cost and scalability in a parallel environment are examined. Section 5.2 will examine an application for nonlinear flexible multi-body dynamics.

5.1 Computational Efficiency for Nonlinear Problem

Before the examination of computational efficiency for the analysis of the CR finite element with the FETI-Local method, geometrically nonlinear characteristic of a cantilevered plate discretized by the CR finite element is evaluated. The geometry and operating condition are described in Fig. 5a. The resulting tip deflection is compared with those predicted by MSC.NASTRAN. Comparison shows excellent correlation between the CR planar element and MSC.NASTRAN prediction and it is illustrated in Fig. 5b. Then, the analysis of the CR finite element with FETI-Local method is performed by using the same condition (Fig. 5a). However, the tip load is chosen to be 150N. The number of the sub-domains is increased from 8 to 60, but the number of DOFs is kept to a total of 39,864. Figure 6 shows benign scalability characteristics exhibited by the CR finite element with FETI-Local method.

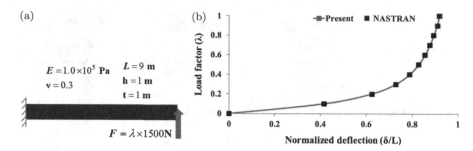

(a)

$E = 1.0 \times 10^5$ Pa $L = 9$ m
$v = 0.3$ $h = 1$ m
                         $t = 1$ m

$F = \lambda \times 1500$N

Fig. 5 Nonlinear analysis regarding a cantilevered plate using the CR finite element. (**a**) Analysis condition. (**b**) Comparison of tip deflection

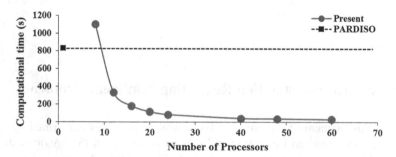

Fig. 6 Computational time and trend of the nonlinear analysis regarding a cantilevered plate

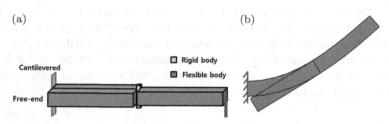

Fig. 7 Analysis condition and deformed configuration of multi-body system. (**a**) Analysis condition. (**b**) Deformed configuration

5.2 Application for Nonlinear Flexible Multi-body Dynamics

In this section, the analysis of the CR finite element with the FETI-Local method is applied to the large scale multi-body system. Analysis condition and resulting deformed configuration is depicted in Fig 7. In parallel computation, the number of the sub-domains is increased from 9 to 36, but the number of DOFs is kept to a total of 32,400. To verify an efficiency of the FETI-Local method in nonlinear flexible multi-body system, equivalent serial analysis employing the classical Lagrange multiplier and PARDISO, is conducted and compared. As the number of processors is increased, the computational time is varied from 2081.09 to 177.03 (s). Figure 8

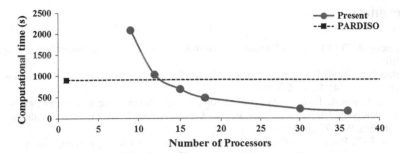

Fig. 8 Computational time and trend of multi-body analysis

shows benign scalability characteristics possessed and exhibited by the analysis of the CR finite element with the FETI-Local method.

6 Conclusion

The development of a nonlinear structural analysis using CR finite element finite element with a domain decomposition algorithm relying on direct solvers only was described. While the FETI-Local method uses the domain decomposition concept that characterizes classical FETI methods, The continuity of the displacement field within sub-domain interfaces is enforced by using a combination of the localized Lagrange multiplier and of the augmented Lagrangian formulation. Therefore, well-conditioned stiffness matrices is derived. Moreover, direct solvers can be used for both sub-domain and interface problems. The FETI-Local method was further improved by employing the sparse matrix solver to handle the sparsity within the governing equation. The computational cost and scalability of the analysis of the CR finite element with the FETI-Local method was compared to those of the sparse linear equation solver, PARDISO. Good scalability characteristics of the analysis of the CR finite element with the FETI-Local method were demonstrated for a general nonlinear analysis and flexible multi-body dynamic analysis.

Acknowledgements This research was supported by the EDISON Program through the National Research Foundation of Korea (NRF) funded by the Ministry of Science, ICT & Future Planning (No. 2014M3C1A6038842) and also be by Advanced Research Center Program (No. 2013073861) through the National Research Foundation of Korea (NRF) grant funded by the Korea government (MSIP) contracted through Next Generation Space Propulsion Research Center at Seoul National University.

References

Anonymous, KISTI TACHYON Userguide, Version 1.2 (KISTI, Supercomputing Center, Daejeon, 2010)

O.A. Bauchau, Parallel computation approaches for flexible multibody dynamics simulations. J. Franklin Inst. **347**(1), 53–68 (2010)

E. Bayo, J. García de Jalón, A. Avello, J. Cuadrado, An efficient computational method for real time multibody dynamic simulation in fully Cartesian coordinates. Comput. Methods Appl. Mech. Eng. **92**, 377–395 (1991)

C. Farhat, F.-X. Roux, A method of finite element tearing and interconnecting and its parallel solution algorithm. Int. J. Numer. Methods Eng. **32**, 1205–1227 (1991)

C. Farhat, M. Lesoinne, K. Pierson, A scalable dual-primal domain decomposition method. Numer. Linear Algebra Appl. **7**, 687–714 (2000)

C. Felippa, B. Haugen, A unified formulation of small-strain corotational finite elements: I. Theory. Comput. Methods Appl. Mech. Eng. **194**(21–24), 2285–2335 (2005)

I. Guèye, S.E. Arem, F. Feyel, R.X. Roux, G. Gailletaud, A new parallel sparse direct solver: presentation and numerical experiments in large-scale structural mechanics parallel computing. Int. J. Numer. Methods Eng. **88**, 370–384 (2011)

P.E. Gill, W. Murray, M.A. Saunders, M.H. Wright, Sequential quadratic programming methods for nonlinear programming, in *Computer-Aided Analysis and Optimization of Mechanical System Dynamics*, ed. by E.J. Haug (Springer, Berlin, Heidelberg, 1984), pp. 679–697

J.Y. Kwak, T.Y. Chun, S.J. Shin, O.A. Bauchau, Domain decomposition approach to flexible multibody dynamics simulation. Comput. Mech. **53**(1), 147–158 (2014)

J.Y. Kwak, H.S. Cho, T.Y. Chun, S.J. Shin, O.A. Bauchau, Domain decomposition approach applied for two- and three-dimensional problems via direct solution methodology. Int. J. Aeronaut. Space Sci. **16**(2), 177–189 (2015)

K.C. Park, C.A. Felippa, U.A. Gumaste, A localized version of the method of Lagrange multipliers and its applications. Comput. Mech. **24**, 476–490 (2000)

C.C. Rankin, A. Brogan, An element-independent co-rotational procedure for the treatment of large rotations. ASME J. Pressure Vessel Technol. **108**(2), 165–175 (1986)

P. Tong, T.H.H. Pian, A hybrid-element approach to crack problems in plane elasticity. Int. J. Numer. Methods Eng. **7**, 297–308 (1973)

An Adaptive Coarse Space for P.L. Lions Algorithm and Optimized Schwarz Methods

Ryadh Haferssas, Pierre Jolivet, and Frédéric Nataf

1 Introduction

Substructuring algorithms such as Balancing Neumann-Neumann (BNN) or Finite Element Tearing and Interconnecting (FETI) are defined for non overlapping domain decompositions but not for overlapping subdomains. Schwarz method (Schwarz, 1870) is defined only for overlapping subdomains. With the help of a coarse space correction, the two-level versions of both type of methods are weakly scalable, see Toselli and Widlund (2005) and references therein.

The domain decomposition method introduced by Lions (1990) can be applied to both overlapping and non overlapping subdomains. It is based on improving Schwarz methods by replacing the Dirichlet interface conditions by Robin interface conditions. This algorithm was extended to Helmholtz problem by Després (1993). Robin interface conditions can be replaced by more general interface conditions that can be optimized (Optimized Schwarz methods, OSM) for a better convergence, see Gander et al. (2002), Gander (2006) and references therein. When the domain is decomposed into a large number of subdomains, these methods are, on a practical point of view, scalable if a second level is added to the algorithm via the introduction of a coarse space Japhet et al. (1998), Farhat et al. (2000), Conen et al. (2014). But there is no systematic procedure to build coarse spaces with a provable efficiency.

The purpose of this article is to define a general framework for building adaptive coarse space for OSM methods for decomposition into overlapping subdomains.

R. Haferssas • F. Nataf (✉)
Laboratoire Jacques-Louis Lions, Paris, France
e-mail: ryadh.haferssas@ljll.math.upmc.fr; https://www.ljll.math.upmc.fr/; nataf@ann.jussieu.fr

P. Jolivet
Toulouse Institute of Computer Science Research, Toulouse, France
e-mail: pierre.jolivet@enseeiht.fr; https://www.irit.fr

© Springer International Publishing AG 2017   43
C.-O. Lee et al. (eds.), *Domain Decomposition Methods in Science and Engineering XXIII*, Lecture Notes in Computational Science and Engineering 116, DOI 10.1007/978-3-319-52389-7_4

We prove that we can achieve the same robustness that what was done for Schwarz (Spillane et al., 2014) and FETI-BDD (Spillane et al., 2013) domain decomposition methods with so called GenEO (Generalized Eigenvalue in the Overlap) coarse spaces. Compared to these previous works, we have to introduce a non standard symmetric variant of the ORAS method as well as two generalized eigenvalue problems. Although theory is valid only in the symmetric positive definite case, the method scales very well for saddle point problems such as highly heterogeneous nearly incompressible elasticity problems as well as the Stokes system.

2 Symmetrized ORAS Method

The problem to be solved is defined via a variational formulation on a domain $\Omega \subset \mathbb{R}^d$ for $d \in \mathbb{N}$:

$$\text{Find } u \in V \text{ such that} : a_\Omega(u, v) = l(v), \quad \forall v \in V,$$

where V is a Hilbert space of functions from Ω with real values. The problem we consider is given through a symmetric positive definite bilinear form that is defined in terms of an integral over any open set $\omega \subset \Omega$. A typical example is the elasticity system (C is the fourth-order stiffness tensor and $\varepsilon(u)$ is the strain tensor of a displacement field u):

$$a_\omega(u, v) := \int_\omega C : \varepsilon(u) : \varepsilon(v)\, dx.$$

The problem is discretized by a finite element method. Let $\mathcal{N}$ denote the set of degrees of freedom and $(\phi_k)_{k \in \mathcal{N}}$ be a finite element basis on a mesh $\mathcal{T}_h$. Let $A \in \mathbb{R}^{\#\mathcal{N} \times \#\mathcal{N}}$ be the associated finite element matrix, $A_{kl} := a_\Omega(\phi_l, \phi_k), k, l \in \mathcal{N}$. For some given right hand side $\mathbf{F} \in \mathbb{R}^{\#\mathcal{N}}$, we have to solve a linear system in $\mathbf{U}$ of the form

$$AU = F.$$

Domain Ω is decomposed into N overlapping subdomains $(\Omega_i)_{1 \le i \le N}$ so that all subdomains are a union of cells of the mesh $\mathcal{T}_h$. This decomposition induces a natural decomposition of the set of indices $\mathcal{N}$ into N subsets of indices $(\mathcal{N}_i)_{1 \le i \le N}$:

$$\mathcal{N}_i := \{k \in \mathcal{N} \mid meas(\text{supp}(\phi_k) \cap \Omega_i) > 0\}, \ 1 \le i \le N. \tag{1}$$

For all $1 \le i \le N$, let R_i be the restriction matrix from $\mathbb{R}^{\#\mathcal{N}}$ to the subset $R^{\#\mathcal{N}_i}$ and D_i be a diagonal matrix of size $\#\mathcal{N}_i \times \#\mathcal{N}_i$, so that we have a partition of unity at the algebraic level, $I_d = \sum_{i=1}^N R_i^T D_i R_i$, where $I_d \in \mathbb{R}^{\#\mathcal{N} \times \#\mathcal{N}}$ is the identity matrix.

For all subdomains $1 \leq i \leq N$, let B_i be a SPD matrix of size $\#\mathcal{N}_i \times \#\mathcal{N}_i$, which comes typically from the discretization of boundary value local problems using optimized transmission conditions, the ORAS preconditioner St-Cyr et al. (2007) is defined as

$$M_{ORAS,1}^{-1} := \sum_{i=1}^{N} R_i^T D_i B_i^{-1} R_i. \tag{2}$$

Due to matrices D_i, this preconditioner is not symmetric. We introduce here a non standard variant of the ORAS preconditioner (2), the symmetrized ORAS (SORAS) algorithm:

$$M_{SORAS,1}^{-1} := \sum_{i=1}^{N} R_i^T D_i B_i^{-1} D_i R_i. \tag{3}$$

More details are given in Dolean et al. (2015).

3   Two-Level SORAS Algorithm

In order to define the two-level SORAS algorithm, we introduce two generalized eigenvalue problems.

First, for all subdomains $1 \leq i \leq N$, we consider the following problem:

Definition 1

$$\text{Find } (\mathbf{U}_{ik}, \mu_{ik}) \in \mathbb{R}^{\#\mathcal{N}_i} \setminus \{0\} \times \mathbb{R} \text{ such that}$$
$$D_i R_i A R_i^T D_i \mathbf{U}_{ik} = \mu_{ik} B_i \mathbf{U}_{ik}. \tag{4}$$

Let $\gamma > 0$ be a user-defined threshold, we define $Z_{geneo}^{\gamma} \subset \mathbb{R}^{\#\mathcal{N}}$ as the vector space spanned by the family of vectors $(R_i^T D_i \mathbf{U}_{ik})_{\mu_{ik} > \gamma, 1 \leq i \leq N}$ corresponding to eigenvalues larger than γ.

In order to define the second generalized eigenvalue problem, we introduce for all subdomains $1 \leq j \leq N$, $\tilde{A}_j$, the $\#\mathcal{N}_j \times \#\mathcal{N}_j$ matrix defined by

$$\mathbf{V}_j^T \tilde{A}_j \mathbf{U}_j := a_{\Omega_j} \left(\sum_{l \in \mathcal{N}_j} \mathbf{U}_{jl} \phi_l, \sum_{l \in \mathcal{N}_j} \mathbf{V}_{jl} \phi_l \right), \quad \mathbf{U}_j, \mathbf{V}_j \in \mathbb{R}^{\mathcal{N}_j}. \tag{5}$$

When the bilinear form a results from the variational solve of a Laplace problem, the previous matrix corresponds to the discretization of local Neumann boundary value problems.

Definition 2 We introduce the generalized eigenvalue problem

$$\text{Find } (\mathbf{V}_{jk}, \lambda_{jk}) \in \mathbb{R}^{\#\mathcal{N}_i} \setminus \{0\} \times \mathbb{R} \text{ such that}$$
$$\tilde{A}^i \mathbf{V}_{ik} = \lambda_{ik} B_i \mathbf{V}_{ik} \ . \tag{6}$$

Let $\tau > 0$ be a user-defined threshold, we define $Z^\tau_{geneo} \subset \mathbb{R}^{\#\mathcal{N}}$ as the vector space spanned by the family of vectors $(R_i^T D_i \mathbf{V}_{ik})_{\lambda_{ik} < \tau, 1 \le i \le N}$ corresponding to eigenvalues smaller than τ.

We are now ready to define the two level SORAS preconditioner

Definition 3 (The SORAS-GenEO-2 Preconditioner) Let P_0 denote the A-orthogonal projection on the coarse space

$$Z_{\text{GenEO-2}} := Z^\gamma_{geneo} \bigoplus Z^\tau_{geneo} \ ,$$

the two-level SORAS-GenEO-2 preconditioner is defined as follows:

$$M^{-1}_{SORAS,2} := P_0 A^{-1} + (I_d - P_0) \sum_{i=1}^{N} R_i^T D_i B_i^{-1} D_i R_i (I_d - P_0^T) . \tag{7}$$

Note that this definition is reminiscent of the balancing domain decomposition preconditioner (Mandel, 1992) introduced for Schur complement based methods as well as of the Broyden-Fletcher-Goldfarb-Shanno (BFGS) update formula, see Nocedal and Wright (2006). We have the following theorem

Theorem 1 (Spectral Estimate for the SORAS-GenEO-2 Preconditioner) *Let k_0 be the maximum number of neighbors of a subdomain (a subdomain is a neighbor of itself) and k_1 be the maximal multiplicity of the subdomain intersections, $\gamma, \tau > 0$ be arbitrary constants used in Definitions 2 and 3.*

Then, the eigenvalues of the two-level preconditioned operator satisfy the following spectral estimate

$$\frac{1}{1 + \dfrac{k_1}{\tau}} \le \lambda(M^{-1}_{SORAS,2} A) \le \max(1, k_0 \gamma)$$

where $\lambda(M^{-1}_{SORAS,2} A)$ is an eigenvalue of the preconditioned operator.

The proof is based on the fictitious space lemma (Nepomnyaschikh, 1991) and is given in Haferssas et al. (2015).

Remark 1 The following heuristic provides an interpretation to both generalized eigenvalues (4) and (6).

We first remark that for the ASM preconditioner we have a very good upper bound for the preconditioned operator that does not depend on the number of subdomains but only on the number of neighbors of a subdomain:

$$\lambda_{max}(M_{ASM}^{-1}A) \leq k_0.$$

Thus from definitions of ASM and SORAS, we can estimate that vectors for which the action of local matrices $(R_i A R^T)^{-1}$ and $D_i B_i^{-1} D_i$ differ notably might lead to a bad upper bound for $M_{SORAS}^{-1}A$. By taking the inverse of both operators this condition means that $R_i A R^T$ and $D_i^{-1} B_i D_i^{-1}$ differ notably. By left and right multiplication by D_i it means we have to look at vectors $\mathbf{V}_i$ for which $D_i R_i A R^T D_i \mathbf{V}_i$ and $B_i \mathbf{V}_i$ have very different values. This a way to interpret the generalized eigenvalue problem (4) which controls the upper bound of the eigenvalues of $M_{SORAS}^{-1} A$.

Second, we introduce the following preconditioner M_{NN}^{-1}

$$M_{NN}^{-1} := \sum_{1 \leq i \leq N} D_i \widetilde{A}_i D_i \tag{8}$$

which is reminiscent of the Neumann-Neumann preconditioner (Tallec et al., 1998) for substructuring methods. We have a very good lower bound for the preconditioned operator $M_{NN}^{-1} A$ that does not depend on the number of subdomains but only on the maximum multiplicity of intersections:

$$\frac{1}{k_1} \leq \lambda_{min}(M_{NN}^{-1} A).$$

If we compare formulas for M_{NN}^{-1} (8) and M_{SORAS}^{-1} (3), we see that we have to look at vectors $\mathbf{V}_i$ for which $D_i \widetilde{A}_i D_i \mathbf{V}_i$ and $B_i \mathbf{V}_i$ have very different values. This is a way to interpret the generalized eigenvalue problem (6) which controls the lower bound of the eigenvalues of $M_{SORAS}^{-1} A$.

4 Nearly Incompressible Elasticity

Although our theory does not apply in a straightforward manner to saddle point problems, we use it for these difficult problems for which it is not possible to preserve both symmetry and positivity of the problem. Note that generalized eigenvalue problems (4) and (6) still make sense if A is the matrix of a saddle point problem and matrices B_i and $\widetilde{A}_i$ are properly defined for each subdomain $1 \leq i \leq N$. The new coarse space was tested quite successfully on Stokes and nearly incompressible elasticity problems with a discretization based on saddle point formulations in order to avoid locking phenomena. The mechanical properties of a solid can be characterized by its Young modulus E and Poisson ratio ν or alternatively by its Lamé coefficients λ and μ. These coefficients relate to each other by the following formulas:

$$\lambda = \frac{E\nu}{(1+\nu)(1-2\nu)} \quad \text{and} \quad \mu = \frac{E}{2(1+\nu)}.$$ (9)

The variational problem consists in finding $(u_h, p_h) \in \mathcal{V}_h := \mathbb{P}_2^d \cap H_0^1(\Omega) \times \mathbb{P}_1$ such that for all $(v_h, q_h) \in \mathcal{V}_h$

$$\begin{cases} \int_\Omega 2\mu\varepsilon(u_h) : \varepsilon(v_h)dx \quad -\int_\Omega p_h \mathrm{div}\,(v_h)dx = \int_\Omega f v_h dx \\ \\ -\int_\Omega \mathrm{div}\,(u_h)q_h dx \quad -\int_\Omega \frac{1}{\lambda}p_h q_h = 0 \end{cases}$$ (10)

$$\implies A U = \begin{bmatrix} H & B^T \\ B & C \end{bmatrix} \begin{bmatrix} u \\ p \end{bmatrix} = \begin{bmatrix} f \\ 0 \end{bmatrix} = \mathbf{F}.$$

Matrix $\widetilde{A}_i$ arises from the variational formulation (10) where the integration over domain Ω is replaced by the integration over subdomain Ω_i and finite element space $\mathcal{V}_h$ is restricted to subdomain Ω_i. Matrix B_i corresponds to a Robin problem and is the sum of matrix $\widetilde{A}_i$ and of the matrix of the following variational formulation restricted to the same finite element space:

$$\int_{\partial\Omega_i \backslash \partial\Omega} \frac{2\alpha\mu(2\mu + \lambda)}{\lambda + 3\mu} u_h \cdot v_h \quad \text{with } \alpha = 10 \text{ in our test.}$$

In Dolean et al. (2015), we tested our method for a heterogeneous beam of eight layers of steel $(E_1, \nu_1) = (210 \cdot 10^9, 0.3)$ and rubber $(E_2, \nu_2) = (0.1 \cdot 10^9, 0.4999)$, see Fig. 1. The beam is clamped on its left and right sides. Table 7.1 of Dolean et al. (2015) shows that our method performs consistently much better than various domain decomposition methods: the one level Additive Schwarz (AS) and SORAS methods, the two level AS and SORAS methods with a coarse space consisting of rigid body motions which are zero energy modes (ZEM) and finally AS with a GenEO coarse space. In our test, the GenEO-2 coarse space defined in Definition 3 was built with $\tau = 0.4$ and $\gamma = 10^3$. Eigenvalue problem (6) accounts for roughly 90% of the GenEO-2 coarse space size. In Figs. 3 and 2, we plot the eigenvectors of the generalized eigenvalue problems (4) and (6) for the linear elasticity case. The domain decomposition is such that all subdomains contain the eight alternating layers of steel and rubber. The GenEO coarse space for lower bound (Fig. 3) will

Fig. 1 2D elasticity: coefficient distribution of steel and rubber

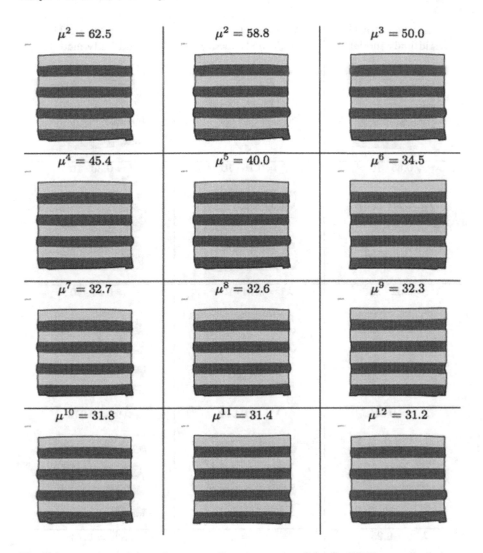

Fig. 2 Largest eigenvalues and corresponding eigenmodes of the GenEO II generalized eigen-problem for the upper bound (4)

consist of the first 12 modes. The first three are known as the rigid body modes. The other nine eigenmodes display very different behaviors for the steel and the rubber. The the 13th eigenvalue and the next ones are larger than 0.25 and are not incorporated into the coarse space. Interestingly enough, steel and rubber have the same deformations in these modes.

In this paragraph, we perform a parametric study of the dependence of the convergence on the thresholds γ and τ of the coarse space. In Fig. 4 we study the influence of the parameter τ alone keeping the parameter $\gamma = 1/0.001$. We see that

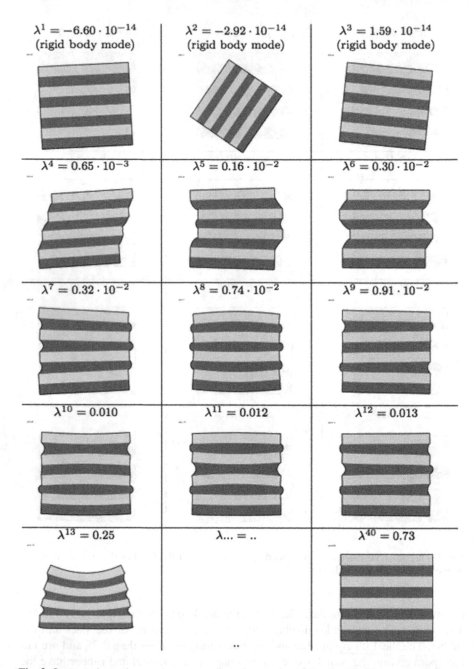

Fig. 3 Lowest eigenvalues and corresponding eigenmodes of the GenEO II generalized eigenproblem for lower bound (6)

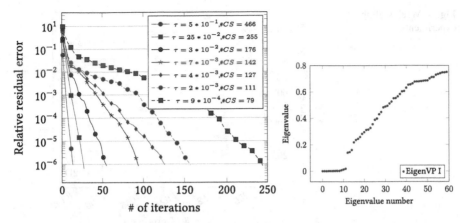

Fig. 4 *Left*: Convergence history vs. threshold τ. *Right*: Eigenvalues for the lower bound eigenvalue problem (6)

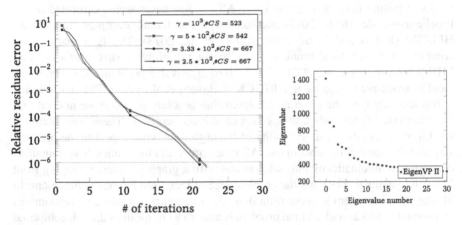

Fig. 5 *Left*: Convergence history vs. threshold γ. *Right*: Eigenvalues for the upper bound eigenvalue problem (4)

for $\tau < 10^{-2}$, there are plateau in the convergence curves. But for larger values of τ, convergence curves are almost straight lines. This is in agreement with the gap in the spectrum of the eigenvalue problem (6), see Fig. 4. A comparable study was made for the impact of the threshold γ. We see on Fig. 5 that this parameter has only a small impact on the iteration count.

We also performed large 3D simulations on 8192 cores of a IBM/Blue Gene Q machine with 1.6 GHz Power A2 processors for both elasticity (200 millions of d.o.f's in 200s) and Stokes (200 millions of d.o.f's in 150s) equations. Computing facilities were provided by an IDRIS-GENCI project. We focus on results for the nearly incompressible elasticity problem. The problem is solved with a geometric overlap of two mesh elements and a preconditioned GMRES is used to solve the resulting linear system where the stopping criteria for the

Fig. 6 Weak scaling
experiments

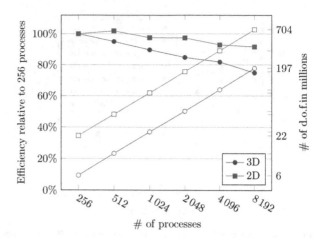

relative residual norm is fixed to 10^{-6}. All the test cases were performed inside
FreeFem++ code (Hecht, 2012) interfaced with the domain decomposition library
HPDDM (Jolivet and Nataf, 2014; Jolivet et al., 2013). The factorizations are
computed for each local problem and also for the global coarse problem using
MUMPS (Amestoy et al., 2001). Generalized eigenvalue problems to generate the
GenEO space are solved using ARPACK (Lehoucq et al., 1998). The coarse space
is formed only with the generalized eigenvalue problem (6) since we noticed that
the other one (4) has only a little effect on the convergence. These computations,
see Fig. 6, assess the weak scalability of the algorithm with respect to the problem
size and the number of subdomains. All times are wall clock times. The domain is
decomposed automatically into subdomains with a graph partitioner, ranging from
256 subdomains to 8192 and the problem size is increased by mesh refinement. In
3D the initial problem is about 6 millions d.o.f decomposed into 256 subdomains
and solved in 145.2s and the final problem is about 197 millions of d.o.f decomposed
into 8192 subdomains and solved in 196s which gives an efficiency near to 75%.

References

P.R. Amestoy, I.S. Duff, J.Y. L'Excellent, J. Koster, A fully asynchronous multifrontal solver using
    distributed dynamic scheduling. SIAM J Matrix Anal. Appl. **23**(1):15–41 (2001)
L. Conen, V. Dolean, R. Krause, F. Nataf, A coarse space for heterogeneous Helmholtz problems
    based on the Dirichlet-to-Neumann operator. J. Comput. Appl. Math. **271**, 83–99 (2014). ISSN:
    0377-0427
B. Després, Domain decomposition method and the Helmholtz problem.II, in *Second International
    Conference on Mathematical and Numerical Aspects of Wave Propagation*, Newark, DE, 1993
    (SIAM, Philadelphia, PA, 1993), pp. 197–206
V. Dolean, P. Jolivet, F. Nataf, *An Introduction to Domain Decomposition Methods: Algorithms,
    Theory and Parallel Implementation* (SIAM, Philadelphia, PA, 2015)

C. Farhat, A. Macedo, M. Lesoinne, A two-level domain decomposition method for the iterative solution of high-frequency exterior Helmholtz problems. Numer. Math. **85**(2), 283–303 (2000)

M.J. Gander, Optimized Schwarz methods. SIAM J. Numer. Anal. 44(2), 699–731 (2006)

M.J. Gander, F. Magoulès, F. Nataf, Optimized Schwarz methods without overlap for the Helmholtz equation. SIAM J. Sci. Comput. **24**(1), 38–60 (2002)

R. Haferssas, P. Jolivet, F. Nataf, An additive Schwarz method type theory for Lions' algorithm and Optimized Schwarz Methods. Working paper or preprint, December 2015. https://hal. archives-ouvertes.fr/hal-01278347

F. Hecht, New development in freefem++. J. Numer. Math. **20**(3–4), 251–265 (2012). ISSN:1570-2820

C. Japhet, F. Nataf, F.-X. Roux, The Optimized Order 2 Method with a coarse grid preconditioner. Application to convection-diffusion problems, in *Ninth International Conference on Domain Decomposition Methods in Science and Engineering*, ed. by P. Bjorstad, M. Espedal, D. Keyes (Wiley, New York, 1998), pp. 382–389

P. Jolivet, F. Nataf, Hpddm: High-Performance Unified framework for Domain Decomposition methods, MPI-C++ library. https://github.com/hpddm/hpddm, 2014

P. Jolivet, F. Hecht, F. Nataf, C. Prud'homme, Scalable domain decomposition preconditioners for heterogeneous elliptic problems, in *Proceedings of the 2013 ACM/IEEE conference on Supercomputing*, SC13. (ACM, New York, 2013), pp. 80:1–80:11. Best paper finalist

R.B. Lehoucq, D.C. Sorensen, C. Yang, *ARPACK Users' Guide: Solution of Large-Scale Eigenvalue Problems with Implicitly Restarted Arnoldi Methods*, vol. 6 (SIAM, Philadelphia, PA, 1998)

P.-L. Lions, On the Schwarz alternating method. III: a variant for nonoverlapping subdomains. In Tony F. Chan, Roland Glowinski, Jacques Périaux, and Olof Widlund, editors, *Third International Symposium on Domain Decomposition Methods for Partial Differential Equations*, Houston, TX, March 20–22, 1989 (SIAM, Philadelphia, PA, 1990)

J. Mandel, Balancing domain decomposition. Comm. Appl. Numer. Methods **9**, 233–241 (1992)

S.V. Nepomnyaschikh, Mesh theorems of traces, normalizations of function traces and their inversions. Sov. J. Numer. Anal. Math. Model. **6**, 1–25 (1991)

J. Nocedal, S.J. Wright, *Numerical Optimization*. Springer Series in Operations Research and Financial Engineering, 2nd edn. (Springer, New York, 2006). ISBN: 978-0387-30303-1; 0-387-30303-0

H.A. Schwarz, Über einen Grenzübergang durch alternierendes Verfahren. Vierteljahrsschrift der Naturforschenden Gesellschaft in Zürich **15**, 272–286 (1870)

N. Spillane, V. Dolean, P. Hauret, F. Nataf, D. Rixen, Solving generalized eigenvalue problems on the interfaces to build a robust two-level FETI method. C. R. Math. Acad. Sci. Paris **351**(5–6), 197–201 (2013) ISSN: 1631-073X. doi: 10.1016/j.crma.2013.03.010. http://dx.doi.org/10. 1016/j.crma.2013.03.010

N. Spillane, V. Dolean, P. Hauret, F. Nataf, C. Pechstein, R. Scheichl, Abstract robust coarse spaces for systems of PDEs via generalized eigenproblems in the overlaps. Numer. Math. **126**(4), 741–770 (2014). ISSN: 0029-599X. doi: 10.1007/s00211-013-0576-y. http://dx.doi.org/10.1007/ s00211-013-0576-y

A. St-Cyr, M.J. Gander, S.J. Thomas, Optimized multiplicative, additive, and restricted additive Schwarz preconditioning. SIAM J. Sci. Comput. **29**(6), 2402–2425 (electronic) (2007). ISSN: 1064-8275. doi: 10.1137/060652610. http://dx.doi.org/10.1137/060652610

P.L. Tallec, J. Mandel, M. Vidrascu, A Neumann-Neumann domain decomposition algorithm for solving plate and shell problems. SIAM J. Numer. Anal. **35**, 836–867 (1998)

A. Toselli, O. Widlund, *Domain Decomposition Methods - Algorithms and Theory*, vol. 34. Springer Series in Computational Mathematics (Springer, Berlin, 2005)

On the Time-Domain Decomposition of Parabolic Optimal Control Problems

Felix Kwok

1 Introduction

The efficient solution of optimal control problems under partial differential equation (PDE) constraints has become an active area of research in the past decade. In this paper, we consider an optimal control problem where the constraint is a large system of linear ordinary differential equations (ODEs) arising from the semi-discretization of a linear PDE in space:

$$\partial_t \mathbf{y} + A\mathbf{y}(t) = B\mathbf{u}(t) + \mathbf{f}(t), \qquad t \in (0, T), \tag{1a}$$

$$\mathbf{y}(0) = \mathbf{y}_0. \tag{1b}$$

The goal is to find a control $\mathbf{u}$ that minimizes the objective functional

$$F(\mathbf{u}) = \frac{1}{2} \int_0^T \|\mathbf{u}(t)\|^2 \, dt + \frac{\alpha_1}{2} \int_0^T \|C\mathbf{y} - \hat{\mathbf{y}}\|^2 \, dt + \frac{\alpha_2}{2} \|D\mathbf{y}(T) - \hat{\mathbf{y}}_T\|^2. \tag{2}$$

In the above, $\hat{\mathbf{y}} = \hat{\mathbf{y}}(t)$ and $\hat{\mathbf{y}}_T$ are the target trajectory and target state, and the functions $\mathbf{u}$ and $\mathbf{y} = \mathbf{y}(t, \mathbf{u})$ are called the control and the state, respectively. (For the purpose of analysis, we will use an appropriate change of variables to subsume any mass matrices that appear into the matrices A, B, C and D.) We will focus on the case where there are no control or state constraints, and where the governing equation is parabolic, i.e., when A is positive semi-definite, but not necessarily symmetric.

F. Kwok (✉)
Department of Mathematics, Hong Kong Baptist University, Kowloon Tong, Hong Kong
e-mail: felix_kwok@hkbu.edu.hk

© Springer International Publishing AG 2017
C.-O. Lee et al. (eds.), *Domain Decomposition Methods in Science and Engineering XXIII*, Lecture Notes in Computational Science and Engineering 116, DOI 10.1007/978-3-319-52389-7_5

A formulation similar to the above has been used for a variety of problems where the goal is to drive a mechanical system to a desired state while minimizing the cost: it has been used for the control of fluid flow modelled by the Navier-Stokes equations (Choi et al., 1999; Unger and Tröltzsch, 2001), boundary control problems for the wave equation (Lagnese and Leugering, 2003) and quantum control (see Maday et al. (2007) and references therein). Recently, medical applications have also been proposed, more specifically in the optimized administration of radiotherapy to control tumour growth (Corwin et al., 2013).

For problems with no control or state constraints, a Lagrange-multiplier argument shows that the optimal control satisfies, in addition to the forward differential equation (1), the adjoint final value problem

$$\dot{\lambda} - A^\mathsf{T}\lambda = \alpha_1 C^\mathsf{T}(C\mathbf{y} - \hat{\mathbf{y}}), \tag{3a}$$

$$\lambda(T) + \alpha_2 D^\mathsf{T} D\mathbf{y}(T) = \alpha_2 D^\mathsf{T}\hat{\mathbf{y}}_T, \tag{3b}$$

where λ, the adjoint state, satisfies $\mathbf{u} = B^\mathsf{T}\lambda$. Together with (1), this leads to a coupled forward-backward ODE system that must be further discretized in time and solved. Alternatively, one can discretize (1) and (2) in time and solve the resulting discrete saddle-point system. Note that the two approaches do not always "commute", even if one chooses compatible time discretizations for (1a) and (3), see Dontchev et al. (2000), Hager (2000). Regardless of the approach taken, the exceedingly large size of the resulting linear system strongly motivates the use of parallel solution strategies. In this paper, we only consider the semi-discrete ODE system; the effect of discretization in time will be studied in a future paper.

There has been much progress in recent years in the development of effective preconditioners for saddle-point systems that arise from PDE-constrained optimal control problems; we only mention two classes of such methods. The first, known as the *all-at-once approach*, uses block preconditioners that are known to be effective for saddle-point systems. Because of its large size, the saddle-point matrix is not formed explicitly; instead, one performs the matrix-vector multiplication and preconditioning steps by solving forward and backward problems similar to (1) and (3). The latter steps can be parallelized in time using e.g. parareal (Lions et al., 2001) or parabolic multigrid (Gander and Neumüller, 2014; Horton and Vandewalle, 1995), or in space by domain decomposition or multigrid methods. We refer the reader to Rees et al. (2010), Pearson et al. (2012), as well as to Schiela and Ulbrich (2014) for an approach in the infinite-dimensional setting which also works for problems with control constraints.

A different idea is to apply parallel methods directly to the optimal control problem itself. One such approach, known as the collective smoothing multigrid (CSMG) scheme, applies multigrid smoothing and coarsening to the coupled system and is analyzed in Borzì (2003). One can also adapt parareal to solve optimal control problems directly, see Maday et al. (2007), Mathew et al. (2010), Maday et al. (2013), Gander et al. (2016). Another approach, which arises from the multiple shooting philosophy, is to create smaller problems by subdividing the time horizon. The problem then consists of finding the intermediate state and adjoint variables that achieve both local optimality on each sub-interval and consistency across neighbouring sub-intervals. The smaller local problems can then be solved independently, and in parallel. This idea has been used in Heinkenschloss (2005) to derive a block preconditioner for parabolic control problems, and in Lagnese and Leugering (2003) to obtain a method with Robin-type consistency conditions in the context of wave equations. In Barker and Stoll (2015), the authors consider an additive Schwarz preconditioner that uses Dirichlet interface conditions in the state and adjoint variables across overlapping sub-intervals.

2   Optimized Schwarz Methods in Time for Control

In Gander and Kwok (2016), we introduced a time-domain decomposition method inspired by the Robin-type interface conditions used in optimized Schwarz methods (OSM) for elliptic problems. In this paper, we consider the natural extension to problems with non-trivial observation and control operators, namely

1. For $k = 1, 2, \ldots$, solve in parallel for $j = 1, 2$

$$\dot{\mathbf{y}}_j^k + A\mathbf{y}_j^k = B\mathbf{u}_j^k + \mathbf{f}(t), \qquad \dot{\boldsymbol{\lambda}}_j^k - A^\top \boldsymbol{\lambda}_j^k = \alpha_1 C^\top (C\mathbf{y}_j^k - \hat{\mathbf{y}}) \qquad (4)$$

on $I_1 = (0, \beta)$ and $I_2 = (\beta, T)$, subject to $\mathbf{u}_j^k = B^\top \boldsymbol{\lambda}_j^k$ and the initial and final conditions

For I_1 : $\mathbf{y}_1^k(0) = \mathbf{y}_0$, $\boldsymbol{\lambda}_1^k(\beta) + p\mathbf{y}_1^k(\beta) = \mathbf{h}^{k-1}$,   (5)

For I_2 : $\mathbf{y}_2^k(\beta) - q\boldsymbol{\lambda}_2^k(\beta) = \mathbf{g}^{k-1}$, $\boldsymbol{\lambda}_2^k(T) + \alpha_2 D^\top D\mathbf{y}_2^k(T) = \alpha_2 D^\top \hat{\mathbf{y}}_T$.   (6)

2. Update traces:

$$\mathbf{g}^k = \mathbf{y}_1^k(\beta) - q\boldsymbol{\lambda}_1^k(\beta), \qquad\qquad \mathbf{h}^k = \boldsymbol{\lambda}_2^k(\beta) + p\mathbf{y}_2^k(\beta). \qquad (7)$$

The parameters p and q are chosen to optimize convergence. In Gander and Kwok (2016), the method is analyzed by assuming $B = C = I, D = 0$ and that A is symmetric. This allows us to diagonalize A and obtain explicit formulas for the contraction factors, but the analysis no longer works when A is non-symmetric. In this paper, we show a different method, based on energy estimates, which allows one to derive optimal parameters for non-symmetric operators A.

In terms of implementation, each iteration of the method (4)–(7) requires the solution of subdomain problems with Robin interface conditions. This may be done using any serial method, such as the all-at-once methods mentioned in

Sect. 1. In the numerical experiments in Sect. 4, we use a Krylov-accelerated iteration based on shooting methods, which are easy to implement and naturally applicable to problems with optimized transmission conditions in time. For example, to solve the local problem on I_2, we consider the mapping $\mathscr{P}_2(\mathbf{y}_\beta, \mathbf{u}) = \left[\mathbf{y}_\beta - q\boldsymbol{\lambda}(\beta) - \mathbf{g}^{k-1}, \mathbf{u} - B^\top \boldsymbol{\lambda}\right]$, where the inputs are the initial state $\mathbf{y}_\beta$ and the control function $\mathbf{u} = \mathbf{u}(t)$, $t \in I_2$, and $\boldsymbol{\lambda}$ is calculated by integrating $\mathbf{y}$ forward in time, obtaining $\boldsymbol{\lambda}(T)$ via the final condition in (6), and integrating $\boldsymbol{\lambda}$ backward in time. Because the differential equations are linear, there exists a linear operator $\mathscr{K}_2$ such that $\mathscr{P}_2(\mathbf{y}_\beta, \mathbf{u}) = \mathscr{K}_2(\mathbf{y}_\beta, \mathbf{u}) + \mathbf{r}_0$ with $\mathbf{r}_0 = \mathscr{P}_2(0, 0)$. To calculate the solution, which satisfies $\mathscr{P}_2(\mathbf{y}_\beta, \mathbf{u}) = 0$, it suffices to solve $\mathscr{K}_2(\mathbf{y}_\beta, \mathbf{u}) = -\mathbf{r}_0$ using a Krylov subspace method such as GMRES. The preconditioning of such systems is an important topic that will be addressed in a future paper. Nonetheless, we have observed in our experiments that the local solves converge within about 20 GMRES iterations, even without preconditioning.

2.1 Energy Estimates

To illustrate the technique for obtaining error estimates, we first consider the simple case of distributed control and observation with no target state (i.e., $B = C = I$, $\alpha_2 = 0$). By linearity, it suffices to consider the problem with zero data (i.e. $\mathbf{f}(t)$, $\mathbf{y}_0$, $\hat{\mathbf{y}}$ and $\hat{\mathbf{y}}_T$ are all taken to be zero) and study how the approximate solution converges to zero. To derive an energy estimate for the first subdomain $\Omega \times I_1$, where $I_1 = (0, \beta)$, we introduce the auxiliary variables $\mathbf{z}_1^k := \mathbf{y}_1^k + r\boldsymbol{\lambda}_1^k$, $\boldsymbol{\mu}_1^k := \boldsymbol{\lambda}_1^k - s\mathbf{y}_1^k$ with $r, s > 0$. Note that the parameters r and s are not the same as the optimization parameters p and q and do not appear in the algorithm; they are introduced for analysis purposes only and must be chosen based on a given (p, q) pair. We now let H and S be the symmetric and skew-symmetric parts of A, such that $A = H + S$, and rewrite the problem (4) for subdomain I_1 in terms of $\mathbf{z}_1^k$ and $\boldsymbol{\mu}_1^k$ to get

$$
\begin{cases}
\partial_t \mathbf{z}_1^k + \dfrac{1}{1+rs}\left[(1-rs)H + (1+rs)S - (\alpha_1 r + s)I\right]\mathbf{z}_1^k \\
\qquad\qquad + \dfrac{1}{1+rs}\left[(\alpha_1 r^2 - 1)I - 2rH\right]\boldsymbol{\mu}_1^k = 0, \\
\partial_t \boldsymbol{\mu}_1^k + \dfrac{1}{1+rs}\left[(s^2 - \alpha_1)I - 2sH\right]\mathbf{z}_1^k \\
\qquad\qquad + \dfrac{1}{1+rs}\left[(\alpha_1 r + s)I - (1-rs)H + (1+rs)S\right]\boldsymbol{\mu}_1^k = 0.
\end{cases}
$$

Note that the matrix multiplying $\mathbf{z}_1^k$ in the first equation is exactly the negative transpose of the matrix multiplying $\boldsymbol{\mu}_1^k$ in the second equation. This means if we multiply the first and second equations by $(\boldsymbol{\mu}_1^k)^\top$ and $(\mathbf{z}_1^k)^\top$ and add the results, the mixed terms cancel. After integrating over $(0, \beta)$, we obtain the energy identity

$$0 = \boldsymbol{\mu}_1^k(\beta)^\top \mathbf{z}_1^k(\beta) - \boldsymbol{\mu}_1^k(0)^\top \mathbf{z}_1^k(0) + \frac{1}{1 + rs} \int_0^\beta (\boldsymbol{\mu}_1^k)^\top (\alpha_1 r^2 - 2rH - 1)\boldsymbol{\mu}_1^k$$

$$+ \frac{1}{1 + rs} \int_0^\beta (\mathbf{z}_1^k)^\top (s^2 - 2sH - \alpha_1)\mathbf{z}_1^k$$

(8)

Similarly, for the second subdomain I_2, we obtain

$$0 = \boldsymbol{\mu}_2^k(T)^\top \mathbf{z}_2^k(T) - \boldsymbol{\mu}_2^k(\beta)^\top \mathbf{z}_2^k(\beta) + \frac{1}{1 + \hat{r}\hat{s}} \int_\beta^T (\boldsymbol{\mu}_2^k)^\top (\alpha_1 \hat{r}^2 - 2\hat{r}H - 1)\boldsymbol{\mu}_2^k$$

$$+ \frac{1}{1 + \hat{r}\hat{s}} \int_\beta^T (\mathbf{z}_2^k)^\top (\hat{s}^2 - 2\hat{s}H - \alpha_1)\mathbf{z}_2^k,$$

(9)

where we used the auxiliary variables $\mathbf{z}_2^k := \mathbf{y}_2^k + \hat{r}\boldsymbol{\lambda}_2^k$ and $\boldsymbol{\mu}_2^k := \boldsymbol{\lambda}_2^k - \hat{s}\mathbf{y}_2^k$, with $\hat{r}, \hat{s}$ possibly different from r, s.

To mimic the energy argument of Lions (1990), we need to ensure that the boundary terms in (8), (9) correspond to differences of incoming and outgoing Robin traces, and that the integral terms never change signs. This motivates the following theorem.

Theorem 1 *Consider the optimized Schwarz method (4)–(7) with $B = C = I$ and $\alpha_2 = 0$. Assume that*

(i) *The parameters $r, s, \hat{r}, \hat{s}$ are non-negative,*
(ii) *The matrices $(1 - \alpha_1 r^2)I + 2rH$, $(1 - \alpha_1 \hat{r}^2)I + 2\hat{r}H$, $(\alpha_1 - s^2)I + 2sH$, $(\alpha_1 - \hat{s}^2)I + 2\hat{s}H$ are all positive definite,*
(iii) *There exist $c_1, c_2 > 0$ such that $(\boldsymbol{\mu}_1^k)^\top \mathbf{z}_1^k = c_1\|\boldsymbol{\lambda}_1^k + p\mathbf{y}_1^k\|^2 - c_2\|\mathbf{y}_1^k - q\boldsymbol{\lambda}_1^k\|^2$,*
(iv) *There exist $\hat{c}_1, \hat{c}_2 > 0$ such that $(\boldsymbol{\mu}_2^k)^\top \mathbf{z}_2^k = \hat{c}_1\|\boldsymbol{\lambda}_2^k + p\mathbf{y}_2^k\|^2 - \hat{c}_2\|\mathbf{y}_2^k - q\boldsymbol{\lambda}_2^k\|^2$.*

Then the method satisfies the two-step error estimates

$$\|\mathbf{y}_1^k(\beta) - q\boldsymbol{\lambda}_1^k(\beta)\|^2 \leq \rho^2 \|\mathbf{y}_1^{k-2}(\beta) - q\boldsymbol{\lambda}_1^{k-2}(\beta)\|^2, \tag{10a}$$

$$\|\boldsymbol{\lambda}_2^k(\beta) + p\mathbf{y}_2^k(\beta)\|^2 \leq \rho^2 \|\boldsymbol{\lambda}_2^{k-2}(\beta) + p\mathbf{y}_2^{k-2}(\beta)\|^2, \tag{10b}$$

with $\rho^2 = \dfrac{c_1 \hat{c}_2}{c_2 \hat{c}_1}$. In particular, the method converges if $\rho^2 < 1$.

Proof Consider the energies

$$E_1^k = \frac{1}{1+rs} \int_0^\beta (\boldsymbol{\mu}_1^k)^\top (1 + 2rH - \alpha_1 r^2)\boldsymbol{\mu}_1^k + (\mathbf{z}_1^k)^\top (\alpha_1 + 2sH - s^2)\mathbf{z}_1^k,$$

$$E_2^k = \frac{1}{1+\hat{r}\hat{s}} \int_\beta^T (\boldsymbol{\mu}_2^k)^\top (1 + 2\hat{r}H - \alpha_1 \hat{r}^2)\boldsymbol{\mu}_2^k + (\mathbf{z}_2^k)^\top (\alpha_1 + 2\hat{s}H - \hat{s}^2)\mathbf{z}_2^k,$$

which must be positive by Assumption (ii) unless $\boldsymbol{\mu}_1^k = \mathbf{z}_1^k = 0$ or $\boldsymbol{\mu}_2^k = \mathbf{z}_2^k = 0$. The energy equality (8) can then be written as

$$\boldsymbol{\mu}_1^k(\beta)^\top \mathbf{z}_1^k(\beta) - \boldsymbol{\mu}_1^k(0)^\top \mathbf{z}_1^k(0) = E_1^k \geq 0.$$

Using Assumption (iii) and the definition of $\boldsymbol{\mu}_1^k$ and $\mathbf{z}_1^k$, we get

$$c_1\|\boldsymbol{\lambda}_1^k(\beta) + p\mathbf{y}_1^k(\beta)\|^2 - c_2\|\mathbf{y}_1^k(\beta) - q\boldsymbol{\lambda}_1^k(\beta)\|^2 - (\boldsymbol{\lambda}_1^k(0) - s\mathbf{y}_1^k(0))^\top (\mathbf{y}_1^k(0) + r\boldsymbol{\lambda}_1^k(0)) = E_1^k.$$

Since $\mathbf{y}_1^k(0) = 0$ by (5), we in fact have

$$c_1\|\boldsymbol{\lambda}_1^k(\beta) + p\mathbf{y}_1^k(\beta)\|^2 - c_2\|\mathbf{y}_1^k(\beta) - q\boldsymbol{\lambda}_1^k(\beta)\|^2 = E_1^k + r\|\boldsymbol{\lambda}_1^k(0)\|^2 \geq 0. \tag{11}$$

But the transmission conditions (7) imply

$$c_1\|\boldsymbol{\lambda}_2^{k-1}(\beta) + p\mathbf{y}_2^{k-1}(\beta)\|^2 \geq c_2\|\mathbf{y}_1^k(\beta) - q\boldsymbol{\lambda}_1^k(\beta)\|^2. \tag{12}$$

A similar calculation on subdomain I_2, using Assumptions (ii), (iv) and the fact that $\boldsymbol{\lambda}_2^k(T) = 0$, yields

$$\hat{c}_2\|\mathbf{y}_2^k(\beta) - q\boldsymbol{\lambda}_2^k(\beta)\|^2 - \hat{c}_1\|\boldsymbol{\lambda}_2^k(\beta) + p\mathbf{y}_2^k(\beta)\|^2 = E_2^k + \hat{s}\|\mathbf{y}_2^k(T)\|^2 \geq 0. \tag{13}$$

The transmission conditions (7) now imply that

$$\hat{c}_2\|\mathbf{y}_1^{k-1}(\beta) - q\boldsymbol{\lambda}_1^{k-1}(\beta)\|^2 \geq \hat{c}_1\|\boldsymbol{\lambda}_2^k(\beta) + p\mathbf{y}_2^k(\beta)\|^2. \tag{14}$$

Combining the inequalities (12) and (14) and shifting indices when necessary leads to the two-step error estimates (10a)–(10b). If $\rho^2 < 1$, then we have

$$\|\mathbf{y}_j^k(\beta) - q\boldsymbol{\lambda}_j^k(\beta)\| \to 0 \quad \text{and} \quad \|\boldsymbol{\lambda}_j^k(\beta) + p\mathbf{y}_j^k(\beta)\| \to 0, \qquad j = 1, 2.$$

We thus conclude from (11) and (13) that $E_j^k \to 0$ for $j = 1, 2$, which implies that $\boldsymbol{\mu}_j^k$ and $\mathbf{z}_j^k$ both go to zero. This in turn shows that the error in the forward and adjoint states $\mathbf{y}_j^k$ and $\boldsymbol{\lambda}_j^k$ converges to zero, as required.   □

In order to prove convergence of the method for a given choice of optimized parameters p and q, we need to show that there exists a choice of $r, s, \hat{r}, \hat{s}$ such that the assumptions in Theorem 1 are satisfied. This is in fact possible if we assume $pq < 1$, together with some mild assumptions on H. For a proof of the following theorem, see Gander and Kwok (2016).

Theorem 2 *Let $B = C = I$ and $\alpha_2 = 0$ (no target state). Assume that $H = \frac{1}{2}(A + A^T)$ is positive semi-definite. If $p, q \geq 0$ satisfy $pq < 1$, then the optimized Schwarz method (4)–(7) converges for any initial guess, provided at least one of p and q is non-zero. Moreover, if H is positive definite, then the method also converges for $p = q = 0$.*

2.2 Choice of Parameters and Convergence Rates

We now show how to choose the parameters p, q in order to minimize the contraction factor ρ in Theorem 1. First, if H is only assumed to be positive semidefinite, then Assumption (ii) is satisfied provided

$$0 \leq r, \hat{r} < 1/\sqrt{\alpha_1}, \qquad 0 \leq s, \hat{s} < \sqrt{\alpha_1}. \tag{15}$$

Now Assumption (iii) says

$$\mu_1^T z_1 = (\lambda_1 - s y_1)^T (y_1 + r\lambda_1) = c_1 \|\lambda_1 + p y_1\|^2 - c_2 \|y_1 - q\lambda_1\|^2, \tag{16}$$

while Assumption (iv) gives a similar relation for $\hat{r}$ and $\hat{s}$. Expanding and equating coefficients for $\lambda_1^T \lambda_1$ and $y_1^T y_1$ in (16) leads to the formulas

$$c_1 = \frac{r + q^2 s}{1 - p^2 q^2}, \qquad c_2 = \frac{s + p^2 r}{1 - p^2 q^2}, \tag{17}$$

where the denominators are non-zero because $pq < 1$, as stated in Theorem 2. Equating coefficients for $\lambda_1^T y_1$ leads to a compatibility condition between r and s:

$$s = \frac{-2pr + (1 - pq)}{2q + r(1 - pq)} \iff r = \frac{-2qs + (1 - pq)}{2p + s(1 - pq)}.$$

For a given pair of optimized parameters (p, q) such that $pq < 1$, there are many ways of choosing r (or, equivalently, s); our task is to choose r to obtain the best estimate for the convergence factor ρ. Using the above expressions to eliminate either r or s from (17) gives

$$\frac{c_1}{c_2} = \frac{q^2 + 2qr + r^2}{1 - 2pr + p^2 r^2} = \left(\frac{q + r}{1 - pr} \right)^2 = \left(\frac{1 - qs}{p + s} \right)^2. \tag{18}$$

After deriving a similar expression for $\hat{c}_1/\hat{c}_2$, we conclude that the contraction factor ρ is

$$\rho = \frac{q+r}{1-pr} \cdot \frac{p+\hat{s}}{1-q\hat{s}}. \tag{19}$$

Theorem 3 *Let $B = C = I$ and $\alpha_2 = 0$ (no target state). If $H = \frac{1}{2}(A + A^\top)$ is positive semidefinite, then the contraction factor ρ in (19) is minimized for*

$$p = \frac{\sqrt{\alpha_1}}{\sqrt{2}+1}, \qquad q = \frac{1}{\sqrt{\alpha_1}(\sqrt{2}+1)}. \tag{20}$$

For these parameters, the two-subdomain OSM converges with the contraction factor

$$\rho = 3 - 2\sqrt{2} \approx 0.1716.$$

Proof Since r is a decreasing function of s (and vice versa), the contraction factor in (19) can be minimized by choosing the smallest possible r and $\hat{s}$ for which the corresponding s and $\hat{r}$ satisfy the upper bounds in (15). Thus, the best choices of r and $\hat{s}$ are given by

$$r = \max\left\{0, \frac{-2q\sqrt{\alpha_1} + (1-pq)}{2p + \sqrt{\alpha_1}(1-pq)}\right\}, \qquad \hat{s} = \max\left\{0, \frac{-2p + \sqrt{\alpha_1}(1-pq)}{2q\sqrt{\alpha_1} + (1-pq)}\right\}.$$

This leads to the following formula for the contraction factor,

$$\rho = \max\left\{q, \frac{1 - q\sqrt{\alpha_1}}{p + \sqrt{\alpha_1}}\right\} \cdot \max\left\{p, \frac{\sqrt{\alpha_1} - p}{q\sqrt{\alpha_1} + 1}\right\},$$

which must be minimized within the region $\{(p,q) : p > 0, q > 0, pq < 1\}$. A somewhat tedious analysis shows that the minimum occurs for the values of p and q shown in (20), with the contraction factor $\rho = 3 - 2\sqrt{2}$. □

Remark Since the contraction estimate is independent of the mesh parameter h and valid for any positive semidefinite matrix H, the above result is robust with respect to spatial and temporal grid refinement.

3 More Convergence Results

We now present two convergence results that hold in more general settings. For a proof of these results, we refer to Gander and Kwok (2016).

Multiple Subdomains It is straightforward to generalize (4)–(7) to the case of many time intervals. Theorem 2 holds for the general case as well. The technique of energy estimates allows us to prove the following result regarding convergence in the multiple subdomain case:

Theorem 4 *Suppose $B = C = I$, $\alpha_2 = 0$. If $H = \frac{1}{2}(A + A^\top)$ is positive semidefinite and h_T is the length of the shortest time sub-interval, then the optimized Schwarz method (4)–(7) converges whenever $pq < 1$ and p, q are not both zero. Moreover, the optimal parameter is given asymptotically by*

$$p_{\text{opt}} = \sqrt{\alpha_1} - \alpha_1^{2/3}(4h_T)^{1/3} + O(h_T^{2/3}), \qquad q_{\text{opt}} = p_{\text{opt}}/\alpha_1,$$

for which we have the contraction factor

$$\rho_{\text{opt}} = 1 - 2h_T\sqrt{\alpha_1} + O((h_T\sqrt{\alpha_1})^{5/3}).$$

Control and Observation Over a Subset Consider a problem with non-trivial control and observation matrices B and C, so that the forcing terms in (4) are restricted to parts of the domain that are controllable or observable. In this case, the quantities inside the integrals in (8) become

$$(\mu_1^k)^\top(\alpha_1 r^2 C^\top C - 2rH - BB^\top)\mu_1^k \quad \text{and} \quad (z_1^k)^\top(s^2 BB^\top - 2sH - \alpha_1 C^\top C)z_1^k,$$

both of which must be zero or negative for all z_1^k and μ_1^k in order for the energy estimates to hold. This restricts the range of allowable parameters r that can be chosen to minimize the contraction factor in (19). Together with a similar criterion on s, we obtain the following theorem.

Theorem 5 *Let $\alpha_2 = 0$ (no target state). Suppose that*

$$\ker(H) \cap \ker(C) \cap \text{range}(B) = \ker(H) \cap \ker(B^\top) \cap \text{range}(C^\top) = \{0\}.$$

Then the method (4)–(7) with two subdomains converges if the non-negative parameters p and q are chosen such that $pq < 1$ and $(1 - pq)(1 - r^*s^*) < 2(pr^* + qs^*)$, where

$$r^* = \min_{\substack{\mu \in \text{range}(C^\top) \\ \mu \neq 0}} \frac{\mu^\top H\mu}{\alpha_1 \|C\mu\|^2} + \sqrt{\left(\frac{\mu^\top H\mu}{\alpha_1 \|C\mu\|^2}\right)^2 + \frac{\|B^\top\mu\|^2}{\alpha_1 \|C\mu\|^2}} > 0,$$

$$s^* = \min_{\substack{z \in \text{range}(B) \\ z \neq 0}} \frac{z^\top Hz}{\|B^\top z\|^2} + \sqrt{\left(\frac{z^\top Hz}{\|B^\top z\|^2}\right)^2 + \frac{\alpha_1 \|Cz\|^2}{\|B^\top z\|^2}} > 0.$$

4 Numerical Experiments

Distributed Control To illustrate Theorem 3, we consider the optimal control problem where the governing PDE is the two-dimensional advection-diffusion equation

$$y_t - \nabla \cdot (\nabla y + \mathbf{b}y) = u \quad \text{on } \Omega = (0,1) \times (0,1)$$

with $\mathbf{b} = \sin \pi x_1 \sin \pi x_2 \left[x_2 - 0.5,\, 0.5 - x_1 \right]^\top$ and no-flow conditions on $\partial\Omega$. The governing PDE is discretized using backward Euler in time and an upwind finite-difference discretization in space, with mesh parameters $h = \frac{1}{16}$ and $h = \frac{1}{32}$ respectively. The adjoint PDE is discretized using "forward" Euler, which is implicit because the adjoint runs backward in time. We solve the optimal control problem (2) over the time horizon $(0, T)$ with $T = 3$, $\alpha_1 = 1$ and $\alpha_2 = 0$, i.e., we do not have a target state. The time window is subdivided into two intervals at $\beta = 1$. At the interface, we use Robin interface conditions with the optimized parameters suggested by Theorem 3, i.e., $p = q = \sqrt{2} - 1$. The convergence history in Fig. 1 shows that the error ratios approach the convergence factor of 0.1716, as predicted by Theorem 3.

Control and Observation Over Subsets For a more realistic example, we consider the problem of pollution tracking, where the goal is to estimate the rate at which a certain pollutant is released based on concentration readings elsewhere in the domain. The governing equation is the 2D advection-diffusion equation, where the domain is as shown in Fig. 2. The flow field is computed by solving the Stokes equation, where the curved part of the domain is a no-flow boundary representing a shoreline, and the straight boundary contains in-flow and out-flow boundary conditions. The source of the pollution is a region near the centre of the domain, and we seek the rate of release that minimizes the discrepancy between the predicted

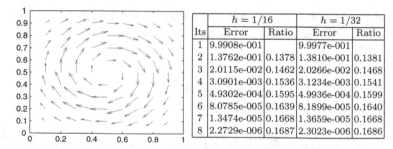

Its	$h = 1/16$ Error	Ratio	$h = 1/32$ Error	Ratio
1	9.9908e-001		9.9977e-001	
2	1.3762e-001	0.1378	1.3810e-001	0.1381
3	2.0115e-002	0.1462	2.0266e-002	0.1468
4	3.0901e-003	0.1536	3.1234e-003	0.1541
5	4.9302e-004	0.1595	4.9936e-004	0.1599
6	8.0785e-005	0.1639	8.1899e-005	0.1640
7	1.3474e-005	0.1668	1.3659e-005	0.1668
8	2.2729e-006	0.1687	2.3023e-006	0.1686

Fig. 1 *Left*: velocity field used in the distributed control problem. *Right*: convergence of OSM for two time sub-intervals

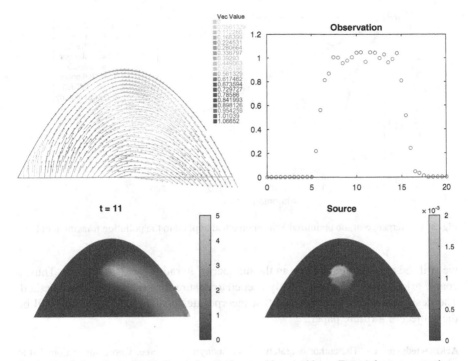

Fig. 2 *Top left*: velocity field for the pollution tracking problem. *Top right*: concentrations observed at one point on the boundary. *Bottom left*: concentration at $t = 11$ that best matches the observations at the boundary point indicated by the *red triangle*. *Bottom right*: release rate that yields the concentration to the left

and observed concentration at the point indicated by the red triangle on the curved boundary.

The advection-diffusion equation that models the concentration of pollutants is discretized using backward Euler in time and a finite volume method in space for unstructured grids, as presented in Bermúdez et al. (1998). The resulting problem has 736 degrees of freedom in space, and the time interval of $(0, T)$ with $T = 20$ is split into 2, 4, 8 and 16 equal sub-intervals to test the optimized Schwarz method. Applying the minimization procedure in Theorem 3 to the bounds on r and s in Theorem 5, we determine the best parameters p and q to be 0.8563. We show in Fig. 2 a snapshot of the concentration and source term that best match the observed concentration shown in the bottom right panel.

In Fig. 3, we show the convergence of the OSM as a stand-alone solver and as a preconditioner used within GMRES. We see that the convergence of the stationary method depends only very weakly on the number of subdomains, even though Theorem 4 suggests that the number of iterations should scale like $O(1/N)$, where N is the number of subdomains. Nonetheless, when Krylov acceleration is used,

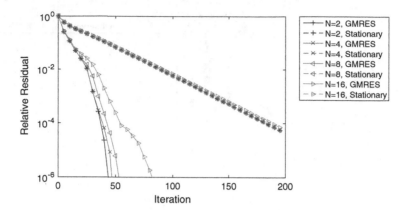

Fig. 3 Convergence of the optimized Schwarz method applied to the pollution tracking problem

we still see a moderate increase in the number of iterations as N increases. Thus, a coarse grid correction is most likely needed to ensure the scalability of the method. The design of a two-level method that incorporates coarse grid correction will be the subject of a future paper.

Acknowledgements The author is grateful to the anonymous referee, whose suggestions led to a better presentation of the paper. We would also like to thank Martin J. Gander for the inspiring collaboration and discussions on this topic. This work is partially supported by Grant No. ECS-22300115 from the Research Grants Council of Hong Kong.

References

A.T. Barker, M. Stoll, Domain decomposition in time for PDE-constrained optimization. Comput. Phys. Commun. **197**, 136–143 (2015)

A. Bermúdez, A. Dervieux, J.-A. Desideri, M.E. Vázquez, Upwind schemes for the two-dimensional shallow water equations with variable depth using unstructured meshes. Comput. Method Appl. Mech. **155**, 49–72 (1998)

A. Borzì, Multigrid methods for parabolic distributed optimal control problems. J. Comput. Appl. Math. **157**, 365–382 (2003)

H. Choi, M. Hinze, K. Kunisch, Instantaneous control of backward-facing step flows. Appl. Numer. Math. **31**(2), 133–158 (1999)

D. Corwin, C. Holdsworth, R. Rockne, A.D. Trister, M.M. Mrugala, J.K. Rockhill, R.D. Stewart, M. Phillips, K.R. Swanson, Toward patient-specific, biologically optimized radiation therapy plans for the treatment of glioblastoma. PLoS ONE **8**(11), e79115 (2013)

A.L. Dontchev, W.W. Hager, V.M. Veliov, Second-order Runge–Kutta approximations in control constrained optimal control. SIAM J. Numer. Anal. **38**, 202–226 (2000)

M.J. Gander, F. Kwok, Optimized Schwarz methods in time for parabolic control problems (in preparation, 2017)

M.J. Gander, F. Kwok, Schwarz methods for the time-parallel solution of parabolic control problems, in *Domain Decomposition Methods in Science and Engineering XXII* (Springer, Switzerland, 2016)

M.J. Gander, F. Kwok, J. Salomon, A parareal algorithm for optimality systems (in preparation, 2017)

M.J. Gander, M. Neumüller, Analysis of a new space-time parallel multigrid algorithm for parabolic problems. SIAM J. Sci. Comput. **38**(4), A2173–A2208 (2016)

W.W. Hager, Runge-Kutta methods in optimal control and the transformed adjoint system. Numer. Math. **87**, 247–282 (2000)

M. Heinkenschloss, A time-domain decomposition iterative method for the solution of distributed linear quadratic optimal control problems. J. Comput. Appl. Math. **173**, 169–198 (2005)

G. Horton, S. Vandewalle, A space-time multigrid method for parabolic partial differential equations. SIAM J. Sci. Comput. **16**, 848–864 (1995)

J.E. Lagnese, G. Leugering, Time-domain decomposition of optimal control problems for the wave equation. Syst. Control Lett. **48**, 229–242 (2003)

J.-L. Lions, Y. Maday, G. Turinici, A parareal in time discretization of PDEs. C.R. Acad. Sci. Paris, Série I **332**, 661–668 (2001)

P.-L. Lions, On the Schwarz alternating method III: a variant for non-overlapping subdomains, in *Third International Symposium on Domain Decomposition Methods for Partial Differential Equations* (SIAM, Philadelphia, PA, 1990), pp. 47–70

Y. Maday, M.-K. Riahi, J. Salomon, Parareal in time intermediate targets methods for optimal control problems, in *Control and Optimization with PDE Constraints* (Springer, Basel, 2013), pp. 79–92

Y. Maday, J. Salomon, G. Turinici, Monotonic parareal control for quantum systems. SIAM J. Numer. Anal. **45**(6), 2468–2482 (2007)

T.P. Mathew, M. Sarkis, Ch.E. Schaerer, Analysis of block parareal preconditioners for parabolic optimal control problems. SIAM J. Sci. Comput. **32**, 1180–1200 (2010)

J.W. Pearson, M. Stoll, A.J. Wathen, Regularization-robust preconditioners for time-dependent PDE-constrained optimization problems. SIAM J. Matrix Anal. A. **33**, 1126–1152 (2012)

T. Rees, M. Stoll, A. Wathen, All-at-once preconditioning in PDE-constrained optimization. Kybernetika **46**, 341–360 (2010)

A. Schiela, S. Ulbrich, Operator preconditioning for a class of inequality constrained optimal control problems. SIAM J. Optimiz. **24**, 435–466 (2014)

A. Unger, F. Tröltzsch, Fast solution of optimal control problems in the selective cooling of steel. Z. Angew. Math. Mech. **81**, 447–456 (2001)

Parallel Solver for $H(\mathrm{div})$ Problems Using Hybridization and AMG

Chak S. Lee and Panayot S. Vassilevski

1 Introduction

This paper is concerned with the $\boldsymbol{H}(\mathrm{div})$ bilinear form acting on vector functions $\boldsymbol{u}, \boldsymbol{v}$:

$$a(\boldsymbol{u}, \boldsymbol{v}) = \int_{\Omega} \alpha \, \nabla \cdot \boldsymbol{u} \, \nabla \cdot \boldsymbol{v} + \beta \, \boldsymbol{u} \cdot \boldsymbol{v} \, d\boldsymbol{x}. \tag{1}$$

Here α, $\beta \in L^{\infty}(\Omega)$ are some positive heterogeneous coefficients, and Ω is a simply-connected polygonal domain in $\mathbb{R}^d$, $d = 2, 3$. Discrete problems associated with $a(\cdot, \cdot)$ arise in many applications, such as first order least squares formulation of second order elliptic problems (Cai et al. 1994), preconditioning of mixed finite element methods (Brezzi and Fortin 1991), Reissner-Mindlin plates (Arnold et al. 1997) and the Brinkman equations (Vassilevski and Villa 2013). Let A be the linear system obtained from discretization of $a(\cdot, \cdot)$ by some $\boldsymbol{H}(\mathrm{div})$-conforming finite elements of arbitrary order on a general unstructured mesh. Our goal is to design a scalable parallel solver for A.

C.S. Lee
Department of Mathematics, Texas A&M University, Mailstop 3368, College Station, TX 77843, USA
e-mail: cslee@math.tamu.edu

P.S. Vassilevski (✉)
Center for Applied Scientific Computing, Lawrence Livermore National Laboratory, P.O. Box 808, L-561, Livermore, CA 94551, USA

Fariborz Maseeh Department of Mathematics and Statistics, Portland State University, Portland, OR 97201, USA
e-mail: panayot@llnl.gov; panayot@pdx.edu

© Springer International Publishing AG 2017
C.-O. Lee et al. (eds.), *Domain Decomposition Methods in Science and Engineering XXIII*, Lecture Notes in Computational Science and Engineering 116, DOI 10.1007/978-3-319-52389-7_6

69

It is well known that finding efficient iterative solvers for A is not trivial because of the "near-null space" of A. The currently available scalable parallel solvers include the auxiliary space divergence solver (ADS) (Kolev and Vassilevski 2012) in the *hypre* library [www.llnl.gov/CASC/hypre/] and PCBDDC (Zampini 2016) in the PETSc library. The former relies on the regular HX-decomposition for $H(\mathrm{div})$ functions proposed in Hiptmair and Xu (2007). The setup of ADS is quite involved and requires additional input from the user, namely, some discrete gradient and discrete curl operators. On the other hand, PCBDDC is based on the Balancing Domain Decomposition by Constraints algorithm (Dohrmann 2003). Its construction requires that the local discrete systems are assembled at subdomain level. To accommodate high contrast and jumps in the coefficients, the primal space in PCBDDC is adaptively enriched by solving some generalized eigenvalue problems, see Zampini and Keyes (2016).

In this paper, we propose an alternative way to solve systems with A. Our approach is based on the traditional hybridization technique used in the mixed finite element method (Brezzi and Fortin 1991), thus reducing the problem to a smaller problem for the respective Lagrange multipliers that are involved in the hybridization. The reduced problem is symmetric positive definite, and as is well-known, is H^1-equivalent. Thus, in principle, one may apply any scalable AMG solver that is suitable for H^1 problems. Unlike ADS, the hybridization approach does not require additional specialized information (such as discrete gradient and discrete curl) from the user. Instead, it requires that the original problem is given in unassembled element-based form.

One main issue that has to be addressed is the choice of the basis of the Lagrange multiplier space. In general, the reduced problem contains the constant function in its near null-space. However, if the basis for the Lagrange multipliers is not properly scaled (i.e., does not provide partition of unity), the coefficient vector of the constant functions is not a constant multiple of the vector of ones. The latter is a main assumption in the design of AMG for H^1-equivalent problems. We resolve this problem in an algebraic way by constructing a diagonal matrix which we use to rescale the reduced system such that the constant vector is the near-null space of the rescaled matrix, so that the respective AMG is correctly designed.

The proposed hybridization with diagonal rescaling is implemented in a parallel code and its scalability is tested in comparison with the state-of-the-art ADS solver. The results demonstrate that the new solver provides a competitive alternative to ADS; it outperforms ADS very clearly for higher order elements.

Although in this paper we focus on finite element problems discretized by Raviart-Thomas elements, the proposed approach can be applied to other $H(\mathrm{div})$ conforming discretizations like Brezzi-Douglas-Marini elements, Arnold-Boffi-Falk elements (Arnold et al. 2005), or numerically upscaled problems (Chung et al. 2015; Kalchev et al. 2016).

The rest of the paper is organized as follows. In Sect. 2, we give a detailed description of the hybridization technique. The properties of the hybridized system are discussed in Sect. 3. After that, we present in Sect. 4 several challenging numerical examples to illustrate the performance of the method comparing it with ADS.

2 Hybridization

We consider the variational problem associated with the bilinear form (1): find $u \in H_0(\text{div};\Omega)$ such that

$$a(u, v) = (f, v), \qquad \forall\, v \in H_0(\text{div}; \Omega). \tag{2}$$

Here, f is a given function in $\left(L^2(\Omega)\right)^d$ and $(\cdot, \cdot)$ is the usual L^2 inner product in Ω. Our following discussion is based on discretization of the variational problem (2) by Raviart-Thomas elements of arbitrary order. We note that other H(div)-conforming finite elements can also be considered. Let $\mathcal{T}_h$ be a general unstructured mesh on Ω. The space of Raviart-Thomas elements of order $k \geq 0$ on $\mathcal{T}_h$ will be denoted by RT_k. For instance, if $\mathcal{T}_h$ is a simplicial mesh, then RT_k is defined to be

$$RT_k = \left\{\, v_h \in H_0(\text{div}; \Omega) \,\big|\, v_h|_\tau \in \left(P_k(\tau)\right)^d + x P_k(\tau) \quad \forall \tau \in \mathcal{T}_h \,\right\},$$

where $P_k(\tau)$ denotes the set of polynomials of degree at most k on τ. For definitions of RT_k on rectangular/cubic meshes, see for example Brezzi and Fortin (1991). Discretization of (2) by RT_k elements results in a linear system of equations

$$Au = f. \tag{3}$$

We are going to formulate an equivalent problem such that the modified problem can be solved more efficiently. We note that RT_k basis functions are either associated with degrees of freedom (dofs) in the interior of elements, on boundary faces, or interior faces of a conforming finite element mesh. Those associated with dofs in the interior of elements or on boundary faces are supported in only one element, while those associated with dofs on interior faces are supported in two elements. In hybridization, the RT_k basis functions that are associated with dofs on interior faces are split into two pieces, each supported in one and only one element. In practice, the splitting can be done by making use of the element-to-dofs relation table to identify the shared dofs between any pair of neighboring elements. This relation table can be constructed during the discretization. The space of Raviart-Thomas element after the splitting will be denoted by $\widehat{RT}_k$. If we discretize $a(\cdot, \cdot)$ with the basis functions in $\widehat{RT}_k$, the resulting system will have a block diagonal matrix $\widehat{A}$. Next, we need to enforce the continuity of the split basis functions in some way such that the solution of the modified system coincides with the original problem. Suppose a RT_k basis function ϕ is split into $\widehat{\phi}_1$ and $\widehat{\phi}_2$. The simplest way is to use Lagrange multiplier space to make the coefficient vectors of the test functions from both sides of an interior interface to be the same. If we set such constraints for all the split basis functions, we obtain a constraint matrix C.

Remark 1 There are other ways to enforce continuity of $\widehat{RT}_k$. For example, when constructing the constraint matrix C, one can also use the normal traces

λ of the original RT_k basis functions as Lagrange multipliers; see Cockburn and Gopalakrishnan (2004).

The modified problem after introducing the Lagrange multipliers takes the saddle–point form

$$\begin{bmatrix} \widehat{A} & C^T \\ C & 0 \end{bmatrix} \begin{bmatrix} \widehat{u} \\ \lambda \end{bmatrix} = \begin{bmatrix} \widehat{f} \\ 0 \end{bmatrix}. \tag{4}$$

Here, $\widehat{u}$ is the coefficient vector of $\widehat{u}_h$. The saddle point problem (4) can be reduced to

$$S\lambda = g, \tag{5}$$

where $S = C\widehat{A}^{-1}C^T$ and $g = C\widehat{A}^{-1}\widehat{f}$. The Schur complement S and the new right-hand side g can be explicitly formed very efficiently because $\widehat{A}$ is block diagonal. In fact, the inversion of $\widehat{A}$ is embarrassingly parallel. Here, each local block of $\widehat{A}$ is invertible, so $\widehat{A}^{-1}$ is well-defined. We will show in the next section that S is actually an s.p.d. system of the Lagrange multipliers, and that it can be solved efficiently by existing parallel linear solvers. After solving for λ, $\widehat{u}$ can be computed by back substitution $\widehat{u} = \widehat{A}^{-1}(\widehat{f} - C^T\lambda)$. Note that the back substitution involves only an action of $\widehat{A}^{-1}$ (already available in the computation of S) and some matrix-vector multiplications, which are inexpensive (local) and scalable computations.

3 Discussion

The hybridization approach described in the previous section can be summarized as follows:

1. Split the RT_k basis to obtain $\widehat{A}$ and $\widehat{f}$.
2. Compute $\widehat{A}^{-1}$ and form $S = C\widehat{A}^{-1}C^T$ and $g = C\widehat{A}^{-1}\widehat{f}$.
3. Solve the system $S\lambda = g$.
4. Recover $\widehat{u}$ by back substitution.

As explained in Sect. 2, steps 2 and 4 are scalable (inexpensive local) computations. In contrast, step 3 involves the main computational cost. Thus, it is important that we can solve S efficiently. In this section, we describe some properties of S. First, we show that S is related to some hybridized mixed discretization of the second order differential operator $-\nabla \cdot (\beta^{-1}\nabla) + \alpha^{-1}I$ (acting on scalar functions). We note that the differential problem associated with (2) is

$$-\nabla(\alpha\nabla \cdot u) + \beta u = f \tag{6}$$

with homogeneous Dirichlet boundary condition $u \cdot n = 0$. The latter operator acts on vector-functions. We now make the following connection between these two operators. If we introduce an additional variable $p = \alpha \nabla \cdot \boldsymbol{u}$, then (6) becomes the following first order system (for $\boldsymbol{u}$ and p)

$$
\begin{aligned}
\beta \boldsymbol{u} - \nabla p &= \boldsymbol{f}, \\
\nabla \cdot \boldsymbol{u} - \alpha^{-1} p &= 0.
\end{aligned}
\tag{7}
$$

It is noteworthy to note that the structure of (7) is the same as the mixed formulation of the differential operator $-\nabla \cdot (\beta^{-1} \nabla) + \alpha^{-1} I$. So we can apply a hybridized mixed discretization (Cockburn and Gopalakrishnan 2004, 2005) for $-\nabla \cdot (\beta^{-1} \nabla) + \alpha^{-1} I$ to discretize (7). To apply the hybridized mixed discretization, we note that the weak form of (7) is to find $(\boldsymbol{u}, p) \in \boldsymbol{H}_0(\text{div}; \Omega) \times L^2(\Omega)$ such that

$$
\begin{aligned}
(\beta \boldsymbol{u}, \boldsymbol{v}) + (p, \nabla \cdot \boldsymbol{v}) &= (\boldsymbol{f}, \boldsymbol{v}) &&\forall \, \boldsymbol{v} \in \boldsymbol{H}_0(\text{div}; \Omega) \\
(\nabla \cdot \boldsymbol{u}, q) - (\alpha^{-1} p, q) &= 0 &&\forall \, q \in L^2(\Omega).
\end{aligned}
\tag{8}
$$

Let $W_h^k \subset L^2(\Omega)$ be a space of piecewise polynomials such that RT_k and W_h^k form a stable pair for the mixed discretization of (8). For instance, for simplicial meshes, we can take

$$
W_h^k = \left\{ q \in L^2(\Omega) \mid q|_\tau \in P_k(\tau) \quad \forall \tau \in \mathcal{T}_h \right\}.
$$

If (8) is discretized by the pair $\widehat{RT}_k$-W_h^k and the continuity of $\widehat{RT}_k$ is enforced by the constraint matrix C as described in Sect. 2, we get a 3 by 3 block system of equations of the form

$$
\begin{bmatrix} \widehat{M} & \widehat{B}^T & C^T \\ \widehat{B} & -W & 0 \\ C & 0 & 0 \end{bmatrix} \begin{bmatrix} \widehat{u} \\ p \\ \lambda \end{bmatrix} = \begin{bmatrix} \widehat{f} \\ 0 \\ 0 \end{bmatrix}.
\tag{9}
$$

As $\widehat{M}$ and W are weighted L^2 mass matrices of the spaces $\widehat{RT}_k$ and W_h^k respectively, they are invertible. Hence, the 2 by 2 block matrix $\begin{bmatrix} \widehat{M} & \widehat{B}^T \\ \widehat{B} & -W \end{bmatrix}$ is invertible, and (9) can be reduced to

$$
\begin{bmatrix} C & 0 \end{bmatrix} \begin{bmatrix} \widehat{M} & \widehat{B}^T \\ \widehat{B} & -W \end{bmatrix}^{-1} \begin{bmatrix} C^T \\ 0 \end{bmatrix} \lambda = \begin{bmatrix} C & 0 \end{bmatrix} \begin{bmatrix} \widehat{M} & \widehat{B}^T \\ \widehat{B} & -W \end{bmatrix}^{-1} \begin{bmatrix} \widehat{f} \\ 0 \end{bmatrix}.
\tag{10}
$$

Since the $(1, 1)$ block of $\begin{bmatrix} \widehat{M} & \widehat{B}^T \\ \widehat{B} & -W \end{bmatrix}^{-1}$ can be written as $(\widehat{M} + \widehat{B}^T W^{-1} \widehat{B})^{-1}$ and $\widehat{A} = \widehat{M} + \widehat{B}^T W^{-1} \widehat{B}$, the reduced problem (10) is in fact identical to (5). Therefore, the Schur complement S in (5) can be characterized by the hybridized mixed discretization for the differential operator $-\nabla \cdot (\beta^{-1}\nabla) + \alpha^{-1}I$.

Remark 2 Actually the hybridized mixed discretization for $-\nabla \cdot (\beta^{-1}\nabla) + \alpha^{-1}I$ in Cockburn and Gopalakrishnan (2004, 2005) gives rise to the reduced system $\widetilde{S}$ for the Lagrange multiplier λ where

$$\widetilde{S} = C\left(\widehat{M}^{-1} - \widehat{M}^{-1}\widehat{B}^T\left(B\widehat{M}^{-1}\widehat{B}^T + W\right)^{-1}B\widehat{M}^{-1}\right)C^T.$$

However, since W is invertible, an application of the Sherman-Morrison-Woodbury formula implies that $\widetilde{S} = S$.

In Cockburn and Gopalakrishnan (2005), the authors proved that S is spectrally equivalent to the norm $\|\|\cdot\|\|$ on the space of Lagrange multipliers defined as

$$\|\|\lambda\|\|^2 = \sum_{\tau \in \mathcal{T}_h} \frac{1}{|\partial \tau|}\|\lambda - m_\tau(\lambda)\|^2_{\partial\tau}$$

where $m_\tau(\lambda) = \frac{1}{|\partial\tau|}\int_{\partial\tau} \lambda \; ds$. More precisely, there are constants C_1 and C_2, depending only on the approximation order k, the coefficients α, β of the operator, and the shape regularity of $\mathcal{T}_h$ such that $C_1\|\|\lambda\|\|^2 \leq \lambda^T S\lambda \leq C_2\|\|\lambda\|\|^2$ for all λ. Consequently, S is symmetric positive definite. Moreover, this shows that the near-null space of S is spanned by the constant functions, which is the main assumption to successfully apply solvers of AMG type. When solving with S, we opt for the parallel algebraic multigrid solver BoomerAMG (Henson and Yang 2002) from the *hypre* library.

Depending on the choice of basis for the Lagrange multipliers space, the coefficient vector of a constant function is not necessarily a constant vector and the latter affects adversely the performance of classical AMG methods such as BoomerAMG from hypre. To resolve this issue, we chose to rescale S by a diagonal matrix D such that the constant vector is now in the near-null space of D^TSD. To achieve this, we solve the homogeneous problem $Sd = 0$ by applying a few smoothing steps to a random initial guess. In our numerical experiments to be presented in the next section, we use five conjugate gradient (CG) iterations preconditioned by the Jacobi smoother in the computation of d, which is fairly inexpensive. Once d is computed, we set $D_{ii} = d_i$ (the i-th entry of d). Noticing that $D\mathbf{1} = d$, so $\mathbf{1}$ is in the near-null space of D^TSD. We can then apply CG preconditioned by BoomerAMG constructed from D^TSD to efficiently solve the system

$$(D^TSD)\lambda_D = D^Tg.$$

Lastly, the original Lagrange multiplier λ is recovered simply by setting $\lambda = D\lambda_D$.

Another useful feature of S is that its size is less than or equal to the size of the original matrix A. This is because there is a one-to-one correspondence between

Lagrange multipliers and Raviart-Thomas basis functions associated with interior faces. For higher order Raviart-Thomas elements, a portion of the basis functions are associated with interior of elements. These basis functions are supported in one element only, so they do not need Lagrange multipliers to enforce their continuity. Hence, for higher order approximations, methods for solving with S are likely to be more efficient and faster than solving with A (using state-of-the-art solvers such as ADS) which is confirmed by our experiments.

4 Numerical Examples

In this section, we present some numerical results regarding the performance of our hybridization AMG solver. The numerical results are generated using MFEM [mfem.org], a scalable C++ library for finite element methods developed in the Lawrence Livermore National Laboratory (LLNL). All the experiments are performed on the cluster Sierra at LLNL. Sierra has a total of 1944 nodes (Intel Xeon EP X5660 clocked at 2.80 GHz), which are connected by InfiniBand QDR. Each node has 12 cores and equipped with 24 GB of memory.

In the solution process, the hybridized system with S is rescaled by the diagonal matrix D as described in the previous section. The rescaled system $D^T SD$ is then solved by the CG method preconditioned with BoomerAMG (constructed from $D^T SD$) from the *hypre* library. As one of our goals is to compare the hybridization AMG solver with ADS, we present also the performance of ADS in all the examples. In order to have fair comparisons, the time to solution for the hybridization AMG solver includes the formation time of the Schur complement S, the computation time to construct the rescaling matrix D, the solve time for the problem with the modified matrix $D^T SD$ by CG preconditioned by BoomerAMG, and the recovery time of the original unknown u. The time to solution for ADS is simply the solve time for the original problem with A by the CG preconditioned by ADS. For the tables in the present section, #proc refers to the number of processors, while #iter refers to the number of PCG iterations.

4.1 Weak Scaling

We first test the weak scaling of the hybridization AMG solver. The problem setting is as follows. We solve problem (3) obtained by RT_k discretization on uniform tetrahedral mesh in 3D. Starting from some initial tetrahedral mesh, we refine the mesh uniformly. The problem size increases by about eight times after one such refinement. At the same time, the number of processors for solving the refined problem is increased eight times so that the problem size per processor is kept roughly the same. Both the lowest order Raviart-Thomas elements RT_0 and a higher order elements, RT_2, are considered. We solve a heterogeneous coefficient problem on the unit cube, i.e. $\Omega = [0, 1]^3$. The boundary conditions are $\boldsymbol{u} \cdot \boldsymbol{n} = 0$ on $\partial\Omega$, and the source function f is the constant vector $[1, 1, 1]^T$. Let $\Omega_i = [\frac{1}{4}, \frac{1}{2}]^3 \cup [\frac{1}{2}, \frac{3}{4}]^3$. We consider β being constant 1 throughout the domain, whereas $\alpha = \begin{cases} 1 & \text{in } \Omega \setminus \Omega_i \\ 10^p & \text{in } \Omega_i \end{cases}$ and we choose $p = -4$, 0, or 4. For RT_2 test case, we first partition Ω into $8 \times 8 \times 4$ parallelepipeds. The initial tetrahedral mesh in this case is then obtained by subdividing each parallelepiped into tetrahedrons, see Fig. 1. The initial mesh of the RT_0 test case is obtained by refining the initial mesh of the RT_2 test case three times. The PCG iterations are stopped when the l_2 norm of the residual is reduced by a factor of 10^{10}. The time to solution (in seconds) of both the hybridization AMG and ADS for the RT_0 case are shown in Table 1. Additionally, we also report the number of PCG iterations in the brackets. We see that the number of iterations of the hybridization solver are very stable against problem size and the heterogeneity of α. The average time to solution of the hybridization approach is about two times faster than that of ADS. The solution time difference between the two solvers is more significant in the high order discretization case. This is due to the fact that size of the hybridized system S is much smaller than the size of the original system A. Indeed, in the case of RT_2, the average time to solution of the hybridization

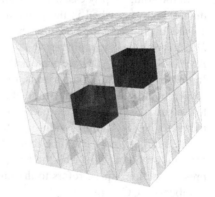

Fig. 1 Initial mesh for the RT_2 weak scaling test case. *Blue* region indicates Ω_i

Table 1 Time to solution (in seconds) in the weak scaling test: RT_0 on tetrahedral meshes, the corresponding number of PCG iterations are the reported in the brackets

#proc	Problem size	$p = -4$	$p = 0$	$p = 4$
Hybridization-BoomerAMG-CG				
3	200, 704	0.97 (24)	0.96 (21)	0.93 (21)
24	1, 589, 248	1.15 (24)	1.15 (23)	1.16 (23)
192	12, 648, 448	1.45 (27)	1.48 (25)	1.43 (24)
1, 536	100, 925, 440	3.31 (29)	3.03 (28)	3.03 (28)
ADS-CG				
3	200, 704	2.68 (21)	1.74 (10)	1.79 (11)
24	1, 589, 248	4.04 (25)	3.53 (13)	3.54 (13)
192	12, 648, 448	7.10 (27)	5.73 (15)	5.61 (14)
1, 536	100, 925, 440	8.30 (28)	6.28 (15)	6.51 (15)

Table 2 Time to solution (in seconds) in the weak scaling test: RT_2 on tetrahedral meshes, the corresponding number of PCG iterations are the reported in the brackets

#proc	Problem size	$p = -4$	$p = 0$	$p = 4$
Hybridization-BoomerAMG-CG				
3	38, 400	0.30 (15)	0.31 (16)	0.31 (16)
24	301, 056	0.48 (18)	0.50 (21)	0.48 (20)
192	2, 383, 872	0.75 (28)	0.89 (29)	0.77 (29)
1, 536	18, 972, 672	1.97 (44)	1.95 (47)	2.10 (47)
ADS-CG				
3	38, 400	4.85 (23)	3.55 (13)	3.80 (14)
24	301, 056	7.24 (29)	5.47 (18)	5.73 (20)
192	2, 383, 872	11.56 (37)	8.89 (25)	9.56 (28)
1, 536	18, 972, 672	24.28 (53)	16.51 (37)	16.37 (39)

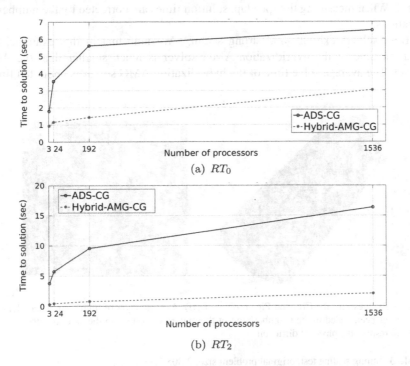

Fig. 2 Weak scaling comparisons between the hybridization AMG solver (*red dotted line*) and ADS (*blue solid line*). (**a**) RT_0. (**b**) RT_2

approach is about eight times faster than that of ADS, see Table 2. In Fig. 2, we plot the solution time of both solvers where $p = 4$ in the definition of α. We can see that the hybridization solver has promising weak scaling over a range of nearly three decades.

4.2 Strong Scaling

In the second example, we investigate the strong scaling of the hybridization AMG
solver. The problem considered in this section is the crooked pipe problem, see
Kolev and Vassilevski (2012) for a detailed description of the problem. The mesh
for this problem is depicted in Fig. 3. The coefficient α and β are piecewise
constants. More precisely, $(\alpha, \beta) = (1.641, 0.2)$ in the red region, and $(\alpha, \beta) =
(0.00188, 2000)$ in the blue region. The difficulties of this problem are the large
jumps of coefficients and the highly stretched elements in the mesh (see Fig. 3). For .
this test, the problem is discretized by RT_1. The size of A is 2,805,520, and we solve
the problem using 4, 8, 16 ,32 and 64 processors. The PCG iterations are stopped
when the l_2 norm of the residual is reduced by a factor of 10^{14}. The number of PCG
iterations and time to solution are reported in Table 3, and we plot the speedup in
Fig. 4. When measuring the speedup, solution times are corrected by the number of
iterations.

Both solvers exhibit good strong scaling. We note that in this example, the
solution time of the hybridization AMG solver is much smaller than the ADS
solver. The average solve time of the hybridization AMG solver is about ten times

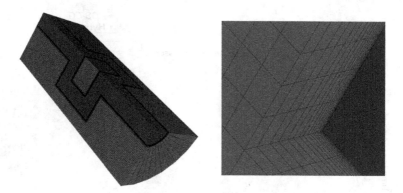

Fig. 3 The mesh for the Crooked Pipe problem (*left*). A dense layer of highly stretched elements
(*right*) has been added to the neighborhood of the material interface in the exterior subdomain in
order to resolve the physical diffusion

Table 3 Strong scaling test, original problem size: 2,805,520

	Hybridization-BoomerAMG-CG		ADS-CG	
#proc	#iter	Time to solution	#iter	Time to solution
4	25	23.46	32	508.66
8	31	14.21	32	251.37
16	28	6.83	33	130.26
32	28	3.98	34	73.47
64	31	2.92	34	54.58

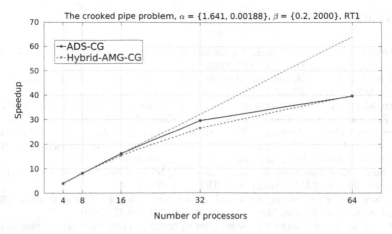

Fig. 4 Strong scaling comparison between the hybridization AMG solver (*red dotted line*) and ADS (*blue solid line*). *Black dotted line* indicates perfect scaling

Table 4 Timing of each component of the new solver

#proc	Formation of S	Computation of D	Setup	PCG solve	Recovery of u
4	7.55	0.22	3.87	11.72	0.092
8	3.95	0.11	2.29	7.81	0.046
16	1.84	0.057	1.4	3.52	0.022
32	1.11	0.034	0.83	2.01	0.012
64	0.68	0.027	0.52	1.7	0.006

smaller than that of ADS. In particular, the hybridization AMG solver with four processors is still two times faster than ADS with 64 processors. The difference in the computation time for this example is highly noticeable.

Lastly, we report the time spent on different components of the hybridization approach in Table 4. We observe that except for solving with S (i.e. setup and PCG solve), the other components scale fairly well. Also, as we point out in Sect. 3, solving with S is the most time consuming part of the hybridization AMG code. We remark that during the formation of S, we stored the inverses of local blocks of $\widehat{A}$. So when we recover u by back substitution, only matrix multiplication is needed. Hence, the recovery of u is extremely cheap and scalable.

Acknowledgements This work was performed under the auspices of the U.S. Department of Energy by Lawrence Livermore National Laboratory under Contract DE-AC52-07NA27344. The work was partially supported by ARO under US Army Federal Grant # W911NF-15-1-0590.

References

D.N. Arnold, R.S. Falk, R. Winther, Preconditioning discrete approximations of the Reissner–Mindlin plate model. **31**(4), 517–557 (1997)

D.N. Arnold, D. Boffi, R.S. Falk, Quadrilateral H(div) finite elements. SIAM J. Numer. Anal. **42**(6), 2429–2451 (2005)

F. Brezzi, M. Fortin, *Mixed and Hybrid Finite Element Methods* (Springer, New York, 1991)

Z. Cai, R. Lazarov, T.A. Manteuffel, S.F. McCormick, First-order system least squares for second-order partial differential equations: Part I. SIAM J. Numer. Anal. **31**(6), 1785–1799 (1994)

E.T. Chung, Y. Efendiev, C.S. Lee, Mixed generalized multiscale finite element methods and applications. SIAM Multiscale Model. Simul. **13**(1), 338–366 (2015)

B. Cockburn, J. Gopalakrishnan, A characterization of hybridized mixed methods for second order elliptic problems. SIAM J. Numer. Anal. **42**(1), 283–301 (2004)

B. Cockburn, J. Gopalakrishnan, Error analysis of variable degree mixed methods for elliptic problems via hybridization. Math. Comput. **74**(252), 1653–1677 (2005)

C.R. Dohrmann, A preconditioner for substructuring based on constrained energy minimization. SIAM J. Sci. Comput. **25**(1), 246–258 (2003)

V.E. Henson, U.M. Yang, BoomerAMG: a parallel algebraic multigrid solver and preconditioner. Appl. Numer. Math. **41**(1), 155–177 (2002)

R. Hiptmair, J. Xu, Nodal auxiliary space preconditioning in H(curl) and H(div) spaces. SIAM J. Numer. Anal. **45**(6), 2483–2509 (2007)

D. Kalchev, C.S. Lee, U. Villa, Y. Efendiev, P.S. Vassilevski, Upscaling of mixed finite element discretization problems by the spectral AMGe method. SIAM J. Sci. Comput. **38**(5), A2912–A2933 (2016)

T.V. Kolev, P.S. Vassilevski, Parallel auxiliary space AMG solver for H(div) problems. SIAM J. Sci. Comput. **34**(6), A3079–A3098 (2012)

P.S. Vassilevski, U. Villa, A block-diagonal algebraic multigrid preconditioner for the Brinkman problem. SIAM J. Sci. Comput. **35**(5), S3–S17 (2013)

S. Zampini, PCBDDC: a class of robust dual-primal methods in PETSc. SIAM J. Sci. Comput. **38**(5), S282–S306 (2016)

S. Zampini, D.E. Keyes, On the robustness and prospects of adaptive BDDC methods for finite element discretizations of elliptic PDEs with high-contrast coefficients, in *Proceedings of the Platform for Advanced Scientific Computing Conference*, PASC '16 (ACM, New York, NY, 2016), pp. 6:1–6:13

Preconditioning for Nonsymmetry and Time-Dependence

Eleanor McDonald, Sean Hon, Jennifer Pestana, and Andy Wathen

1 Introduction

Preconditioning, whether by domain decomposition or other methods, is well understood for symmetric (or Hermitian) matrices at least in the sense that guaranteed convergence bounds based on eigenvalues alone describe convergence of iterative methods. Establishing spectral properties of preconditioned operators or matrices is thus all that is required to reliably predict the number of steps of an appropriate Krylov subspace method—it would be Conjugate Gradients (CG) (Hestenes and Stiefel, 1952) in the case of positive definite matrices and MINRES (Paige and Saunders, 1975) for indefinite matrices—in the symmetric case. Faster convergence than that predicted by these bounds occurs in rare cases when only few eigenspaces are important; thus in the rare cases that the convergence bounds fail to be descriptive, it is because they overestimate the number of iterations required for convergence—a good thing! Put another way, we know what we're trying to achieve in the construction of preconditioners in the case of symmetric coefficient matrices.

By contrast, in the nonsymmetric case, no generally descriptive convergence bounds are known. In specialist situations, the field of values or other sets can occasionally be usefully employed (Loghin and Wathen, 2004), but it is known that GMRES can converge in any (monotone) specified manner whatever the

E. McDonald (✉) • S. Hon • A. Wathen
Mathematical Institute, Oxford University, Radcliffe Observatory Quarter, Oxford OX2 6GG, England, UK
e-mail: mcdonalde@maths.ox.ac.uk; hon@maths.ox.ac.uk; wathen@maths.ox.ac.uk

J. Pestana
Department of Mathematics and Statistics, University of Strathclyde, Glasgow G1 1XH, Scotland, UK
e-mail: jennifer.pestana@strath.ac.uk

© Springer International Publishing AG 2017
C.-O. Lee et al. (eds.), *Domain Decomposition Methods in Science and Engineering XXIII*, Lecture Notes in Computational Science and Engineering 116, DOI 10.1007/978-3-319-52389-7_7

eigenvalues for the coefficient matrix; precisely, it is proved in Greenbaum et al. (1996) [and the results extended in Tebbens and Meurant (2014)] that given any set of n eigenvalues and any monotonic convergence curve terminating at or before the nth iteration, then for any b there exists an $n \times n$ matrix B having those eigenvalues and an initial guess x_0 such that GMRES (Saad and Schultz, 1986) for $Bx = b$ with x_0 as starting vector will give that convergence curve. More negative results than this exist [see for example Tebbens and Meurant (2012)].

Thus, one can for example have an $n \times n$ nonsymmetric matrix with all eigenvalues equal to 1 for which GMRES gives no reduction in the norm of the residual vectors—that is, no convergence—for $n - 1$ iterations. For any of the range of other nonsymmetric Krylov subspace methods, convergence theory is extremely limited. Thus, even though there is often consideration of eigenvalues when considering possible preconditioners even in the nonsymmetric case, this is not well-founded. It is not however foolish, since poor convergence can certainly in general be associated with problems with widely spread eigenvalues!

The important point nevertheless remains that the construction of preconditioners for nonsymmetric problems is of necessity currently heuristic.

In this short paper, we describe at least one simple and frequently arising situation—that of nonsymmetric real Toeplitz (constant diagonal) matrices—where we can *guarantee* rapid convergence of the appropriate iterative method by manipulating the problem into a symmetric form without recourse to the normal equations. This trick can be applied regardless of the nonnormality of the Toeplitz matrix. We also propose a symmetric and positive definite preconditioner for this situation which is proved to cluster eigenvalues and is by consequence guaranteed to ensure convergence in a number of iterations independent of the matrix dimension. This is described in Sect. 2 and more fully in Pestana and Wathen (2015).

We then go on to exploit these observations in considering time-stepping problems for ordinary differential equations. The result we establish in this setting is guaranteed convergence of an iterative method for an all-at-once formulation in a number of iterations independent of the number of time-steps. This is described in Sect. 3.

2   Real Nonsymmetric Toeplitz Matrices

If B is a real Toeplitz matrix then

$$
\underbrace{\begin{bmatrix} a_0 & a_{-1} & \cdots & a_{-n+2} & a_{-n+1} \\ a_1 & a_0 & a_{-1} & & a_{-n+2} \\ \vdots & a_1 & a_0 & \ddots & \vdots \\ a_{n-2} & & \ddots & \ddots & a_{-1} \\ a_{n-1} & a_{n-2} & \cdots & a_1 & a_0 \end{bmatrix}}_{B}
\underbrace{\begin{bmatrix} 0 & 0 & \cdots & 0 & 1 \\ 0 & & 0 & 1 & 0 \\ \vdots & \cdots & 1 & 0 & \vdots \\ 0 & \cdots & \cdots & & 0 \\ 1 & 0 & \cdots 0 & 0 \end{bmatrix}}_{Y}
$$

is the real *symmetric* matrix

$$
\underbrace{\begin{bmatrix}
a_{-n+1} & a_{-n+2} & \cdots & a_{-1} & a_0 \\
a_{-n+2} & & a_{-1} & a_0 & a_1 \\
\vdots & \ddots & a_0 & a_1 & \vdots \\
a_{-1} & \ddots & \ddots & & a_{n-2} \\
a_0 & a_1 & \cdots & a_{n-2} & a_{n-1}
\end{bmatrix}}_{\hat{B}}.
$$

Thus the simple trick of reversing the order of the unknowns which is effected by multiplication with Y yields a matrix for which we can get theoretical a priori convergence bounds for MINRES based only on eigenvalues. We comment that the (Hankel) matrix $\hat{B}$ is most likely indefinite, but it is clearly symmetric. Premultiplication by Y leads to similar conclusions: see Pestana and Wathen (2015).

It is quite likely that MINRES applied to any linear system involving $\hat{B}$ would converge slowly, but fortunately it is well-known that Toeplitz matrices are well preconditioned by related circulant matrices in many cases [see Chan (1988), Strang (1986), Tyrtyshnikov (1996), Tyrtyshnikov et al. (1997)]. Any circulant matrix $C \in \mathbb{R}^{n \times n}$ is diagonalised as $C = U^* \Lambda U$ by a Fast Fourier Transform (FFT) (Cooley and Tukey, 1965) in $O(n \log n)$ operations and so matrix multiplication by a vector or solution of equations with a circulant is computationally achieved in $O(n \log n)$ operations. For many Toeplitz matrices which have sufficient decay in the entries in the first row and column moving away from the diagonal it is known that

$$
C^{-1}B = I + R + E
$$

where R is of small rank and E is of small norm. This implies that the eigenvalues of the preconditioned matrix $C^{-1}B$ are clustered around 1 except for a few outliers. Precise statements about the decay of entries are usually expressed in terms of the smoothness of the generating function associated with the Toeplitz matrix which relates to the decay of Fourier coefficients and thus the speed of convergence of Fourier series.

Now, for use with MINRES a symmetric and positive definite preconditioner is required [see Wathen (2015)]. Fortunately via the FFT diagonalisation this is easily achieved by taking the absolute value

$$
|C| = U^* |\Lambda| U \tag{1}
$$

where $|\Lambda|$ is just the diagonal matrix of absolute values of the eigenvalues for an appropriate (e.g. Strang or Chan) circulant, C. For a nonsymmetric Toeplitz matrix with decay of entries as above, there now follows.

Theorem 2.1 (Pestana and Wathen 2015)

$$|C|^{-1}\hat{B} = J + \hat{R} + \hat{E}$$

where J is a real symmetric and orthogonal matrix with eigenvalues ± 1, $\hat{R}$ is of small rank and $\hat{E}$ is of small norm.

The eigenvalues of $|C|^{-1}\hat{B}$ are thus clustered around ± 1 together with a few outliers and guaranteed rapid convergence follows (Elman et al., 2014, Chap. 4).

A very simple example demonstrates the point: let

$$B = \begin{bmatrix} 1 & 0.01 & & & \\ 1 & 1 & 0.01 & & \\ & \ddots & \ddots & \ddots & \\ & & 1 & 1 & 0.01 \\ & & & 1 & 1 \end{bmatrix} \in \mathbb{R}^{n \times n} \tag{2}$$

with preconditioning via the Strang preconditioner (which simply takes C as B but with an additional 1 in the nth entry of the first row and 0.01 in the first entry of the nth row). The result of (implicitly) reordering/multiplying by Y and preconditioning with $|C|$ are shown in the MINRES iteration counts in Table 1 for a randomly generated right hand side vector. Convergence is accepted when the preconditioned residual vector has norm less than 10^{-10} for the results shown. The eigenvalues of the preconditioned matrix are shown in Table 2.

In fact for this example one can prove these and simpler results via consideration of low rank updates and the degree of the minimal polynomial so it is also possible to prove that GMRES will terminate in just a few iterations.

Table 1 Condition numbers $\kappa(B)$ for the Toeplitz matrix B described in (2) and iteration counts for *MINRES* applied to the symmetrized matrix $\widehat{B}$ with preconditioner $|C|$

n	$\kappa(B)$	Iterations
10	14	6
100	207	6
1000	2.6×10^6	6

Table 2 Eigenvalues of the Toeplitz matrix as described in (2) preconditioned with absolute value circulant (to four decimal places)

n	Eigenvalues of $|C|^{-1}\hat{B}$
10	$\{-9.9107, -1.0002, (-1 \times 2), -0.9640, 0.9893, (1 \times 4)\}$
100	$\{-2.2803, -1.0007, (-1 \times 47), -0.2536, 0.9919, (1 \times 49)\}$
1000	$\{-2.1626, -1.0008, (-1 \times 497), -1.8309e\text{-}5, 0.9929, (1 \times 499)\}$

Repeated eigenvalues are shown in brackets with the number of repeated eigenvalues indicated

Table 3 Preconditioned *MINRES* convergence for dense nonsymmetric Toeplitz matrices of Wiener class with absolute value circulant preconditioner

n	Eigenvalue inclusion	Iterations
10	$[-1.018, -0.710] \cup [0.981, 1.804]$	10
100	$[-1.092, -0.856] \cup [0.912, 1.160]$	14
1000	$[-1.154, -0.708] \cup [0.864, 1.381]$	20
10000	$[-1.078, -0.980] \cup [0.922, 1.017]$	12

For a dense Toeplitz with sufficient decay of entries in the first row and column this is not the case however, so the results presented in Table 3 for random nonsymmetric Toeplitz matrices of so-called Wiener class [see e.g. Ng (2004, p. 51)] are not explained by any other means as far as we know, but are a demonstration of the theory presented here. The matrices for these numerical experiments were generated by initially selecting independently the entries of two n-vectors, r and c with $r_1 = c_1$ from a normal distribution with mean zero and variance 1 (using the randn command in Matlab), then setting $r_i \leftarrow r_i/(i^2), c_i \leftarrow c_i/(i^2)$ and using these vectors as the first row and column of the nonsymmetric Toeplitz matrix, B.

3 Preconditioning for Time-Dependence

3.1 Theta Method

Here, we consider only a scalar linear ordinary differential equation,

$$\frac{dy}{dt} = ay + f, \qquad y(0) = y_0$$

on the time interval $[0, T]$. For the solution of systems of ODE and PDE problems via the method of lines, see McDonald et al. (in preparation). Likewise to begin with for simplicity we consider only the simple two-level θ-method, which gives,

$$\frac{y^{n+1} - y^n}{\tau} = \theta a y^{n+1} + (1 - \theta) a y^n + f^n, \qquad y^0 = y_0,$$

where τ is the constant time step with $N\tau = T$. The discrete equations to be solved are

$$(1 - a\theta\tau)y^{n+1} = (1 + a(1 - \theta)\tau)y^n + \tau f^n, \qquad n = 0, 1, 2, \ldots, N - 1, \qquad (3)$$

with $y^0 = y_0$.

The usual approach would be to solve the equations (3) sequentially for $n = 0, 1, 2, \ldots$ which is exactly forward substitution for the all-at-once system

$$
B \underbrace{\begin{bmatrix} y^1 \\ y^2 \\ y^3 \\ \vdots \\ y^N \end{bmatrix}}_{y} = \underbrace{\begin{bmatrix} \tau f^1 + (1 + a(1 - \theta)\tau)y^0 \\ \tau f^2 \\ \tau f^3 \\ \vdots \\ f^N \end{bmatrix}}_{f}
$$

where

$$
B = \begin{bmatrix} 1 - a\theta\tau \\ -1 - a(1 - \theta)\tau & 1 - a\theta\tau \\ & -1 - a(1 - \theta)\tau & 1 - a\theta\tau \\ & & \ddots & \ddots \\ & & & -1 - a(1 - \theta)\tau & 1 - a\theta\tau \end{bmatrix}.
$$

$$(4)$$

However, we can note that the coefficient matrix B, in the all-at-once system is real Toeplitz, hence solution using the idea in the section above is possible. MINRES for $\widehat{B}y = BYy = f, x = Yy$ converges in 4 iterations independently of N as can be seen from the results in Tables 4 and 5 below. The parameter values for the presented results are $a = -0.3, \tau = 0.2, \theta = 0.8$; similar behaviour has been observed for many other sets of parameter values. The eigenvalues of the preconditioned matrix for this problem are shown in Table 5.

For such a bidiagonal Toeplitz matrix, with Strang circulant preconditioning, one can show that the minimal polynomial is quadratic, hence this is a rare situation in which it is possible to deduce that GMRES must terminate with the solution after two iterations.

Table 4 Condition numbers $\kappa(B)$ for a time-dependent linear ODE using the θ-method, i.e. for B given by (4) and *MINRES* iteration counts with absolute value Strang circulant preconditioner described by (1) applied to the symmetrized matrix $\widehat{B}$

N	$\kappa(B)$	Iterations
10	10.474	4
100	30.852	4
1000	33.887	4

Table 5 Eigenvalues of the preconditioned system (to four decimal places)

N	Eigenvalues of $|C|^{-1}\hat{B}$
10	$\{-0.7206, (-1 \times 4), (1 \times 4), 3.1155\}$
100	$\{-0.4975, (-1 \times 49), (1 \times 49), 2.0157\}$
1000	$\{-0.4966, (-1 \times 499), (1 \times 499), 2.0139\}$

Repeated eigenvalues are shown in brackets with the number of repeated eigenvalues indicated

Theorem 3.1 *Let α and $\beta \neq 0 \in \mathbb{C}$. If*

$$B = \begin{bmatrix} \alpha & & & \\ \beta & \alpha & & \\ & \ddots & \ddots & \\ & & \beta & \alpha \\ & & & \beta & \alpha \end{bmatrix} \in \mathbb{C}^{n \times n}$$

is preconditioned by

$$C = \begin{bmatrix} \alpha & & & \beta \\ \beta & \alpha & & \\ & \ddots & \ddots & \\ & & \beta & \alpha \\ & & & \beta & \alpha \end{bmatrix},$$

the minimal polynomial of the preconditioned system $T = C^{-1}B$ is quadratic provided that both B and C are nonsingular.

Proof A simple calculation gives

$$T = C^{-1}B = \begin{bmatrix} 1 & & & \frac{-\alpha^{n-1}\beta}{\det C} \\ & 1 & & \frac{\alpha^{n-2}\beta^2}{\det C} \\ & & \ddots & \vdots \\ & & & 1 & \frac{(-1)^{n-1}\alpha\beta^{n-1}}{\det C} \\ & & & & \frac{\alpha^n}{\det C} \end{bmatrix},$$

where

$$\det C = \begin{cases} \alpha^n + \beta^n & \text{when } n \text{ is odd} \\ \alpha^n - \beta^n & \text{when } n \text{ is even} \end{cases}.$$

We can now easily show that T satisfies

$$(T - I)(T - \frac{\alpha^n}{\det C} I) = 0.$$

Since $(T - I) \neq 0$ and $(T - \frac{\alpha^n}{\det C} I) \neq 0$, $(T - I)(T - \frac{\alpha^n}{\det C} I)$ is the minimal polynomial of the preconditioned system T.

Since the minimal polynomial for the preconditioned coefficient matrix is in this case quadratic we must therefore have that the Krylov subspace is of dimension 2 and so because of its minimisation property, GMRES termination must occur within two iterations.

3.2 Multi-Step Method

In order to examine a slightly more complex system where the minimal polynomial is not as trivial as with the theta method above, we examine also a 2-step BDF time stepping method. We now require two initial conditions y_{-1} and y_0. For the BDF2 method we have

$$\frac{y^{n+1} - \frac{4}{3} y^n + \frac{1}{3} y^{n-1}}{\tau} = \frac{2}{3} a y^{n+1} + \frac{2}{3} f^{n+1}, \qquad y^0 = y_0, \quad y^{-1} = y_{-1}$$

where τ is the constant time step with $N\tau = T$. The discrete equations to be solved are

$$(1 - \frac{2}{3} a\tau) y^{n+1} = \frac{4}{3} y^n - \frac{1}{3} y^{n-1} + \frac{2}{3} \tau f^{n+1}, \qquad n = 0, 1, 2, \ldots, N - 1$$

with $y^0 = y_0$ and $y^{-1} = y_{-1}$. The corresponding all-at-once system is

$$B \underbrace{\begin{bmatrix} y^1 \\ y^2 \\ y^3 \\ \vdots \\ y^N \end{bmatrix}}_{y} = \underbrace{\begin{bmatrix} \frac{2}{3} \tau f^1 + \frac{4}{3} y^0 - \frac{1}{3} y^{-1} \\ \frac{2}{3} \tau f^2 - \frac{1}{3} y^0 \\ \frac{2}{3} \tau f^3 \\ \vdots \\ \frac{2}{3} \tau f^N \end{bmatrix}}_{f}$$

Table 6 Condition numbers $\kappa(B)$ for a time-dependent linear ODE using the BDF2 method, i.e. for B given by (5) and *MINRES* iteration counts with absolute value Strang circulant preconditioner described by (1) applied to the symmetrized matrix $\widehat{B}$

N	$\kappa(B)$	Iterations
10	29.33	6
100	67.49	6
1000	67.67	6

Table 7 Eigenvalues of the preconditioned system (to four decimal places)

N	Eigenvalues of $|C|^{-1}\widehat{B}$
10	$\{-1.0442, (-1 \times 3), -0.6781, 0.9219, (1 \times 3), 3.3921\}$
100	$\{-1.0610, (-1 \times 48), -0.4410, 0.9424, (1 \times 48), 2.2736\}$
1000	$\{-1.0610, (-1 \times 498), -0.4401, 0.9425, (1 \times 498), 2.2720\}$

Repeated eigenvalues are shown in brackets with the number of repeated eigenvalues indicated

where

$$B = \begin{bmatrix} 1 - \frac{2}{3}a\tau & & & & \\ -\frac{4}{3} & 1 - \frac{2}{3}a\tau & & & \\ \frac{1}{3} & -\frac{4}{3} & 1 - \frac{2}{3}a\tau & & \\ & \ddots & \ddots & \ddots & \\ & & \frac{1}{3} & -\frac{4}{3} & 1 - \frac{2}{3}a\tau \end{bmatrix}. \tag{5}$$

The coefficient matrix B in (5) has an additional subdiagonal but is still Toeplitz and the method above therefore still applies. Applying MINRES to solve the equation $\widehat{B}y = BYy = f, x = Yy$ with a random starting vector, we see convergence in 6 iterations independently of N as can be seen from the results in Table 6. The parameter values for the presented results are again chosen as $a = -0.3$ and $\tau = 0.2$ with zero forcing but the behaviour does not change for many other choices of a and τ; this apparent insensitivity is just an observation, for which we do not have a mathematical explanation. As we have used implicit time-stepping we have no restrictions on the value of τ to maintain stability and, as Theorem 3.1 seems to indicate, it is only the lower diagonal Toeplitz structure of B which ensures the number of unique eigenvalues of $C^{-1}B$ so it is not surprising that other parameter values behaviour in the same manner for the symmetrized system. The eigenvalues of the preconditioned matrix in this case are shown in Table 7.

This approach for time-dependent problems may not seem of any advantage for such a simple problems as considered here because MINRES requires matrix vector multiplication with B (and Y) as well as solution of a system with $|C|$ at each iteration. Its potential for time-dependent PDEs is however more intriguing [see McDonald et al. (in preparation)].

4 Conclusions

Preconditioning for nonsymmetric linear systems is generally heuristic with no guarantee of the speed of convergence from a priori spectral estimation. This is in stark contrast to the case of real symmetric or complex Hermitian matrices. We have shown that for nonsymmetric real Toeplitz matrices the use of a simple trick gives symmetry so that convergence estimates for MINRES which are based only on eigenvalues rigorously apply. Further, we propose the use of an absolute value circulant matrix as preconditioner: the action of this preconditioner is effected in $O(n \log n)$ operations via use of the FFT as originally suggested in Strang (1986). These constructions apply independently of nonnormality and rapid, n-independent convergence is guaranteed and hence observed.

It is further observed how this preconditioning can be applied in the context of time-stepping problems and that convergence is achieved in a small number of iterations independent of the number of time-steps.

References

T.F. Chan, An optimal circulant preconditioner for Toeplitz systems. J. Sci. Stat. Comput. **9**, 766–771 (1988)

J.W Cooley, J.W Tukey, An algorithm for the machine calculation of complex Fourier series. Math. Comput. **19**, 297–301 (1965)

H.C. Elman, D.J. Silvester, A.J. Wathen, *Finite Elements and Fast Iterative Solvers with Applications in Incompressible Fluid Dynamics*, 2nd edn. Numerical Mathematics and Scientific Computation (Oxford University Press, Oxford, 2014)

A. Greenbaum, V. Ptak, Z. Strakos, Any nonincreasing convergence curve is possible for GMRES. SIAM J. Matrix Anal. **17**(3), 465–469 (1996)

M.R. Hestenes, E. Stiefel, Methods of conjugate gradients for solving linear systems. J. Res. Natl. Bur. Stand. **49**, 409–435 (1952). nvl.nist.gov/pub/nistpubs/jres/049/6/V49.N06.A08.pdf?

D. Loghin, A.J. Wathen, Analysis of preconditioners for saddle-point problems. SIAM J. Sci. Comp. **25**(6), 2029–2049 (2004). doi:10.1137/S1064827502418203. http://epubs.siam.org/doi/abs/10.1137/S1064827502418203.

E. McDonald, J. Pestana, A.J. Wathen, Preconditioning and iterative solution of all-at-once systems for evolutionary partial differential equations. Technical Report (2016)

M.K. Ng, *Iterative Methods for Toeplitz Systems (Numerical Mathematics and Scientific Computation)* (Oxford University Press, New York, NY, 2004). ISBN: 0198504209

C. Paige, M. Saunders, Solution of sparse indefinite systems of linear equations. SIAM J. Numer. Anal **12**, 617–629 (1975)

J. Pestana, A.J. Wathen, A preconditioned MINRES method for nonsymmetric Toeplitz matrices. SIAM J. Matrix Anal. Appl. **36**(1), 273–288 (2015)

Y. Saad, M.H. Schultz, GMRES: a generalized minimal residual algorithm for solving nonsymmetric linear systems. SIAM J. Sci. Stat. Comput. **7**(3), 856–869 (1986). ISSN: 0196-5204. doi: 10.1137/0907058. http://dx.doi.org/10.1137/0907058

G. Strang, A proposal for toeplitz matrix calculations. Stud. Appl. Math. **74**, 171–176 (1986)

J.D. Tebbens, G. Meurant, Any Ritz value behavior is possible for Arnoldi and for GMRES. SIAM J. Matrix Anal. Appl. **33**, 958–978 (2012)

J.D. Tebbens, G. Meurant, Prescribing the behavior of early terminating GMRES and Arnoldi iterations. Numer. Algorithms **65**, 69–90 (2014)

E.E. Tyrtyshnikov, A unifying approach to some old and new theorems on distribution and clustering. Linear Algebra Appl. **232**, 1–43 (1996)

E.E. Tyrtyshnikov, A.Y. Yeremin, N.L. Zamarashkin, Clusters, preconditioners, convergence. Linear Algebra Appl. **263**, 25–48 (1997)

A.J. Wathen, Preconditioning. Acta Numer. **24**, 329–376 (2015)

Algebraic Adaptive Multipreconditioning Applied to Restricted Additive Schwarz

Nicole Spillane

In 2006 the Multipreconditioned Conjugate Gradient (MPCG) algorithm was introduced by Bridson and Greif (2006). It is an iterative linear solver, adapted from the Preconditioned Conjugate Gradient (PCG) algorithm (Saad, 2003), which can be used in cases where several preconditioners are available or the usual preconditioner is a sum of operators. In Bridson and Greif (2006) it was already pointed out that Domain Decomposition algorithms are ideal candidates to benefit from MPCG. This was further studied in Greif et al. (2014) which considers Additive Schwarz preconditioners in the Multipreconditioned GMRES (MPGMRES) Greif et al. (2016) setting. In 1997, Rixen had proposed in his thesis (Rixen, 1997) the Simultaneous FETI algorithm which turns out to be MPCG applied to FETI. The algorithm is more extensively studied in Gosselet et al. (2015) where its interpretation as an MPCG solver is made explicit.

The idea behind MPCG is that if at a given iteration N preconditioners are applied to the residual, then the space spanned by all of these directions is a better minimization space than the one spanned by their sum. This can significantly reduce the number of iterations needed to achieve convergence, as we will observe in Sect. 3, but comes at the cost of loosing the short recurrence property in CG. This means that at each iteration the new search directions must be orthogonalized against all previous ones. For this reason, in Spillane (2016) it was proposed to make MPCG into an Adaptive MPCG (AMPCG) algorithm where, at a given iteration, only the contributions that will accelerate convergence are kept, and all others are added into a global contribution (as they would be in classical PCG). This works very well for FETI and BDD but the theory in that article does not apply to Additive Schwarz. Indeed, the assumption is made that the smallest eigenvalue

N. Spillane (✉)
CMAP, Ecole polytechnique, CNRS, Université Paris-Saclay, 91128, Palaiseau, France
e-mail: nicole.spillane@cmap.polytechnique.fr

© Springer International Publishing AG 2017
C.-O. Lee et al. (eds.), *Domain Decomposition Methods in Science and Engineering XXIII*, Lecture Notes in Computational Science and Engineering 116, DOI 10.1007/978-3-319-52389-7_8

of the (globally) preconditioned operator is known. The test (called the τ-test), which chooses at each iteration which contributions should be kept, heavily relies on it. More precisely, the quantity that is examined by the τ-test can be related to a Rayleigh quotient, and the vectors that are selected to form the next minimization space correspond to large frequencies of the (globally) preconditioned operator. These are exactly the ones that are known to slow down convergence of BDD and FETI. Moreover, they are generated by the first few iterations of PCG (van der Sluis and van der Vorst, 1986). These two reasons make BDD and FETI ideal for the AMPCG framework.

The question posed by the present work is whether an AMPCG algorithm can be developed for Additive Schwarz type preconditioners. The goal is to design an adaptive algorithm that is robust at a minimal cost. One great feature of Additive Schwarz is that it is algebraic (all the components in the preconditioner can be computed from the knowledge of the matrix $\mathbf{A}$), and we will aim to preserve this property. The algorithms will be presented in an abstract framework. Since the short recurrence property is lost anyway in the MPCG setting, we will consider the more efficient (Efstathiou and Gander, 2003) Restricted Additive Schwarz preconditioner (RAS) (Cai and Sarkis, 1999) in our numerical experiments, instead of its symmetric counterpart the Additive Schwarz preconditioner [see Toselli and Widlund (2005)]. RAS is a non symmetric preconditioner but, provided that full recurrence is used, conjugate gradient based algorithms apply and still have nice properties (in particular the global minimization property). We will detail this in the next section where we briefly introduce the problem at hand, the Restricted Additive Schwarz preconditioner, and the MPCG solver. Then in Sect. 2, we propose two ways to make MPCG adaptive. Finally, Sect. 3 presents some numerical experiments on matrices arising from the finite element discretization of two dimensional elasticity problems. Three types of difficulties will be considered: heterogeneous coefficients, automatic (irregular) partitions into subdomains and almost incompressible behaviour.

These are sources of notoriously hard problems that have been, and are still, at the heart of much effort in the domain decomposition community, in particular by means of choosing an adequate coarse spaces (see Sarkis (2002), Nataf et al. (2011), Spillane et al. (2014), Efendiev et al. (2012), Brezina et al. (1999), Sousedík et al. (2013), Spillane and Rixen (2013), Haferssas et al. (2015), Klawonn et al. (2015), Cai et al. (2015), Dohrmann and Widlund (2010), Klawonn et al. (2016) and many more).

1   Preliminaries

Throughout this work, we consider the problem of solving the linear system

$$\mathbf{A}\mathbf{x}_* = \mathbf{b},$$

where $\mathbf{A} \in \mathbb{R}^{n \times n}$ is a sparse symmetric positive definite matrix, $\mathbf{b} \in \mathbb{R}^n$ is a given right hand side, and $\mathbf{x}_* \in \mathbb{R}^n$ is the unknown. We consider Conjugate Gradient type solvers preconditioned by the Restricted Additive Schwarz (RAS) preconditioner.

To construct the RAS preconditioner, a non overlapping partition of the degrees of freedom into N subsets, or subdomains, must first be chosen and then overlap must be added to each subset to get an overlapping partition. Denoting for each $s = 1, \ldots, N$, by $\widetilde{\mathbf{R}}^s$ and $\mathbf{R}^s$, the restriction operators from $[\![1, n]\!]$ into the s-th non overlapping and overlapping subdomains, respectively, the preconditioner is defined as:

$$\mathbf{H} = \sum_{s=1}^{N} \mathbf{H}^s \text{ with } \mathbf{H}^s = \widetilde{\mathbf{R}}^{s\top} \mathbf{A}^{s-1} \mathbf{R}^s \text{ and } \mathbf{A}^s = \mathbf{R}^s \mathbf{A} \mathbf{R}^{s\top}.$$

In Algorithm 1 the MPCG iterations are defined. Each contribution $\mathbf{H}^s$ to $\mathbf{H}$ is treated separately. This corresponds to the non adaptive algorithm, i.e., the condition in line 8 is not satisfied and N search directions are added to the minimization space at each iteration (namely the columns in $\mathbf{Z}_{i+1}$). We have denoted by $\boldsymbol{\Delta}_i^{\dagger}$ the pseudo inverse of $\boldsymbol{\Delta}_i$ to account for the fact that some search directions may be linearly dependent [see Gosselet et al. (2015), Spillane (2016)].

Although RAS is a non symmetric preconditioner the following properties hold:

- $\|\mathbf{x}_* - \mathbf{x}_i\|_A = \min\left\{\|\mathbf{x}_* - \mathbf{x}\|_A; \mathbf{x} \in \mathbf{x}_0 + \sum_{j=0}^{i-1} \text{range}(\mathbf{P}_j)\right\}$,
- $\mathbf{P}_j^\top \mathbf{A} \mathbf{P}_i = 0$ $(i \neq j)$, $\mathbf{r}_i^\top \mathbf{P}_j = 0$ $(i > j)$, and $\mathbf{r}_i^\top \mathbf{H} \mathbf{r}_j = 0$ $(i > j)$.

This can be proved easily following similar proofs in Spillane (2016) and the textbook Saad (2003). The difference from the symmetric case is that the two last properties only hold for $i > j$, and not for every pair $i \neq j$.

Algorithm 1: Adaptive Multipreconditioned Conjugate Gradient Algorithm for $\mathbf{A}\mathbf{x}_* = \mathbf{b}$. Preconditioners: $\{\mathbf{H}^s\}_{s=1,\ldots,N}$. Initial guess: $\mathbf{x}_0$.

---

1   $\mathbf{r}_0 = \mathbf{b} - \mathbf{A}\mathbf{x}_0$; $\mathbf{Z}_0 = [\mathbf{H}^1\mathbf{r}_0 | \ldots | \mathbf{H}^N\mathbf{r}_0]$; $\mathbf{P}_0 = \mathbf{Z}_0$;
2   **for** $i = 0, 1, \ldots,$ *convergence* **do**
3      $\mathbf{Q}_i = \mathbf{A}\mathbf{P}_i$;
4      $\boldsymbol{\Delta}_i = \mathbf{Q}_i^\top \mathbf{P}_i$; $\gamma_i = \mathbf{P}_i^\top \mathbf{r}_i$; $\alpha_i = \boldsymbol{\Delta}_i^\dagger \gamma_i$;
5      $\mathbf{x}_{i+1} = \mathbf{x}_i + \mathbf{P}_i\alpha_i$;
6      $\mathbf{r}_{i+1} = \mathbf{r}_i - \mathbf{Q}_i\alpha_i$;
7      $\mathbf{Z}_{i+1} = [\mathbf{H}^1\mathbf{r}_{i+1} | \ldots | \mathbf{H}^N\mathbf{r}_{i+1}]$;   // Generate N search directions.
8   **if** *Adaptive Algorithm* **then**
9 | Reduce number of columns in $\mathbf{Z}_{i+1}$ (see Sect. 2) ;
10   **end**
11     $\boldsymbol{\Phi}_{i,j} = \mathbf{Q}_j^\top \mathbf{Z}_{i+1}$; $\beta_{i,j} = \boldsymbol{\Delta}_j^\dagger \boldsymbol{\Phi}_{i,j}$ for each $j = 0, \ldots, i$;
12     $\mathbf{P}_{i+1} = \mathbf{Z}_{i+1} - \sum_{j=0}^{i} \mathbf{P}_j\beta_{i,j}$;
13 **end**
14 Return $\mathbf{x}_{i+1}$;

Multipreconditioning significantly improves convergence as has already been observed (Bridson and Greif, 2006; Gosselet et al., 2015; Greif et al., 2014; Spillane, 2016) and as will be illustrated in the numerical result section. The drawback is that a dense matrix $\Delta_i \in \mathbb{R}^{N \times N}$ must be factorized at each iteration and that N search directions per iteration need to be stored. In the next section, we will try to remove these limitations by reducing the number of search directions at every iteration. We aim to do this without having too strong a negative impact on the convergence.

2 An Adaptive Algorithm

There is definitely a balance to be found between the number of iterations, the cost of each iteration, and the memory required for storage. Here, we do not claim that we have achieved a perfect balance, but we introduce some ways to influence it. More precisely, we propose two methods of reducing the number of columns in Z_{i+1} (or in other words how to fill in line 9 in Algorithm 1). In Sect. 2.1, we propose a τ-test that measures the relevance of every *candidate* $H^s r_{i+1}$ and only keeps the most relevant contributions. In Sect. 2.2, we propose to form m coarser subdomains (which are *agglomerates* of the initial N subdomains) and *aggregate* the N candidates into only m search directions. Note that there is a definite connection with multigrid studies from where we have borrowed some vocabulary [see Vassilevski (2008), Chartier et al. (2003), Brandt et al. (2011) and many references therein].

2.1 Select Search Directions with a τ-Test

The τ-test in the original AMPCG publication (Spillane, 2016) is based on the assumption that the smallest eigenvalue for the globally preconditioned operator $\mathbf{HA}$ is known (Toselli and Widlund, 2005). This allows for an error estimate inspired by those in Axelsson and Kaporin (2001), and the choice of the τ-test is a direct consequence of it. Here, the largest eigenvalue is known and it is the presence of small eigenvalues that is responsible for slow convergence. Unfortunately, we have failed to produce an estimate similar to that in Spillane (2016) in this case. Note that there is no such estimate in Axelsson and Kaporin (2001) either, and we believe that this is inherent to the properties of the conjugate gradient algorithm.

The approach that we propose here to select local contributions is different. It is well known by now [see, e.g., Saad (2003)] that, at each iteration, the approximate solution returned by the conjugate gradient algorithm is the $\mathbf{A}$-orthogonal projection of the exact solution $\mathbf{x}_*$ onto the minimization space. Here, the property satisfied by the update between in iteration $i + 1$ is

$$\|\mathbf{x}_* - \mathbf{x}_{i+1}\|_{\mathbf{A}} = \min\{\|\mathbf{x}_* - \mathbf{x}\|_{\mathbf{A}}; \ \mathbf{x} \in \mathbf{x}_i + \mathrm{range}(\mathbf{P}_i)\},$$

where $\mathbf{P}_i$ forms a basis of range($\mathbf{Z}_i$) after orthogonalization against previous search spaces (line 12 in Algorithm 1).

For this reason, the τ-test that we propose aims at evaluating, for each $s = 1, \ldots, N$, the ratio between the norm of the error projected onto the global vector $\mathbf{H}\mathbf{r}_{i+1}$ and the norm of the error projected onto the local candidate $\mathbf{H}^s\mathbf{r}_{i+1}$. More precisely, we compute (with $\langle \cdot, \cdot \rangle$ denoting the ℓ_2 inner product)

$$t_i^s = \frac{\langle \mathbf{r}_{i+1}, \mathbf{H}\mathbf{r}_{i+1} \rangle^2}{\langle \mathbf{H}\mathbf{r}_{i+1}, \mathbf{A}\mathbf{H}\mathbf{r}_{i+1} \rangle} \times \frac{\langle \mathbf{H}^s\mathbf{r}_{i+1}, \mathbf{A}\mathbf{H}^s\mathbf{r}_{i+1} \rangle}{\langle \mathbf{r}_{i+1}, \mathbf{H}^s\mathbf{r}_{i+1} \rangle^2}. \tag{1}$$

This is indeed the announced quantity, since the square of the $\mathbf{A}$-norm of the $\mathbf{A}$ orthogonal projection of $\mathbf{x}_* - \mathbf{x}_i$ onto any vector $\mathbf{v}$ is

$$\| \mathbf{v}(\mathbf{v}^\top \mathbf{A}\mathbf{v})^{-1}\mathbf{v}^\top \underbrace{\mathbf{A}(\mathbf{x}_* - \mathbf{x}_i)}_{=\mathbf{r}_i} \|_\mathbf{A}^2 = \frac{\langle \mathbf{r}_i, \mathbf{v} \rangle^2}{\langle \mathbf{v}, \mathbf{A}\mathbf{v} \rangle}.$$

Then, given a threshold τ, the number of columns in $\mathbf{Z}_{i+1}$ is reduced by eliminating all those for which $t_i^s > \tau$. In order for the global preconditioned residual $\sum_{s=1}^{N} \mathbf{H}^s\mathbf{r}_{i+1}$ to be included in the search space (as is always the case in PCG), we add it to $\mathbf{Z}_{i+1}$ in a separate column. This way we obtain a minimization space range($\mathbf{P}_i$) of dimension anywhere between 1 and N

An important question is of course how to choose τ. Considering that t_i^s measures the (inverse of the) impact of one of N contributions compared to the impact of the sum of the N contributions, it is quite natural to choose $\tau \approx N$. In the next section, we illustrate the behaviour of the adaptive algorithm with the τ-test for values of τ ranging between $N/10$ and $10N$ with satisfactory results.

In order to determine whether or not $t_i^s \leq \tau$ (i.e., perform the τ-test) it is necessary to compute t_i^s. Here, we will not discuss how to do this at the smallest cost but it is of course an important consideration (that was discussed for the AMPCG algorithm applied to BDD in Spillane (2016)). One noteworthy observation is that if $\mathbf{H}$ were either the Additive Schwarz (AS), or the Additive Schwarz with Harmonic overlap [ASH Cai and Sarkis (1999)] preconditioner (i.e., $\mathbf{H} = \sum_{s=1}^{N} \mathbf{R}^{s\top}\mathbf{A}^{s-1}\mathbf{R}^s$ or $\mathbf{H} = \sum_{s=1}^{N} \mathbf{R}^{s\top}\mathbf{A}^{s-1}\widetilde{\mathbf{R}}^s$) then all terms involving $\mathbf{H}^{s\top}\mathbf{A}\mathbf{H}^s$ would simplify since, obviously, $\mathbf{A}^{s-1}\mathbf{R}^s\mathbf{A}\mathbf{R}^{s\top}\mathbf{A}^{s-1} = \mathbf{A}^{s-1}$.

Another option is to prescribe a number m of vectors to be selected at each iteration instead of a threshold τ, and keep the m vectors with smallest values of t_i^s. Then, only the second factor in (1) would be required. We leave a more in depth study of these questions for future work.

2.2 Aggregate Search Directions

Here, we propose a completely different, and much simpler, way of reducing the number of vectors in $\mathbf{Z}_{i+1}$. This is to choose a prescribed number m, with $m \leq N$, of search directions per iteration, and a partition of $[\![1, N]\!]$ into m subsets. Then, the columns of $\mathbf{Z}_{i+1}$ that correspond to the same subset are simply replaced by their sum, leaving m vectors. We refer to this as aggregation as it is the same as assembling coarse domains from the original subdomains and computing coarse search directions as sums of the $\mathbf{H}_{i+1}^s$. The question of how to choose m is of course important. It can be a fraction of N or the maximal size of the dense matrix that the user is prepared to factorize. In the next section, we consider values ranging from $N/20$ to N.

3 Numerical Results with FreeFem++ (Hecht, 2013) and GNU Octave (Eaton et al., 2009)

In this section, we consider the linear elasticity equations posed in $\Omega = [0, 1]^2$ with mixed boundary conditions. We search for $\mathbf{u} = (u_1, u_2)^\top \in H^1(\Omega)^2$ such that

$$
\begin{cases}
-\mathrm{div}(\boldsymbol{\sigma}(\mathbf{u})) = (0, 0)^\top, \text{ in } \Omega, \\
\mathbf{u} = (1/2(y(1 - y)), 0)^\top, \text{ on } \{(x, y) \in \partial\Omega : x = 0\}, \\
\mathbf{u} = (-1/2(y(1 - y)), 0)^\top, \text{ on } \{(x, y) \in \partial\Omega : x = 1\}, \\
\quad\quad \sigma(\mathbf{u}) \cdot \mathbf{n} = 0, \text{ on the rest of } \partial\Omega\,(\mathbf{n} : \text{ outward normal}).
\end{cases}
$$

The stress tensor $\sigma(\mathbf{u})$ is defined by $\sigma_{ij}(\mathbf{u}) = 2\mu\varepsilon_{ij}(\mathbf{u}) + \lambda\delta_{ij}\mathrm{div}(\mathbf{u})$ for $i, j = 1, 2$ where $\varepsilon_{ij}(\mathbf{u}) = \frac{1}{2}\left(\frac{\partial u_i}{\partial x_j} + \frac{\partial u_j}{\partial x_i}\right)$, δ_{ij} is the Kronecker symbol and the Lamé coefficients are functions of Young's modulus E and Poisson's ratio ν : $\mu = \frac{E}{2(1+\nu)}$, $\lambda = \frac{E\nu}{(1+\nu)(1-2\nu)}$. In all test cases, ν is uniform and equal either to 0.4 (compressible test case) or 0.49999 (almost incompressible test case) while E varies between 10^6 and 10^{12} in a pattern presented in Fig. 1-left. The geometries of the solutions are also presented in this figure.

The computational domain is discretized into a uniform mesh with mesh size: $h = 1/60$, and partitioned into $N = 100$ subdomains by the automatic graph partitioner METIS (Karypis and Kumar, 1998). One layer of overlap is added to each subdomain. In the compressible case, the system is discretized by piecewise second order polynomial ($\mathbb{P}_2$) Lagrange finite elements. In the almost incompressible setting it is known that the locking phenomenon occurs rendering the solution unreliable. To remedy this, the problem is rewritten in a mixed formulation with an additional unknown $p = \mathrm{div}(u)$, and then discretized. Although the $\mathbb{P}_2 - \mathbb{P}_0$ mixed finite element does not satisfy the discrete inf-sup condition it is often used

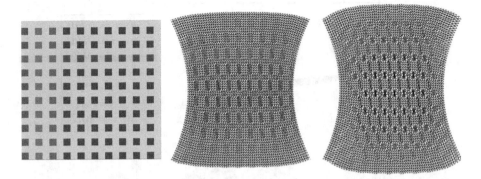

Fig. 1 Test case setup (all three configurations are drawn to scale). *Left*: Young's modulus—$E = 10^6$ with square inclusions of larger E, up to 10^{12}. *Middle*: Solution for $\nu = 0.4$. *Right*: Solution for $\nu = 0.49999$

Table 1 Summary of all numerical results presented

Compressible						Incompressible					
τ-test (see Fig. 2)			Aggregates (see Fig. 4)			τ-test (see Fig. 3)			Aggregates (see Fig. 5)		
τ	*iter.*	*# vec.*	*m*	*iter.*	*# vec.*	τ	*iter.*	*# vec.*	*m*	*iter.*	*# vec.*
10	104	6059	1	889	890	10	124	4865	1	> 999	>1000
25	85	5769	5	381	1910	25	99	4889	5	512	2565
50	91	6625	10	277	2780	50	79	4621	10	345	3460
100	82	6339	20	186	3740	100	72	4521	20	194	3900
200	84	6876	40	111	4480	200	68	4593	40	125	5040
400	78	6817	100	60	6100	400	65	4552	100	56	5700
1000	69	6153				1000	68	5156			

iter.: number of iterations needed to reduce the initial error by a factor 10^{-7}
 # vec.: size of the minimization space. There are two test cases: Compressible and Incompressible, and for each there are two ways of reducing the number of search directions at each iteration: with the τ-test (as proposed in Sect. 2.1) or by aggregating into m directions (as proposed in Sect. 2.2)

in practice, and we choose it here. Finally, the pressure unknowns are eliminated by static condensation.

In both cases the problem has 28798 degrees of freedom (once degrees of freedom corresponding to Dirichlet boundary conditions have been eliminated). As an initial guess, we first compute a random vector $\mathbf{v}$ and then scale it to form $\mathbf{x}_0 = \frac{\mathbf{b}^\top \mathbf{v}}{\|\mathbf{v}\|_A^2}$, according to what is proposed in Strakoš and Tichý (2005). This guarantees that $\|\mathbf{x}_* - \mathbf{x}_0\|_A \leq \|\mathbf{x}_*\|_A$: the initial error is at most as large as it would be with a zero initial guess.

In Table 1, we report on the number of iterations needed to reduce the initial error $\|\mathbf{x}_* - \mathbf{x}_0\|_A$ by a factor 10^{-7} and on the size of the minimization space that was constructed to do this, which is $\sum_i \mathrm{rank}(\mathbf{P}_i)$. Note that, although they are presented in the same table, we cannot compare the compressible and incompressible test cases

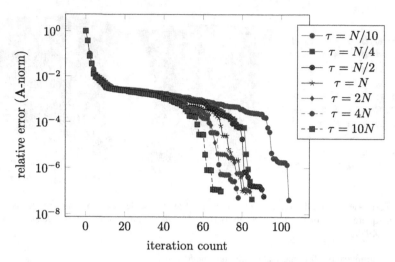

Fig. 2 Compressible test case—reducing the number of directions with the τ-test—error norm versus iteration count for different values of τ

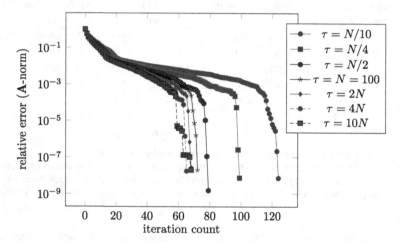

Fig. 3 Incompressible test case—reducing the number of directions with the τ-test—error norm versus iteration count for different values of τ

as they are simply not the same problem. Figures 2, 3, 4 and 5 show in more detail the convergence behaviour of each method.

The first point to be made is that the MPCG algorithm does an excellent job at reducing the number of iterations. This can be observed by looking at the data for $m = 100 = N$ directions per iteration in Figs. 4 and 5. The iteration counts are reduced from 889 to 60 and from over 999 to 56 compared to the classical PCG iterations ($m = 1$ direction per iteration). Secondly the adaptation steps that we introduced seem to do their job since they ensure fast convergence with smaller minimization spaces. In particular, all of these adaptive methods converged in less

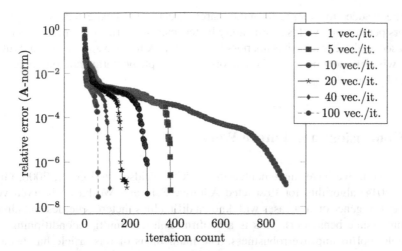

Fig. 4 Compressible test case—reducing the number of directions by aggregating them into m vectors—error norm versus iteration count for different values of m

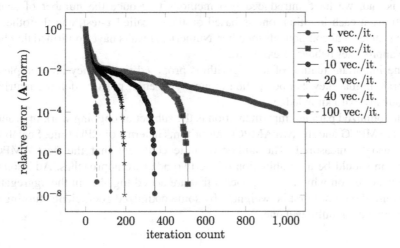

Fig. 5 Incompressible test case—reducing the number of directions by aggregating them into m vectors—error norm versus iteration count for different values of m

than 512 iterations even for the incompressible case (for which the usual PCG still has a relative error of $8 \cdot 10^{-4}$ after 999 iterations).

With the τ-test, the number of iterations is always reduced by a factor at least 8 compared to PCG even with the smallest threshold $\tau = 10 = N/10$. With $\tau = 10N$ the number of iterations is almost the same as with the full MPCG. For these test cases the choice $\tau = N$ advocated in Sect. 2 seems to be a good compromise.

With the aggregation procedure, convergence is achieved even when the coarsening is quite aggressive (5 vectors per iteration means that 20 local contributions have been added together to form the search direction). As expected, keeping more vectors per iteration yields significantly better results in terms of iteration count.

Based on these results, it is not possible to compare the two approaches and future work will definitely be focused on an optimized implementation and on decreasing the CPU time.

4   Conclusions and Future Work

In this work, we have implemented the MPCG (Bridson and Greif, 2006; Greif et al., 2014) algorithm for Restricted Additive Schwarz. We have observed very good convergence on test cases with known difficulties (heterogeneities and almost incompressible behaviour). This is a confirmation that multipreconditioning is a valuable tool to improve robustness. The main focus of this article has been to propose an adaptive version of the algorithm so that, when possible, the cost of each iteration and the cost of storage can be reduced while maintaining fast convergence. To this end, we have introduced two methods to reduce the number of search directions at each iteration: one is based on the so called τ-test, and the other on adding some local components together. Numerical results have confirmed that both these approaches behave as expected.

One important feature of the algorithms proposed is that they are completely algebraic in that they can be applied to any symmetric, positive definite matrix **A** without any extra knowledge.

An optimized parallel implementation is the subject of ongoing work in order to compare MPCG and the two AMPCG algorithms in terms of CPU time. Scalability must also be measured. The author is quite confident that the *best* AMPCG algorithm should be a combination of the two adaptive approaches. Additionally there is no reason why the components that are added together in the aggregation procedure should not first be weighted by some optimized coefficients, turning the algorithm into a multilevel one.

References

O. Axelsson, I. Kaporin, Error norm estimation and stopping criteria in preconditioned conjugate gradient iterations. Numer. Linear Algebra Appl. **8**(4), 265–286 (2001)

A. Brandt, J. Brannick, K. Kahl, I. Livshits, Bootstrap AMG. SIAM J. Sci. Comput. **33**(2), 612–632 (2011)

M. Brezina, C. Heberton, J. Mandel, P. Vaněk, An iterative method with convergence rate chosen a priori. Technical Report 140, University of Colorado Denver, April 1999

R. Bridson, C. Greif, A multipreconditioned conjugate gradient algorithm. SIAM J. Matrix Anal. Appl. **27**(4), 1056–1068 (electronic) (2006)

M. Cai, L.F. Pavarino, O.B. Widlund, Overlapping Schwarz methods with a standard coarse space for almost incompressible linear elasticity. SIAM J. Sci. Comput. **37**(2), A811–A830 (2015)

X.-C. Cai, M. Sarkis, A restricted additive Schwarz preconditioner for general sparse linear systems. SIAM J. Sci. Comput. **21**(2), 792–797 (electronic) (1999)

T. Chartier, R.D. Falgout, V.E. Henson, J. Jones, T. Manteuffel, S. McCormick, J. Ruge, P.S. Vassilevski, Spectral AMGe (ρAMGe). SIAM J. Sci. Comput. **25**(1), 1–26 (2003)

C.R. Dohrmann, O.B. Widlund, Hybrid domain decomposition algorithms for compressible and almost incompressible elasticity. Int. J. Numer. Methods Eng. **82**(2), 157–183 (2010)

W.E. John, B. David, H. Søren, W. Rik, GNU Octave version 4.0.0 manual: a high-level interactive language for numerical computations. http://www.gnu.org/software/octave/doc/interpreter/ (2015)

Y. Efendiev, J. Galvis, R. Lazarov, J. Willems, Robust domain decomposition preconditioners for abstract symmetric positive definite bilinear forms. ESAIM Math. Model. Numer. Anal. **46**(5), 1175–1199 (2012)

E. Efstathiou, M.J. Gander, Why restricted additive Schwarz converges faster than additive Schwarz. BIT **43**(suppl.), 945–959 (2003)

P. Gosselet, D. Rixen, F.-X. Roux, N. Spillane, Simultaneous FETI and block FETI: robust domain decomposition with multiple search directions. Int. J. Numer. Methods Eng. **104**(10), 905–927 (2015)

C. Greif, T. Rees, D. Szyld, Additive Schwarz with variable weights, in *Domain Decomposition Methods in Science and Engineering XXI* (Springer, 2014)

C. Greif, T. Rees, D.B. Szyld, GMRES with multiple preconditioners. SeMA J. 1–19 (2016) ISSN: 2281-7875, doi:10.1007/s40324-016-0088-7, http://dx.doi.org/10.1007/s40324-016-0088-7

R. Haferssas, P. Jolivet, F. Nataf, A robust coarse space for optimized Schwarz methods: SORAS-GenEO-2. C. R. Math. Acad. Sci. Paris, 353(10), 959–963 (2015)

F. Hecht, *FreeFem++*. Numerical Mathematics and Scientific Computation. Laboratoire J.L. Lions, Université Pierre et Marie Curie, http://www.freefem.org/ff++/, 3.23 edition, 2013

G. Karypis, V. Kumar, A fast and high quality multilevel scheme for partitioning irregular graphs. SIAM J. Sci. Comput. **20**(1), 359–392 (electronic) (1998)

A. Klawonn, M. Kühn, O. Rheinbach, Adaptive coarse spaces for FETI-DP in three dimensions. SIAM J. Sci. Comput. **38**(5), A2880–A2911 (2016). ISSN:1064-8275, MRCLASS:65N30 (65N25 65N50 65N55 74E05),MRNUMBER:3546980, MRREVIEWER:Carlos A. de Moura, doi:10.1137/15M1049610, http://dx.doi.org/10.1137/15M1049610

A. Klawonn, P. Radtke, O. Rheinbach, FETI-DP methods with an adaptive coarse space. SIAM J. Numer. Anal. **53**(1), 297–320 (2015)

F. Nataf, H. Xiang, V. Dolean, N. Spillane, A coarse space construction based on local Dirichlet-to-Neumann maps. SIAM J. Sci. Comput. **33**(4), 1623–1642 (2011)

D. Rixen, Substructuring and Dual Methods in Structural Analysis. PhD thesis, Université de Liège, Collection des Publications de la Faculté des Sciences appliquées, n.175, 1997

Y. Saad, *Iterative Methods for Sparse Linear Systems*, 2nd edn. (Society for Industrial and Applied Mathematics (SIAM), Philadelphia, PA, 2003)

M. Sarkis, Partition of unity coarse spaces, in *Fluid Flow and Transport in Porous Media: Mathematical and Numerical Treatment (South Hadley, MA, 2001)*. Contemporary Mathematics, vol. 295 (American Mathematical Society, Providence, RI, 2002), pp. 445–456

B. Sousedík, J. Šístek, J. Mandel, Adaptive-multilevel BDDC and its parallel implementation. Computing **95**(12), 1087–1119 (2013)

N. Spillane, An adaptive multipreconditioned conjugate gradient algorithm. SIAM J. Sci. Comput. **38**(3), A1896–A1918 (2016)

N. Spillane, V. Dolean, P. Hauret, F. Nataf, C. Pechstein, R. Scheichl, Abstract robust coarse spaces for systems of PDEs via generalized eigenproblems in the overlaps. Numer. Math. **126**(4), 741–770 (2014)

N. Spillane, D.J. Rixen, Automatic spectral coarse spaces for robust FETI and BDD algorithms. Int. J. Numer. Meth. Eng. **95**(11), 953–990 (2013)

Z. Strakoš, P. Tichý, Error estimation in preconditioned conjugate gradients. BIT **45**(4), 789–817 (2005)

A. Toselli, O. Widlund, *Domain Decomposition Methods—Algorithms and Theory*. Springer Series in Computational Mathematics, vol. 34 (Springer, Berlin, 2005)

A. van der Sluis, H.A. van der Vorst, The rate of convergence of conjugate gradients. Numer. Math. **48**(5), 543–560 (1986)

P.S. Vassilevski, Multilevel block factorization preconditioners, in *Matrix-Based Analysis and Algorithms for Solving Finite Element Equations* (Springer, New York, 2008), xiv+529 pp. ISBN: 978-0-387-71563-6

Part II
Talks in Minisymposia (MT)

Closed Form Inverse of Local Multi-Trace Operators

Alan Ayala, Xavier Claeys, Victorita Dolean, and Martin J. Gander

1 Introduction

Local multi-trace operators arise when one uses a particular integral formulation for a transmission problem. A transmission problem for a second order elliptic operator is a problem defined on a domain which is decomposed into non-overlapping subdomains, but instead of imposing the continuity of the traces of the solution and their normal derivative along the interfaces between the subdomains, given jumps are imposed along the interfaces. The solution of a transmission problem is thus naturally discontinuous along the interfaces, and hence a domain decomposition formulation is imposed by the problem.

A local multi-trace formulation represents the solution in each subdomain using an integral formulation, and couples these solutions imposing the given jumps in the traces of the solution and the normal derivatives along the interfaces (hence the name multi-trace). This formulation was introduced in Hiptmair and Jerez-Hanckes (2012) to tackle transmission problems for the Helmholtz equation, where the material properties are constant in each subdomain, see also Claeys and Hiptmair (2012, 2013), and Claeys et al. (2015) for associated boundary integral methods.

A. Ayala • X. Claeys (✉)
Université Pierre et Marie Curie, Paris, France

INRIA Paris, Paris, France
e-mail: claeys@ann.jussieu.fr

V. Dolean
University of Strathclyde, Glasgow, UK
e-mail: Victorita.Dolean@strath.ac.uk

M.J. Gander
University of Geneva, Geneva, Switzerland
e-mail: martin.gander@unige.ch

© Springer International Publishing AG 2017
C.-O. Lee et al. (eds.), *Domain Decomposition Methods in Science and Engineering XXIII*, Lecture Notes in Computational Science and Engineering 116, DOI 10.1007/978-3-319-52389-7_9

Multi-trace formulations lead naturally to block preconditioners, see Hiptmair et al. (2014). In Dolean and Gander (2015), a simple introduction to local multi-trace formulations is given in the language of domain decomposition, and it is shown that these block preconditioners are equivalent to the simultaneous application of a Dirichlet-Neumann and a Neumann-Dirichlet method to the transmission problem. Block preconditioners based on multi-trace formulations have also the potential to lead to nil-potent iterations, a more recent area of research in domain decomposition (Chaouqui et al. 2016), and it was shown that for two subdomains, they correspond to optimal Schwarz methods, see Claeys et al. (2015).

We are interested here in the inverse of local multi-trace operators. We exhibit a closed form of this inverse for a model problem with three subdomains in the special case where the coefficients are homogeneous. An essential ingredient to obtain this closed form inverse are several remarkable identities which were recently discovered, see Claeys et al. (2015). We illustrate our findings with a numerical experiment that shows that discretizing the closed form inverse gives indeed and approximate inverse of the discretized local multi-trace operator.

2   Local Multi-Trace Formulation

We start by introducing the local multi-trace formulation for a model problem. Consider a partition of the space $\mathbb{R}^d = \overline{\Omega}_0 \cup \overline{\Omega}_1 \cup \overline{\Omega}_2$ as shown in Fig. 1. We assume that $\Omega_j, j = 0, 1, 2$ are Lipschitz domains such that $\Omega_j \cap \Omega_k = \emptyset$ for $j \neq k$. Denoting by $\Gamma_j := \partial \Omega_j$, we assume in addition that $\Gamma_1 \cap \Gamma_2 = \emptyset$ and $\Gamma_0 = \Gamma_1 \cup \Gamma_2$. Let $\boldsymbol{n}_j$ be the unit outer normal for Ω_j on its boundary Γ_j. For a sufficiently regular function v we denote by $v|_{\Gamma_j}^+$ the trace of v and by $\partial_{n_j} v|_{\Gamma_j}^+$ the trace of $\boldsymbol{n}_j \cdot \nabla v$ on Γ_j taken from inside of Ω_j. Similarly we define $v|_{\Gamma_j}^-$ and $\partial_{n_j} v|_{\Gamma_j}^-$ but with traces from outside of Ω_j.

Fig. 1 Geometrical configuration we consider in the analysis

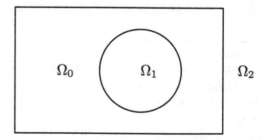

The elliptic transmission problem for which we want to study the local multi-trace formulation and its inverse is: find $u \in H^1(\mathbb{R}^d)$ such that

$$
\begin{aligned}
-\Delta u + a_j^2 u &= 0 \quad \text{in } \Omega_j, \ j = 0, 1, 2, \\
[u]_{\Gamma_1} &= g_1, \quad [u]_{\Gamma_2} = g_2, \\
[\partial_n u]_{\Gamma_1} &= h_1, \quad [\partial_n u]_{\Gamma_2} = h_2,
\end{aligned}
\tag{1}
$$

where $a_j > 0$ for $j = 0, 1, 2$, $g_j \in H^{+1/2}(\Gamma_j)$ and $h_j \in H^{-1/2}(\Gamma_j)$ are given data of the transmission problem, and we used the classical jump notation for the Dirichlet and Neumann traces of the solution across the interfaces $\Gamma_j, j = 1, 2$, i.e. $[u]_{\Gamma_j} := u|_{\Gamma_j}^+ - u|_{\Gamma_j}^-$ and $[\partial_n u]_{\Gamma_j} := \partial_{n_j} u|_{\Gamma_j}^+ - \partial_{n_j} u|_{\Gamma_j}^-$.

Following Hiptmair et al. (2014), this problem can be rewritten as a boundary integral local multi-trace formulation, using the Calderón projector: let $\mathbb{H}(\Gamma_j) := H^{1/2}(\Gamma_j) \times H^{-1/2}(\Gamma_j)$; then for $(g, h) \in \mathbb{H}(\Gamma_j)$, the Calderón projector $\mathbb{P}_j : \mathbb{H}(\Gamma_j) \to \mathbb{H}(\Gamma_j)$ interior to Ω_j associated to the operator $-\Delta + a_j^2$ is defined by

$$
\begin{aligned}
\mathbb{P}_j(g, h) &:= (v|_{\Gamma_j}^+, \partial_{n_j} v|_{\Gamma_j}^+) \quad \text{where } v \text{ satisfies} \\
-\Delta v + a_j^2 v &= 0 \quad \text{in } \Omega_j \text{ and in } \mathbb{R}^d \setminus \overline{\Omega}_j, \\
[v]_{\Gamma_j} &= g \quad \text{and} \quad [\partial_n v]_{\Gamma_j} = h, \quad \text{and} \\
\limsup_{|x| \to \infty} &|v(x)| < +\infty,
\end{aligned}
$$

and $\mathbb{P}_j$ is known to be a continuous map, see Sauter and Schwab (2011). The decomposition $\Gamma_0 = \Gamma_1 \cup \Gamma_2$ induces a natural decomposition of $\mathbb{P}_0$ in the following manner: for any $U \in \mathbb{H}(\Gamma_0)$ set $\rho_j(U) := U|_{\Gamma_j} \in \mathbb{H}(\Gamma_j), j = 1, 2$. In addition, for any $V \in \mathbb{H}(\Gamma_j), j = 1, 2$, define $\rho_j^*(V) \in \mathbb{H}(\Gamma_0)$ by $\rho_j^*(V) = V$ on Γ_j and $\rho_j^*(V) = 0$ on $\Gamma_0 \setminus \Gamma_j$. Then the projector $\mathbb{P}_0$ can be decomposed as

$$
\mathbb{P}_0 = \begin{bmatrix} \tilde{\mathbb{P}}_1 & R_{1,2}/2 \\ R_{2,1}/2 & \tilde{\mathbb{P}}_2 \end{bmatrix}, \quad \text{where} \quad \begin{cases} \tilde{\mathbb{P}}_j := \rho_j \cdot \mathbb{P}_0 \cdot \rho_j^*, \\ R_{j,k}/2 := \rho_j \cdot \mathbb{P}_0 \cdot \rho_k^*. \end{cases}
$$

The operators $\tilde{\mathbb{P}}_j : \mathbb{H}(\Gamma_j) \to \mathbb{H}(\Gamma_j)$ and $R_{j,k} : \mathbb{H}(\Gamma_k) \to \mathbb{H}(\Gamma_j)$ are continuous. Following this decomposition, we identify $\mathbb{H}(\Gamma_0)$ with $\mathbb{H}(\Gamma_1) \times \mathbb{H}(\Gamma_2)$. We also introduce the sign switching operator $X(v, q) := (v, -q)$, and a relaxation parameter $\sigma \in \mathbb{C} \setminus \{0\}$. The local multi-trace formulation of problem (1) is then: find $(U_1, U_1^{(0)}, U_2^{(0)}, U_2) \in \mathbb{H}(\Gamma_1)^2 \times \mathbb{H}(\Gamma_2)^2$ such that

$$
\begin{bmatrix}
(1+\sigma)\text{Id} - \mathbb{P}_1 & -\sigma X & 0 & 0 \\
-\sigma X & (1+\sigma)\text{Id} - \tilde{\mathbb{P}}_1 & -R_{1,2}/2 & 0 \\
0 & -R_{2,1}/2 & (1+\sigma)\text{Id} - \tilde{\mathbb{P}}_2 & -\sigma X \\
0 & 0 & -\sigma X & (1+\sigma)\text{Id} - \mathbb{P}_2
\end{bmatrix}
\cdot
\begin{bmatrix}
U_1 \\
U_1^{(0)} \\
U_2^{(0)} \\
U_2
\end{bmatrix}
= F,
\tag{2}
$$

where $F \in \mathbb{H}(\Gamma_1)^2 \times \mathbb{H}(\Gamma_2)^2$ is some right-hand side depending on g_j, h_j, σ whose precise expression is not important for our present study, where we want to obtain an explicit expression for the operator in (2) and its inverse for the special case

$$a_0 = a_1 = a_2. \tag{3}$$

To simplify the calculations when working with the entries of the operator in (2), we set $A_j := -\mathrm{Id} + 2\mathbb{P}_j$ and $\tilde{A}_j := -\mathrm{Id} + 2\tilde{\mathbb{P}}_j$. The following remarkable identities were established in Claeys et al. (2015, §4.4) for the special case (3): $\mathbb{P}_j^2 = \mathbb{P}_j$, $\tilde{\mathbb{P}}_j^2 = \tilde{\mathbb{P}}_j$, $\tilde{\mathbb{P}}_1 R_{1,2} = \tilde{\mathbb{P}}_2 R_{2,1} = 0$, $X\mathbb{P}_j X = \mathrm{Id} - \tilde{\mathbb{P}}_j$, and finally $R_{1,2}R_{2,1} = R_{2,1}R_{1,2} = 0$. These five properties can be reformulated in terms of the operators A_j, namely

$$
\begin{aligned}
&i) \quad A_j^2 = \tilde{A}_j^2 = \mathrm{Id}, \\
&ii) \quad \tilde{A}_1 R_{1,2} = -R_{1,2} \text{ and } \tilde{A}_2 R_{2,1} = -R_{2,1}, \\
&iii) \quad X \cdot A_j \cdot X = -\tilde{A}_j, \\
&iv) \quad R_{1,2}R_{2,1} = R_{2,1}R_{1,2} = 0, \\
&v) \quad R_{1,2}\tilde{A}_2 = R_{1,2} \text{ and } R_{2,1}\tilde{A}_1 = R_{2,1}.
\end{aligned}
\tag{4}
$$

Let us introduce auxiliary operators $\mathbb{A}, \Pi : \mathbb{H}(\Gamma_1)^2 \times \mathbb{H}(\Gamma_2)^2$ defined by

$$
\mathbb{A} := \begin{bmatrix} A_1 & 0 & 0 & 0 \\ 0 & \tilde{A}_1 & R_{1,2} & 0 \\ 0 & R_{2,1} & \tilde{A}_2 & 0 \\ 0 & 0 & 0 & A_2 \end{bmatrix}, \quad \Pi := \begin{bmatrix} 0 & X & 0 & 0 \\ X & 0 & 0 & 0 \\ 0 & 0 & 0 & X \\ 0 & 0 & X & 0 \end{bmatrix}. \tag{5}
$$

According to property $i)$ in (4), we have $(\mathrm{Id} + \mathbb{A})^2/4 = (\mathrm{Id} + \mathbb{A})/2$, which implies the well known Calderón identity from the boundary integral equation literature, i.e.

$$\mathbb{A}^2 = \mathrm{Id}, \tag{6}$$

see for example Nédélec (2001, §4.4). The local multi-trace operator on the left-hand side of Eq. (2) can then be rewritten as

$$\mathrm{MTF}_{\mathrm{loc}} := -\frac{1}{2}\mathbb{A} - \sigma\Pi + (\sigma + \frac{1}{2})\mathrm{Id}. \tag{7}$$

In (2), the terms associated with the relaxation parameter σ, namely $\mathrm{Id} - \Pi$, enforce the transmission conditions of problem (1). For $\sigma = 0$, we have $\mathrm{MTF}_{\mathrm{loc}} = \frac{1}{2}(\mathrm{Id} - \mathbb{A})$, which is a projector, and $\mathrm{MTF}_{\mathrm{loc}}$ is thus not invertible. For $\sigma \neq 0$ however, $\mathrm{MTF}_{\mathrm{loc}}$ was proved to be invertible in Claeys (2016, Corollary 6.3). The goal of the present contribution is to derive an explicit formula for the inverse of $\mathrm{MTF}_{\mathrm{loc}}$, and we will thus assume $\sigma \neq 0$.

3 Inverse of the Local Multi-Trace Operator

We now derive a closed form inverse of the local multi-trace operator in (7) for the special case (3). Using that $\Pi^2 = \text{Id}$ and (6), we obtain

$$
\begin{aligned}
[-\mathbb{A}/2 - \sigma\Pi + (\sigma + 1/2)\text{Id}] \, [-\mathbb{A}/2 - \sigma\Pi - (\sigma + 1/2)\text{Id}] \\
= (\mathbb{A}/2 + \sigma\Pi)^2 - (\sigma + 1/2)^2 \, \text{Id} \\
= (\sigma^2 + 1/4 - \sigma^2 - \sigma - 1/4)\text{Id} + \sigma(\mathbb{A}\Pi + \Pi\mathbb{A})/2 \\
= -\sigma\text{Id} + \sigma(\mathbb{A}\Pi + \Pi\mathbb{A})/2.
\end{aligned} \tag{8}
$$

Inspired by the calculations in Claeys et al. (2015, §4.4) as well as Claeys (2016, Proposition 6.1), we examine more closely $\mathbb{A}\Pi + \Pi\mathbb{A}$. We start by comparing $\mathbb{A}\Pi$ and $\Pi\mathbb{A}$:

$$
\mathbb{A}\Pi = \begin{bmatrix} 0 & A_1X & 0 & 0 \\ \tilde{A}_1X & 0 & 0 & R_{1,2}X \\ R_{2,1}X & 0 & 0 & \tilde{A}_2X \\ 0 & 0 & A_2X & 0 \end{bmatrix}, \quad \Pi\mathbb{A} = \begin{bmatrix} 0 & X\tilde{A}_1 & XR_{1,2} & 0 \\ XA_1 & 0 & 0 & 0 \\ 0 & 0 & 0 & XA_2 \\ 0 & XR_{2,1} & X\tilde{A}_2 & 0 \end{bmatrix}. \tag{9}
$$

According to Property $iii)$ in (4), we have $X\tilde{A}_j + A_jX = 0$ and $XA_j + \tilde{A}_jX = 0$, and thus from (9) we obtain

$$
\Pi\mathbb{A} + \mathbb{A}\Pi = \begin{bmatrix} 0 & 0 & XR_{1,2} & 0 \\ 0 & 0 & 0 & R_{1,2}X \\ R_{2,1}X & 0 & 0 & 0 \\ 0 & XR_{2,1} & 0 & 0 \end{bmatrix}.
$$

Computing the square of this operator, and taking into account Property $iv)$ from (4), we obtain

$$
(\Pi\mathbb{A} + \mathbb{A}\Pi)^2 = \begin{bmatrix} XR_{1,2}R_{2,1}X & 0 & 0 & 0 \\ 0 & R_{1,2}R_{2,1} & 0 & 0 \\ 0 & 0 & R_{2,1}R_{1,2} & 0 \\ 0 & 0 & 0 & XR_{2,1}R_{1,2}X \end{bmatrix} = 0.
$$

From this we conclude that $(-\text{Id} + (\mathbb{A}\Pi + \Pi\mathbb{A})/2)^{-1} = -\text{Id} - (\mathbb{A}\Pi + \Pi\mathbb{A})/2$. Coming back to (8), we obtain a first expression for the inverse of the local multi-trace operator, namely

$$
\begin{aligned}
[-\mathbb{A}/2 - \sigma\Pi + (\sigma + 1/2)\text{Id}]^{-1} \\
= \sigma^{-1}[\mathbb{A}/2 + \sigma\Pi + (\sigma + 1/2)\text{Id}] \, [\text{Id} + (\mathbb{A}\Pi + \Pi\mathbb{A})/2] \\
= \sigma^{-1}[\tfrac{1}{2}(1 + \sigma)\mathbb{A} + (\sigma + 1/4)\Pi + (\sigma + 1/2)(\text{Id} + (\mathbb{A}\Pi + \Pi\mathbb{A})/2)] \\
+ \sigma^{-1}[\tfrac{\sigma}{2}\Pi\mathbb{A}\Pi + \tfrac{1}{4}\mathbb{A}\Pi\mathbb{A}].
\end{aligned} \tag{10}
$$

The only terms that are not explicitly known yet in (10) are the last two, $\Pi A \Pi$ and $A \Pi A$. Combining (9) with Definition (5), direct calculation yields

$$\Pi A \Pi = \begin{bmatrix} -A_1 & 0 & 0 & XR_{1,2}X \\ 0 & -\tilde{A}_1 & 0 & 0 \\ 0 & 0 & -\tilde{A}_2 & 0 \\ XR_{2,1}X & 0 & 0 & -A_2 \end{bmatrix},$$

and similarly, we also obtain

$$A \Pi A = \begin{bmatrix} 0 & -X & XR_{1,2} & 0 \\ -X & 0 & 0 & -R_{1,2}X \\ -R_{2,1}X & 0 & 0 & -X \\ 0 & XR_{2,1} & -X & 0 \end{bmatrix}.$$

We have now derived an explicit expression for each term in (10), which leads to a close form matrix expression for the inverse of the local multi-trace operator, namely

$$\mathrm{MTF}_{\mathrm{loc}}^{-1} = (1 + \frac{1}{2\sigma})\mathrm{Id} + \frac{1}{\sigma} \begin{bmatrix} \frac{1}{2}A_1 & \sigma X & \frac{\sigma+1}{2}XR_{1,2} & \frac{\sigma}{2}XR_{1,2}X \\ \sigma X & \frac{1}{2}\tilde{A}_1 & \frac{\sigma+1}{2}R_{1,2} & \frac{\sigma}{2}R_{1,2}X \\ \frac{\sigma}{2}R_{2,1}X & \frac{\sigma+1}{2}R_{2,1} & \frac{1}{2}\tilde{A}_2 & \sigma X \\ \frac{\sigma}{2}XR_{2,1}X & \frac{\sigma+1}{2}XR_{2,1} & \sigma X & \frac{1}{2}A_2 \end{bmatrix}.$$

$$(11)$$

The expression $\mathrm{MTF}_{\mathrm{loc}} \cdot \mathrm{MTF}_{\mathrm{loc}}^{-1} = \mathrm{Id}$ should not be mistaken for the Calderón identity (6). The primary difference is that (11) involves coupling terms between Ω_1 and Ω_2, whereas in (6), all three subdomains are decoupled.

4 Numerical Experiment

We now illustrate the closed form inversion formula (11) for the local multi-trace formulation by a numerical experiment. We consider a three dimensional version of the geometrical setting described at the beginning in Fig. 1. Here $\Omega_1 := B(0, 0.5)$ is the open ball centered at 0 with radius 0.5, $\Omega_2 := \mathbb{R}^3 \backslash [-1, +1]^3$, and $\Omega_0 := \mathbb{R}^3 \setminus \overline{\Omega}_1 \cup \overline{\Omega}_2$, see Fig. 2.

For our numerical results, we discretize both $\mathrm{MTF}_{\mathrm{loc}}$ given by (7) leading to a matrix we denote by $[\mathrm{MTF}_{\mathrm{loc}}]$, and $\mathrm{MTF}_{\mathrm{loc}}^{-1}$ given by (11) leading to a matrix denoted by $[\mathrm{MTF}_{\mathrm{loc}}^{-1}]$. Our discretization using the code BEMTOOL[1] is based on a Galerkin method where both Dirichlet and Neumann traces are approximated by means of continuous piece-wise linear functions on the same mesh. We use a triangulation

---

[1] Available on https://github.com/xclaeys/bemtool under Lesser Gnu Public License.

Fig. 2 3D geometry for the numerical experiment

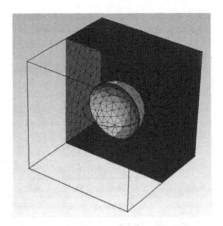

with a mesh width $h = 0.35$, and generated the mesh using GMSH, see Geuzaine and Remacle (2009).

Let M_h be the mass matrix associated with the duality pairing used to write (2) in variational form. We represent the spectrum of the matrix $M_h^{-1} \cdot [\mathrm{MTF_{loc}}] \cdot M_h^{-1} \cdot [\mathrm{MTF_{loc}^{-1}}]$ in Fig. 3. We see that the eigenvalues are clustered around 1, which agrees well with our analysis at the continuous level.

5 Conclusions

We have shown in this paper that it is possible for the local multi-trace operator of a model transmission problem to obtain a closed form for the inverse. This would therefore be an ideal preconditioner for local multi-trace formulations. We are currently investigating if such closed form inverses are also possible for more general situations, where the coefficients are only constant in each subdomain, and in the presence of more subdomains. The closed form inverse seems to be inherent to the formulation, and not dependent on the specific form of the partial differential equation.

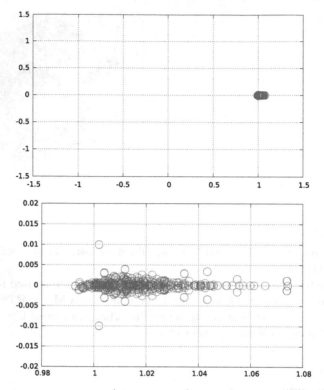

Fig. 3 Eigenvalues of the matrix $M_h^{-1} \cdot [\text{MTF}_{\text{loc}}] \cdot M_h^{-1} \cdot [\text{MTF}_{\text{loc}}^{-1}]$ for $\sigma = -\frac{1}{2}$, with a zoom below around 1

Acknowledgements This work received support from the ANR research Grant ANR-15-CE23-0017-01.

References

F. Chaouqui, M.J. Gander, K. Santugini-Repiquet, On nilpotent subdomain iterations, in *Domain Decomposition Methods in Science and Engineering XXIII*, ed. by C. Cai, D. Keyes, H.H. Kim, A. Klawonn, C.-O. Lee, E.-J. Park, O. Widlund (Springer, Berlin, 2016)

X. Claeys, Essential spectrum of local multi-trace boundary integral operators. IMA J. Appl. Math. **81**, 961–983 (2016)

X. Claeys, V. Dolean, M.J. Gander, An introduction to multitrace formulations and associated domain decomposition solvers. https://arxiv.org/abs/1605.04422 (2015, preprint)

X. Claeys, R. Hiptmair, Electromagnetic scattering at composite objects: a novel multi-trace boundary integral formulation. ESAIM Math. Model. Numer. Anal. **46**(6), 1421–1445 (2012)

X. Claeys, R. Hiptmair, Multi-trace boundary integral formulation for acoustic scattering by composite structures. Commun. Pure Appl. Math. **66**(8), 1163–1201 (2013)

X. Claeys, R. Hiptmair, E. Spindler, A second-kind Galerkin boundary element method for scattering at composite objects. BIT **55**(1), 33–57 (2015)

V. Dolean, M.J. Gander, Multitrace formulations and Dirichlet-Neumann algorithms, in *Domain Decomposition Methods in Science and Engineering XXII*, ed. by J. Erhel, M.J. Gander, L. Halpern, G. Pichot, T. Sassi, O. Widlund (Springer, Berlin, 2015)

C. Geuzaine, J.-F. Remacle, Gmsh: a 3-D finite element mesh generator with built-in pre- and post-processing facilities. Int. J. Numer. Methods Eng. **79**(11), 1309–1331 (2009)

R. Hiptmair, C. Jerez-Hanckes, Multiple traces boundary integral formulation for Helmholtz transmission problems. Adv. Comput. Math. **37**(1), 39–91 (2012)

R. Hiptmair, C. Jerez-Hanckes, J.-F. Lee, Z. Peng, Domain decomposition for boundary integral equations via local multi-trace formulations, in *Domain Decomposition Methods in Science and Engineering XXI*, ed. by J. Erhel, M.J. Gander, L. Halpern, G. Pichot, T. Sassi, O. Widlund (Springer, Berlin, 2014), pp. 43–57

J.-C. Nédélec, *Acoustic and Electromagnetic Equations*. Applied Mathematical Sciences, vol. 144 (Springer, New York, 2001). Integral representations for harmonic problems

S.A. Sauter, C. Schwab, *Boundary Element Methods*. Springer Series in Computational Mathematics, vol. 39 (Springer, Berlin, 2011)

Schwarz Preconditioning for High Order Edge Element Discretizations of the Time-Harmonic Maxwell's Equations

Marcella Bonazzoli, Victorita Dolean, Richard Pasquetti, and Francesca Rapetti

1 Introduction

High order discretizations of PDEs for wave propagation can provide a highly accurate solution with very low dispersion and dissipation errors. The resulting linear systems can however be ill conditioned, so that preconditioning becomes mandatory. Moreover, the time-harmonic Maxwell's equations with high frequency are known to be difficult to solve by classical iterative methods, like the Helmholtz equation (Ernst and Gander 2012). Domain decomposition methods are currently the most promising techniques for this class of problems (see Dolean et al. 2009, 2015).

In order to simulate propagation in waveguide structures, we consider the *second order time-harmonic Maxwell's equation*:

$$\nabla \times \left(\frac{1}{\mu} \nabla \times \mathbf{E} \right) + (i\omega\sigma - \omega^2\varepsilon)\mathbf{E} = -i\omega\mathbf{J}, \tag{1}$$

in the domain $\mathcal{D} \subset \mathbb{R}^3$ contained between two infinite parallel metallic plates $y = 0$ and $y = Y$. The wave propagates in the x-direction and all physical parameters

M. Bonazzoli (✉) • R. Pasquetti • F. Rapetti
Laboratoire J.A. Dieudonné, University of Nice Sophia Antipolis, Nice Cedex 02, 06108 Parc Valrose, France
e-mail: marcella.bonazzoli@unice.fr; victorita.dolean@unice.fr; victorita.dolean@strath.ac.uk; richard.pasquetti@unice.fr; francesca.rapetti@unice.fr

V. Dolean
Laboratoire J.A. Dieudonné, University of Nice Sophia Antipolis, Nice Cedex 02, 06108 Parc Valrose, France

Department of Mathematics and Statistics, University of Strathclyde, Glasgow, UK

© Springer International Publishing AG 2017
C.-O. Lee et al. (eds.), *Domain Decomposition Methods in Science and Engineering XXIII*, Lecture Notes in Computational Science and Engineering 116, DOI 10.1007/978-3-319-52389-7_10

(magnetic permeability μ, electrical conductivity σ, and electric permittivity ε) are invariant in the z-direction. Equation (1) assumes that the electric field $\mathcal{E}(\mathbf{x}, t) = \mathrm{Re}(\mathbf{E}(\mathbf{x})e^{i\omega t})$ has harmonic dependence on time enforced by the imposed current source $\mathcal{J}(\mathbf{x}, t) = \mathrm{Re}(\mathbf{J}(\mathbf{x})e^{i\omega t})$, ω being the angular frequency. We work in a bounded section $\Omega = (0, X) \times (0, Y)$ of $\mathcal{D}$ and solve the boundary value problem given by Eq. (1), where we set $\mathbf{J} = \mathbf{0}$, with metallic boundary conditions on the waveguide walls:

$$\mathbf{E} \times \mathbf{n} = \mathbf{0}, \text{ on } \Gamma_{\mathrm{w}} = \{y = 0, y = Y\},$$

and impedance boundary conditions at the waveguide entrance and exit:

$$(\nabla \times \mathbf{E}) \times \mathbf{n} + i\kappa\mathbf{n} \times (\mathbf{E} \times \mathbf{n}) = \mathbf{g}^{\mathrm{in}}, \text{ on } \Gamma_{\mathrm{in}} = \{x = 0\},$$

$$(\nabla \times \mathbf{E}) \times \mathbf{n} + i\kappa\mathbf{n} \times (\mathbf{E} \times \mathbf{n}) = \mathbf{g}^{\mathrm{out}}, \text{ on } \Gamma_{\mathrm{out}} = \{x = X\},$$

$\kappa = \omega\sqrt{\varepsilon\mu}$ being the wavenumber and $\mathbf{n} = (n_x, n_y, 0)$ the outward normal to $\Gamma = \partial\Omega$. The assumptions on Ω and on the physical parameters distribution are such that $\mathbf{E} = (E_x, E_y, 0)$, which yields $\nabla \times \mathbf{E} = (0, 0, \partial_x E_y - \partial_y E_x)$.

The variational formulation of the problem is: find $\mathbf{E} \in V$ such that

$$\int_{\Omega} \left[\mu\vartheta\mathbf{E} \cdot \mathbf{v} + (\nabla \times \mathbf{E}) \cdot (\nabla \times \mathbf{v}) \right] + \int_{\Gamma_{\mathrm{in}} \cup \Gamma_{\mathrm{out}}} i\kappa(\mathbf{E} \times \mathbf{n}) \cdot (\mathbf{v} \times \mathbf{n})$$

$$= \int_{\Gamma_{\mathrm{in}}} \mathbf{g}^{\mathrm{in}} \cdot \mathbf{v} + \int_{\Gamma_{\mathrm{out}}} \mathbf{g}^{\mathrm{out}} \cdot \mathbf{v}, \quad \forall \mathbf{v} \in V,$$

with $V = \{\mathbf{v} \in H(\mathrm{curl}, \Omega), \mathbf{v} \times \mathbf{n} = 0 \text{ on } \Gamma_{\mathrm{w}}\}$, where $H(\mathrm{curl}, \Omega)$ is the space of square integrable functions whose curl is also square integrable, $\vartheta = i\omega\sigma - \omega^2\varepsilon$, and μ is supposed constant. To write a finite element discretization of this problem we introduce a triangulation $\mathcal{T}_h$ of Ω and a finite dimensional subspace $V_h \subset H(\mathrm{curl}, \Omega)$. The simplest possible conformal discretization for the space $H(\mathrm{curl}, \Omega)$ is given by the low order *Nédélec edge finite elements* (Nédélec 1980): the local basis functions are associated with the oriented edges $E = \{v_i, v_j\}$ of a given triangle T of $\mathcal{T}_h$ and they are given by

$$\mathbf{w}^E = \lambda_i\nabla\lambda_j - \lambda_j\nabla\lambda_i,$$

where the λ_ℓ are the barycentric coordinates of a point w.r.t. the node v_ℓ.

2   High Order Edge Finite Elements

We adopt here the *high order extension* of Nédélec elements presented in Rapetti (2007) and Rapetti and Bossavit (2009). The definition of the basis functions is rather simple since it only involves the barycentric coordinates of the simplex. Given a multi-index $\mathbf{k} = (k_1, k_2, k_3)$ of weight $k = k_1 + k_2 + k_3$ (where k_1, k_2, k_3 are non negative integers), we denote by $\lambda^{\mathbf{k}}$ the product $\lambda_1^{k_1} \lambda_2^{k_2} \lambda_3^{k_3}$. The *basis functions* of *polynomial degree* $r = k + 1$ over the triangle T are defined as

$$\mathbf{w}^e = \lambda^{\mathbf{k}} \mathbf{w}^E, \tag{2}$$

for all edges E of the triangle T, and for all multi-indices $\mathbf{k}$ of weight k. Notice that these high order elements still yield a conformal discretization of $H(\mathrm{curl}, \Omega)$. Indeed, they are products between Nédélec elements, which are curl-conforming, and the continuous functions $\lambda^{\mathbf{k}}$.

An interesting point of the proposed construction is the possible *geometrical localization* of the basis functions: the couples $\{\mathbf{k}, E\}$ appearing in (2) are in one-to-one correspondence with *small edges* e in the principal lattice of degree r of T (see Fig. 1). More precisely, the small edge $e = \{\mathbf{k}, E\}$ is the small edge parallel to E that belongs to the small triangle of barycentre G of coordinates $\lambda_i(G) = \frac{1/3 + k_i}{k + 1}, i = 1, 2, 3$. Thanks to the definition of the basis the circulation of each basis function along a small edge is a constant that does not depend on the triangle T of the mesh.

Even if the described basis functions are very easy to generate, they don't really form a basis as they are *not linearly independent*. Indeed, for each small triangle which is not homothetic to the big one (the white ones in Fig. 1) one can check that the sum of the basis functions associated with its small edges is zero. Hence a redundant function should be eliminated for each 'reversed' small triangle.

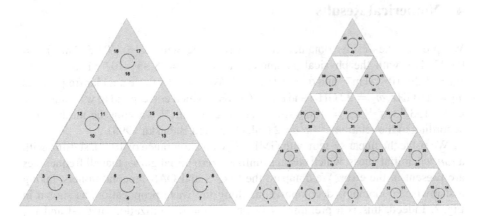

Fig. 1 The small triangles (*shaded regions*) and their small edges in the principal lattice of degree $r = 3$ (*left*) and $r = 5$ (*right*)

3 Schwarz Preconditioning

As shown numerically in Rapetti (2007), the matrix of the linear system resulting from the described high order discretization is ill conditioned. Therefore, we use and compare two domain decomposition preconditioners, the *Optimized Additive Schwarz* (OAS) and the *Optimized Restricted Additive Schwarz* (ORAS)

$$M_{\text{OAS}}^{-1} = \sum_{s=1}^{N_{\text{sub}}} R_s^T A_s^{-1} R_s, \quad M_{\text{ORAS}}^{-1} = \sum_{s=1}^{N_{\text{sub}}} \widetilde{R}_s^T A_s^{-1} R_s,$$

where N_{sub} is the number of *overlapping* subdomains Ω_s into which the domain Ω is decomposed. The matrices A_s are the local matrices of the *subproblems* with impedance boundary conditions $(\nabla \times \mathbf{E}) \times \mathbf{n} + i\kappa \mathbf{n} \times (\mathbf{E} \times \mathbf{n})$ as transmission conditions between subdomains.

In order to describe the matrices $R_s, \widetilde{R}_s$, let $\mathcal{N}$ be the set of degrees of freedom and $\mathcal{N} = \bigcup_{s=1}^{N_{\text{sub}}} \mathcal{N}_s$ its decomposition into the subsets corresponding to different subdomains. The matrix R_s is a $\#\mathcal{N}_s \times \#\mathcal{N}$ boolean matrix, which is the restriction matrix from Ω to the subdomain Ω_s. Its (i, j) entry is equal to 1 if the i-th degree of freedom in Ω_s is the j-th one in the whole Ω. Notice that R_s^T is then the extension matrix from the subdomain Ω_s to Ω. The matrix $\widetilde{R}_s$ is a $\#\mathcal{N}_s \times \#\mathcal{N}$ restriction matrix, like R_s, but with some of the unit entries associated with the overlap replaced by zeros: this would correspond to a decomposition into *non overlapping* subdomains $\widetilde{\Omega}_s \subset \Omega_s$ (completely non overlapping, not even on their border!) (see Gander 2008). This way $\sum_{s=1}^{N_{\text{sub}}} \widetilde{R}_s^T R_s = I$, that is the matrices $\widetilde{R}_s$ give a discrete partition of unity (which is made only of 1 and 0).

4 Numerical Results

We present the results obtained for a waveguide with $X = 0.0502\,\text{m}$, $Y = 0.00254\,\text{m}$, with the physical parameters: $\varepsilon = \varepsilon_0 = 8.85 \cdot 10^{-12}\,\text{F}\,\text{m}^{-1}$, $\mu = \mu_0 = 1.26 \cdot 10^{-6}\,\text{H}\,\text{m}^{-1}$ and $\sigma = 0.15\,\text{S}\,\text{m}^{-1}$. We consider three angular frequencies $\omega_1 = 16\,\text{GHz}$, $\omega_2 = 32\,\text{GHz}$, and $\omega_3 = 64\,\text{GHz}$, which correspond to wavenumbers $\kappa_1 = 153.43\,\text{m}^{-1}$, $\kappa_2 = 106.86\,\text{m}^{-1}$, $\kappa_3 = 213.72\,\text{m}^{-1}$, varying the mesh size h according to the relation $h^2 \cdot \kappa^3 = 2$ (Ihlenburg and Babuška 1995).

We solve the linear system with GMRES (with a tolerance of 10^{-6}), starting with a *random* initial guess, which ensures, unlike a zero initial guess, that all frequencies are present in the error. We compare the ORAS and OAS preconditioners, taking a stripwise subdomains decomposition, along the wave propagation, as shown in Fig. 2. Indeed, this is a preliminary testing of the discretization method and the preconditioner on a simple geometry which is the two-dimensional rectangular waveguide propagating only one mode; in this case, it is not necessary to consider more complicated or general decompositions.

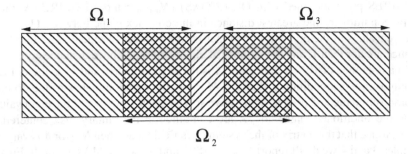

Fig. 2 The stripwise decomposition of the domain

Table 1 Influence of k ($\omega = \omega_2$, $N_{\text{sub}} = 2$, $\delta_{\text{ovr}} = 2h$)

k	N_{dofs}	N_{iterNp}	N_{iter}	$\max|\lambda - 1|$	$\#\{\lambda : |\lambda - 1| > 1\}$	$\#\{\lambda : |\lambda - 1| = 1\}$
0	282	179	5 (10)	$1.04e{-}1(1.38e{+}1)$	0 (4)	0 (12)
1	884	559	6 (15)	$1.05e{-}1(1.63e{+}1)$	0 (8)	0 (40)
2	1806	1138	6 (17)	$1.05e{-}1(1.96e{+}1)$	0 (12)	0 (84)
3	3048	1946	6 (21)	$1.05e{-}1(8.36e{+}2)$	0 (16)	0 (144)
4	4610	2950	6 (26)	$1.05e{-}1(1.57e{+}3)$	0 (20)	0 (220)

Table 2 Influence of ω ($k = 2$, $N_{\text{sub}} = 2$, $\delta_{\text{ovr}} = 2h$)

κ	N_{dofs}	N_{iterNp}	N_{iter}	$\max|\lambda - 1|$	$\#\{\lambda : |\lambda - 1| > 1\}$	$\#\{\lambda : |\lambda - 1| = 1\}$
153.43	339	232	5 (11)	$2.46e{-}1(1.33e{+}1)$	0 (6)	0 (45)
106.86	1806	1138	6 (17)	$1.05e{-}1(1.96e{+}1)$	0 (12)	0 (84)
213.72	7335	4068	9 (24)	$3.03e{-}1(2.73e{+}1)$	0 (18)	0 (123)

Table 3 Influence of N_{sub} ($k = 2$, $\omega = \omega_2$, $\delta_{\text{ovr}} = 2h$)

N_{sub}	N_{iter}	$\max|\lambda - 1|$	$\#\{\lambda : |\lambda - 1| > 1\}$	$\#\{\lambda : |\lambda - 1| = 1\}$
2	6 (17)	1.05e-1 (1.96e+1)	0 (12)	0 (84)
4	10 (27)	5.33e-1 (1.96e+1)	0 (38)	0 (252)
8	19 (49)	7.73e-1 (1.96e+1)	0 (87)	0 (588)

Table 4 Influence of δ_{ovr} ($k = 2$, $\omega = \omega_2$, $N_{\text{sub}} = 2$)

δ_{ovr}	N_{iter}	$\max|\lambda - 1|$	$\#\{\lambda : |\lambda - 1| > 1\}$	$\#\{\lambda : |\lambda - 1| = 1\}$
1h	10 (20)	1.95e+1 (1.96e+1)	3 (12)	0 (39)
2h	6 (17)	1.05e-1 (1.96e+1)	0 (12)	0 (84)
4h	5 (14)	1.06e-1 (1.96e+1)	0 (12)	0 (174)

In our tests we vary the polynomial degree $r = k + 1$, the angular frequency ω and so the wavenumber κ, the number of subdomains N_{sub}, and finally the overlap size δ_{ovr}. Here, $\delta_{\text{ovr}} = h, 2h, 4h$ means that we consider an overlap of $1, 2, 4$ mesh triangles along the horizontal direction. Tables 1, 2, 3, 4 show the total number of degrees of freedom N_{dofs}, the number of iterations N_{iter} for convergence of GMRES preconditioned with ORAS(OAS) (N_{iterNp} refers to GMRES without any preconditioner), the greatest distance in the complex plane between $(1,0)$ and the eigenvalues of the preconditioned matrix, the number of eigenvalues that have distance greater than 1, and the number of eigenvalues that have distance equal to 1 (up to a tolerance of 10^{-10}). Indeed, if A is the system matrix and M is the domain decomposition preconditioner, then $I - M^{-1}A$ is the iteration matrix of the domain decomposition method used as an iterative solver. So, here we see if the eigenvalues of the preconditioned matrix $M^{-1}A$ are contained in the unitary disk centered at $(1,0)$. Notice that the matrix of the system doesn't change when N_{sub} or δ_{ovr} vary, so in Tables 3 and 4 we don't report $N_{\text{dofs}} = 1806$ and $N_{\text{iterNp}} = 1138$ again. In Figs. 3 and 4 we show for certain values of the parameters the whole spectrum of the matrix preconditioned with ORAS and OAS respectively (notice that many eigenvalues are multiple).

We can see that the non preconditioned GMRES is very slow, and the ORAS preconditioning gives much faster convergence than the OAS preconditioning.

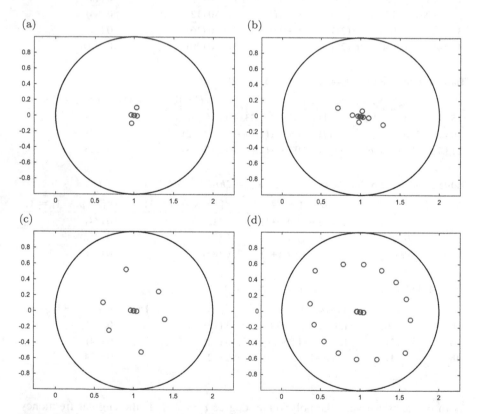

Fig. 3 Spectrum in the complex plane of the ORAS-preconditioned matrix. (**a**) $k = 2, \omega_2, N_{\text{sub}} = 2, \delta_{\text{ovr}} = 2h$. (**b**) $k = 2, \omega_3, N_{\text{sub}} = 2, \delta_{\text{ovr}} = 2h$. (**c**) $k = 2, \omega_2, N_{\text{sub}} = 4, \delta_{\text{ovr}} = 2h$. (**d**) $k = 2, \omega_2, N_{\text{sub}} = 8, \delta_{\text{ovr}} = 2h$

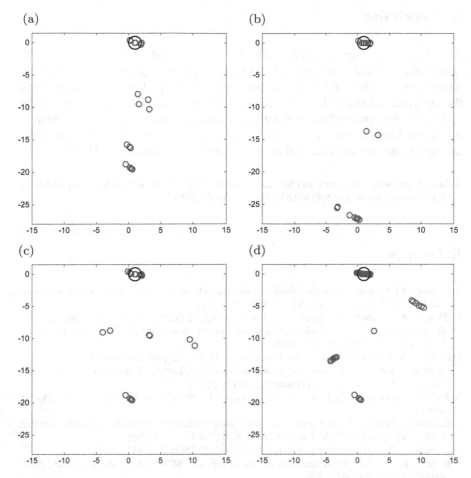

Fig. 4 Spectrum in the complex plane of the OAS-preconditioned matrix. **(a)** $k = 2, \omega_2, N_{\text{sub}} = 2, \delta_{\text{ovr}} = 2h$. **(b)** $k = 2, \omega_3, N_{\text{sub}} = 2, \delta_{\text{ovr}} = 2h$. **(c)** $k = 2, \omega_2, \underline{N_{\text{sub}} = 4}, \delta_{\text{ovr}} = 2h$. **(d)** $k = 2, \omega_2, \underline{N_{\text{sub}} = 8}, \delta_{\text{ovr}} = 2h$

Moreover, convergence becomes slower when k, ω or N_{sub} increase, or when the overlap size decreases; actually, when varying k, the number of iterations for convergence using the ORAS preconditioner is equal to 5 for $k = 0$ and then it stays equal to 6 for $k > 0$.

Notice also that for 2 subdomains the spectrum is well clustered inside the unitary disk with the ORAS preconditioner, except for the case with $\delta_{\text{ovr}} = h$, in which 3 eigenvalues are outside with distances from $(1, 0)$ equal to 19.5, 19.4, 14.4. Then, for 4 and 8 subdomains the spectrum is not so well clustered. With the OAS preconditioner there are always eigenvalues outside the unitary disk. For all the considered cases, the less clustered the spectrum, the slower the convergence.

5   Conclusion

Numerical experiments have shown that Schwarz preconditioning improves significantly the GMRES convergence for different values of physical and numerical parameters, and that the ORAS preconditioner always performs much better than the OAS preconditioner. The only advantage of the OAS method is to preserve the symmetry of the preconditioner. Finally, it has been pointed out that the spectrum of the preconditioned matrix reflects the convergence qualities, which improve when the eigenvalues are well clustered inside the unitary disk centered at $(1, 0)$.

Acknowledgements This work was financed by the French National Research Agency (ANR) in the framework of the project MEDIMAX, ANR-13-MONU-0012.

References

V. Dolean, M.J. Gander, L. Gerardo-Giorda, Optimized Schwarz methods for Maxwell's equations. SIAM J. Sci. Comput. **31**(3), 2193–2213 (2009)

V. Dolean, M.J. Gander, S. Lanteri, J.-F. Lee, Z. Peng, Effective transmission conditions for domain decomposition methods applied to the time-harmonic curl-curl Maxwell's equations. J. Comput. Phys. **280**, 232–247 (2015)

O.G. Ernst, M.J. Gander, Why it is difficult to solve Helmholtz problems with classical iterative methods, in *Numerical Analysis of Multiscale Problems*. Lecture Notes in Computer Science Engineering, vol. 83 (Springer, Heidelberg, 2012), pp. 325–363

M.J. Gander, Schwarz methods over the course of time. Electron. Trans. Numer. Anal. **31**, 228–255 (2008)

F. Ihlenburg, I. Babuška, Finite element solution of the Helmholtz equation with high wave number. I. The *h*-version of the FEM. Comput. Math. Appl. **30**(9), 9–37 (1995)

J.-C. Nédélec. Mixed finite elements in $\mathbf{R}^3$. Numer. Math. **35**(3), 315–341 (1980)

F. Rapetti, High order edge elements on simplicial meshes. M2AN Math. Model. Numer. Anal. **41**(6), 1001–1020 (2007)

F. Rapetti, A. Bossavit, Whitney forms of higher degree. SIAM J. Numer. Anal. **47**(3), 2369–2386 (2009)

On Nilpotent Subdomain Iterations

Faycal Chaouqui, Martin J. Gander, and Kévin Santugini-Repiquet

1 Introduction and Model Problem

Subdomain iterations which lead to a nilpotent iteration operator converge in a finite number of steps, and thus are equivalent to direct solvers. Such methods have led to very powerful new algorithms over the last few years, like the sweeping preconditioner of Engquist and Ying (2011a,b), or the source transfer domain decomposition method of Chen and Xiang (2013a,b). Their underlying mathematical structure are optimal Schwarz methods, see Gander (2006, 2008); Nataf et al. (1994) and references therein.[1]

We study here under which conditions the classical Neumann-Neumann, Dirichlet-Neumann and optimal Schwarz method can be nilpotent for the model problem

$$\eta u - \partial_{xx} u = f \text{ in } \Omega := (0, 1), \quad u(0) = u(1) = 0, \tag{1}$$

and a decomposition of the domain into J subdomains, $\Omega_j := (x_{j-1}, x_j)$, with $0 = x_0 < x_1 < \ldots < x_J = 1$ and subdomain length $\ell_j := x_j - x_{j-1}$. For two

---

[1]Optimal here is not in the sense of scalable, but really optimal: faster convergence is not possible.

F. Chaouqui • M.J. Gander (✉)
Section de mathématiques, Université de Genève, Geneva, Switzerland
e-mail: Faycal.Chaouqui@unige.ch; Martin.Gander@unige.ch

K. Santugini-Repiquet
Université Bordeaux, IMB, CNRS UMR5251, MC2, INRIA Bordeaux - Sud-Ouest, Bordeaux Cedex, France
e-mail: Kevin.Santugini@math.u-bordeaux1.fr

© Springer International Publishing AG 2017
C.-O. Lee et al. (eds.), *Domain Decomposition Methods in Science and Engineering XXIII*, Lecture Notes in Computational Science and Engineering 116, DOI 10.1007/978-3-319-52389-7_11

subdomains, we show that they all can be made nilpotent. For three subdomains, Neumann-Neumann can not be made nilpotent any more, but Dirichlet-Neumann can. For four subdomains, also Dirichlet-Neumann can not be made nilpotent any more for general decompositions, but for decompositions with subdomains of equal size, Dirichlet-Neumann can be made nilpotent for an arbitrary number of subdomains. Optimal Schwarz methods are always nilpotent for an arbitrary number of subdomains, even unequal ones. Our results indicate that for more general problems and more than two subdomains, only the optimal Schwarz method will be nilpotent.

2   The Neumann-Neumann Algorithm

For two subdomains, $J = 2$, the Neumann-Neumann algorithm applied to (1) is

$$\begin{cases} \eta u_j^{(n)} - \partial_{xx} u_j^{(n)} = f \text{ in } \Omega_j, \\ u_j^{(n)}(x_1) = h^{(n)}, \end{cases} \quad \begin{cases} \eta \psi_j^{(n)} - \partial_{xx} \psi_j^{(n)} = 0 \text{ in } \Omega_j, \\ \partial_{n_j} \psi_j^{(n)}(x_1) = \partial_{n_1} u_1^{(n)}(x_1) + \partial_{n_2} u_2^{(n)}(x_1), \end{cases}$$
$$h^{(n+1)} := h^{(n)} - \theta(\psi_1^{(n)}(x_1) + \psi_2^{(n)}(x_1)),$$
$$(2)$$

with $h^{(0)}$ an initial guess, θ a relaxation parameter, and in each iteration $u_1^{(n)}(0) = u_2^{(n)}(1) = 0$ and $\psi_1^{(n)}(0) = \psi_2^{(n)}(1) = 0$.

Since the problem is linear, it suffices to consider the homogeneous case of equation (1) and analyze the convergence of (2) to the zero solution. For $\eta > 0$ and $f = 0$, the differential equations in (2) can readily be solved,[2] and we obtain for the relaxation after a short calculation the relation

$$h^{(n+1)} = (1 - \theta(2 + \varphi(\eta)))h^{(n)}, \quad \varphi(t) := \frac{\tanh(\sqrt{t}\ell_1)}{\tanh(\sqrt{t}\ell_2)} + \frac{\tanh(\sqrt{t}\ell_2)}{\tanh(\sqrt{t}\ell_1)}, t > 0. \quad (3)$$

Proposition 1 *For two subdomains, the Neumann-Neumann algorithm (2) is convergent iff $0 < \theta < \theta_\eta^*$, $\theta_\eta^* := \frac{2}{2+\varphi(\eta)}$. Moreover, convergence is reached after two iterations for $\theta := \frac{\theta_\eta^*}{2}$, which in the symmetric case (i.e. $x_1 = \frac{1}{2}$) becomes $\theta := \frac{1}{4}$, i.e. the method is then nilpotent.*

Proof The convergence factor of the Neumann-Neumann algorithm (2) is $\rho_{\theta,\eta} := |1 - \theta(2 + \varphi(\eta))|$, and thus the algorithm is convergent iff $\rho_{\theta,\eta} < 1$, which is equivalent to requiring that $0 < \theta < \theta_\eta^*$. Moreover, $\rho_{\theta,\eta}$ vanishes when $\theta := \frac{\theta_\eta^*}{2}$, which makes the algorithm nilpotent.

---

[2]All our results remain valid also for $\eta = 0$ by taking limits.

Proposition 2 *For three subdomains, it is not possible to make the Neumann-Neumann algorithm nilpotent in general.*

Proof We consider the analogous definition of the Neumann-Neumann algorithm from (2) for three equal subdomains, i.e. $x_0 = 0$, $x_1 = \frac{1}{3}$, $x_2 = \frac{2}{3}$, $x_3 = 1$, and obtain after a short calculation as in Proposition 1 with explicit subdomain solutions

$$\begin{pmatrix} h_1^{(n+1)} \\ h_2^{(n+1)} \end{pmatrix} = \begin{pmatrix} 1 - \theta_1(4 + \frac{1}{s^2}) & -\frac{\theta_1}{cs^2} \\ -\frac{\theta_2}{cs^2} & 1 - \theta_2(4 + \frac{1}{s^2}) \end{pmatrix} \begin{pmatrix} h_1^{(n)} \\ h_2^{(n)} \end{pmatrix}, \tag{4}$$

where $s := \sinh(\sqrt{\eta}/3)$ and $c := \cosh(\sqrt{\eta}/3)$. Convergence in a finite number of iterations is possible iff the spectral radius of the iteration matrix in (4) vanishes, which means that the characteristic polynomial must be a monomial of degree 2. The fact that the other coefficients must vanish implies that the relaxation parameters θ_1 and θ_2 must satisfy the system of equations

$$(4 + \frac{1}{s^2})\theta_1 + (4 + \frac{1}{s^2})\theta_2 = 2 \quad \text{and} \quad (4 + \frac{1}{s^2})^2 \theta_1 \theta_2 = \alpha, \tag{5}$$

where $\alpha := \frac{(4+\frac{1}{s^2})^2}{(4+\frac{1}{s^2})^2 - (\frac{1}{s^2 c})^2} > 1$. Now (5) has no real solution, since the associated characteristic equation $\lambda^2 - 2\lambda + \alpha = 0$ does not admit one. It is thus not possible in general to obtain a nilpotent iteration for the Neumann-Neumann algorithm with three subdomains. $\qquad \blacksquare$

We will see in the numerical section that also for more than three subdomains, it is not possible in general to make the Neumann-Neumann algorithm nilpotent, and we will even get divergent iterations.

3 The Dirichlet-Neumann Algorithm

The Dirichlet-Neumann algorithm applied to (1) for two subdomains is

$$\begin{cases} \eta u_1^{(n)} - \partial_{xx} u_1^{(n)} = f \text{ in } \Omega_1, \\ u_1^{(n)}(x_1) = h^{(n)}, \end{cases} \quad \begin{cases} \eta u_2^{(n)} - \partial_{xx} u_2^{(n)} = f \text{ in } \Omega_2, \\ \partial_x u_2^{(n)}(x_1) = \partial_x u_1^{(n)}(x_1), \end{cases} \tag{6}$$
$$h^{(n+1)} := (1 - \theta)h^{(n)} + \theta u_2^{(n)}(x_1),$$

with $h^{(0)}$ an initial guess, θ a relaxation parameter, and $u_1^{(n)}(0) = u_2^{(n)}(1) = 0$. As for the Neumann-Neumann algorithm, we study the homogeneous part of 1, and obtain after a short calculation using the explicitly available subdomain solutions

$$h^{(n+1)} = (1 - \theta(1 + \psi(\eta)))h^{(n)}, \quad \psi(t) := \frac{\tanh(\sqrt{t}\ell_2)}{\tanh(\sqrt{t}\ell_1)}, \quad t > 0. \tag{7}$$

Proposition 3 *The Dirichlet-Neumann algorithm (6) is convergent for two subdomains iff* $0 < \theta < \theta_\eta^*$, $\theta_\eta^* := \frac{2}{1+\psi(\eta)}$. *Moreover, convergence is reached after two iterations for* $\theta := \frac{\theta_\eta^*}{2}$, *which in the symmetric case (i.e.* $x_1 = \frac{1}{2}$*) becomes* $\theta := \frac{1}{2}$, *i.e. the algorithm is then nilpotent.*

Proof The proof is similar to the proof of Proposition 1.

Proposition 4 *For three subdomains, the Dirichlet-Neumann algorithm converges in three iterations if either*

$$(\theta_1^*, \theta_2^*) = \left(\frac{1-\sqrt{1-\alpha}}{1+\frac{c_1 s_2}{s_1 c_2}}, \frac{1+\sqrt{1-\alpha}}{1+\frac{s_2 s_3}{c_2 c_3}} \right) \quad or \quad (\theta_1^*, \theta_2^*) = \left(\frac{1+\sqrt{1-\alpha}}{1+\frac{c_1 s_2}{s_1 c_2}}, \frac{1-\sqrt{1-\alpha}}{1+\frac{s_2 s_3}{c_2 c_3}} \right), \quad (8)$$

where $s_i := \sinh\left(\sqrt{\eta}\ell_i\right)$, $c_i := \cosh\left(\sqrt{\eta}\ell_i\right)$, $i = 1, \ldots, 3$, *and* $\alpha := \frac{(1+\frac{c_1 s_2}{s_1 c_2})(1+\frac{s_2 s_3}{c_2 c_3})}{1+\frac{c_1 s_2}{s_1 c_2}+\frac{s_2 s_3}{c_2 c_3}+\frac{c_1 s_3}{s_1 c_3}}$.

Proof With the analogously to (6) defined Dirichlet-Neumann algorithm for three subdomains, and solving the subdomain problems explicitly, we obtain after a short calculation

$$\begin{pmatrix} h_1^{(n+1)} \\ h_2^{(n+1)} \end{pmatrix} = \begin{pmatrix} 1 - \theta_1(1 + \frac{c_1 s_2}{s_1 c_2}) & \frac{\theta_1}{c_2} \\ -\theta_2 \frac{c_1 s_3}{s_1 c_2 c_3} & 1 - \theta_2(1 + \frac{s_2 s_3}{c_2 c_3}) \end{pmatrix} \begin{pmatrix} h_1^{(n)} \\ h_2^{(n)} \end{pmatrix}, \quad (9)$$

and the matrix is nilpotent iff its spectral radius vanishes, i.e.

$$\theta_1(1 + \frac{c_1 s_2}{s_1 c_2}) + \theta_2(1 + \frac{s_2 s_3}{c_2 c_3}) = 2, \quad (1 + \frac{c_1 s_2}{s_1 c_2})(1 + \frac{s_2 s_3}{c_2 c_3})\theta_1\theta_2 = \alpha. \quad (10)$$

This system admits the real solutions given in (8), since $0 < \alpha < 1$.

Proposition 5 *For four subdomains, convergence of the Dirichlet-Neumann algorithm in a finite number of iterations can not always be achieved.*

Proof We focus for simplicity on the case $\eta = 0$ and obtain for the analogously to (6) defined Dirichlet-Neumann algorithm for four subdomains after a short calculation

$$\begin{pmatrix} h_1^{(n+1)} \\ h_2^{(n+1)} \\ h_3^{(n+1)} \end{pmatrix} = \begin{pmatrix} 1 - \left(\frac{\ell_2}{\ell_1} + 1\right)\theta_1 & \theta_1 & 0 \\ -\frac{\theta_2 \ell_3}{\ell_1} & 1 - \theta_2 & \theta_2 \\ -\frac{\theta_3 \ell_4}{\ell_1} & 0 & 1 - \theta_3 \end{pmatrix} \begin{pmatrix} h_1^{(n)} \\ h_2^{(n)} \\ h_3^{(n)} \end{pmatrix}. \quad (11)$$

For nilpotence, the spectral radius of (11) must vanish, which means that the characteristic polynomial must be a monomial of degree 3. The fact that the other coefficients must vanish implies after a short calculation that θ_1, θ_2 and θ_3 must satisfy the system of equations $(1 + \frac{\ell_2}{\ell_1})\theta_1 + \theta_2 + \theta_3 = 3$, $(1 + \frac{\ell_2+\ell_3}{\ell_1})\theta_1\theta_2 +$

$(1 + \frac{\ell_2}{\ell_1})\theta_1\theta_3 + \theta_2\theta_3 = 3$, $(1 + \frac{\ell_2+\ell_3+\ell_4}{\ell_1})\theta_1\theta_2\theta_3 = 1$. Substituting the first equation into the second one we obtain $\frac{\ell_1+\ell_2+\ell_3}{\ell_1}\theta_1\theta_2 + \theta_3(3 - \theta_3) = 3 \implies \frac{(1-\ell_4)}{\ell_1}\theta_1\theta_2 + \theta_3(3 - \theta_3) = 3$, and replacing $\theta_1\theta_2$ by $\frac{\ell_1}{\theta_3}$ yields $1 - \ell_4 + \theta_3^2(3 - \theta_3) = 3\theta_3 \implies (\theta_3 - 1)^3 = -\ell_4 \implies \theta_3^* = 1 - \sqrt[3]{\ell_4}$. We therefore get

$$(1 + \frac{\ell_2}{\ell_1})\theta_1 + \theta_2 = 3 - \theta_3^*, \qquad (1 + \frac{\ell_2}{\ell_1})\theta_1\theta_2 = (1 + \frac{\ell_2}{\ell_1})\frac{\ell_1}{\theta_3^*}. \qquad (12)$$

The system (12) has real solutions if and only if the discriminant is non negative,

$$\Delta := \left(-3\ell_4 - 4\ell_3 + 3\ell_4^{2/3}\right)\left(\sqrt[3]{\ell_4} - 1\right)^{-1} \geq 0, \qquad (13)$$

which is equivalent to $-3\ell_4 - 4\ell_3 + 3\ell_4^{2/3} \leq 0$, and hence if this condition is not satisfied, the algorithm can not be made nilpotent.

We will see in Sect. 5 that for subdomains of equal size, Dirichlet-Neumann can be made nilpotent also for a larger number of subdomains.

4   The Optimal Schwarz Algorithm

A non-overlapping Schwarz algorithm for (1) with two subdomains is

$$\begin{cases} \eta u_1^{(n+1)} - \partial_{xx}u_1^{(n+1)} = f \text{ in } \Omega_1, \\ (\partial_x + p_1^+)u_1^{(n+1)}(x_1) = (\partial_x + p_1^+)u_2^{(n)}(x_1), \end{cases} \begin{cases} \eta u_2^{(n+1)} - \partial_{xx}u_2^{(n+1)} = f \text{ in } \Omega_2, \\ (\partial_x - p_2^-)u_2^{(n+1)}(x_1) = (\partial_x - p_2^-)u_1^{(n)}(x_1), \end{cases}$$
$$(14)$$

with $p_1^+, p_2^- > 0$ and $u_1^{(n)}(0) = u_2^{(n)}(1) = 0$. A direct computations shows that an optimal Schwarz method converging in two iterations is obtained for an arbitrary initial guess if $p_1^+ = \sqrt{\eta}\coth(\sqrt{\eta}\ell_2)$ and $p_2^- = \sqrt{\eta}\coth(\sqrt{\eta}\ell_1)$, and we even have

Proposition 6   *For J subdomains, let $\ell_j^+ := x_J - x_j, j = 1\dots,J-1$ and $\ell_j^- := x_{j-1} - x_0, j = 2,\dots,J$. Then setting $p_j^- := \sqrt{\eta}\coth(\sqrt{\eta}\ell_j^-)$ and $p_j^+ := \sqrt{\eta}\coth(\sqrt{\eta}\ell_j^+)$ in an analogously to (14) defined algorithm with $J \geq 2$ subdomains, an optimal Schwarz method converging in J iterations is obtained.*

Proof   By linearity, we again study convergence to the zero solution. Let $u_j^{(n)}$ be the approximate solution in each Ω_j at iteration n. First we prove that if

$$\begin{aligned} \partial_x u_j^{(n)} + p_j^+ u_j^{(n)} &= 0 \text{ at } x = x_j \implies \partial_x u_j^{(n)} + p_{j-1}^+ u_j^{(n)} = 0 \text{ at } x = x_{j-1}, \\ \partial_x u_j^{(n)} - p_j^- u_j^{(n)} &= 0 \text{ on } x = x_{j-1} \implies \partial_x u_j^{(n)} - p_{j+1}^- u_j^{(n)} = 0 \text{ on } x = x_j. \end{aligned}$$
$$(15)$$

To see this, suppose that $\partial_x u_j^{(n)} + p_j^+ u_j^{(n)} = 0$ on $x = x_j$, and let v be defined by $v(x) := u_j^{(n)}(x_{j-1}) \frac{\sinh(\sqrt{\eta}(x_J - x))}{\sinh(\sqrt{\eta}\ell_{j-1}^+)}$. Then $\partial_x v + p_j^+ v = 0$ at $x = x_j$, and by construction $v(x_{j-1}) = u_j^{(n)}(x_{j-1})$. Hence v satisfies

$$(\eta - \partial_{xx})(u_j^{(n)} - v) = 0 \text{ in } (x_{j-1}, x_j),$$
$$(\partial_x + p_j^+)(u_j^{(n)} - v) = 0 \text{ at } x = x_j, \quad u_j^{(n)} - v = 0 \text{ at } x = x_{j-1}. \tag{16}$$

Therefore, by uniqueness of the solution we must have $u_j^{(n)} = v$ on (x_{j-1}, x_j) and thus $\partial_x u_j^{(n)} + p_{j-1}^+ u_j^{(n)}$ at $x = x_{j-1}$, as it holds for v. The proof for the second line in (15) is similar.

Now since $\partial_x u_1^{(1)} - p_2^- u_1^{(1)} = 0$, we have from the transmission condition $\partial_x u_2^{(2)} - p_2^- u_2^{(2)} = \partial_x u_1^{(1)} - p_2^- u_1^{(1)} = 0$, which gives $\partial_x u_2^{(2)} - p_3^- u_2^{(2)} = 0$, and using the transmission condition again we get $\partial_x u_3^{(3)} - p_3^- u_3^{(3)} = \partial_x u_2^{(2)} - p_3^- u_2^{(2)} = 0$, and so on, until $\partial_x u_J^{(j)} - p_J^- u_J^{(j)} = 0$ and a similar argument holds for p_j^+. Hence, after J iterations the interior iterates $u_j^{(j)}$ satisfy

$$(\eta - \partial_{xx})(u_j^{(j)}) = 0 \text{ in } (x_{j-1}, x_j),$$
$$(\partial_x + p_j^+)u_j^{(j)} = 0 \text{ at } x = x_j, \quad (\partial_x - p_j^-)u_j^{(j)} = 0 \text{ at } x = x_{j-1}, \tag{17}$$

and on the domains on the left and right, we get

$$(\eta - \partial_{xx})(u_1^{(j)}) = 0 \text{ in } (x_0, x_1), \quad (\eta - \partial_{xx})(u_J^{(j)}) = 0 \text{ in } (x_{J-1}, x_J),$$
$$(\partial_x + p_1^+)u_1^{(j)} = 0 \text{ at } x = x_1, \quad (\partial_x - p_J^-)u_J^{(j)} = 0 \text{ at } x = x_{J-1}. \tag{18}$$
$$u_1^{(j)} = 0 \text{ at } x = x_0. \qquad u_J^{(j)} = 0 \text{ at } x = x_J,$$

Hence, $u_j^{(j)} = 0$, for all $j = 1, \ldots, J$, which concludes the proof.

One can show that this result still holds in higher dimensions for a decomposition into strips, provided one uses the then non-local Dirichlet to Neumann operators in the transmission conditions, see Nataf et al. (1994). One can however also obtain a nilpotent iteration with less restrictions, which also holds for higher dimensions just by replacing the transmission parameters below by the Dirichlet to Neumann operators again.

Proposition 7 *For J subdomains and $1 < d < J$,[3] choosing p_j^- for $j = 2, \ldots, d$ and p_j^+ for $j = d, \ldots J - 1$ as in Proposition 6, optimal Schwarz will converge in*

---

[3]Even the case $d = 1$ and $d = J$ can be handled by changing one of the Robin conditions into a Dirichlet one.

$2J^* - 1$ *iterations where* $J^* := \max(d, J - d + 1)$, *independently of the choice of the remaining* p_j^-, p_j^+.

Proof Following the proof of Proposition 6, after $j^* := \max(d, J-d+1)$ iterations, the $u_d^{(j^*)}$ satisfy

$$
\begin{aligned}
(\eta - \partial_{xx})(u_d^{(j^*)}) &= 0 \text{ in } (x_{d-1}, x_d), \\
(\partial_x - p_d^-)u_d^{(j^*)} = 0 \text{ at } x = x_{d-1}, \quad (\partial_x + p_d^+)u_d^{(j^*)} &= 0 \text{ at } x = x_d.
\end{aligned}
\tag{19}
$$

Hence $u_d^{(j^*)}$ vanishes in (x_{d-1}, x_d) and it follows that $u_j^{(j^*+j-d)} = 0$ for $j = d + 1, \ldots J$, and $u_{d-j}^{(j^*+j)} = 0$ for $j = 1, \ldots d - 1$. Thus optimal Schwarz will converge after $j^* + \max(d - 1, J - d) = 2 \max(d, J - d + 1) - 1$ iterations, which concludes the proof.

5 Numerical Experiments

We discretize our model problem (1) using finite differences with a mesh size $\Delta x = 10^{-5}$ and chose the right hand side such that the exact solution is $\sin(\pi x)$ for the parameter $\eta = 1$. We decompose the domain into $J = 2, 3, \ldots, 10$ equal subdomains, and start the iterations with a random initial guess. For each algorithm, we use the best possible relaxation parameters, i.e. the ones that minimize the spectral radius of the iteration operator, and we plot the error versus iteration on a semi-log scale. In Fig. 1 we see on the left that Neumann-Neumann is nilpotent for 2 subdomains, as shown in Proposition 1. For 3, 4 and 5 subdomains, Neumann-Neumann still converges, but is not nilpotent, see Proposition 2, and for more than 5 subdomains, the iterations even diverge. One can show that the convergence factor of Neumann-Neumann for this model problem with optimized relaxation parameters

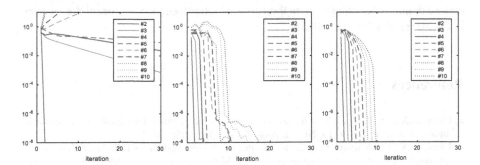

Fig. 1 Error versus number of iterations for Neumann-Neumann (*left*), Dirichlet-Neumann (*middle*), and optimal Schwarz (*right*) for different numbers of subdomains $J = 2, 3, \ldots, 10$ using the best possible relaxation parameters at the interfaces

behaves like $\mathcal{O}(\frac{1}{\ell^2})$ where ℓ is the subdomain size, so divergence will always set in at some point. For Dirichlet-Neumann in the middle of Fig. 1, we see nilpotence for all J in this special case of equal sized subdomains, but this would not be the case for general decompositions, see Proposition 5. The optimal Schwarz method on the right of Fig. 1 always converges in J iterations, as expected from Proposition 6.

6   Conclusion

We showed for a one dimensional model problem that the Neumann-Neumann method can only be nilpotent for a decomposition into two general subdomains; the Dirichlet-Neumann method can be nilpotent also for a decomposition into 3 general subdomains, but not any more for a decomposition into four general subdomains. We expect that for subdomains of equal size, Dirichlet-Neumann can be made nilpotent for an arbitrary number of subdomains. The optimal Schwarz method is nilpotent for a decomposition into an arbitrary number of subdomains, also of unequal size and in higher spatial dimensions, and this even if one does not use systematically the Dirichlet to Neumann operators, see our new result in Proposition 7. Our negative results for Neumann-Neumann and Dirichlet-Neumann methods in one spatial dimension imply that these algorithms can not be nilpotent in higher spatial dimensions either. For the Dirichlet-Neumann method and equal subdomains, our result indicates that nilpotence is also possible in higher dimensions for a strip decomposition, provided that the relaxation parameters become non-local operators. Optimal Schwarz methods are nilpotent in higher dimensions without any restrictions. Such nilpotent iterations have led to some of the best solvers for Helmholtz problems recently, see Gander and Nataf (2000), Gander and Nataf (2005), Engquist and Ying (2011a), Engquist and Ying (2011b), Chen and Xiang (2013a), Chen and Xiang (2013b), Zepeda-Núñez and Demanet (2016), and have been important in the development of optimized Schwarz methods (Côté et al. 2005; Gander 2006, 2008; Japhet and Nataf 2001). Well chosen coarse corrections can make a domain decomposition method also nilpotent, see the very recent discoveries in Gander et al. (2014a,b, 2015).

References

Z. Chen, X. Xiang, A source transfer domain decomposition method for Helmholtz equations in unbounded domain. SIAM J. Numer. Anal. **51**(4), 2331–2356 (2013)

Z. Chen, X. Xiang, A source transfer domain decomposition method for Helmholtz equations in unbounded domain part II: extensions. Numer. Math. Theory Methods Appl. **6**(03), 538–555 (2013)

J. Côté, M.J. Gander, L. Laayouni, S. Loisel, Comparison of the Dirichlet-Neumann and optimal Schwarz method on the sphere, in *Domain Decomposition Methods in Science and Engineering* (Springer, Berlin, 2005), pp. 235–242

B. Engquist, L. Ying, Sweeping preconditioner for the Helmholtz equation: hierarchical matrix representation. Commun. Pure Appl. Math. **64**(5), 697–735 (2011)

B. Engquist, L. Ying, Sweeping preconditioner for the Helmholtz equation: moving perfectly matched layers. SIAM Multiscale Model. Simul. **9**(2), 686–710 (2011)

M.J. Gander, Optimized Schwarz methods. SIAM J. Numer. Anal. **44**(2), 699–731 (2006)

M.J. Gander, Schwarz methods over the course of time. Electron. Trans. Numer. Anal. **31**(5), 228–255 (2008)

M.J. Gander, L. Halpern, K. Santugini-Repiquet, Discontinuous coarse spaces for DD-methods with discontinuous iterates, in *Domain Decomposition Methods in Science and Engineering XXI* (Springer, Berlin, 2014), pp. 607–615

M.J. Gander, L. Halpern, K. Santugini-Repiquet, A new coarse grid correction for RAS/AS, in *Domain Decomposition Methods in Science and Engineering XXI* (Springer, Berlin, 2014), pp. 275–283

M.J. Gander, A. Loneland, T. Rahman, Analysis of a new harmonically enriched multiscale coarse space for domain decomposition methods. arXiv preprint arXiv:1512.05285 (2015)

M.J. Gander, F. Nataf, AILU: a preconditioner based on the analytic factorization of the elliptic operator. Numer. Linear Alg. **7**(7–8), 543–567 (2000)

M.J. Gander, F. Nataf, An incomplete LU preconditioner for problems in acoustics. J. Comput. Acoust. **13**(03), 455–476 (2005)

C. Japhet, F. Nataf, The best interface conditions for domain decomposition methods: absorbing boundary conditions, in *Absorbing Boundaries and Layers, Domain Decomposition Methods: Applications to Large Scale Computers* (CMAP (Ecole Polytechnique), Palaiseau, France, 2001), p. 348

F. Nataf, F. Rogier, E. de Sturler, Optimal interface conditions for domain decomposition methods. CMAP (Ecole Polytechnique) **301**, 1–18 (1994)

L. Zepeda-Núnez, L. Demanet, The method of polarized traces for the 2d Helmholtz equation. J. Comput. Phys. **308**, 347–388 (2016)

A Direct Elliptic Solver Based on Hierarchically Low-Rank Schur Complements

Gustavo Chávez, George Turkiyyah, and David E. Keyes

1 Introduction

Cyclic reduction was conceived in 1965 for the solution of tridiagonal linear systems, such as the one-dimensional Poisson equation (Hockney 1965). Generalized to higher dimensions by recursive blocking, it is known as block cyclic reduction (BCR) (Buzbee et al. 1970). It can be used for general (block) Toeplitz and (block) tridiagonal linear systems; however, it is not competitive for large problems, because its arithmetic complexity grows superlinearly. Cyclic reduction can be thought of as a direct Gaussian elimination that recursively computes the Schur complement of half of the system. The complexity of Schur complement computations is dominated by the inverse. By considering a tridiagonal system and an even/odd ordering, cyclic reduction decouples the system such that the inverse of a large block is the block-wise inverse of a collection of independent smaller blocks. This addresses the most expensive step of the Schur complement computation in terms of operation complexity and does so in a way that launches concurrent subproblems. Its concurrency feature, in the form of recursive bisection, makes it interesting for parallel environments, provided that its arithmetic complexity can be improved.

We address the time and memory complexity growth of the traditional cyclic reduction algorithm by approximating dense blocks as they arise with hierarchical matrices ($\mathcal{H}$-Matrices). The effectiveness of the block approximation relies on the rank structure of the original matrix. Many relevant operators are known to have blocks of low rank off the diagonal. This philosophy follows recent work

G. Chávez • G. Turkiyyah • D.E. Keyes (✉)
Extreme Computing Research Center, King Abdullah University of Science and Technology,
23955 Thuwal, Saudi Arabia
e-mail: gustavo.chavezchavez@kaust.edu.sa; george.turkiyyah@kaust.edu.sa;
david.keyes@kaust.edu.sa

© Springer International Publishing AG 2017
C.-O. Lee et al. (eds.), *Domain Decomposition Methods in Science
and Engineering XXIII*, Lecture Notes in Computational Science
and Engineering 116, DOI 10.1007/978-3-319-52389-7_12

discussed below, but to our knowledge this is the first demonstration of the utility of complexity-reducing hierarchical substitution in the context of cyclic reduction.

The synergy of cyclic reduction and hierarchical matrices leads to a parallel fast direct solver of log-linear arithmetic complexity, $O(N \log^2 N)$, with controllable accuracy. The algorithm is purely algebraic, depending only on a block tridiagonal structure. We call it Accelerated Cyclic Reduction (ACR). Using a well-known implementation of $\mathcal{H}$-LU (Grasedyck et al. 2009), we demonstrate the range of applicability of ACR over a set of model problems including the convection-diffusion equation with recirculating flow and the wave Helmholtz equation, problems that cannot be tackled with the traditional FFT enabled version of cyclic reduction, FACR (Swarztrauber 1977). We show that ACR is competitive in time to solution as compared with a global $\mathcal{H}$-LU factorization that does not exploit the cyclic reduction structure. The fact that ACR is completely algebraic expands its range of applicability to problems with arbitrary coefficient structure within the block tridiagonal sparsity structure, subject to their amenability to rank compression. This gives the method robustness in some applications that are difficult for multigrid. The concurrency and flexibility to tune the accuracy of individual matrix block approximations makes it interesting for emerging many-core architectures. Finally, as with other direct solvers, there are complexity-accuracy tradeoffs that would naturally lead to the development of a new scalable preconditioner based on ACR.

2 Related Work

Exploiting underlying low-rank structure is a trending strategy for improving the performance of sparse direct solvers.

Nested dissection based clustering of an $\mathcal{H}$-Matrix is known as $\mathcal{H}$-Cholesky by Ibragimov et al. (2007) and $\mathcal{H}$-LU by Grasedyck et al. (2009), the main idea being to introduce $\mathcal{H}$-Matrix approximation on Schur complements based on domain decomposition. This is accomplished by a nested dissection ordering of the unknowns, and the advantage is that large blocks of zeros are preserved after factorization. The non-zero blocks are replaced with low-rank approximations, and an LU factorization is performed, using hierarchical matrix arithmetics. Recently, Kriemann (2013) demonstrated that $\mathcal{H}$-LU implemented with a task-based scheduling based on a directed acyclic graph is well suited for modern many-core systems when compared with the conventional recursive algorithm. A similar line of work by Xia and Gu (2010) also proposes the construction of a rank-structured Cholesky factorization via the HSS hierarchical format (Chandrasekaran et al. 2006). Figure 1 illustrates the differences between nested dissection ordering and the even/odd (or red/black) ordering of cyclic reduction.

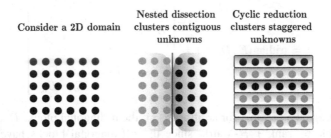

Fig. 1 The nested dissection ordering recursively clusters contiguous unknowns by bisection, whereas the red/black ordering recursively clusters staggered unknowns, allowing isolation of a new readily manipulated diagonal block

Multifrontal factorization, with low-rank approximations of frontal matrices, as in the work of Xia et al. (2010) also relies on nested dissection as the permutation strategy, but it uses the multifrontal method as a solver. Frontal matrices are approximated with the HSS format, while the solver relies on the corresponding HSS algorithms for elimination (Xia et al. 2010). A similar line of work is the generalization of this method to 3D problems and general meshes by Schmitz and Ying (2012, 2014). More recently, Ghysels et al. (2015) introduced a method based on a fast ULV decomposition and randomized sampling of HSS matrices in a many-core environment, where HSS approximations are used to approximate fronts of large enough size, as the complexity constant in building an HSS approximation is only convenient for large matrices.

This strategy is not limited to any specific hierarchical format. Aminfar et al. (2016) proposed the use of the HODLR matrix format Ambikasaran and Darve (2013), also in the context of the multifrontal method. The well known solver MUMPS now also exploits the low-rank property of frontal matrices to accelerate its multifrontal implementation, as described in Amestoy et al. (2014).

3   Accelerated Cyclic Reduction

Consider the two-dimensional linear variable-coefficient Poisson equation (1) and its corresponding block tridiagonal matrix structure resulting from a second order finite difference discretization, as shown in (2):

$$-\nabla \cdot \kappa(\mathbf{x})\nabla u = f(\mathbf{x}), \tag{1}$$

$$A = \text{tridiag}(E_i, D_i, F_i) = \begin{bmatrix} D_1 & F_1 & & & \\ E_2 & D_2 & F_2 & & \\ & \ddots & \ddots & \ddots & \\ & & E_{n-1} & D_{n-1} & F_{n-1} \\ & & & E_n & D_n \end{bmatrix}. \qquad (2)$$

We leverage the fact that for arbitrary $\kappa(\mathbf{x})$, the tridiagonal blocks D_i are *exactly* representable by rank 1 $\mathcal{H}$-Matrix since the off-diagonal blocks have only one entry regardless of their coefficient, and the blocks E_i and F_i are diagonal. As cyclic reduction progresses, the resulting blocks will have a bounded increase in the numerical ranks of their off-diagonal blocks. This numerical off-diagonal rank may be tuned to accommodate for a specified accuracy. We choose the $\mathcal{H}$-Matrix format proposed in Hackbusch (1999) by Hackbusch, although ACR is not limited to a specific hierarchical format. In terms of admissibility condition, we choose weak admissibility, as the sparsity structure is known beforehand and it proved effective in our numerical experiments.

Approximating each block as an $\mathcal{H}$-Matrix, we use the corresponding hierarchical arithmetic operations as cyclic reduction progresses, instead of the conventional linear algebra arithmetic operations. The following table summarizes the complexity estimates in terms of time and memory while dealing with a $n \times n$ block in a typical dense format and as a block-wise approximation with a rank-r $\mathcal{H}$-Matrix.

	Inverse	Storage
Dense Block	$\mathcal{O}(n^3)$	$\mathcal{O}(n^2)$
$\mathcal{H}$ Block	$\mathcal{O}(r^2 n \log^2 n)$	$\mathcal{O}(rn \log n)$

The following table summarizes the complexity estimates of the methods discussed so far in a two-dimensional square mesh where N is the total number of unknowns, neglecting the dependence upon rank. The derivation of the complexity estimates for $\mathcal{H}$-LU can be found in Bebendorf (2008).

	Operations	Memory
BCR	$\mathcal{O}(N^2)$	$\mathcal{O}(N^{1.5})$
$\mathcal{H}$-LU	$\mathcal{O}(N \log^2 N)$	$\mathcal{O}(N \log N)$
ACR	$\mathcal{O}(N \log^2 N)$	$\mathcal{O}(N \log N)$

With block-wise approximations in place, block cyclic reduction becomes ACR. BCR consists of two phases: reduction and back-substitution. The reduction phase is equivalent to block Gaussian elimination without pivoting on a permuted system $(PAP^T)(Pu) = Pf$. Permutation decouples the system, and the computation of the Schur complement reduces the problem size by half. This process is recursive and finishes when a single block is reached, although the recursion can be stopped when the system is small enough to be solved directly.

As an illustration, consider a system of $n = 8$ points per dimension, which translates into a $N \times N$ sparse matrix, with $N = n^2$. The first step is to permute the system, which with an even/odd ordering becomes:

$$
\begin{bmatrix}
D_0 & & & & F_0 & & & \\
& D_2 & & & E_2 & F_2 & & \\
& & D_4 & & & E_4 & F_4 & \\
& & & D_6 & & & E_6 & F_6 \\
\hline
E_1 & F_1 & & & D_1 & & & \\
& E_3 & F_3 & & & D_3 & & \\
& & E_5 & F_5 & & & D_5 & \\
& & & E_7 & & & & D_7
\end{bmatrix}
\begin{bmatrix}
u_0 \\ u_2 \\ u_4 \\ u_6 \\ u_1 \\ u_3 \\ u_5 \\ u_7
\end{bmatrix}
=
\begin{bmatrix}
f_0 \\ f_2 \\ f_4 \\ f_6 \\ f_1 \\ f_3 \\ f_5 \\ f_7
\end{bmatrix}.
\tag{3}
$$

Consider the above 2×2 partitioned system (3) as H. The upper-left block is block-diagonal, which means that its inverse can be computed as the inverse of each individual block (D_0, D_2, D_4, and D_6), in parallel and with hierarchical matrix arithmetics. The Schur complement of the upper-left partition may then be computed as follows:

$$
\begin{bmatrix}
H_{11} & H_{12} \\
H_{21} & H_{22}
\end{bmatrix}
\begin{bmatrix}
u_{even} \\
u_{odd}
\end{bmatrix}
=
\begin{bmatrix}
f_{even} \\
f_{odd}
\end{bmatrix}.
\tag{4}
$$

$$
(H_{22} - H_{21}H_{11}^{-1}H_{12})u_{odd} = f^{(1)}, \qquad f^{(1)} = f_{odd} - H_{21}H_{11}^{-1}f_{even}.
\tag{5}
$$

Superscripts indicates algorithmic steps. A key property of the Schur complement of a block tridiagonal matrix is that it yields another block tridiagonal matrix, as can been seen in the resulting permuted matrix system (5):

$$
\begin{bmatrix}
D_0^{(1)} & & F_0^{(1)} & \\
& D_2^{(1)} & E_2^{(1)} & F_2^{(1)} \\
\hline
E_1^{(1)} & F_1^{(1)} & D_1^{(1)} & \\
& E_3^{(1)} & & D_3^{(1)}
\end{bmatrix}
\begin{bmatrix}
u_0^{(1)} \\ u_2^{(1)} \\ u_1^{(1)} \\ u_3^{(1)}
\end{bmatrix}
=
\begin{bmatrix}
f_0^{(1)} \\ f_2^{(1)} \\ f_1^{(1)} \\ f_3^{(1)}
\end{bmatrix}.
\tag{6}
$$

One step further, the computation of the Schur complement of the permuted system (6), results in:

$$
\begin{bmatrix}
D_0^{(2)} & F_0^{(2)} \\
E_1^{(2)} & D_1^{(2)}
\end{bmatrix}
\begin{bmatrix}
u_0^{(2)} \\
u_1^{(2)}
\end{bmatrix}
=
\begin{bmatrix}
f_0^{(2)} \\
f_1^{(2)}
\end{bmatrix}.
\tag{7}
$$

A last round of permutation and Schur complement computation leads to the $D_0^{(3)}$ block, which is the last step of the reduction phase of Cyclic Reduction. A back-substitution phase to recover the solution also consists of $\log n$ steps. Each step involves matrix-vector products involving the off-diagonal blocks $E^{(i)}$ and $F^{(i)}$ and the inverses of the diagonal $D^{(i)}$ blocks computed during the elimination phase. These matrix-vector operations are done efficiently with hierarchical matrix arithmetics.

4 Numerical Results in 2D

We select two test cases to provide a baseline of performance and robustness as compared with the $\mathcal{H}$-LU implementation in HLIBpro Hackbusch et al. (xxxx), and with the AMG implementation in Hypre Lawrence Livermore National Laboratory (2017). Tests are performed in the shared memory environment of a 36-core Intel Haswell processor.

The first test is the wave Helmholtz equation.

$$\nabla^2 u + k^2 u = f(\mathbf{x}), \qquad \mathbf{x} \in \Omega = [0,1]^2 \quad u(\mathbf{x}) = 0,\ x \in \Gamma$$
$$f(\mathbf{x}) = 100e^{-100((x-0.5)^2+(y-0.5)^2)}. \tag{8}$$

For large values of kh, where h is the mesh spacing, discretization leads to an indefinite matrix. Performance over a range of k is shown in Fig. 2, for $h = 2^{-10}$. We compare ACR and $\mathcal{H}$-LU with AMG as a direct solver and as a preconditioner in combination with GMRES. For small α AMG outperforms the direct methods, but AMG loses robustness with rising indefiniteness.

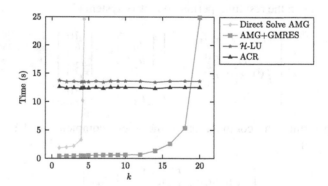

Fig. 2 Runtime versus wavenumber for fixed mesh size in the Wave Helmhotz equation. AMG is the method of choice for small k, but loses robustness with indefiniteness

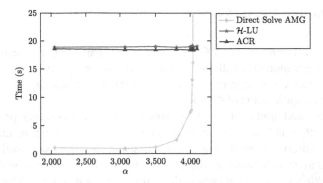

Fig. 3 Runtime versus velocity magnitude in convection-diffusion. AMG is the method of choice in the diffusion dominated limit, but loses robustness with skew-symmetry

The second test is convection-diffusion equation with recirculating flow.

$$- \nabla^2 u + \alpha b(\mathbf{x}) \cdot \nabla u = f(\mathbf{x}), \qquad \mathbf{x} \in \Omega = [0,1]^2 \ u(\mathbf{x}) = 0, \ x \in \Gamma$$

$$b(\mathbf{x}) = \begin{pmatrix} \sin(4\pi x) \sin(4\pi y) \\ \cos(4\pi x) \cos(4\pi y) \end{pmatrix} \qquad f(\mathbf{x}) = 100 e^{-100((x-0.5)^2 + (y-0.5)^2)}. \tag{9}$$

Discretization of this equation, again with $h = 2^{-10}$, leads to a nonsymmetric matrix, whose eigenvalues go complex (with central differencing) when the cell Peclet number exceeds 2. Direct algebraic methods are unaffected.

We progressively increase the convection dominance with α. For small α AMG outperforms the direct methods, but AMG is not robust with respect to the rising skew-symmetry. ACR maintains its performance for any α, as shown in Fig. 3.

5 Extensions

The discretization of 3D elliptic operators also leads to a block tridiagonal structure, with the difference that each block is of size $n^2 \times n^2$, instead of $n \times n$, as in the 2D discretization. A similar reduction strategy in the outermost dimension is possible, and leads to a solver with log-linear complexity in N and similar parallel structure, except that ranks grow.

The controllable accuracy feature of hierarchical matrices suggests the possibility of using ACR as a preconditioner, with rank becoming a tuning parameter balancing the cost per and the number of iterations, while preserving the rich concurrency features of the method.

6 Concluding Remarks

We present a fast direct solver, ACR, for structured sparse linear systems that arise from the discretization of 2D elliptic operators. The solver approximates every block using an $\mathcal{H}$-Matrix, resulting in a log-linear arithmetic complexity of $\mathcal{O}(N \log^2 N)$ with memory requirements of $\mathcal{O}(N \log N)$.

Robustness and applicability are demonstrated on model scalar problems and contrasted with established solvers based on the $\mathcal{H}$-LU factorization and algebraic multigrid. Multigrid maintains superiority in scalar problems with sufficient definiteness and symmetry, whereas hierarchical matrix-based replacements of direct methods tackle some problems where these properties are lacking. Although being of the same asymptotic complexity as $\mathcal{H}$-LU, ACR has fundamentally different algorithmic roots which produce a novel alternative for a relevant class of problems with competitive performance, and concurrency that grows with the problem size.

In Chávez et al. (2016) we expand on the consideration of cyclic reduction as a fast direct solver for 3D elliptic operators.

References

S. Ambikasaran, E. Darve, An $\mathcal{O}(N \log N)$ fast direct solver for partial hierarchically semiseparable matrices. J. Sci. Comput. **57**(3), 477–501 (2013)

P. Amestoy, A. Buttari, G. Joslin, J.-Y. L'Excellent, M. Sid-Lakhdar, C. Weisbecker, M. Forzan, C. Pozza, R. Perrin, V. Pellissier, Shared-memory parallelism and low-rank approximation techniques applied to direct solvers in FEM simulation. IEEE Trans. Mag. **50**(2), 517–520 (2014)

A. Aminfar, S. Ambikasaran, E. Darve, A fast block low-rank dense solver with applications to finite-element matrices. J. Comput. Phys. **304**, 170–188 (2016)

M. Bebendorf, *Hierarchical Matrices: A Means to Efficiently Solve Elliptic Boundary Value Problems*. Lecture Notes in Computational Science and Engineering, vol. 63 (Springer, Berlin, 2008)

B.L. Buzbee, G.H. Golub, C.W. Nielson, On direct methods for solving Poisson equation. SIAM J. Numer. Anal. **7**(4), 627–656 (1970)

S. Chandrasekaran, M. Gu, T. Pals, A fast *ULV* decomposition solver for hierarchically semiseparable representations. SIAM J. Matrix Anal. Appl. **28**(3), 603–622 (2006)

G. Chavez, G. Turkiyyah, S. Zampini, H. Ltaief, D. Keyes, Accelerated cyclic reduction: a distributed-memory fast direct solver for structured linear systems (2016). Preprint, arXiv:1701.00182

P. Ghysels, X.S. Li, F.-H. Rouet, S. Williams, A. Napov, An efficient multi-core implementation of a novel HSS-structured multifrontal solver using randomized sampling. arXiv:1502.07405 [cs.MS], pp. 1–26, 2015

L. Grasedyck, R. Kriemann, S. Le Borne, Domain decomposition based $\mathcal{H}$-LU preconditioning. Numer. Math. **112**(4), 565–600 (2009)

W. Hackbusch, A sparse matrix arithmetic based on $\mathcal{H}$-Matrices. Part I: introduction to $\mathcal{H}$-Matrices. Computing **62**(2), 89–108 (1999)

W. Hackbusch, S. Börm, L. Grasedyck, HLib 1.4. http://hlib.org, 1999–2012. Max-Planck-Institut, Leipzig

R.W. Hockney, A fast direct solution of Poisson's equation using Fourier analysis. J. ACM **12**(1), 95–113 (1965)

I. Ibragimov, S. Rjasanow, K. Straube, Hierarchical Cholesky decomposition of sparse matrices arising from curl-curl-equation. J. Numer. Math. **15**(1), 31–57 (2007)

R. Kriemann, Parallel $\mathcal{H}$-Matrix arithmetics on shared memory systems. Computing **74**(3), 273–297 (2005). ISSN=0010-485X

R.D. Falgout, U.M. Yang, Hypre: a library of high performance preconditioners, in *Computational Science - ICCS 2002: International Conference Amsterdam*, ed. by P.M. A. Sloot, A.G. Hoekstra, C.J.K. Tan, J.J. Dongarra. 2002 Proceedings, Part III (Springer, Berlin/Heidelberg, 2002), pp. 632–641. ISBN:978-3-540-47789-1

P.G. Schmitz, L. Ying, A fast direct solver for elliptic problems on general meshes in 2D. J. Comput. Phys. **231**(4), 1314–1338 (2012)

P.G. Schmitz, L. Ying, A fast nested dissection solver for Cartesian 3D elliptic problems using hierarchical matrices. J. Comput. Phys. **258**, 227–245 (2014)

P. Swarztrauber, The methods of cyclic reduction, Fourier analysis and the FACR algorithm for the discrete solution of Poisson equation on a rectangle. SIAM Rev. **19**(3), 490–501 (1977)

J. Xia, S. Chandrasekaran, M. Gu, X. Li, Superfast multifrontal method for large structured linear systems of equations. SIAM J. Matrix Anal. Appl. **31**(3), 1382–1411 (2010)

J. Xia, S. Chandrasekaran, M. Gu, X.S. Li, Fast algorithms for hierarchically semiseparable matrices. Numer. Lin. Alg. Appl. **17**(6), 953–976 (2010)

J. Xia, M. Gu, Robust approximate Cholesky factorization of rank-structured symmetric positive definite matrices. SIAM J. Matrix Anal. Appl. **31**(5), 2899–2920 (2010)

Optimized Schwarz Methods for Heterogeneous Helmholtz and Maxwell's Equations

Victorita Dolean, Martin J. Gander, Erwin Veneros, and Hui Zhang

1 Introduction

The Helmholtz equation is very difficult to solve by iterative methods (Ernst and Gander 2012), and the time harmonic Maxwell's equations inherit these difficulties. Optimized Schwarz methods are among the most promising iterative techniques. For the Helmholtz equation, they have their roots in the seminal work of Deprés (Després 1990; Després et al. 1992), which led to the development of optimized transmission conditions (Boubendir et al. 2012; Chevalier and Nataf 1997; Gander 2001; Gander et al. 2002, 2007), and these techniques were independently rediscovered for the sweeping preconditioner (Engquist and Ying 2011) and the source transfer domain decomposition method (Chen and Xiang 2013). For the time harmonic Maxwell's equations, optimized transmission conditions were developed and tested for problems without conductivity in Alonso-Rodriguez and Gerardo-Giorda (2006), Dolean et al. (2009), Peng and Lee (2010), Peng et al. (2010), El Bouajaji et al. (2012), and with conductivity in Dolean et al. (2011a), Dolean et al. (2011b). Particular Galerkin discretizations of transmission conditions were studied

V. Dolean (✉)
University of Strathclyde, Glasgow, UK

University Cote d'Azur, CNRS, LJAD, France
e-mail: victorita.dolean@strath.ac.uk

M.J. Gander • E. Veneros
Section de mathématiques, Université de Genève, 1211 Genève 4, Geneva, Switzerland
e-mail: martin.gander@unige.ch; erwin.veneros@unige.ch

H. Zhang
Key Laboratory of Oceanographic Big Data Mining & Application of Zhejiang Province, 316022
Zhoushan, China
e-mail: huiz@zjou.edu.cn

© Springer International Publishing AG 2017
C.-O. Lee et al. (eds.), *Domain Decomposition Methods in Science and Engineering XXIII*, Lecture Notes in Computational Science and Engineering 116, DOI 10.1007/978-3-319-52389-7_13

in Dolean et al. (2008a,b), and for scattering applications, see Peng and Lee (2010), Peng et al. (2010).

In Dubois (2007), Gander and Dubois (2015), it was discovered that heterogeneous media can actually improve the convergence of optimized Schwarz methods, provided that the coefficient jumps are aligned with the interfaces, and the jumps are taken into account in an appropriate way in the transmission conditions. Similar results were found for Maxwell's equations in Veneros et al. (2014) and Veneros et al. (2015); it is even possible to obtain convergence independently of the mesh size in certain situations. We present and study here transmission conditions for the Helmholtz equation with heterogeneous media, and establish a relation to the results of Veneros et al. (2014), Veneros et al. (2015) written for Maxwell's equations. We then study improved convergence behavior for specific choices of the discretization parameters related to the pollution effect.

2 Optimized Schwarz Methods for Helmholtz and Maxwell's Equations

We consider the two dimensional Helmholtz equation in discontinuous media with piece-wise constant density ρ and wave-speed c. The Helmholtz equation in $\Omega = \mathbb{R}^2$ is defined by

$$\nabla(\frac{1}{\rho}\nabla \cdot u) + \frac{\omega^2}{c^2\rho}u = f, \text{ in } \Omega, \tag{1}$$

with

$$\rho =: \begin{cases} \rho_1 & \text{in } \Omega_1, \\ \rho_2 & \text{in } \Omega_2, \end{cases} \qquad c := \begin{cases} c_1 & \text{in } \Omega_1, \\ c_2 & \text{in } \Omega_2, \end{cases}$$

where $\Omega_1 = \mathbb{R}^- \times \mathbb{R}$, $\Omega_2 = \mathbb{R}^+ \times \mathbb{R}$ and the Sommerfeld radiation condition is imposed at infinity,

$$\lim_{|x|\to\infty} \sqrt{|x|}\left(\partial_{|x|}u + i\omega u\right) = 0, \tag{2}$$

for every possible direction $\frac{x}{|x|}$.

We can naturally define a Schwarz algorithm for Eq. (1) with Robin transmission conditions at the interface aligned with the discontinuity between the coefficients, and parameters $s_1, s_2 \in \mathbb{C}$,

$$\begin{cases} \nabla(\frac{1}{\rho_1}\nabla \cdot u_1^n) + \frac{\omega^2}{c_1^2\rho_1}u_1^n = f, & \text{in } \Omega_1, \\ (\frac{1}{\rho_1}\partial_{n_1} + \frac{1}{\rho_2}s_2)u_1^n = (\frac{1}{\rho_2}\partial_{n_1} + \frac{1}{\rho_2}s_2)u_2^{n-1}, & \text{on } \Gamma, \\ \nabla(\frac{1}{\rho_2}\nabla \cdot u_2^n) + \frac{\omega^2}{c_2^2\rho_1}u_2^n = f, & \text{in } \Omega_2, \\ (\frac{1}{\rho_2}\partial_{n_2} + \frac{1}{\rho_1}s_1)u_2^n = (\frac{1}{\rho_1}\partial_{n_2} + \frac{1}{\rho_1}s_1)u_1^{n-1}, & \text{on } \Gamma. \end{cases} \tag{3}$$

Proposition 1 *The convergence factor of algorithm (3) is given by*

$$\rho_{opt}(k, \rho_1, \rho_2, \omega, c_1, c_2, s_1, s_2) = \left| \frac{(\lambda_1 - s_1)(\lambda_2 - s_2)}{(\lambda_1 + s_2 \frac{\rho_1}{\rho_2})(\lambda_2 + s_1 \frac{\rho_2}{\rho_1})} \right|^{1/2}, \tag{4}$$

with $\lambda_j = \sqrt{k^2 - \omega_j^2}$, $\omega_j = \frac{\omega}{c_j}$ for $j = 1, 2$.

The proof of Proposition 1 is based in Fourier analysis, see Veneros (2015) for details.

In order to obtain an efficient algorithm, we have to choose s_1 and s_2 such that ρ_{opt} becomes as small as possible for all relevant numerical frequencies $k \in K := [k_{min}, k_{max}]$, where k_{min} is the lowest relevant frequency (k_{min} depends on the geometry of the media) and $k_{max} = \frac{c_{max}}{h}$ is the highest numerical frequency supported by the numerical grid with mesh size h.

In what follows, we only consider $s_1 = P_1(1 + i)$ and $s_2 = P_2(1 + i)$, a choice that has been justified in Gander et al. (2002), and thus study the min-max problem

$$\rho_{opt}^* = \min_{P_1, P_2 > 0} \max_{k \in K} |\rho_{opt}(k, \rho_1, \rho_2, \omega, c_1, c_2, P_1(1 + i), P_2(1 + i))|. \tag{5}$$

Similarly we can define a Schwarz algorithm for the time-harmonic Maxwell equations in a given domain $\Omega = \mathbb{R}^3$

$$-i\omega\varepsilon\mathbf{E} + \nabla \times \mathbf{H} = \mathbf{J}, \quad i\omega\mu\mathbf{H} + \nabla \times \mathbf{E} = \mathbf{0}, \tag{6}$$

with the Silver Müller radiation condition

$$\lim_{r \to \infty} r(\mathbf{H} \times e_{\mathbf{r}} + \frac{1}{Z_j}\mathbf{E}) = 0, \tag{7}$$

where $r := |\mathbf{x}|$ and $e_{\mathbf{r}} = \mathbf{x}/r$ for any vector $\mathbf{x} \in \mathbb{R}^3$.

We also consider the heterogeneous case where the domain Ω consists of two non-overlapping subdomains $\Omega_1 := \mathbb{R}^- \times \mathbb{R}^2$ and $\Omega_2 := \mathbb{R}^+ \times \mathbb{R}^2$ with interface Γ, with piece-wise constant parameters ε_j and μ_j in $\Omega_j, j = 1, 2$. A general Schwarz algorithm for this configuration is

$$\begin{aligned} -i\omega\varepsilon_1\mathbf{E}^{1,n} + \nabla \times \mathbf{H}^{1,n} &= \mathbf{J}, \quad i\omega\mu_1\mathbf{H}^{1,n} + \nabla \times \mathbf{E}^{1,n} = \mathbf{0} \quad \text{in } \Omega_1, \\ (\mathcal{B}_{\mathbf{n}_1} + \mathcal{S}_1\mathcal{B}_{\mathbf{n}_2})(\mathbf{E}^{1,n}, \mathbf{H}^{1,n}) &= (\mathcal{B}_{\mathbf{n}_1} + \mathcal{S}_1\mathcal{B}_{\mathbf{n}_2})(\mathbf{E}^{2,n-1}, \mathbf{H}^{2,n-1}) \text{ on } \Gamma, \\ -i\omega\varepsilon_2\mathbf{E}^{2,n} + \nabla \times \mathbf{H}^{2,n} &= \mathbf{J}, \quad i\omega\mu_2\mathbf{H}^{2,n} + \nabla \times \mathbf{E}^{2,n} = \mathbf{0} \quad \text{in } \Omega_2, \\ (\mathcal{B}_{\mathbf{n}_2} + \mathcal{S}_2\mathcal{B}_{\mathbf{n}_1})(\mathbf{E}^{2,n}, \mathbf{H}^{2,n}) &= (\mathcal{B}_{\mathbf{n}_2} + \mathcal{S}_2\mathcal{B}_{\mathbf{n}_1})(\mathbf{E}^{1,n-1}, \mathbf{H}^{1,n-1}) \text{ on } \Gamma, \end{aligned} \tag{8}$$

where $\mathcal{S}_j, j = 1, 2$ are tangential, possibly pseudo-differential operators, and

$$\mathcal{B}_{\mathbf{n}_j}(\mathbf{E}^{j,n}, \mathbf{H}^{j,n}) = \frac{\mathbf{E}^{j,n}}{Z_j} \times \mathbf{n}_j + \mathbf{n}_j \times (\mathbf{H}^{j,n} \times \mathbf{n}_j)$$

are the characteristic conditions, with $Z_j = \sqrt{\mu_j/\epsilon_j}$, $j = 1, 2$. Different choices of $S_j, j = 1, 2$ lead to different Schwarz methods, see Dolean et al. (2009).

Remark 1 A direct computation shows that algorithms (3) and (8) have the same convergence factor, when setting $\rho_j := \mu_j$ and $c_j := \frac{1}{\sqrt{\epsilon_j \mu_j}}$ for $j = 1, 2$. Hence we can use all the results presented in Veneros et al. (2014) for Maxwell's equations for the case of the Helmholtz equation (3). We thus focus in the remainder on the Helmholtz case, but keep in mind that all results we will obtain hold mutatis mutandis also for the Maxwell case.

Using Remark 1, we obtain from Veneros et al. (2014) and Veneros et al. (2015)

Corollary 1 *The solution of (5) for $c_1 \neq c_2$ is asymptotically*

$$\rho_{opt}^* = \begin{cases} 1 - \mathcal{O}(h^{1/4}) & \text{if } \rho_1 = \rho_2, \\ \sqrt{\frac{\rho_{min}}{\rho_{max}}} + \mathcal{O}(h^{1/2}) & \text{if } \frac{1}{\sqrt{2}} \leq \frac{\rho_1}{\rho_2} \leq \sqrt{2}, \\ \sqrt[4]{\frac{1}{2}} + \mathcal{O}(h^{1/2}) & \text{if } \frac{\rho_1}{\rho_2} < \frac{1}{\sqrt{2}} \text{ or } \frac{\rho_1}{\rho_2} > \sqrt{2}. \end{cases} \tag{9}$$

If $\rho_1 \neq \rho_2$ and $c_1 = c_2$, we obtain after excluding the resonance frequency (Dolean et al. 2009)

$$\rho_{opt}^* = \sqrt{\frac{\rho_{min}}{\rho_{max}}} + \mathcal{O}(h^{1/2}), \tag{10}$$

with $\rho_{min} = \min\{\rho_1, \rho_2\}$ and $\rho_{max} = \max\{\rho_1, \rho_2\}$.

The detailed proof of Corollary 1 and the values of P_j can be found in Veneros (2015). We see from Corollary 1 that in most of the cases the optimized convergence factor ρ_{opt}^* has an asymptotic behavior independent of the mesh size h.

3 Scaling Results When Controlling the Pollution Effect

The core of our study is the asymptotic analysis of algorithms (3) and (8) when the mesh size h is related to the wave number ω to control the pollution effect. We will focus on the first case of Corollary 1, because this is the only case where the convergence can deteriorate in the mesh size h, see the first line in (9). We will consider three particular relationships between ω and h: $\omega h = C_\omega$, C_ω a constant, where the pollution effect is not controlled, $\omega^2 h = C_\omega$ where the pollution effect is provably controlled, and finally $\omega^{3/2} h = C_\omega$ which is widely believed to suffice to control the pollution effect.

Theorem 1 *Let $\rho_1 = \rho_2$, $c_1 \neq c_2$ and $\omega h = C_\omega$. If $|\rho_{opt}|$ defined in (4) is maximal for the frequencies $k = \omega_1$, $k = \omega_2$ and $k = k_{\max}$, and $s_j = (1 + i)P_j$, then the* *solution of the min-max problem (5) is*

$$P_1^* = \frac{\bar{p}_1}{h}, \quad P_2^* = \frac{\bar{p}_2}{h}, \quad \rho_{opt}^* = \left(\frac{\bar{p}_1^2 (2\bar{p}_2^2 - 2\bar{p}_2 c_r + c_r^2)}{\bar{p}_2^2 (2\bar{p}_1^2 + 2\bar{p}_1 c_r + c_r^2)} \right)^{\frac{1}{4}}, \tag{11}$$

where $\{\bar{p}_1, \bar{p}_2\}$ is solution of the system of equations

$$\frac{p_1^2 (2p_2^2 - 2p_2 c_r + c_r^2)}{p_2^2 (2p_1^2 + 2p_1 c_r + c_r^2)} = \frac{\rho^2 p_2^2 (2p_1^2 - 2p_1 c_r + c_r^2)}{p_1^2 (2p_2^2 + 2p_2 c_r + c_r^2)},$$
$$\frac{p_1^2 (2p_2^2 - 2p_2 c_r + c_r^2)}{p_2^2 (2p_1^2 + 2p_1 c_r + c_r^2)} = \frac{\rho^2 (2p_2^2 - 2p_2 c_{max_2} + c_{max_2}^2)(2p_1^2 - 2p_1 c_{max_1} + c_{max_1}^2)}{(2p_2^2 + 2p_2 c_{max_2} + c_{max_2}^2)(2p_1^2 + 2p_1 c_{max_1} + c_{max_1}^2)},$$

$c_r := rh := \sqrt{|\omega_1^2 - \omega_2^2|} h, \ c_{max_1} := \sqrt{c_{max}^2 - C_\omega^2 / c_1^2}, \ c_{max_2} := \sqrt{c_{max}^2 - C_\omega^2 / c_2^2}.$

Proof Evaluating $|\rho_{opt}|^4$ from (4) at $s_j := \frac{p_j}{h}(1+i)$ for $k = \omega_1, k = \omega_2$ and $k = k_{max}$ yields

$$R_1 = \frac{(h^2 r^2 - 2p_2 hr + 2p_2^2) p_1^2}{p_2^2 (h^2 r^2 + 2p_1 hr + 2p_1^2)}, \quad R_2 = \frac{\rho^2 p_2^2 (h^2 r^2 - 2p_1 hr + 2p_1^2)}{(2p_2^2 + 2p_2 hr + h^2 r^2) p_1^2},$$

$$R_3 = \frac{\left(h^2 (\frac{c_{max}^2}{h^2} - \frac{C_\omega^2}{c_2^2 h^2}) - 2p_2 h \sqrt{\frac{c_{max}^2}{h^2} - \frac{C_\omega^2}{c_2^2 h^2}} + 2p_2^2 \right) \left(h^2 (\frac{c_{max}^2}{h^2} - \frac{C_\omega^2}{c_1^2 h^2}) - 2p_1 h \sqrt{\frac{c_{max}^2}{h^2} - \frac{C_\omega^2}{c_1^2 h^2}} + 2p_1^2 \right)}{\left(h^2 (\frac{c_{max}^2}{h^2} - \frac{C_\omega^2}{c_2^2 h^2}) - 2p_1 h \sqrt{\frac{c_{max}^2}{h^2} - \frac{C_\omega^2}{c_2^2 h^2}} + 2p_1^2 \right) \left(h^2 (\frac{c_{max}^2}{h^2} - \frac{C_\omega^2}{c_1^2 h^2}) - 2p_2 h \sqrt{\frac{c_{max}^2}{h^2} - \frac{C_\omega^2}{c_1^2 h^2}} + 2p_2^2 \right)}.$$

Replacing rh by c_r, $c_{max_1} = \sqrt{c_{max}^2 - C_\omega^2 / c_1^2}$ and $c_{max_2} = \sqrt{c_{max}^2 - C_\omega^2 / c_2^2}$, the expressions can be simplified to

$$R_1 = \frac{p_1^2 (2p_2^2 - 2p_2 c_r + c_r^2)}{p_2^2 (2p_1^2 + 2p_1 c_r + c_r^2)}, \quad R_2 = \frac{\rho^2 p_2^2 (2p_1^2 - 2p_1 c_r + c_r^2)}{p_1^2 (2p_2^2 + 2p_2 c_r + c_r^2)},$$

$$R_3 = \frac{(2p_2^2 - 2p_2 c_{max_2} + c_{max_2}^2)(2p_1^2 - 2p_1 c_{max_1} + c_{max_1}^2)}{(2p_2^2 + 2p_2 c_{max_2} + c_{max_2}^2)(2p_1^2 + 2p_1 c_{max_1} + c_{max_1}^2)}.$$

Equioscillation between R_1, R_2 and R_3 then gives the result.

Remark 2 Note that Theorem 1 gives a closed form solution of the min-max problem (5), not just an asymptotic one.

For the special case of equal transmission conditions, we have

Corollary 2 *Under the same assumptions as in Theorem 1, if $s_j = (1 + i)P_j$ with $P_1 = P_2$, then the solution of the min-max problem (5) is given by*

$$P_1^* = P_2^* = \frac{\bar{p}}{h}, \quad \rho_{opt}^* = \left(\frac{(2\bar{p}^2 - 2\bar{p} c_r + c_r^2)}{(2\bar{p}^2 + 2\bar{p} c_r + c_r^2)} \right)^{\frac{1}{4}},$$

with $\bar{p}$ the solution of the equation

$$\frac{(2p^2 - 2pc_r + c_r^2)}{(2p^2 + 2pc_r + c_r^2)} = \frac{(2p^2 - 2pc_{\max_2} + c_{\max_2}^2)(2p^2 - 2pc_{\max_1} + c_{\max_1}^2)}{(2p^2 + 2pc_{\max_2} + c_{\max_2}^2)(2p^2 + 2pc_{\max_1} + c_{\max_1}^2)}.$$

Proof The proof follows along the same lines as the proof of Theorem 1.

Theorem 2 *Let* $\rho_1 = \rho_2$, $c_1 \neq c_2$ *and* $\omega^2 h = C_\omega$. *If* $|\rho_{opt}|$ *defined in (4) is maximal for the frequencies* $k = \omega_1$, $k = \omega_2$, $k = k_m := \frac{c_m}{h^{3/4}}$ *and* $k = k_{\max}$, *and* $s_j = (1+i)P_j$, $P_1 = \frac{p_1}{h}$ *and* $P_2 = \frac{p_2}{\sqrt{h}}$, *then the asymptotic solution of the min-max problem (5) for h small is given by*

$$P_1^* = \frac{c_{\max}^{3/4} c_r^{1/4}}{2^{1/4} h^{7/8}}, \quad P_2^* = \frac{1}{2} \frac{c_{\max}^{1/4} c_r^{3/4}}{2^{3/4} h^{5/8}}, \quad \rho_{opt}^* = 1 - \frac{r^{1/4}}{2^{1/4} c_{\max}^{1/4}} h^{1/8} + \mathcal{O}(h^{1/4}).$$

Interchanging the role of P_1 *and* P_2 *leads to the same result.*

Proof The proof is based again on equioscillation.

Theorem 3 *Let* $\rho_1 = \rho_2$, $c_1 \neq c_2$ *and* $\omega^{3/2} h = C_\omega$. *If the frequencies* $k = \omega_1$, $k = \omega_2$, $k = k_m := \frac{c_m}{h^{5/6}}$ *and* $k = k_{\max}$ *are the local maxima of the convergence factor* ρ_{opt} *from (4), and if* $s_1 = (1 + i)P_1$, $s_2 = (1 + i)P_2$, *with* $P_1 = \frac{p_1}{h^{11/12}}$ *and* $P_2 = \frac{p_2}{h^{3/4}}$, *then the asymptotic solution of the min-max problem (5) for h small is given by*

$$P_1^* = \frac{c_{\max}^{3/4} c_r^{1/4}}{2^{1/4} h^{11/12}}, \quad P_2^* = \frac{1}{2} \frac{c_{\max}^{1/4} c_r^{3/4}}{2^{3/4} h^{3/4}}, \quad \rho_{opt}^* = 1 - \frac{r^{1/4}}{2^{1/4} c_{\max}^{1/4}} h^{1/12} + \mathcal{O}(h^{1/6}).$$

Interchanging the role of P_1 *and* P_2 *leads to the same result.*

Proof The proof is similar to the proof of Theorem 2.

One can justify the choice of the frequencies $k = \omega_1$, $k = \omega_2$, $k = k_m$ and $k = k_{\max}$ as the correct candidates for the $|\rho_{opt}|$ using asymptotic analysis, but this exceeds the space available, see Veneros (2015) for more details.

Remark 3 One can obtain similar results also for the cases $\rho_1 \neq \rho_2$ but this will only reduce the order of the second asymptotic term, as in Theorems 2 and 3. For the relationship $\omega h = C_\omega$ one can also obtain a similar result to Theorem 1.

We give a summary of all these results in Table 1.

Table 1 Comparison of the convergence factors with different relationships between ω and h

	$\omega = C_\omega$	$\omega^2 h = C_\omega$	$\omega^{3/2} h = C_\omega$	$\omega h = C_\omega$
$\rho_1 = \rho_2,\ c_1 \neq c_2$	$1 - \mathcal{O}(h^{1/4})$ (Corollary 1)	$1 - \mathcal{O}(h^{1/8})$ (Theorem 2)	$1 - \mathcal{O}(h^{1/12})$ (Theorem 3)	< 1 (Theorem 1)
$\rho_1 \neq \rho_2,\ c_1 \neq c_2$	$\max\{\sqrt[4]{\frac{1}{2}}, \sqrt{\frac{\rho_{min}}{\rho_{max}}}\}$ (Corollary 1)	$\max\{\sqrt[4]{\frac{1}{2}}, \sqrt{\frac{\rho_{min}}{\rho_{max}}}\}$ (Remark 3)	$\max\{\sqrt[4]{\frac{1}{2}}, \sqrt{\frac{\rho_{min}}{\rho_{max}}}\}$ (Remark 3)	< 1 (Remark 3)
$\rho_1 \neq \rho_2,\ c_1 = c_2$	$\sqrt{\frac{\rho_{min}}{\rho_{max}}}$ (Corollary 1)	$\sqrt{\frac{\rho_{min}}{\rho_{max}}}$ (Remark 3)	$\sqrt{\frac{\rho_{min}}{\rho_{max}}}$ (Remark 3)	< 1 (Remark 3)

4 Conclusions

We studied the performance of optimized Schwarz methods for Helmholtz and Maxwell's equations for heterogeneous media. Using Fourier analysis, we showed that the convergence factor of the optimized Schwarz methods for the Helmholtz equation and the Maxwell's equations are the same, and it suffices therefore to study the algorithms only for the Helmholtz equation. We then studied in detail the performance for three different choices of the relationship between the wave number and the mesh size to control the pollution effect, and showed that increasing the resolution improves the performance of the optimized Schwarz methods. It was not possible to show all the proofs in detail in this short manuscript, but more information can be found in the PhD thesis (Veneros 2015).

Acknowledgements Hui Zhang was supported by Research Start Funding of Zhejiang Ocean University

References

A. Alonso-Rodriguez, L. Gerardo-Giorda, New nonoverlapping domain decomposition methods for the harmonic Maxwell system. SIAM J. Sci. Comput. **28**(1), 102–122 (2006)

Y. Boubendir, X. Antoine, C. Geuzaine, A quasi-optimal non-overlapping domain decomposition algorithm for the Helmholtz equation. J. Comput. Phys. **231**(2), 262–280 (2012)

Z. Chen, X. Xiang, A source transfer domain decomposition method for Helmholtz equations in unbounded domain. SIAM J. Numer. Anal. **51**(4), 2331–2356 (2013)

P. Chevalier, F. Nataf, An OO2 (Optimized Order 2) method for the Helmholtz and Maxwell equations, in *10th International Conference on Domain Decomposition Methods in Science and in Engineering* (American Mathematical Society, Providence, RI, 1997), pp. 400–407

B. Deprés, Décomposition de domaine et problème de Helmholtz. C. R. Acad. Sci. Paris **1**(6), 313–316 (1990)

B. Deprés, P. Joly, J.E. Roberts, A domain decomposition method for the harmonic Maxwell equations, in *Iterative Methods in Linear Algebra* (North-Holland, Amsterdam, 1992), pp. 475–484

V. Dolean, S. Lanteri, R. Perrussel, A domain decomposition method for solving the three-dimensional time-harmonic Maxwell equations discretized by discontinuous Galerkin methods. J. Comput. Phys. **227**(3), 2044–2072 (2008a)

V. Dolean, S. Lanteri, R. Perrussel, Optimized Schwarz algorithms for solving time-harmonic Maxwell's equations discretized by a discontinuous Galerkin method. IEEE. Trans. Magn. **44**(6), 954–957 (2008b)

V. Dolean, L. Gerardo-Giorda, M.J. Gander, Optimized Schwarz methods for Maxwell equations. SIAM J. Sci. Comput. **31**(3), 2193–2213 (2009)

V. Dolean, M. El Bouajaji, M.J. Gander, S. Lanteri, Optimized Schwarz methods for Maxwell's equations with non-zero electric conductivity, in *Domain Decomposition Methods in Science and Engineering XIX*. Lecture Notes in Computational Science and Engineering, vol. 78 (Springer, Heidelberg, 2011a), pp. 269–276

V. Dolean, M. El Bouajaji, M.J. Gander, S. Lanteri, R. Perrussel, Domain decomposition methods for electromagnetic wave propagation problems in heterogeneous media and complex domains, in *Domain Decomposition Methods in Science and Engineering XIX*, vol. 78 Lecture Notes in Computational Science Engineering (Springer, Heidelberg, 2011b), pp. 15–26

O. Dubois, Optimized Schwarz methods for the advection-diffusion equation and for problems with discontinuous coefficients. PhD thesis, McGill University, June (2007)

M. El Bouajaji, V. Dolean, M.J. Gander, S. Lanteri, Optimized Schwarz methods for the time-harmonic Maxwell equations with damping. SIAM J. Sci. Comput. **34**(4), A2048–A2071 (2012)

B. Engquist, L. Ying, Sweeping preconditioner for the Helmholtz equation: hierarchical matrix representation. Commun. Pure Appl. Math. **64**(5), 697–735 (2011)

O.G. Ernst, M.J. Gander, Why it is difficult to solve Helmholtz problems with classical iterative methods. in *Numerical Analysis of Multiscale Problems*, vol. 83 Lecture Notes in Computational Science and Engineering (Springer, Heidelberg, 2012), pp. 325–363

M.J. Gander, Optimized Schwarz methods for Helmholtz problems, in *Thirteenth International Conference on Domain Decomposition*, pp. 245–252, 2001

M.J. Gander, O. Dubois, Optimized Schwarz methods for a diffusion problem with discontinuous coefficient. Numer. Alg. **69**(1), 109–144 (2015)

M.J. Gander, F. Magoulès, F. Nataf, Optimized Schwarz methods without overlap for the Helmholtz equation. SIAM J. Sci. Comput. **24**(1), 38–60 (2002)

M.J. Gander, L. Halpern, F. Magoulès, An optimized Schwarz method with two-sided Robin transmission conditions for the Helmholtz equation. Int. J. Numer. Methods Fluids **55**(2), 163–175 (2007)

Z. Peng, J.F. Lee, Non-conformal domain decomposition method with second-order transmission conditions for time-harmonic electromagnetics. J. Comput. Phys. **229**(16), 5615–5629 (2010)

Z. Peng, V. Rawat, J.F. Lee, One way domain decomposition method with second order transmission conditions for solving electromagnetic wave problems. J. Comput. Phys. **229**(4), 1181–1197 (2010)

E. Veneros, Méthodes des décomposition de domaines pour des problèmes de propagation d'ondes heterogènes. PhD thesis, University of Geneva (2015)

E. Veneros V. Dolean, M.J. Gander, Optimized Schwarz methods for Maxwell equations with discontinuous coefficients. in *Domain Decomposition Methods in Science and Engineering XXI*. Lecture Notes in Computational Science and Engineering (Springer, Berlin, 2014), pp. 517–526

E. Veneros, V. Dolean, M.J. Gander, Schwarz methods for second order Maxwell equations in 3d with coefficient jumps, in *Domain Decomposition Methods in Science and Engineering XXII*. Lecture Notes in Computational Science and Engineering (Springer, Berlin, 2015)

On the Origins of Linear and Non-linear Preconditioning

Martin J. Gander

1 Linear Preconditioning

On December 26, 1823, Gauss (1903) sent a letter to his friend Gerling to explain how he computed an approximate least squares solution based on angle measurements between the locations Berger Warte, Johannisberg, Taufstein and Milseburg. The system is symmetric, see Fig. 1; it comes from the normal equations, and Gauss explains [translation by Forsythe (1951)]:

> In order to eliminate indirectly, I note that, if 3 of the quantities a, b, c, d are set to 0, the fourth gets the largest value when d is chosen as the fourth. Naturally, every quantity must be determined from its own equation, and hence d from the fourth. I therefore set $d = -201$ and substitute this value. The absolute terms then become: $+5232$, -6352, $+1074$, $+46$; the other terms remain the same.

With the new right hand side, Gauss then chooses again the variable to update which gives the largest value, and we recognize the well known Gauss-Seidel method, with the extra feature that at each step a particular variable is chosen to be updated, instead of just cycling through all the variables. Note also that the matrix is singular, but consistent (summing all equations gives zero, as indicated by Gauss' comment 'Summe=0' in Fig. 1), and the method gives one particular solution. Gauss concludes his letter with the statement in Fig. 2 [translation by Forsythe (1951)]:

> Almost every evening I make a new edition of the tableau, wherever there is easy improvement. Against the monotony of the surveying business, this is always a pleasant entertainment; one can also see immediately whether anything doubtful has crept in, what still remains to be desired, etc. I recommend this method to you for imitation. You will

M.J. Gander (✉)
Section de mathématiques, Université de Genève, Geneva, Switzerland
e-mail: martin.gander@unige.ch

© Springer International Publishing AG 2017

C.-O. Lee et al. (eds.), *Domain Decomposition Methods in Science and Engineering XXIII*, Lecture Notes in Computational Science and Engineering 116, DOI 10.1007/978-3-319-52389-7_14

Die Bedingungsgleichungen sind also:

$$0 = +\quad\ \ 6 + 67a - 13b - \ \ 28c - \ 26d$$
$$0 = -\ \ 7558 - 13a + 69b - \ \ 50c - \ \ 6d$$
$$0 = -14604 - 28a - 50b + 156c - \ 78d$$
$$0 = +22156 - 26a - \ \ 6b - \ \ 78c + 110d;$$

Summe $= 0$.

Um nun indirect zu eliminiren, bemerke ich, dass, wenn 3 der Grössen a, b, c, d gleich 0 gesetzt werden, die vierte den grössten Werth bekommt, wenn d dafür gewählt wird. Natürlich muss jede Grösse aus ihrer eigenen Gleichung, also d aus der vierten, bestimmt werden. Ich setze also $d = -201$ und substituire diesen Werth. Die absoluten Theile werden dann: $+5232$, -6352, $+1074$, $+46$; das Übrige bleibt dasselbe.

Fig. 1 Letter of Gauss from 1823 explaining what is now known as the Gauss-Seidel method

Fast jeden Abend mache ich eine neue Auflage des Tableaus, wo immer leicht nachzuhelfen ist. Bei der Einförmigkeit des Messungsgeschäfts gibt dies immer eine angenehme Unterhaltung; man sieht dann auch immer gleich, ob etwas zweifelhaftes eingeschlichen ist, was noch wünschenswerth bleibt, etc. Ich empfehle Ihnen diesen Modus zur Nachahmung. Schwerlich werden Sie je wieder direct eliminiren, wenigstens nicht, wenn Sie mehr als 2 Unbekannte haben. Das indirecte Verfahren lässt sich halb im Schlafe ausführen, oder man kann während desselben an andere Dinge denken.

Fig. 2 Gauss explains how relaxing these relaxations are

hardly ever again eliminate directly, at least not when you have more than 2 unknowns. The indirect procedure can be done while half asleep, or while thinking about other things.

A general description of the method was then given by Seidel (1874), who also proved convergence of the method for the case of the normal equations, proposed to do the relaxations cyclically, and also to distribute them to two computers (humans) to do parallel computing.[1]

In 1845, Jacobi (1845) presented the variant of Gauss' method now known as the Jacobi method, where one simultaneously relaxes all the variables. He acknowledges the computations that were performed by his friend Dr. Seidel. Realizing that the method can be slow or even fail if the system is not diagonally dominant enough, Jacobi then presents the groundbreaking idea of preconditioning using Jacobi rotations, see Fig. 3:

As an example we use the method for the equations from Theoria motus p. 219. The original equations are (see Fig. 3). If we remove the coefficient 6 in front of q in the first equation, the angle of rotation is $\alpha = 22^0 30'$, and the new equations are...

---

[1] "... sich unter zwei Rechner so vertheilen lässt ...".

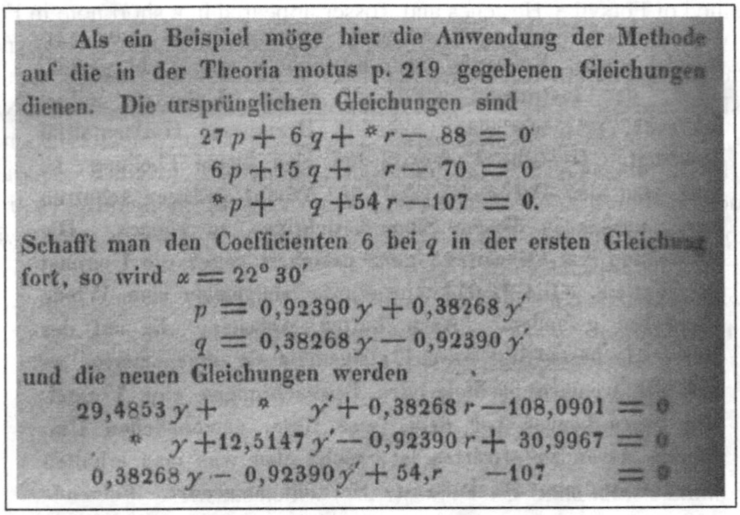

> Als ein Beispiel möge hier die Anwendung der Methode
> auf die in der Theoria motus p. 219 gegebenen Gleichungen
> dienen. Die ursprünglichen Gleichungen sind
>
> $$27\,p + 6\,q + {}^*r - 88 = 0$$
> $$6\,p + 15\,q + r - 70 = 0$$
> $${}^*p + q + 54\,r - 107 = 0.$$
>
> Schafft man den Coefficienten 6 bei q in der ersten Gleichung
> fort, so wird $\alpha = 22° 30'$
>
> $$p = 0{,}92390\,y + 0{,}38268\,y'$$
> $$q = 0{,}38268\,y - 0{,}92390\,y'$$
>
> und die neuen Gleichungen werden
>
> $$29{,}4853\,y + {}^* \quad y' + 0{,}38268\,r - 108{,}0901 = 0$$
> $${}^* \quad y + 12{,}5147\,y' - 0{,}92390\,r + 30{,}9967 = 0$$
> $$0{,}38268\,y - 0{,}92390\,y' + 54{,}r \quad -107 \quad = 0$$

Fig. 3 Jacobi's idea of preconditioning the linear system using Jacobi rotations

After preconditioning, it takes then only three Jacobi iterations to obtain three accurate digits!

In modern notation, a stationary iterative method for the linear system

$$A\mathbf{u} = \mathbf{f} \tag{1}$$

is obtained from a splitting of the matrix $A = M - N$, followed by the iteration

$$M\mathbf{u}^{n+1} = N\mathbf{u}^n + \mathbf{f}. \tag{2}$$

For Jacobi, we would have $M = \mathrm{diag}(A)$, for Gauss-Seidel $M = \mathrm{tril}(A)$, a Schwarz domain decomposition method with minimal overlap would have M block diagonal, and for multigrid, M represents a V-cycle or W-cycle. Rewriting the stationary iterative method (2) as

$$\mathbf{u}^{n+1} = M^{-1}N\mathbf{u}^n + M^{-1}\mathbf{f} = (I - M^{-1}A)\mathbf{u}^n + M^{-1}\mathbf{f},$$

we see that the method converges fast if the spectral radius $\rho(I - M^{-1}A)$ is small, and it is cheap, if systems with M can easily be solved.

In 1951, Stiefel and Rosser[2] gave both a presentation at a symposium on simultaneous linear equations and the determination of eigenvalues at the National Bureau of Standards (UCLA), and realized that they presented the same method.

---

[2]Rosser was working with Forsythe and Hestenes at that time.

The method of Forsythe, Hestenes and Rosser appeared in a short note in Forsythe et al. (1951), and the method of Stiefel in a comprehensive and elegant exposition on iterative methods in Stiefel (1952). Hestenes, who was also present at the symposium, and Stiefel then wrote together during Stiefel's stay at the National Bureau of Standards the famous 1952 conjugate gradient paper (Hestenes and Stiefel 1952).[3] Independently in 1952, Lanczos had also invented essentially the same method (Lanczos 1952), based on his earlier work on eigenvalues problems (Lanczos 1950), where he already pointed out that solving linear systems with this method was just a special case.

So what is this famous conjugate gradient (CG) method? To solve approximately $A\mathbf{u} = \mathbf{f}$, A symmetric and positive definite, CG finds at step n using the Krylov space[4]

$$\mathscr{K}_n(A, \mathbf{r}^0) := \{\mathbf{r}^0, A\mathbf{r}^0, \ldots, A^{n-1}\mathbf{r}^0\}, \quad \mathbf{r}^0 := \mathbf{f} - A\mathbf{u}^0$$

an approximate solution $\mathbf{u}^n \in \mathbf{u}^0 + \mathscr{K}_n(A, \mathbf{r}^0)$ which satisfies

$$||\mathbf{u} - \mathbf{u}^n||_A \longrightarrow \min, \qquad ||\mathbf{u}||_A^2 := \mathbf{u}^T A \mathbf{u}.$$

Using Chebyshev polynomials, one can prove the following convergence estimate for CG:

Theorem 1 *With $\kappa(A) := \frac{\lambda_{\max}(A)}{\lambda_{\min}(A)}$ the condition number of A, the iterate $\mathbf{u}^n$ of CG satisfies the convergence estimate*

$$||\mathbf{u} - \mathbf{u}^n||_A \leq 2 \left(\frac{\sqrt{\kappa(A)} - 1}{\sqrt{\kappa(A)} + 1} \right)^n ||\mathbf{u} - \mathbf{u}^0||_A.$$

We see that the conjugate gradient method converges very fast, if the condition number $\kappa(A)$ is not very large.

The success of CG motivated researchers to design similar methods searching in a Krylov space for solutions when the system matrix is not symmetric and positive definite. There are two classes of such methods: the first class are the Minimum Residual methods (MR) which search for $\mathbf{u}^n \in \mathbf{u}_0 + \mathscr{K}_n(A, \mathbf{r}^0)$ such that

$$||\mathbf{f} - A\mathbf{u}^n||_2 \longrightarrow \min.$$

MINRES (Paige and Saunders 1975) is such an algorithm, designed for symmetric systems which are not positive definite. GMRES (Saad and Schultz 1986) does the

---

[3]"An iterative algorithm is given for solving a system $Ax = k$ of n linear equations in n unknowns. The solution is given in n steps."

[4]The name is going back to Krylov (1931) studying the solution of systems of second order ordinary differential equations, and the now called Krylov space only appears implicitly there.

same for arbitrary systems, and QMR (Freund and Nachtigal 1991) tries to solve the minimization problem approximately. The second class of methods is based on orthogonalization (OR): they search for $\mathbf{u}^n \in \mathbf{u}_0 + \mathscr{K}_n(A, \mathbf{r}^0)$ such that

$$\mathbf{f} - A\mathbf{u}^n \perp \mathscr{K}_n(A, \mathbf{r}^0).$$

SymmLQ (Paige and Saunders 1975) does this for symmetric indefinite systems, FOM (Saad 1981) for general systems, and BiCGstab (Van der Vorst 1992) does it approximately. All these methods converge well, if the spectrum of the matrix A is clustered around 1 provided the matrices are normal ($AA^T = A^TA$).

If the spectrum of A is not clustered around 1, the old idea of Jacobi can be used: find a preconditioner, a matrix M, such that the preconditioned system

$$M^{-1}A\mathbf{u} = M^{-1}\mathbf{f}$$

has a spectrum which clusters much better around 1 than the spectrum of the matrix A itself. For CG, using Theorem 1 one can even say more specifically that M should make the condition number $\kappa(M^{-1}A)$ much smaller than $\kappa(A)$. In all cases however it should be inexpensive to apply M^{-1}.

It is sometimes possible to directly design preconditioners with good properties: excellent examples in domain decomposition are the additive Schwarz method (Dryja and Widlund 1987), FETI (Farhat and Roux 1991) and Balancing Domain Decomposition (Mandel and Brezina 1993), but it takes a lot of experience and intuition to do so.

A systematic approach for constructing preconditioners is to recall what we have seen for stationary iterative methods: we needed M such that the spectral radius $\rho(I - M^{-1}A)$ is small, and it is inexpensive to apply M^{-1}. The last point is identical with preconditioning, and note that

$$\rho(I - M^{-1}A)\text{small} \iff \text{the spectrum of } M^{-1}A \text{ is close to one!}$$

It is therefore natural to first design a good M for a stationary iterative method, and then use it as a preconditioner for a Krylov method.

Theorem 2 *Using an MR Krylov method with preconditioner M never gives worse (and usually much better) residual reduction than just using the stationary iteration.*

Proof The stationary iterative method computes

$$\mathbf{u}^n = (I - M^{-1}A)\mathbf{u}^{n-1} + M^{-1}\mathbf{f} = \mathbf{u}^{n-1} + \mathbf{r}_{stat}^{n-1},$$

where we introduced $\mathbf{r}_{stat}^n := M^{-1}\mathbf{f} - M^{-1}A\mathbf{u}^n$. Multiplying this equation by $-M^{-1}A$ and adding $M^{-1}\mathbf{f}$ on both sides then gives

$$\mathbf{r}_{stat}^n = (I - M^{-1}A)\mathbf{r}_{stat}^{n-1} = (I - M^{-1}A)^n\mathbf{r}^0. \tag{3}$$

The preconditioned Krylov method will use the Krylov space

$$\mathscr{K}_n(M^{-1}A, \mathbf{r}^0) := \{\mathbf{r}^0, M^{-1}A\mathbf{r}^0, \ldots, (M^{-1}A)^{n-1}\mathbf{r}^0\}$$

to search for $\mathbf{u}^n \in \mathbf{u}^0 + \mathscr{K}_n(M^{-1}A, \mathbf{r}^0)$, i.e. it will determine coefficients α_i s.t.

$$\mathbf{u}^n = \mathbf{u}^0 + \sum_{i=1}^{n} \alpha_i (M^{-1}A)^{i-1}\mathbf{r}^0.$$

Multiplying this equation by $-M^{-1}A$ and adding $M^{-1}\mathbf{f}$ on both sides then gives

$$\mathbf{r}_{kry}^n = p^n(M^{-1}A)\mathbf{r}^0, \tag{4}$$

p^n a polynomial of degree n with $p^n(0) = 1$. Since the MR Krylov method finds the polynomial which minimizes the residual in norm, it is at least as good as the specific polynomial $(I - M^{-1}A)^n$ chosen by the stationary iterative method in (3).

The classical alternating and parallel Schwarz methods are such stationary iterative methods, and also RAS (Cai and Sarkis 1999) and optimized Schwarz methods (Gander 2006), and the Dirichlet-Neumann and Neumann-Neumann methods (Quarteroni and Valli 1999). They all are convergent as stationary iterative methods, while for example additive Schwarz is not (Efstathiou and Gander 2003; Gander 2008).

2   Non-linear Preconditioning

In contrast to linear preconditioning, non-linear preconditioning is a much less explored area of research. In the context of domain decomposition, a seminal contribution for non-linear preconditioning was made by Cai, Keyes and Young at DD13 Cai et al. (2001), namely the Additive Schwarz Preconditioned Inexact Newton method (ASPIN), see also Cai and Keyes (2002). The idea is:

> The nonlinear system is transformed into a new nonlinear system, which has the same solution as the original system. For certain applications the nonlinearities of the new function are more balanced and, as a result, the inexact Newton method converges more rapidly.

Instead of solving $F(\mathbf{u}) = \mathbf{0}$, one solves instead $G(F(\mathbf{u})) = \mathbf{0}$ where according to the authors the function G should have the properties: 1) if $G(\mathbf{v}) = \mathbf{0}$ then $\mathbf{v} = \mathbf{0}$, 2) $G \approx F^{-1}$ in some sense, 3) $G(F(\mathbf{v}))$ is easy to compute, and 4) applying Newton, $(G(F(\mathbf{v})))'\mathbf{w}$ should also be easy to compute. The authors then define the ASPIN preconditioner as follows: for $F : \mathbb{R}^m \to \mathbb{R}^m$, define J (overlapping) subsets Ω_j for the indices $\{1, 2, \ldots, m\}$, such that $\bigcup_j \Omega_j = \{1, 2, \ldots, m\}$, and corresponding restriction matrices R_j, e.g. $\Omega_1 = \{1, 2, 3\} \implies R_1 = [I\ 0]_{3 \times m}$, I the 3×3 identity matrix. Define the solution operator $T_j : \mathbb{R}^m \to \mathbb{R}^{|\Omega_j|}$ such that

$$R_j F(\mathbf{v} - R_j^T T_j(\mathbf{v})) = 0. \tag{5}$$

Then ASPIN solves using inexact Newton

$$\sum_{j=1}^{J} R_j^T T_j(\mathbf{u}) = 0. \tag{6}$$

It is not easy to understand where this transformation comes from.[5] Let us first look at a fixed point iteration like Gauss-Seidel or Jacobi for this nonlinear problem. If we denote the unknowns corresponding to the subsets Ω_j by $\mathbf{u}_j$, the corresponding block Jacobi fixed point iteration would be to solve for $n = 0, 1, 2, \ldots$

$$
\begin{array}{ll}
F_1(\mathbf{u}_1^{n+1}, \mathbf{u}_2^n, \ldots, \mathbf{u}_J^n) = 0 & \mathbf{u}_1^{n+1} = G_1(\mathbf{u}_2^n, \ldots, \mathbf{u}_J^n) \\
F_2(\mathbf{u}_1^n, \mathbf{u}_2^{n+1}, \ldots, \mathbf{u}_J^n) = 0 & \mathbf{u}_2^{n+1} = G_2(\mathbf{u}_1^n, \mathbf{u}_3^n, \ldots, \mathbf{u}_J^n) \\
\quad \vdots & \quad \vdots \\
F_J(\mathbf{u}_1^n, \mathbf{u}_2^n, \ldots, \mathbf{u}_J^{n+1}) = 0 & \mathbf{u}_J^{n+1} = G_J(\mathbf{u}_1^n, \mathbf{u}_2^n, \ldots, \mathbf{u}_{J-1}^n)
\end{array}
\tag{7}
$$

where we denoted the solutions of the non-linear equation F_j by G_j. At the fixed point, which solves $F(\mathbf{u}) = 0$, we must have $\mathbf{u} = G(\mathbf{u})$, and thus instead of solving $F(\mathbf{u}) = 0$ using Newton's method, one can instead solve $\mathbf{u} - G(\mathbf{u}) = 0$ using Newton's method. This gives us a very general idea of non-linear preconditioning: one first designs a fixed point iteration (like the stationary iterative method in the linear case); but then one does not use this method directly, one applies Newton's method to the equation at the fixed point (like one applies a Krylov method to the fixed point of the stationary iterative method).

Theorem 3 *ASPIN in the case of no algebraic overlap (which means minimal geometric overlap of one mesh size) is identical to solving with an inexact Newton method the non-linear block Jacobi iteration equations at the fixed point.*

Proof The definition of the solution operator in (5) shows that we can use it to replace G_j in (7), namely

$$\mathbf{u}_j^{n+1} = R_j \mathbf{u}^n - T_j(\mathbf{u}^n).$$

Now in the case of no algebraic overlap (minimal geometric overlap), the sum in (6) just composes the operators T_j in a large vector, there is never actually a

---

[5] "ASPIN may look a bit complicated ..." (Cai and Keyes 2002).

sum computed, and thus (6) represents precisely (7) at the fixed point, i.e.

$$0 = \mathbf{u} - G(\mathbf{u}) = \mathbf{u} - \sum_{j=1}^{J} R_j^T (R_j \mathbf{u} - T_j(\mathbf{u})) = \sum_{j=1}^{J} R_j^T T_j(\mathbf{u}),$$

where we used that $\mathbf{u} - \sum_{j=1}^{J} R_j^T R_j \mathbf{u} = 0$ in the case of zero algebraic overlap.

Remark 1 In the case of more overlap, ASPIN has the same problem as the additive Schwarz method in the overlap, it is inconsistent and can only be used as a preconditioner (Efstathiou and Gander 2003; Gander 2008), where a Krylov method must correct this inconsistency. In the case of ASPIN, Newton must to the same; ASPIN then does not correspond to a consistent fixed point iteration in the case of more than minimal overlap.

3 Conclusion

We have explained how first stationary iterative methods were invented for linear systems of equations by Gauss and Jacobi, and how Jacobi had already the idea of preconditioning in 1845. With the invention of Krylov methods, stationary iterations have lost their importance as solvers, but good splittings from stationary iterative methods found great use as preconditioners for Krylov methods. In the case of non-linear problems, one can follow the same principle: one first conceives a fixed point iteration for the non-linear problem, like a non-linear iterative domain decomposition method, or the full approximation scheme from multigrid. One then however does not use this fixed point iteration as a solver, one solves instead the equations at the fixed point: *this is the meaning of non-linear preconditioning.* This observation allowed the authors in Dolean et al. (2016) to devise a new non-linear preconditioner called RASPEN, which avoids the problem ASPIN has in the overlap, and also introduces the coarse grid correction in a consistent way by using the full approximation scheme from multigrid. It is also shown in Dolean et al. (2016) that one can actually use the exact Jacobian, since the non-linear subdomain solvers provide this information already, and extensive numerical experiments in Dolean et al. (2016) show that RASPEN performs significantly better as non-linear preconditioner than ASPIN.

References

X.-C. Cai, D.E. Keyes, Nonlinearly preconditioned inexact Newton algorithms. SIAM J. Sci. Comput. **24**(1), 183–200 (2002)

X.-C. Cai, M. Sarkis, A restricted additive Schwarz preconditioner for general sparse linear systems. SIAM J. Sci. Comput. **21**(2), 792–797 (1999)

X.-C. Cai, D.E. Keyes, D.P. Young, A nonlinear additive Schwarz preconditioned inexact Newton method for shocked duct flow, in *Proceedings of the 13th International Conference on Domain Decomposition Methods*, pp. 343–350, 2001. DDM.org

V. Dolean, M.J. Gander, W. Kheriji, F. Kwok, R. Masson, Nonlinear preconditioning: How to use a nonlinear Schwarz method to precondition Newton's method. SIAM J. Sci. Comput. **38**(6), A3357–A3380 (2016)

M. Dryja, O. Widlund, An Additive Variant of the Schwarz Alternating Method for the Case of Many Subregions (Ultracomputer Research Laboratory, Courant Institute of Mathematical Sciences, Division of Computer Science, 1987)

E. Efstathiou, M.J. Gander, Why Restricted Additive Schwarz converges faster than Additive Schwarz. BIT Numer. Math. **43**(5), 945–959 (2003)

C. Farhat, F.-X. Roux, A method of finite element tearing and interconnecting and its parallel solution algorithm. Int. J. Numer. Methods Eng. **32**(6), 1205–1227 (1991)

G.E. Forsythe, Notes. Math. Tables Other Aids Comput. **5**(36), 255–258 (1951)

G.E. Forsythe, M.R. Hestenes, J.B. Rosser, Iterative methods for solving linear equations. Bull. Am. Math. Soc. **57**(6), 480–480 (1951)

R.W. Freund, N. Nachtigal, QMR: a quasi-minimal residual method for non-Hermitian linear systems. Numer. Math. **60**(1), 315–339 (1991)

M.J. Gander, Optimized Schwarz methods. SIAM J. Numer. Anal. **44**(2), 699–731 (2006)

M.J. Gander, Schwarz methods over the course of time. Electron. Trans. Numer. Anal. **31**(5), 228–255 (2008)

C.F. Gauss, Letter to Gerling, Dec 26, 1823, in *Werke*, vol. 9 (Göttingen, Berlin, 1903), pp. 278–281

M.R. Hestenes, E. Stiefel, Methods of conjugate gradients for solving linear systems, vol. 49. NBS (1952)

C.G.J. Jacobi, Ueber eine neue Auflösungsart der bei der Methode der kleinsten Quadrate vorkommenden lineären Gleichungen. Astron. Nachr. **22**(20), 297–306 (1845)

A.N. Krylov, On the numerical solution of the equation by which in technical questions frequencies of small oscillations of material systems are determined. Izvestija AN SSSR (News of Academy of Sciences of the USSR), Otdel. mat. i estest. nauk **7**(4), 491–539 (1931)

C. Lanczos, An iteration method for the solution of the eigenvalue problem of linear differential and integral operators. United States Government Press Office Los Angeles, CA (1950)

C. Lanczos, Solution of systems of linear equations by minimized iterations. J. Res. Nat. Bur. Standards **49**(1), 33–53 (1952)

J. Mandel, M. Brezina, Balancing domain decomposition: theory and computations in two and three dimensions. Technical Report UCD/CCM 2, Center for Computational Mathematics, University of Colorado at Denver (1993)

C.C. Paige, M.A. Saunders, Solution of sparse indefinite systems of linear equations. SIAM J. Numer. Anal. **12**(4), 617–629 (1975)

A. Quarteroni, A. Valli, *Domain Decomposition Methods for Partial Differential Equations* (Oxford Science Publications, New York, 1999)

Y. Saad, Krylov subspace methods for solving large unsymmetric linear systems. Math. Comput. **37**(155), 105–126 (1981)

Y. Saad, M.H. Schultz, GMRES: a generalized minimal residual algorithm for solving nonsymmetric linear systems. SIAM J. Sci. Stat. Comput. **7**(3), 856–869 (1986)

L. Seidel, Über ein Verfahren, die Gleichungen, auf welche die Methode der kleinsten Quadrate führt, sowie lineäre Gleichungen überhaupt, durch successive Annäherung aufzulösen, in *Abhandlungen der Mathematisch-Physikalischen Klasse der Königlich Bayerischen Akademie der Wissenschaften, Band 11, III. Abtheilung*, pp. 81–108 (1874)

E. Stiefel, Über einige Methoden der Relaxationsrechnung. Z. Angew. Math. Phys. **3**(1), 1–33 (1952)

H.A. Van der Vorst, Bi-CGSTAB: A fast and smoothly converging variant of Bi-CG for the solution of nonsymmetric linear systems. SIAM J. Sci. Stat. Comput. **13**(2), 631–644 (1992)

Time Parallelization for Nonlinear Problems Based on Diagonalization

Martin J. Gander and Laurence Halpern

1 Introduction

Over the last decade, an intensive research effort has been devoted to investigate the time direction in evolution problems for parallelization. This is because modern supercomputers have now so many processors that often space parallelization strategies for evolution problems saturate before all available processors can be used. In the relatively recent field of time parallelization, there are four main algorithmic techniques that have been investigated: methods based on multiple shooting (Chartier and Philippe 1993), like the parareal algorithm (Lions et al. 2001) for which a detailed convergence analysis can be found in Gander and Vandewalle (2007) for the linear case and in Gander and Hairer (2008) for the nonlinear case; methods based on space-time decomposition, like classical Schwarz waveform relaxation (Bjørhus 1995; Gander and Stuart 1998; Giladi and Keller 2002) and optimized variants (Bennequin et al. 2009; Gander and Halpern 2005, 2007; Gander et al. 2003), and Dirichlet-Neumann and Neumann-Neumann waveform relaxation (Gander et al. 2016b; Kwok 2014; Mandal 2014); space-time multigrid methods (Emmett and Minion 2012; Gander and Neumüller 2016; Hackbusch 1984; Horton and Vandewalle 1995); and direct time parallelization methods like tensor product methods (Maday and Rønquist 2008), RIDC (Christlieb et al. 2010), and ParaExp (Gander and Güttel 2013); for an up to date overview and a historical perspective of these approaches, see Gander (2015).

M.J. Gander
University of Geneva, Geneva, Switzerland
e-mail: martin.gander@unige.ch

L. Halpern (✉)
University Paris 13, Paris, France
e-mail: halpern@math.univ-paris13.fr

© Springer International Publishing AG 2017
C.-O. Lee et al. (eds.), *Domain Decomposition Methods in Science and Engineering XXIII*, Lecture Notes in Computational Science and Engineering 116, DOI 10.1007/978-3-319-52389-7_15

We have recently proposed and analyzed a new approach to make the tensor product time parallelization technique from Maday and Rønquist (2008) robust. For linear problems of diffusion type, we have derived in Gander et al. (2014) asymptotic estimates of the best choice of the main parameter in these methods, balancing truncation error and roundoff error, and the study for wave equations is in preparation (Gander et al. 2016a). These methods are however only applicable to linear problems. We propose here a new idea which permits these techniques also to be used for nonlinear problems.

2 Scalar Model Problem

We start with the nonlinear scalar model problem

$$u_t = f(u), \quad u(0) = u_0. \tag{1}$$

Discretization using a backward Euler method with variable time step leads to

$$\frac{u_n - u_{n-1}}{\Delta t_n} = f(u_n), \tag{2}$$

and writing this system over several time steps, we obtain

$$B\mathbf{u} := \begin{pmatrix} \frac{1}{\Delta t_1} & & & \\ -\frac{1}{\Delta t_2} & \frac{1}{\Delta t_2} & & \\ & \ddots & \ddots & \\ & & -\frac{1}{\Delta t_n} & \frac{1}{\Delta t_n} \end{pmatrix} \begin{pmatrix} u_1 \\ u_2 \\ \vdots \\ u_n \end{pmatrix} = \begin{pmatrix} f(u_1) + \frac{1}{\Delta t_1} u_0 \\ f(u_2) \\ \vdots \\ f(u_n) \end{pmatrix} =: \mathbf{f(u)}. \tag{3}$$

Parallelization in time based on diagonalization uses the assumption that B can be diagonalized, $B = SAS^{-1}$, which is possible if all the time steps are different. One then diagonalizes the system (3) in time,

$$\Lambda\hat{\mathbf{u}} := S^{-1}BSS^{-1}\mathbf{u} = S^{-1}\mathbf{f(u)}. \tag{4}$$

If the right-hand side is linear, $f(u) = au$, we get with $\mathbf{e}_1 := (1, 0, \ldots, 0)^T$

$$S^{-1}\mathbf{f(u)} = S^{-1}(a\mathbf{u} + \frac{u_0}{\Delta t_1}\mathbf{e}_1) = a\hat{\mathbf{u}} + \frac{u_0}{\Delta t_1}S^{-1}\mathbf{e}_1,$$

and the system is indeed diagonalized in time, and all time steps can be solved in parallel by a diagonal solve,

$$(\Lambda - aI)\hat{\mathbf{u}} = \frac{u_0}{\Delta t_1}S^{-1}\mathbf{e}_1.$$

The solution is then obtained by simply applying S,

$$\mathbf{u} = S\hat{\mathbf{u}}.$$

Since our problem is nonlinear however, it is not possible to directly diagonalize (4).

Since the discretized system (3) is nonlinear, we will have to apply an iterative method to solve it, e.g. we can apply Newton's method to

$$\mathbf{F}(\mathbf{u}) := B\mathbf{u} - \mathbf{f}(\mathbf{u}) = 0.$$

This leads with some initial guess $\mathbf{u}^0$ to the iteration

$$\mathbf{u}^m = \mathbf{u}^{m-1} - (F'(\mathbf{u}^{m-1}))^{-1}\mathbf{F}(\mathbf{u}^{m-1}).$$

Now the Jacobian is

$$F'(\mathbf{u}) = B - \mathrm{diag}(f'(u_1), f'(u_2), \ldots, f'(u_n)) =: B - D(\mathbf{u}).$$

The Newton iteration can thus be rewritten as

$$(B - D(\mathbf{u}^{m-1}))\mathbf{u}^m = (B - D(\mathbf{u}^{m-1}))\mathbf{u}^{m-1} - (B\mathbf{u}^{m-1} - \mathbf{f}(\mathbf{u}^{m-1}))$$
$$= \mathbf{f}(\mathbf{u}^{m-1}) - D(\mathbf{u}^{m-1})\mathbf{u}^{m-1}, \qquad (5)$$

and for a given iteration step $m - 1$, $\mathbf{u}^{m-1}$ is known. Denoting by $\tilde{B}^{m-1} := B - D(\mathbf{u}^{m-1})$ and $\tilde{\mathbf{f}}^{m-1} := \mathbf{f}(\mathbf{u}^{m-1}) - D(\mathbf{u}^{m-1})\mathbf{u}^{m-1}$, we have to solve at each iteration step of Newton the evolution problem

$$\tilde{B}^{m-1}\mathbf{u}^m = \tilde{\mathbf{f}}^{m-1}.$$

This can be done by diagonalization now, since it is a linear problem: having $\tilde{B}^{m-1} = \tilde{S}\tilde{\Lambda}\tilde{S}^{-1}$, we can solve

$$\tilde{\Lambda}\hat{\mathbf{u}}^m := \tilde{S}^{-1}\tilde{B}^{m-1}\tilde{S}\tilde{S}^{-1}\mathbf{u}^m = \tilde{S}^{-1}\tilde{\mathbf{f}}^{m-1}$$

for all $\hat{u}_j^m$, $j = 1, 2, \ldots, n$ in parallel.

A major disadvantage that is brought in by the nonlinear term is that one has to compute a factorization of the time stepping matrix $\tilde{B}^{m-1}$ at each Newton iteration. This could be avoided if we do not use the exact Jacobian at each Newton iteration, but an approximation which uses for example a scalar approximation of the diagonal matrix by averaging,

$$D(\mathbf{u}) \approx \frac{1}{n}\sum_{j=1}^{n} f'(u_j)I.$$

Now we can use the old factorization of the time stepping matrix B and solve in parallel at each quasi Newton step

$$(\Lambda - \frac{1}{n}\sum_{j=1}^{n} f'(u_j^{m-1})I)\hat{\mathbf{u}}^m = \tilde{S}^{-1}\mathbf{f}(\mathbf{u}^{m-1}) - \frac{1}{n}\sum_{j=1}^{n} f'(u_j^{m-1})\mathbf{u}^{m-1}. \tag{6}$$

Using this approximate Jacobian, the quasi Newton method will then however only converge linearly in general, and we will compare in the numerical section the two approaches to see how much is lost due to this approximation.

3   A PDE Model Problem

Suppose we want to solve the time dependent semi-linear heat equation

$$u_t = \Delta u + f(u), \quad u(0, x) = u^0(x), \tag{7}$$

with homogeneous Dirichlet boundary conditions. Using a standard five point finite difference discretization in space over a rectangular grid of size $J = J_1 J_2$, we obtain the discrete problem

$$\frac{\mathbf{u}_n - \mathbf{u}_{n-1}}{\Delta t_n} = \Delta_h \mathbf{u}_n + f(\mathbf{u}_n), \tag{8}$$

where now $\mathbf{u}_n$ and $\mathbf{u}_{n-1}$ are vectors in $\mathbb{R}^J$. As in the scalar case, we need to introduce an iteration to solve this nonlinear problem, but here the system has to be treated also by tensor products to separate space and time. Let I_t be the $N \times N$ identity matrix associated with the time domain and I_x be the $J \times J$ identity matrix associated with the spatial domain. Setting $\mathbf{u} := (\mathbf{u}_1, \ldots, \mathbf{u}_N)$, $\mathbf{f}(\mathbf{u}) := (f(\mathbf{u}_1) + \frac{1}{\Delta t_1}\mathbf{u}_0, f(\mathbf{u}_2), \cdots, f(\mathbf{u}_N))$, and using the Kronecker symbol, we can rewrite (8) as one large nonlinear system,

$$(B \otimes I_x)\mathbf{u} = (I_t \otimes \Delta_h)\mathbf{u} + \mathbf{f}(\mathbf{u}). \tag{9}$$

To solve (9) with an iterative method, one could for example apply Newton's method to solve

$$\mathbf{F}(\mathbf{u}) := (B \otimes I_x - I_t \otimes \Delta_h)\mathbf{u} - \mathbf{f}(\mathbf{u}) = 0.$$

To obtain the Jacobian needed, we define the diagonal matrix function

$$J(\mathbf{u}) := \begin{pmatrix} J_s(\mathbf{u}_1) & & \\ & \ddots & \\ & & J_s(\mathbf{u}_N) \end{pmatrix}, \tag{10}$$

where $J_s(\mathbf{u}_n) := \mathrm{diag}(f'(u_n^1), \cdots, f'(u_n^J)) \in \mathcal{M}_J(\mathbb{R})$. We can then write the Jacobian of $\mathbf{F}$ in compact form,

$$\mathbf{F}'(\mathbf{u}) = B \otimes I_x - I_t \otimes \Delta_h - J(\mathbf{u}).$$

Newton's method corresponds then to computing for $m = 1, 2, \ldots$

$$\left(B \otimes I_x - I_t \otimes \Delta_h - J(\mathbf{u}^{m-1})\right)(\mathbf{u}^m - \mathbf{u}^{m-1}) = f(\mathbf{u}^{m-1}) - (B \otimes I_x - I_t \otimes \Delta_h)\mathbf{u}^{m-1},$$

and we see that the linear terms cancel, so we can simplify to obtain

$$\left(B \otimes I_x - I_t \otimes \Delta_h - J(\mathbf{u}^{m-1})\right)\mathbf{u}^m = f(\mathbf{u}^{m-1}) - J(\mathbf{u}^{m-1})\mathbf{u}^{m-1}. \tag{11}$$

In contrast to the scalar case, where one could simply diagonalize at each Newton iteration a modified time stepping matrix $\tilde{B}^{m-1}$ to keep Newton's method without any approximation, this modified $\tilde{B}^{m-1}$ would here also depend on the space dimension now, and one would have to diagonalize a $\tilde{B}^{m-1}$ matrix at each spatial discretization point, which becomes prohibitive. So we perform a similar approximation as in the scalar case: we define

$$\tilde{J}(\mathbf{u}) := \frac{1}{N} \sum_{n=1}^{N} J_s(\mathbf{u}_n),$$

and obtain with this approximation the quasi-Newton algorithm

$$\left(B \otimes I_x - I_t \otimes (\Delta_h + \tilde{J}(\mathbf{u}^{m-1}))\right)\mathbf{u}^m = \mathbf{f}(\mathbf{u}^{m-1}) - (I_t \otimes \tilde{J}(\mathbf{u}^{m-1}))\mathbf{u}^{m-1}. \tag{12}$$

Now we can use the factorization $B = S\Lambda S^{-1}$, and defining

$$\tilde{\mathbf{f}}^{m-1} := \mathbf{f}(\mathbf{u}^{m-1}) - (I_t \otimes \tilde{J}(\mathbf{u}^{m-1}))\mathbf{u}^{m-1},$$

the quasi-Newton step (12) over all time steps can be parallelized in time by solving

$$(\Lambda \otimes I_x - I_t \otimes (\Delta_h + \tilde{J}(\mathbf{u}^{m-1})))\hat{\mathbf{u}}^m = (S^{-1} \otimes I_x)\tilde{\mathbf{f}}^{m-1}, \tag{13}$$

followed by computing $\mathbf{u}^m = (S \otimes I_x)\hat{\mathbf{u}}^m$.

4 Numerical Experiments

We first show a numerical experiment for the scalar model problem (1) where we chose either $f(u) = -u^2$ or $f(u) = \sqrt{u}$. We solve these problems on the time interval $(0, T)$ using N time steps on a geometrically stretched grid (Gander et al. 2014)

$$\Delta t_n := \frac{(1 + \varepsilon)^n}{\sum_{n=1}^{N}(1 + \varepsilon)^n} T,$$

with $T = 1$, $N = 10$, and initial condition $u(0) = 1$. We show in Fig. 1 on the left how the time parallel Newton method (5) and the Quasi-Newton method (6) converge for $\varepsilon = 0.05$. Although the approximation leads only to linear convergence, the first few steps lead already to a high accuracy approximation, like for the true Newton method. On the right in Fig. 1, we show how the accuracy at the end of the time interval is influenced by the stretching of the time grid determined by ε. For a highly anisotropic time grid, ε close to 1, the truncation error is bigger than for a time grid with equal time steps (Gander et al. 2014). When ε becomes too small however, then roundoff errors due to the diagonalization process lead to large errors, and an optimal choice has been determined asymptotically for linear problems in Gander et al. (2014). We can see on the right in Fig. 1 that there is also an optimal choice in the nonlinear case, and it seems to be very similar for the two examples we considered.

We next test the algorithm for the PDE model problem (7) using the same two nonlinear functions as for the scalar model problem, homogeneous boundary conditions and initial condition $u(0, x) = 1$. We discretize the Laplacian using a five point finite difference stencil with mesh size $h = 1/20$ and use the same time grid as for the scalar model problem. We show in Fig. 2 on the left how the

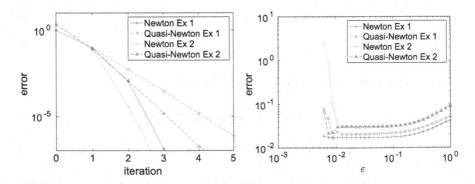

Fig. 1 *Left*: quadratic and linear convergence of the time parallel Newton and Quasi-Newton methods for two scalar model problems. *Right*: accuracy for different choices of the time grid stretching ε

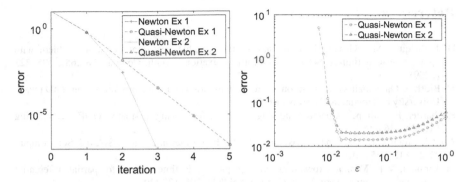

Fig. 2 *Left*: linear convergence of the time parallel Quasi-Newton method for two PDE model problems. *Right*: accuracy for different choices of the time grid stretching ε

Newton method (11) which can only be time parallelized at the cost of many time stepping matrix factorizations, and the Quasi-Newton method (13) that is easily time parallelized converge. Again the approximation still leads to a rapidly converging method. On the right in Fig. 2, we show how the accuracy at the end of the time interval is influenced by the stretching of the time grid in the PDE case, and again we see that there is an optimal choice for the stretching parameter.

5 Conclusion

We have introduced a new method which allows us to use diagonalization for time parallelization also for nonlinear problems. We have shown two variants for nonlinear scalar problems, and one for a nonlinear PDE. Numerical experiments show that the methods converge rapidly, and there is also an optimal choice of the geometric time grid stretching, like in the original algorithm for linear problems (Gander et al. 2014, 2016a). The geometric stretching is only one way to make diagonalization possible: random or adaptive time steps could also be used, but they must be determined for the entire time window before its parallel solve, and they must all be different, otherwise the diagonalization is not possible. In an adaptive setting, one could adaptively determine a macro time step with a larger tolerance as time window, before parallelizing its solve with smaller geometric or random time steps. We are currently investigating such variants, and also the generalization to nonlinear hyperbolic problems.

References

D. Bennequin, M.J. Gander, L. Halpern, A homographic best approximation problem with application to optimized Schwarz waveform relaxation. Math. Comput. **78**(265), 185–223 (2009)

M. Bjørhus, On domain decomposition, subdomain iteration and waveform relaxation. PhD thesis, University of Trondheim, Norway (1995)

P. Chartier, B. Philippe, A parallel shooting technique for solving dissipative ODEs. Computing **51**, 209–236 (1993)

A.J. Christlieb, C.B. Macdonald, B.W. Ong, Parallel high-order integrators. SIAM J. Sci. Comput. **32**(2), 818–835 (2010)

M. Emmett, M.L. Minion, Toward an efficient parallel in time method for partial differential equations. Commun. Appl. Math. Comput. Sci **7**(1), 105–132 (2012)

M.J. Gander, 50 years of time parallel time integration, in *Multiple Shooting and Time Domain Decomposition Methods* (Springer, Berlin, 2015), pp. 69–113

M.J. Gander, S. Güttel, Paraexp: a parallel integrator for linear initial-value problems. SIAM J. Sci. Comput. **35**(2), C123–C142 (2013)

M.J. Gander, E. Hairer, Nonlinear convergence analysis for the parareal algorithm. in *Domain Decomposition Methods in Science and Engineering XVII*, vol. 60 (Springer, Berlin, 2008), pp. 45–56

M.J. Gander, L. Halpern, Absorbing boundary conditions for the wave equation and parallel computing. Math. Comput. **74**, 153–176 (2005)

M.J. Gander, L. Halpern, Optimized Schwarz waveform relaxation methods for advection reaction diffusion problems. SIAM J. Numer. Anal. **45**(2), 666–697 (2007)

M.J. Gander, M. Neumüller, Analysis of a new space-time parallel multigrid algorithm for parabolic problems. SIAM J. Sci. Comput. **38**(4), A2173–A2208 (2016)

M.J. Gander, A.M. Stuart, Space-time continuous analysis of waveform relaxation for the heat equation. SIAM J. Sci. Comput. **19**(6), 2014–2031 (1998)

M.J. Gander, S. Vandewalle, Analysis of the parareal time-parallel time-integration method. SIAM J. Sci. Comput. **29**(2), 556–578 (2007)

M.J. Gander, L. Halpern, F. Nataf, Optimal Schwarz waveform relaxation for the one dimensional wave equation. SIAM J. Numer. Anal. **41**(5), 1643–1681 (2003)

M.J. Gander, L. Halpern, J. Ryan, T.T.B. Tran, A direct solver for time parallelization, in *22nd International Conference of Domain Decomposition Methods* (Springer, Berlin, 2014)

M.J. Gander, L. Halpern, J. Rannou, J. Ryan, A direct solver for time parallelization of the wave equation. (2016a, in preparation)

M.J. Gander, F. Kwok, B. Mandal, Dirichlet-Neumann and Neumann-Neumann waveform relaxation algorithms for parabolic problems. Electron. Trans. Numer. Anal. **45**, 424–456 (2016b)

E. Giladi, H.B. Keller, Space time domain decomposition for parabolic problems. Numer. Math. **93**(2), 279–313 (2002)

W. Hackbusch, Parabolic multi-grid methods, in *Computing Methods in Applied Sciences and Engineering, VI*, (North-Holland, Amsterdam, 1984), pp. 189–197

G. Horton, S. Vandewalle, A space-time multigrid method for parabolic partial differential equations. SIAM J. Sci. Comput. **16**(4), 848–864 (1995)

F. Kwok, Neumann–Neumann waveform relaxation for the time-dependent heat equation, in *Domain Decomposition Methods in Science and Engineering XXI* (Springer, Berlin, 2014), pp. 189–198

J.L. Lions, Y. Maday, G. Turinici, A parareal in time discretization of PDE's. C.R. Acad. Sci. Paris Ser. I **332**, 661–668 (2001)

Y. Maday, E.M. Rønquist, Parallelization in time through tensor-product space-time solvers. C. R. Math. Acad. Sci. Paris **346**(1–2), 113–118 (2008)

B. Mandal, A time-dependent Dirichlet-Neumann method for the heat equation, in *Domain Decomposition Methods in Science and Engineering, DD21* (Springer, Berlin, 2014)

The Effect of Irregular Interfaces on the BDDC Method for the Navier-Stokes Equations

Martin Hanek, Jakub Šístek, and Pavel Burda

1 Introduction

The Balancing Domain Decomposition based on Constraints (BDDC) was introduced by Dohrmann (2003) as an efficient method to solve large systems of linear equations arising from the finite element method on parallel computers. Dohrmann (2003) applied BDDC to elliptic problems, namely Poisson equation and linear elasticity. Li and Widlund (2006) extended the method to the Stokes equations. However, the approach requires a discontinuous approximation of the pressure. An attempt to apply the BDDC method in connection to a continuous approximation of the pressure was presented by Šístek et al. (2011) employing Taylor-Hood finite elements. Another construction of the BDDC preconditioner for the Stokes problem with a continuous approximation of the pressure was proposed by Li and Tu (2013).

Hanek et al. (2015) combined the approach to building the interface problem by Šístek et al. (2011) with the extension of BDDC to nonsymmetric problems from Yano (2009). The algorithm has been applied to linear systems obtained by Picard linearisation of the Navier-Stokes equations. One step of BDDC is applied as a preconditioner for the BiCGstab method. These generalizations have

M. Hanek • P. Burda
Faculty of Mechanical Engineering, Czech Technical University in Prague, Karlovo náměstí 13, CZ - 121 35 Prague 2, Czech Republic
e-mail: martin.hanek@fs.cvut.cz; pavel.burda@fs.cvut.cz

J. Šístek (✉)
Institute of Mathematics of the Czech Academy of Sciences, Žitná 25, CZ - 115 67 Prague 1, Czech Republic

School of Mathematics, The University of Manchester, M13 9PL Manchester, UK
e-mail: sistek@math.cas.cz

© Springer International Publishing AG 2017
C.-O. Lee et al. (eds.), *Domain Decomposition Methods in Science and Engineering XXIII*, Lecture Notes in Computational Science and Engineering 116, DOI 10.1007/978-3-319-52389-7_16

been implemented to our open-source parallel multilevel BDDC solver *BDDCML* described by Sousedík et al. (2013).

The main focus of this study is an investigation of the robustness of the algorithm of Hanek et al. (2015) with respect to interface irregularities and element aspect ratios. The motivation comes from simulations of hydrostatic bearings, where very bad element aspect ratios appear. A benchmark problem of a narrowing channel is proposed in two dimensions (2D) and three dimensions (3D), and numerical results for this problem are presented.

2   BDDC for Navier-Stokes Equations

In this section, we briefly recall our approach to using BDDC for steady Navier-Stokes problems. Details of the method can be found in Hanek et al. (2015).

A steady flow of an incompressible fluid in a two-dimensional (2-D) or three-dimensional (3-D) domain Ω is governed by the Navier-Stokes equations without body forces

$$(\mathbf{u} \cdot \nabla)\mathbf{u} - \nu \Delta \mathbf{u} + \nabla p = \mathbf{0} \quad \text{in } \Omega, \tag{1}$$

$$\nabla \cdot \mathbf{u} = 0 \quad \text{in } \Omega, \tag{2}$$

where $\mathbf{u}$ is an unknown velocity vector, p is an unknown pressure normalised by (constant) density, and ν is a given kinematic viscosity. In addition, the usual 'no-slip' boundary conditions $\mathbf{u} = \mathbf{g}$ on Γ_D and 'do-nothing' boundary conditions $-\nu(\nabla \mathbf{u})\mathbf{n} + p\mathbf{n} = 0$ on Γ_N are considered.

Applying the finite element method leads to a nonlinear system of algebraic equations [see e.g. Elman et al. (2005)]. For its linearisation, we use the Picard iteration and get the system

$$\begin{bmatrix} \nu A + N(\mathbf{u}^k) & B^T \\ B & 0 \end{bmatrix} \begin{bmatrix} \mathbf{u}^{k+1} \\ \mathbf{p}^{k+1} \end{bmatrix} = \begin{bmatrix} \mathbf{f} \\ \mathbf{g} \end{bmatrix}, \tag{3}$$

where $\mathbf{u}^{k+1}$ is the vector of unknown coefficients of velocity in the $(k + 1)$-th iteration, $\mathbf{p}^{k+1}$ is the vector of unknown coefficients of the pressure, A is the matrix of diffusion, $N(\mathbf{u}^k)$ is the matrix of the advection where we substitute velocity from the previous step, B is the matrix from the continuity equation, and $\mathbf{f}$ and $\mathbf{g}$ are discrete right-hand side vectors arising from the Dirichlet boundary conditions. This already linear nonsymmetric system is solved by means of iterative substructuring.

To this end, we decompose Ω into N_S nonoverlapping subdomains. Degrees of freedom shared by several subdomains form the *interface*, whereas the rest are in the interior of subdomains. Importantly, for the Taylor–Hood elements employed in this work, parts of both velocity and pressure unknowns form the interface, denoted $\mathbf{u}_\Gamma$ and $\mathbf{p}_\Gamma$, respectively (superscript $^{k+1}$ will be omitted).

By eliminating interior unknown coefficients for velocity and pressure on each subdomain, the local Schur complement S_i can be formed. Finally, a global Schur complement can be assembled as $S = \sum_{i=1}^{N_S} R_i^{\Gamma T} S_i R_i^{\Gamma}$, where R_i^{Γ} is the 0–1 matrix selecting the interface unknowns of the i-th subdomain from the global vector of interface unknowns. We then solve the problem

$$S \begin{bmatrix} \mathbf{u}_\Gamma \\ \mathbf{p}_\Gamma \end{bmatrix} = g, \tag{4}$$

where g is the reduced right-hand side vector. In our implementation, Schur complements are not actually constructed. Instead, only their actions on vectors are evaluated within each iteration of a Krylov method.

Problem (4) is solved by the BiCGstab method and one step of BDDC is used as a preconditioner. As usual, a coarse correction is combined with independent subdomain corrections in each action of the preconditioner. The main difference of the employed approach from the standard BDDC preconditioner as introduced by Dohrmann (2003) is the need of the *adjoint* coarse basis functions for mapping fine residuals to the coarse problem, following Yano (2009). This involves solving two saddle-point systems in the set-up phase of the preconditioner,

$$\begin{bmatrix} S_i & C_i^T \\ C_i & 0 \end{bmatrix} \begin{bmatrix} \Psi_i \\ \Lambda_i \end{bmatrix} = \begin{bmatrix} 0 \\ I \end{bmatrix}, \quad \begin{bmatrix} S_i^T & C_i^T \\ C_i & 0 \end{bmatrix} \begin{bmatrix} \Psi_i^* \\ \Lambda_i^T \end{bmatrix} = \begin{bmatrix} 0 \\ I \end{bmatrix},$$

where C_i is the matrix defining the local coarse degrees of freedom, which has as many rows as coarse degrees of freedom located in the subdomain. Finally, Ψ_i and Ψ_i^* are the matrices of standard and adjoint coarse basis functions.

As coarse degrees of freedom, we consider components of the velocity and the pressure at several *corners* selected according to Šístek et al. (2012), and arithmetic averages over edges and faces of subdomains. Constraints on their continuity in the coarse space are enforced component-wise on the velocities as well as on the pressure. The averaging at the interface unknowns applies diagonal matrix of weights to satisfy the partition of unity. The weights correspond to the inverse of the number of subdomains containing an interface unknown in this work.

3 Mesh Partitioning

We compare two approaches to partitioning the computational domain and the mesh into subdomains. A standard approach is based on a conversion of the computational mesh into a graph. In the so-called dual graph, the finite elements represent vertices of the graph and if two elements share an edge (in 2D) or a face (in 3D), the corresponding graph vertices are connected by a graph edge. The task of partitioning a mesh is translated into a problem of dividing a graph into subgraphs, with the goal that the subgraphs contain approximately the same number of vertices and the number of edges connecting the subgraphs is minimized. We make use of the *METIS* library (version 4.0) for this purpose.

Graph partitioning provides an automated way for dividing the computational mesh into subdomains of well-balanced sizes even for complex geometries and meshes. However, information about the geometry of the interface is lost during the conversion into a graph, and the resulting interface can be very irregular. This is a known issue studied mathematically for elliptic problems e.g. by Klawonn et al. (2008).

Another, somewhat opposite, strategy is based on the geometry of the domain. The domain can be enclosed into its cuboidal bounding box $[x_{min}, x_{max}] \times [y_{min}, y_{max}] \times [z_{min}, z_{max}]$. Two subdomains are created by bisecting the box into halves, with the cutting plane perpendicular to the longest edge. In the recursive bisection (RCB) algorithm, the longest subdomain edge is found as the maximum over subdomains, and one of the adjacent subdomains is bisected. This process is repeated until the given number of subdomains is reached.

This algorithm does not work well for complex unstructured meshes, since the strategy ignores numbers of elements in each block, and it can even create 'empty subdomains' with no elements. Nevertheless, for simple cuboidal domains, it is straightforward to produce a partition avoiding irregular interfaces. For a suitable number of subdomains and regular meshes, subdomain sizes are well-balanced in addition. In the rest of the paper, we refer to this strategy as the *geometric* partitioner.

Many geometries, including those of the hydrostatic bearings we aim at, are not completely general and can be decomposed into several cuboidal blocks in the first stage. In the second stage, each of these blocks can be partitioned as above.

4 Numerical Results

Our computations aim at the influence of interface irregularities on the BDDC solver for Navier-Stokes equations. In particular, we investigate the effect of the aspect ratio of the finite elements at the interface on convergence. This is motivated by our target application—simulations of oil flow in hydrostatic bearings with very narrow throttling gaps. In order to study this phenomenon, a benchmark problem suitable for such a study is proposed and the partitioning strategies described in Sect. 3 are compared.

The computations are performed by a parallel finite element package written in C++ and described by Šístek and Cirak (2015), with the *BDDCML* library being used for solving the arising systems of linear equations. The Picard iteration is terminated based on the change of subsequent solutions when $\left\| u^k - u^{k-1} \right\|_2 \leq 10^{-5}$ or after performing 100 iterations. The BiCGstab method is stopped based on the relative residual if $\left\| r^k \right\|_2 / \left\| g \right\|_2 \leq 10^{-6}$, with the limit of 1000 iterations.

As a measure of convergence, we monitor the number of BiCGstab iterations needed in one Picard iteration. Two matrix-vector multiplications are needed in each

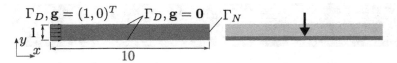

Fig. 1 The narrowing channel 2-D benchmark; original channel (*left*) and narrowing along the *y*-axis (*right*)

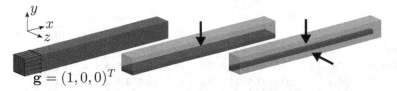

Fig. 2 The narrowing channel 3-D benchmark; original channel (*left*), narrowing along the *y*-axis (*centre*), and narrowing along both *y* and *z*-axes (*right*)

iteration of BiCGstab, and after each of them, the terminal condition is evaluated. Correspondingly, inspired by the Matlab `bicgstab` function, termination after the first matrix-vector multiplication is reported by a half iteration in the BiCGstab iteration counts. Numbers of iterations are presented as minimum, maximum, and mean over all nonlinear iterations for a given case.

The benchmark problem consists of a sequence of simple channels in 2D (Fig. 1) and 3D (Fig. 2). The dimension of the channels along one or two (in 3D) coordinates is gradually decreased, with the initial dimensions 10×1×1 along the *x*, *y*, and *z* axes.

The computational mesh is based on rectangular (in 2D) or cuboidal (in 3D) finite elements uniformly distributed along each direction. The number of elements is $100 \times 10 \times 10$ along the *x*, *y* and *z* coordinates. In total, the 3-D problem contains 10,000 elements, 88,641 nodes, and 278,144 unknowns.

The *aspect ratio of elements* $\mathcal{R} = h_{max}/h_{min}$ is defined as the ratio of the longest edge of the element h_{max} to its shortest counterpart h_{min}. The $\mathcal{R} = 1$ corresponds to square (or cubic) elements. We test the sequence of narrowing channels for $\mathcal{R} \in \{1, 2, 4, 10, 20, 40, 100\}$.

The velocity at the inlet starts from $\mathbf{g} = (1, 0, 0)^T$ for $x = 0$, the velocity at the walls is fixed to $\mathbf{g} = \mathbf{0}$, and the face of the channel for $x = 10$ corresponds to Γ_N. We have considered two scenarios for the inflow velocity during the narrowing. The first is simply keeping the magnitude of the velocity fixed throughout the sequence. In the second scenario, the magnitude of the velocity is increased proportionally to the decrease of the height, so that the Reynolds number, defined as $Re = \frac{|\mathbf{u}|D}{\nu}$, is kept constant for the decreasing channel height D. However, results for both scenarios of the inlet boundary condition have been almost identical, and we present only the results for fixed Reynolds number for brevity. We use $\nu = 1$ for our computations. The channel is divided into four subdomains by the graph and the geometric partitioners described in Sect. 3.

Fig. 3 Detail of the interface between two subdomains in 2D for graph (*left*) and geometric (*right*) partitioner

Table 1 Numbers of iterations for graph and geometric partitioners for 2-D narrowing channel

Partitioner		Graph							Geometric						
$\mathcal{R}$		1	2	4	10	20	40	100	1	2	4	10	20	40	100
Picard its.		4	4	5	5	7	6	40	3	4	5	5	6	6	5
BiCGstab its.	Min	9	10.5	13.5	13.5	15	16.5	17.5	4.5	4.5	4.5	4	3	3	3
	Max	9.5	10.5	13.5	15	16	17.5	19.5	4.5	4.5	4.5	4	3	3	3
	Mean	9.4	10.5	13.5	14.2	15.2	16.7	18.1	4.5	4.5	4.5	4	3	3	3

Fig. 4 Detail of the interface between two subdomains for narrowing along the *y*-coordinate in 3D for graph (*left*) and geometric (*right*) partitioner

First we look at the two-dimensional problem. For the graph partitioner, the interface contains both long and short edges of elements. On the other hand, the interface is composed solely from short edges for the geometric partitioner (see Fig. 3). Corresponding results are in Table 1.

For the 3-D case, we consider two kinds of problems. First we decrease only the *y*-dimension of the channel, while in the second case, we shrink both *y* and *z* dimensions of the cross-section (see Fig. 2). The graph partitioner produces rough interface in both cases, while the geometric partitioner leads to rectangular faces at the interface in the first case (see Fig. 4) and square faces in the second case. Resulting numbers of iterations are presented in Tables 2 and 3. Numbers in italic are runs that did not converge due to reaching the maximal number of iterations or time restrictions. A solution of the problem for the initial channel geometry is presented in Fig. 5.

From Tables 1, 2, and 3 we can conclude that $\mathcal{R}$ of faces at the interface has a remarkable influence on the number of BiCGstab iterations in each Picard iteration.

Using the graph partitioner results in a rough interface combining long and short edges. This has a large impact on the efficiency of the BDDC preconditioner and the number of linear iterations increases significantly.

Table 2 Numbers of iterations for graph and geometric partitioners for 3-D channel narrowed along the y-coordinate

Partitioner		Graph							Geometric						
$\mathcal{R}$		1	2	4	10	20	40	100	1	2	4	10	20	40	100
Picard its.		4	5	5	42	5	*100*	*100*	4	5	5	5	5	5	99
BiCGstab its.	Min	17.5	20	25.5	44.5	84.5	145	400	5.5	6.5	7.5	11.5	16	19.5	19.5
	Max	18.5	20.5	25.5	51	113.5	858	*1000*	5.5	6.5	7.5	12	17.5	19.5	21
	Mean	18.3	20.4	25.5	46.2	93.9	209	761	5.5	6.5	7.5	11.9	17.2	19.5	19.5

Table 3 Numbers of iterations for graph and geometric partitioners for 3-D channel narrowed along both y and z-coordinates

Partitioner		Graph							Geometric						
$\mathcal{R}$		1	2	4	10	20	40	100	1	2	4	10	20	40	100
Picard its.		4	4	4	5	8	19	28	4	4	4	5	5	5	4
BiCGstab its.	Min	17.5	19.5	27.5	36	51	80	197	5.5	5.5	6	5	4.5	4.5	4.5
	Max	18.5	20.5	28	41.5	53	92.5	*1000*	5.5	6	6	5.5	5	5	4.5
	Mean	18.3	19.8	27.9	39.5	51.8	87.7	590	5.5	5.9	6	5.1	4.9	4.6	4.5

Fig. 5 Solution in the initial 3-D channel geometry; magnitude of velocity (*left*) and pressure in the plane of symmetry (*right*)

Employing the geometric partitioner leads to straight cuts between subdomains aligned with layers of elements. In 2D, this is sufficient to achieve convergence of the linear solver independent of $\mathcal{R}$. In 3D, the situation is more subtle. For the case of narrowing the channel only along the y-axis, the aspect ratio of the rectangular element faces at the interface also worsens during contracting the channel. This is translated into a slight growth of the number of BiCGstab iterations in Table 2 even in this case, although the convergence is much more favourable than for the graph partitioner. If we narrow the channel along both y and z coordinates, the shape of the element faces at the interface does not deteriorate from squares, and we observe fast convergence independent of $\mathcal{R}$ in Table 3.

5 Conclusion

We have investigated the influence of an irregular interface on the performance of the BDDC method for Navier-Stokes equations. A benchmark problem of a narrowing channel in 2D and 3D has been proposed to evaluate the impact of aspect ratios of finite elements on the convergence of iterative solvers for the arising system of equations. A simple partitioning strategy based on an application of a regular geometric division of simple sub-blocks of the computational mesh has been presented. This approach was applied to the benchmark channel problems. The number of BiCGstab iterations required when using the geometric partitioner has been compared to the number of iterations required when using a graph partitioner. This rather simple idea has dramatically improved convergence of our *BDDCML* solver. Our next aim is to apply the idea to real geometries of hydrostatic bearings with block structured meshes. The preliminary results in this direction are very promising.

Acknowledgements We are grateful to Fehmi Cirak for valuable discussions on domain partitioning by the recursive bisection algorithm, and to Santiago Badia for discussing the impact of element aspect ratios on convergence of BDDC. This work was supported by the Czech Technical University in Prague through the student project SGS16/206/OHK2/3T/12, by the Czech Science Foundation through grant 14-02067S, and by the Czech Academy of Sciences through RVO:67985840.

References

C.R. Dohrmann, A preconditioner for substructuring based on constrained energy minimization. SIAM J. Sci. Comput. **25**(1), 246–258 (2003)

H.C. Elman, D.J. Silvester, A.J. Wathen, *Finite Elements and Fast Iterative Solvers: With Applications in Incompressible Fluid Dynamics*. Numerical Mathematics and Scientific Computation (Oxford University Press, New York, 2005)

M. Hanek, J. Šístek, P. Burda, An application of the BDDC method to the Navier-Stokes equations in 3-D cavity, in *Proceedings of Programs and Algorithms of Numerical Mathematics 17, Dolní Maxov, June 8–13, 2014* ed. by J. Chleboun, P. Přikryl, K. Segeth, J. Šístek, T. Vejchodský (Institute of Mathematics AS CR, 2015), pp. 77–85

A. Klawonn, O. Rheinbach, O.B. Widlund, An analysis of a FETI-DP algorithm on irregular subdomains in the plane. SIAM J. Numer. Anal. **46**(5), 2484–2504 (2008)

J. Li, X. Tu, A nonoverlapping domain decomposition method for incompressible Stokes equations with continuous pressures. SIAM J. Numer. Anal. **51**(2), 1235–1253 (2013)

J. Li, O.B. Widlund, BDDC algorithms for incompressible Stokes equations. SIAM J. Numer. Anal. **44**(6), 2432–2455 (2006)

J. Šístek, F. Cirak, Parallel iterative solution of the incompressible Navier-Stokes equations with application to rotating wings. Comput. Fluids **122**, 165–183 (2015)

J. Šístek, B. Sousedík, P. Burda, J. Mandel, J. Novotný, Application of the parallel BDDC preconditioner to the Stokes flow. Comput. Fluids **46**, 429–435 (2011)

J. Šístek, M. Čertíková, P. Burda, J. Novotný, Face-based selection of corners in 3D substructuring. Math. Comput. Simul. **82**(10), 1799–1811 (2012)

B. Sousedík, J. Šístek, J. Mandel, Adaptive-multilevel BDDC and its parallel implementation. Computing **95**(12), 1087–1119 (2013)

M. Yano, Massively parallel solver for the high-order Galerkin least-squares method. Master's thesis, Massachusests Institute of Technology (2009)

BDDC and FETI-DP Methods with Enriched Coarse Spaces for Elliptic Problems with Oscillatory and High Contrast Coefficients

Hyea Hyun Kim, Eric T. Chung, and Junxian Wang

1 Introduction

BDDC (Balancing Domain Decomposition by Constraints) and FETI-DP (Dual-Primal Finite Element Tearing and Interconnecting) algorithms with adaptively enriched coarse spaces are developed and analyzed for second order elliptic problems with high contrast and random coefficients. Among many approaches to form adaptive coarse spaces, we consider an approach using eigenvectors of generalized eigenvalues problems defined on each subdomain interface, see Mandel and Sousedík (2007), Galvis and Efendiev (2010), Spillane et al. (2011), Spillane et al. (2013), Klawonn et al. (2015).

The main contribution of the current work is to extend the methods in Dohrmann and Pechstein (2013), Klawonn et al. (2014) to three-dimensional problems. In three dimensions, there are three types of equivalence classes on the subdomain interfaces, i.e., faces, edges, and vertices. A face is shared by two subdomains. An edge is shared by more than two subdomains. Vertices are end points of edges.

H.H. Kim (✉)
Department of Applied Mathematics and Institute of Natural Sciences, Kyung Hee University, Yongin, Korea
e-mail: hhkim@khu.ac.kr

E.T. Chung
Department of Mathematics, The Chinese University of Hong Kong, Sha Tin, Hong Kong SAR
e-mail: tschung@math.cuhk.edu.hk

J. Wang
Department of Mathematics, The Chinese University of Hong Kong, Sha Tin, Hong Kong SAR

School of Mathematics and Computational Science, Xiangtan University, 411105 Xiangtan, Hunan, China
e-mail: wangjunxian@xtu.edu.cn

© Springer International Publishing AG 2017   179
C.-O. Lee et al. (eds.), *Domain Decomposition Methods in Science
and Engineering XXIII*, Lecture Notes in Computational Science
and Engineering 116, DOI 10.1007/978-3-319-52389-7_17

In addition to the generalized eigenvalue problems on faces, which are already considered in Dohrmann and Pechstein (2013), Klawonn et al. (2014) for two-dimensional problems, generalized eigenvalues problems on edges are proposed.

Equipped with the coarse space formed by using the selected eigenvectors, the condition numbers of the resulting algorithms are determined by the user defined tolerance value λ_{TOL} that is used to select the eigenvectors. An estimate of condition numbers is obtained as $C\lambda_{TOL}$, where the constant C is independent of coefficients and any mesh parameters. We note that a full version of the current paper was submitted to a journal. We also note that an adaptive BDDC algorithm for three-dimensional problems was considered and numerically tested in Mandel et al. (2012) for difficult engineering applications.

This paper is organized as follows. A brief description of BDDC and FETI-DP algorithms is given in Sect. 2. Adaptive selection of coarse spaces is presented in Sect. 3 and the estimate of condition numbers of the both algorithms is provided in Sect. 4.

2   BDDC and FETI-DP Algorithms

To present BDDC and FETI-DP algorithms, we introduce a finite element space $\widehat{X}$ for a given domain Ω, where the model elliptic problem is defined as

$$- \nabla \cdot (\rho(x)\nabla u(x)) = f(x) \tag{1}$$

with a boundary condition on $u(x)$ and with $\rho(x)$ highly varying and random. The domain is then partitioned into non-overlapping subdomains $\{\Omega_i\}$ and X_i are the restrictions of $\widehat{X}$ to Ω_i. The subdomain interfaces are assumed to be aligned to the given triangles in X. In three dimensions, the subdomain interfaces consist of faces, edges, and vertices. We introduce W_i as the restriction of X_i to the subdomain interface unknowns, W and X as the product of the local finite element spaces W_i and X_i, respectively. We note that functions in W or X are decoupled across the subdomain interfaces. We then select some primal unknowns among the decoupled unknowns on the interfaces and enforce continuity on them and denote the corresponding spaces $\widetilde{W}$ and $\widetilde{X}$.

The preconditioners in BDDC and FETI-DP algorithms will be developed based on the partially coupled space $\widetilde{W}$ and appropriate scaling matrices. We refer to Dohrmann (2003), Farhat et al. (2001), Li and Widlund (2006) for general introduction of these algorithms. The unknowns at subdomain vertices will first be included in the set of primal unknowns. Additional set of primal unknowns will be selected by solving generalized eigenvalue problems on faces and edges. In the BDDC algorithm, they are enforced just like unknowns at subdomain vertices after a change of basis, while in the FETI-DP algorithm they are enforced by using a projection, see Klawonn et al. (2015).

We next define the matrices K_i and S_i. The matrices K_i are obtained from the Galerkin approximation of

$$a(u, v) = \int_{\Omega_i} \rho(x) \nabla u \cdot \nabla v \, dx$$

by using finite element spaces X_i and S_i are the Schur complements of K_i, that are obtained after eliminating unknowns interior to Ω_i. Let $\widetilde{R}_i : \widetilde{W} \to W_i$ be the restriction into $\partial \Omega_i \setminus \partial \Omega$ and let $\widetilde{S}$ be the partially coupled matrix defined by

$$\widetilde{S} = \sum_{i=1}^{N} \widetilde{R}_i^T S_i \widetilde{R}_i.$$

Let $\widehat{R}$ be the restriction from $\widehat{W}$ to $\widetilde{W}$. The discrete problem of (1) is then written as

$$\widehat{R}^T \widetilde{S} \widehat{R} = \widehat{R}^T \widetilde{g},$$

where $\widetilde{g}$ is the vector given by the right hand side $f(x)$. The above matrix equation can be solved iteratively by using preconditioners. The BDDC preconditioner is then given by

$$M_{BDDC}^{-1} = \widehat{R}^T \widetilde{D} \widetilde{S}^{-1} \widetilde{D}^T \widehat{R},$$

where $\widetilde{D}$ is a scaling matrix of the form

$$\widetilde{D} = \sum_{i=1}^{N} \widetilde{R}_i^T D_i \widetilde{R}_i.$$

Here the matrices D_i are defined for unknowns in W_i and they are introduced to resolve heterogeneity in $\rho(x)$ across the subdomain interface. In more detail, D_i consists of blocks $D_F^{(i)}$, $D_E^{(i)}$, $D_V^{(i)}$, where F denotes an equivalence class shared by two subdomains, i.e., Ω_i and its neighboring subdomain Ω_j, E denotes an equivalence class shared by more than two subdomains, and V denotes the end points of E, respectively. We note that those blocks should satisfy the partition of unity for a given F, E, and V, respectively, and call them faces, edges, and vertices, respectively. We refer to Klawonn and Widlund (2006) for these definitions.

The FETI-DP preconditioner is a dual form of the BDDC preconditioner. In our case, the unknowns at subdomain vertices are chosen as the initial set of primal unknowns and the algebraic system of the FETI-DP algorithm is obtained as

$$B \widetilde{S}^{-1} B^T \lambda = d,$$

where $\widetilde{S}$ is the partially coupled matrix at subdomain vertices and B is a matrix with entries 0, -1, and 1, which is used to enforce continuity at the decoupled interface unknowns. The above algebraic system is then solved by an iterative method with the following projected preconditioner

$$M_{FETI}^{-1} = (I - P)B_D\widetilde{S}B_D^T(I - P^T),$$

where B_D is defined by

$$B_D = \left(B_{D,\Delta}\ 0\right) = \left(B_{D,\Delta}^{(1)} \cdots B_{D,\Delta}^{(i)}\ 0\right).$$

In the above, $B_{D,\Delta}^{(i)}$ is a scaled matrix of $B_\Delta^{(i)}$ where rows corresponding to Lagrange multipliers to the unknowns $w^{(i)} \in W_i$ are multiplied with a scaling matrix $(D_C^{(j)})^T$ when the Lagrange multipliers connect $w^{(i)}$ to $w^{(j)} \in W_j$ and Ω_j is the neighboring subdomain sharing the interface C of $\partial\Omega_i$. The interface C can be F, faces, or E, edges. The matrix P is a projection operator related to the additional primal constraints and it is given by

$$P = U(U^T F_{DP} U)^{-1} U^T F_{DP},$$

where $F_{DP} = B\widetilde{S}^{-1}B^T$ and U consists of columns related to the additional primal constraints on the decoupled interface unknowns.

3   Adaptively Enriched Coarse Spaces

With the standard choice of primal unknowns, values at subdomain vertices, edge averages, and face averages, the performance of BDDC and FETI-DP preconditioners can often deteriorate for bad arrangements of the coefficient $\rho(x)$. The preconditioner can be enriched by using adaptively chosen primal constraints. The adaptive constraints will be selected by considering generalized eigenvalue problems on each equivalence class. The idea is originated from the upper bound estimate of BDDC and FETI-DP preconditioners. In the estimate of condition numbers of BDDC and FETI-DP preconditioners, the average and jump operators are defined as

$$E_D = \widetilde{R}\widetilde{R}^T\widetilde{D}, \quad P_D = B_D^T B.$$

When adaptive constraints are introduced, they are enforced strongly just like unknowns at vertices after a change of basis formulation in the BDDC algorithm. In contrast, in the FETI-DP algorithm the additional constraints are enforced weakly by using a projection P. In general, $E_D + P_D = I$ does not hold when adaptively enriched constraints are included in the preconditioners. Thus the analysis of BDDC and FETI-DP algorithms requires the following estimates, respectively,

$$\langle \widetilde{S}(I - E_D)\widetilde{w}_a, (I - E_D)\widetilde{w}_a \rangle \leq C \langle \widetilde{S}\widetilde{w}_a, \widetilde{w}_a \rangle,$$

$$\langle \widetilde{S}P_D\widetilde{w}, P_D\widetilde{w} \rangle \leq C \langle \widetilde{S}\widetilde{w}, \widetilde{w} \rangle.$$

In the above, $\widetilde{w}_a$ is strongly coupled at the initial set of primal unknowns and the adaptively enriched primal unknowns after the change of basis while $\widetilde{w}$ is strongly coupled at the initial set of primal unknowns and satisfies the adaptive constraints across the subdomain interfaces, $v^T(w_i - w_j) = 0$ with v a vector of an adaptive constraint.

For a face F, shared by two subdomains Ω_i and Ω_j, we restrict the operator $I - E_D$ to $F \subset \partial\Omega_i$ and obtain

$$((I - E_D)\widetilde{w}_a)|_F = D_F^{(j)}(\widetilde{w}_{F,\Delta}^{(i)} - \widetilde{w}_{F,\Delta}^{(j)}), \tag{2}$$

where $\widetilde{w}_{F,\Delta}^{(i)}$ denotes the vector of unknowns on $F \subset \partial\Omega_i$ with zero primal unknowns and the dual unknowns identical to $\widetilde{w}_a$. Similarly, for an edge $E \subset \partial\Omega_i$,

$$((I - E_D)\widetilde{w}_a)|_E = \sum_{m \in E(i)} D_E^{(m)}(\widetilde{w}_{E,\Delta}^{(i)} - \widetilde{w}_{E,\Delta}^{(m)}),$$

where $E(i)$ denotes the set of subdomain indices sharing the edge E with Ω_i. We now introduce a Schur complement matrix $\widetilde{S}_C^{(i)}$ of S_i, which are obtained after eliminating unknowns except those interior to C. Here C can be an equivalence class, F or E. For semi-positive definite matrices A and B, we introduce a parallel sum defined as, see Anderson and Duffin (1969),

$$A : B = A(A + B)^+ B,$$

where $(A + B)^+$ denotes a pseudo inverse. The parallel sum satisfies the following properties

$$A : B = B : A, \quad A : B \leq A, \quad A : B \leq B, \tag{3}$$

and it was first used in forming generalized eigenvalues problems by Dohrmann and Pechstein (2013). We note that a similar approach was considered by Klawonn et al. (2014) in a more general form. Both are limited to the two-dimensional problems with only face equivalence classes. In this work, generalized eigenvalue problems for edge equivalence classes will be introduced to extend the previous approaches to three dimensions.

For a face F, the following generalized eigenvalue problem is considered

$$A_F v_F = \lambda \widetilde{A}_F v_F,$$

where

$$A_F = (D_F^{(j)})^T S_F^{(i)} D_F^{(j)} + (D_F^{(i)})^T S_F^{(j)} D_F^{(i)}, \; \widetilde{A}_F = \widetilde{S}_F^{(i)} : \widetilde{S}_F^{(j)},$$

and $S_F^{(i)}$ denote block matrix of S_i to the unknowns interior to F. The eigenvalues are all positive and we select eigenvectors $v_{F,l}$, $l \in N(F)$ with associated eigenvalues λ_l larger than a given λ_{TOL}. The following constraints will then be enforced on the unknowns in F,

$$(A_F v_{F,l})^T (w_F^{(i)} - w_F^{(j)}) = 0, \; l \in N(F).$$

After a change of unknowns, the above constraints can be transformed into explicit unknowns and they are added to the initial set of primal unknowns and denoted by $w_{F,\Pi}^{(i)}$. The remaining unknowns are called dual unknowns and denoted by $w_{F,\Delta}^{(i)}$. Using (2), for the two-dimensional case we obtain that

$$\langle \widetilde{S}(I - E_D)\widetilde{w}_a, (I - E_D)\widetilde{w}_a \rangle \leq C \sum_F (\langle A_F \widetilde{w}_{F,\Delta}^{(i)}, \widetilde{w}_{F,\Delta}^{(i)} \rangle + \langle A_F \widetilde{w}_{F,\Delta}^{(j)}, \widetilde{w}_{F,\Delta}^{(j)} \rangle)$$

$$\leq C\lambda_{TOL} \sum_F (\langle \widetilde{A}_F \widetilde{w}_{F,\Delta}^{(i)}, \widetilde{w}_{F,\Delta}^{(i)} \rangle + \langle \widetilde{A}_F \widetilde{w}_{F,\Delta}^{(j)}, \widetilde{w}_{F,\Delta}^{(j)} \rangle)$$

$$\leq C\lambda_{TOL} \sum_F (\langle S^{(i)} w_i, w_i \rangle + \langle S^{(j)} w_j, w_j \rangle),$$

where the estimate on the dual unknowns are bounded by λ_{TOL} in the second inequality, and (3) and the minimum energy property of $\widetilde{S}_F^{(i)}$ are used in the last inequality.

For an edge E, shared by more than two subdomains, we introduce the following generalized eigenvalue problem,

$$A_E v_E = \lambda \widetilde{A}_E v_E,$$

where

$$A_E = \sum_{m \in I(E)} \sum_{l \in I(E) \setminus \{m\}} (D_E^{(l)})^T S_E^{(m)} D_E^{(l)}, \; \widetilde{A}_E = \prod_{m \in I(E)} \widetilde{S}_E^{(m)},$$

and $I(E)$ denotes the set of subdomain indices sharing E in common, and $\prod_{m \in I(E)} \widetilde{S}_E^{(m)}$ is the parallel sum applied to those matrices $\widetilde{S}_E^{(m)}$. For a given λ_{TOL}, the eigenvectors with their eigenvalues larger than λ_{TOL} will be selected and denoted by $v_{E,l}$, $l \in N(E)$. The following constraints will then be enforced on the unknowns in E,

$$(A_E v_{E,l})^T (w_E^{(i)} - w_E^{(m)}) = 0, \; l \in N(E), \; m \in I(E) \setminus \{i\}.$$

Using the adaptively selected primal unknowns on each face F and edge E, we can obtain the following estimate

$$\langle \widetilde{S}(I - E_D)\widetilde{w}_a, (I - E_D)\widetilde{w}_a \rangle \leq C\lambda_{TOL} \langle \widetilde{S}\widetilde{w}_a, \widetilde{w}_a \rangle,$$

where C is a constant depending on the maximum number of edges and faces per subdomain, and the maximum number of subdomains sharing an edge but is independent of the coefficient $\rho(x)$.

4 Condition Number Estimate

Using the adaptively enriched primal constraints described in Sect. 3, we can obtain the following bound of the condition numbers for the given λ_{TOL}:

Theorem 1 *The BDDC algorithm with the change of basis formulation for the adaptively chosen set of primal unknowns with a given tolerance λ_{TOL} has the following bound of condition numbers,*

$$\kappa(M_{BDDC,a}^{-1}\widetilde{R}^T \widetilde{S}_a\widetilde{R}) \leq C\lambda_{TOL},$$

and the FETI-DP algorithm with the projector preconditioner M_{FETI}^{-1} has the bound

$$\kappa(M_{FETI}^{-1}F_{DP}) \leq C\lambda_{TOL},$$

where C is a constant depending only on $N_{F(i)}$, $N_{E(i)}$, $N_{I(E)}$, which are the number of faces per subdomain, the number of edges per subdomain, and the number of subdomains sharing an edge E, respectively.

In the above $M_{BDDC,a}$ and $\widetilde{S}_a$ denote the BDDC preconditioner and the partially assembled matrix of S_i after the change of unknowns for the adaptive primal constraints. We refer to Kim et al. (2015) for detailed proofs of the above theorem. We note that for the FETI-DP algorithm with the projector preconditioner the approaches in Toselli and Widlund (2005) can be used to obtain the upper bound estimate

$$\langle \widetilde{S}P_D\widetilde{w}, P_D\widetilde{w} \rangle \leq C\lambda_{TOL} \langle \widetilde{S}\widetilde{w}, \widetilde{w} \rangle,$$

where $\widetilde{w}$ is strongly coupled at vertices and the adaptive primal constraints on F and E are enforced on $\widetilde{w}$ by using the projection P.

References

W.N. Anderson Jr., R.J. Duffin, Series and parallel addition of matrices. J. Math. Anal. Appl. **26**, 576–594 (1969)

C.R. Dohrmann, A preconditioner for substructuring based on constrained energy minimization. SIAM J. Sci. Comput. 25(1), 246–258 (2003)

C.R. Dohrmann, C. Pechstein, Modern domain decomposition solvers: BDDC, deluxe scaling, and an algebraic approach (2013). http://people.ricam.oeaw.ac.at/c.pechstein/pechstein-bddc2013.pdf

C. Farhat, M. Lesoinne, P. LeTallec, K. Pierson, D. Rixen, FETI-DP: a dual-primal unified FETI method. I. A faster alternative to the two-level FETI method. Int. J. Numer. Methods Eng. **50**(7), 1523–1544 (2001)

J. Galvis, Y. Efendiev, Domain decomposition preconditioners for multiscale flows in high-contrast media. Multiscale Model. Simul. **8**(4), 1461–1483, (2010)

H.H. Kim, E.T. Chung, J. Wang, BDDC and FETI-DP algorithms with adaptive coarse spaces for three-dimensional elliptic problems with oscillatory and high contrast coefficients. http://arxiv.org/abs/1606.07560 (2016)

A. Klawonn, O.B. Widlund, Dual-primal FETI methods for linear elasticity. Commun. Pure Appl. Math. **59**(11), 1523–1572 (2006)

A. Klawonn, P. Radtke, O. Rheinbach, FETI-DP with different scalings for adaptive coarse spaces, in *Proceedings in Applied Mathematics and Mechanics* (2014)

A. Klawonn, P. Radtke, O. Rheinbach, FETI-DP methods with an adaptive coarse space. SIAM J. Numer. Anal. **53**(1), 297–320 (2015)

J. Li, O.B. Widlund, FETI-DP, BDDC, and block Cholesky methods. Int. J. Numer. Methods Eng. **66**(2), 250–271 (2006)

J. Mandel, B. Sousedík, Adaptive selection of face coarse degrees of freedom in the BDDC and the FETI-DP iterative substructuring methods. Comput. Methods Appl. Mech. Eng. **196**(8), 1389–1399 (2007)

J. Mandel, B. Sousedík, J. Šístek, Adaptive BDDC in three dimensions. Math. Comput. Simul. **82**(10), 1812–1831 (2012)

N. Spillane, V. Dolean, P. Hauret, F. Nataf, C. Pechstein, R. Scheichl, A robust two-level domain decomposition preconditioner for systems of PDEs. C. R. Math. Acad. Sci. Paris **349**(23–24), 1255–1259 (2011)

N. Spillane, V. Dolean, P. Hauret, F. Nataf, D.J. Rixen, Solving generalized eigenvalue problems on the interfaces to build a robust two-level FETI method. C. R. Math. Acad. Sci. Paris **351**(5–6), 197–201 (2013)

A. Toselli, O. Widlund, *Domain Decomposition Methods—Algorithms and Theory*, vol. 34. Springer Series in Computational Mathematics (Springer, Berlin, 2005)

Adaptive Coarse Spaces for FETI-DP in Three Dimensions with Applications to Heterogeneous Diffusion Problems

Axel Klawonn, Martin Kühn, and Oliver Rheinbach

1 Introduction

We consider an adaptive coarse space for FETI-DP or BDDC methods in three dimensions. We have user-given tolerances for certain eigenvalue problems which determine the computational overhead needed to obtain fast convergence. Similar adaptive strategies are available for many kinds of domain decomposition methods; see, e.g., Galvis and Efendiev (2010), Dolean et al. (2012), Spillane and Rixen (2013), Kim and Chung (2015), Klawonn et al. (2015), Mandel and Sousedík (2007), Dohrmann and Pechstein.

We will give numerical results for our algorithm for the diffusion equation on a bounded polyhedral domain Ω, i.e., for the weak formulation of

$$
\begin{aligned}
-\nabla \cdot (\rho \nabla u) &= f \quad \text{in } \Omega, \\
u &= 0 \quad \text{on } \partial\Omega_D, \\
\rho \nabla u \cdot \mathbf{n} &= 0 \quad \text{on } \partial\Omega_N.
\end{aligned}
\tag{1}
$$

A. Klawonn (✉) • M. Kühn
Mathematisches Institut, Universität zu Köln, Weyertal 86-90, 50931 Köln, Germany
e-mail: axel.klawonn@uni-koeln.de; martin.kuehn@uni-koeln.de

O. Rheinbach
Institut für Numerische Mathematik und Optimierung, Fakultät für Mathematik und Informatik, Technische Universität Bergakademie Freiberg, Akademiestr. 6, 09596 Freiberg, Germany
e-mail: oliver.rheinbach@math.tu-freiberg.de

© Springer International Publishing AG 2017
C.-O. Lee et al. (eds.), *Domain Decomposition Methods in Science and Engineering XXIII*, Lecture Notes in Computational Science and Engineering 116, DOI 10.1007/978-3-319-52389-7_18

Here, $\partial\Omega_D \subset \partial\Omega$ is a subset with positive surface measure where Dirichlet boundary conditions are prescribed. Furthermore, $\partial\Omega_N := \partial\Omega \setminus \partial\Omega_D$ is the part of the boundary where Neumann boundary conditions are given and $\mathbf{n}$ is the outward pointing unit normal on $\partial\Omega_N$. The function $\rho = \rho(x)$ will be called coefficient (distribution).

2 FETI-DP with Projector Preconditioning and Balancing

Due to space limitation, we will only provide the most important FETI-DP operators and the FETI-DP system. For a more detailed description of FETI-DP; see, e.g., Farhat et al. (2000), Toselli and Widlund (2005). The FETI-DP system is given by $F\lambda = d$ where

$$F = B_B K_{BB}^{-1} B_B^T + B_B K_{BB}^{-1} \widetilde{K}_{\Pi B}^T \widetilde{S}_{\Pi\Pi}^{-1} \widetilde{K}_{\Pi B} K_{BB}^{-1} B_B^T,$$

$$d = B_B K_{BB}^{-1} f_B + B_B K_{BB}^{-1} \widetilde{K}_{\Pi B}^T \widetilde{S}_{\Pi\Pi}^{-1} \left(\left(\sum_{i=1}^{N} R_{\Pi}^{(i)T} f_{\Pi}^{(i)} \right) - \widetilde{K}_{\Pi B} K_{BB}^{-1} f_B \right).$$

Here, $\widetilde{S}_{\Pi\Pi}$ defines the primal coarse space which, in our case, will be given by all vertex variables being primal. We now present Projector Preconditioning and Balancing in a very short form; for a more detailed description see Klawonn and Rheinbach (2012), and for a semidefinite matrix F, (Klawonn et al. 2016a). Given a matrix U representing constraints $U^T B w = 0$, we define $P := U(U^T F U)^+ U^T F$ and solve the preconditioned system

$$M_{PP}^{-1} F\lambda := (I - P)M_D^{-1}(I - P)^T F\lambda = (I - P)M_D^{-1}(I - P)^T d.$$

Here, M_D^{-1} is the Dirichlet preconditioner. In our computations, we exclusively use patch-ρ-scaling (see Klawonn and Rheinbach 2007) but other scalings are possible. We can also use the balancing preconditioner $M_{BP}^{-1} = M_{PP}^{-1} + U(U^T F U)^+ U^T$ instead of M_{PP}^{-1}.

3 Adaptive Constraints and Condition Number Bound

We now present our adaptive approach that is based on modifications of the approach in Mandel and Sousedík (2007); see also Klawonn et al. (2016b) and Klawonn et al. (2016a). In Klawonn et al. (2016b), for two dimensions, a complete theory including a condition number bound for the coarse space introduced by

Mandel and Sousedík (2007) was given. However, this coarse space turns out not to be sufficient in three dimensions. In Klawonn et al. (2016a), we therefore have added certain edge eigenvalue problems to prove a condition number bound also in three dimensions and in the numerical experiments, we have focussed on elasticity. In the present paper, we consider scalar second-order elliptic problems.

For a given subdomain Ω_i, we assume that it shares an edge $\mathcal{E}$ and an adjacent face with Ω_j and Ω_k, respectively, while it only shares the edge $\mathcal{E}$ with Ω_l. More general cases can be treated analogously. In the following we will use the index $s \in \{j, l\}$ to describe simultaneously eigenvalue problems and their operators defined on faces ($s = j$) and edges ($s = l$), respectively. Note that eigenvalue problems on faces are defined on the closure of the face.

Let G be a face or an edge shared by Ω_i and Ω_s. Then, we define $B_{G_{is}} = [B_{G_{is}}^{(i)} B_{G_{is}}^{(s)}]$ as all the rows of $[B^{(i)} B^{(s)}]$ that contain exactly one $+1$ and one -1. In the same manner, we define the scaled matrix $B_{D,G_{is}} = [B_{D,G_{is}}^{(i)} B_{D,G_{is}}^{(s)}]$ as the submatrix of $[B_D^{(i)} B_D^{(s)}]$. Furthermore, define $S_{is} := \begin{pmatrix} S^{(i)} & 0 \\ 0 & S^{(s)} \end{pmatrix}$ and $P_{D_{is}} := B_{D,G_{is}}^T B_{G_{is}}$.

The space of functions in $W_i \times W_s$ that are continuous in the primal variables shared by Ω_i and Ω_s will be denoted by $\widetilde{W}_{is}$. Then, we introduce the ℓ_2-orthogonal projection Π_{is} from $W_i \times W_s$ to $\widetilde{W}_{is}$ as well as a second ℓ_2-orthogonal projection $\overline{\Pi}_{is}$ from $W_i \times W_s$ to $\text{range}(\Pi_{is}S_{is}\Pi_{is} + \sigma(I - \Pi_{is}))$. There, σ is a possibly large positive constant, e.g., the maximum of the diagonal entries of S_{ij}, to avoid numerical instabilities. Without loss of generality we can assume that the projections are symmetric.

Then, we build and solve the generalized eigenvalue problems

$$\overline{\Pi}_{is}\Pi_{is}P_{D_{is}}^T S_{is}P_{D_{is}}\Pi_{is}\overline{\Pi}_{is}w_{is}^k$$
$$= \mu_{is}^k(\overline{\Pi}_{is}(\Pi_{is}S_{is}\Pi_{is} + \sigma(I - \Pi_{is}))\overline{\Pi}_{is} + \sigma(I - \overline{\Pi}_{is}))w_{is}^k, \qquad (2)$$

for $\mu_{is}^k \geq \text{TOL}$. Let us note that the projections are built such that the right hand side of the eigenvalue problem (2) is symmetric positive definite; cf. Mandel and Sousedík (2007). For an eigenvalue problem defined on (the closure of) a face (i.e. $s = j$), we split the computed constraint columns $u_{ij}^k := B_{D,G_{ij}}S_{ij}P_{D_{ij}}w_{ij}^k$ into several edge constraints $u_{ij,\mathcal{E}_m}^k$ and a constraint on the open face $u_{ij,\mathcal{F}}^k$, all extended by zero to the closure of the face. The splitting avoids coupling of the constraints and preserves a block structure of the constraint matrix; cf. Mandel et al. (2012). We then enforce all the constraints

$$u_{ij,\mathcal{E}_m}^{kT}B_{F_{ij}}w_{ij} = 0, \ m = 1, 2, \ldots, \quad u_{ij,\mathcal{F}}^{kT}B_{F_{ij}}w_{ij} = 0.$$

For a given edge with corresponding edge eigenvalue problem, we enforce

$$w_{il}^{kT} P_{D_{il}}^T S_{il} P_{D_{il}} w_{il} = 0.$$

For $w_{is} \in W_i \times W_s$ satisfying the constraints, we have the local estimate

$$w_{is}^T \overline{\Pi}_{is} \Pi_{is} P_{D_{is}}^T S_{is} P_{D_{is}} \Pi_{is} \overline{\Pi}_{is} w_{is} \leq \text{TOL} \, w_{is}^T \overline{\Pi}_{is} \Pi_{is} S_{is} \Pi_{is} \overline{\Pi}_{is} w_{is};$$

cf. (Klawonn et al. 2016b). For $w \in \widetilde{W}$ we have $\begin{pmatrix} R^{(i)} w \\ R^{(s)} w \end{pmatrix} \in \widetilde{W}_{is}$ and therefore

$\Pi_{is} \begin{pmatrix} R^{(i)} w \\ R^{(s)} w \end{pmatrix} = \begin{pmatrix} R^{(i)} w \\ R^{(s)} w \end{pmatrix}$. As argued in Klawonn et al. (2016b) we have $\Pi_{is}(I - \overline{\Pi}_{is}) w_{is} = (I - \overline{\Pi}_{is}) w_{is}$. This gives $P_{D_{is}} \Pi_{is}(I - \overline{\Pi}_{is}) w_{is} = 0$ and $S_{is} \Pi_{is}(I - \overline{\Pi}_{is}) w_{is} = 0$. Therefore, for all $w_{is} \in \widetilde{W}_{is}$ with $w_{is}^{kT} P_{D_{is}}^T S_{is} P_{D_{is}} w_{is} = 0$, $\mu_{is}^k \geq \text{TOL}$ we obtain

$$w_{is}^T \Pi_{is} P_{D_{is}}^T S_{is} P_{D_{is}} \Pi_{is} w_{is} \leq \text{TOL} w_{is}^T \Pi_{is} S_{is} \Pi_{is} w_{is}; \tag{3}$$

cf. (Mandel and Sousedík 2007).

Let $U = (u_1, \ldots, u_k)$ be the matrix where the adaptive constraints are stored in its columns. Then, $\widetilde{W}_U := \{w \in \widetilde{W} \mid U^T B w = 0\}$ will be the subspace of $\widetilde{W}$ which contains all elements $w \in \widetilde{W}$ satisfying the adaptively computed constraints, i.e., $Bw \in \ker U^T$. We are now ready to give the following lemma.

Lemma 1 *Let $N_{\mathcal{F}}$ denote the maximum number of faces of a subdomain, $N_{\mathcal{E}}$ the maximum number of edges of a subdomain, $M_{\mathcal{E}}$ the maximum multiplicity of an edge and TOL a given tolerance for solving the local generalized eigenvalue problems. If all vertices are chosen to be primal, for $w \in \widetilde{W}_U$ it holds*

$$|P_D w|_{\widetilde{S}}^2 \leq 4 \max\{N_{\mathcal{F}}, N_{\mathcal{E}} M_{\mathcal{E}}\}^2 \text{TOL}|w|_{\widetilde{S}}^2.$$

Proof See Klawonn et al. (2016a).

We can now provide a condition number estimate for the preconditioned FETI-DP algorithm with all vertex constraints being primal and additional, adaptively chosen, edge and face constraints.

Theorem 1 *Let $N_{\mathcal{F}}$ denote the maximum number of faces of a subdomain, $N_{\mathcal{E}}$ the maximum number of edges of a subdomain, $M_{\mathcal{E}}$ the maximum multiplicity of an edge and TOL a given tolerance for solving the local generalized eigenvalue problems.*

If all vertices are chosen to be primal, the condition number $\kappa(\widehat{M}^{-1}F)$ of the FETI-DP algorithm with adaptive constraints as described, e.g., enforced by the projector preconditioner $\widehat{M}^{-1} = M_{PP}^{-1}$ or the balancing preconditioner $\widehat{M}^{-1} = M_{BP}^{-1}$, satisfies

$$\kappa(\widehat{M}^{-1}F) \leq 4 \max\{N_{\mathcal{F}}, N_{\mathcal{E}}M_{\mathcal{E}}\}^2 \text{TOL}.$$

Proof See Klawonn et al. (2016a). $\square$

4 Heuristic Modifications

In this section we introduce two modifications of our algorithm. We will test the performance of the heuristically reduced coarse spaces along with the algorithm presented before.

Reducing the Number of Edge Eigenvalue Problems Our first modification consists of discarding edge eigenvalue problems on edges where no coefficient jumps occur. Therefore, we traverse the corresponding edge nodes and check for coefficient jumps. If no jumps occur we will not solve the corresponding edge eigenvalue problem and discard it with all possible constraints. Let us note that the condition number bound mentioned before might no longer hold if we use this strategy. However, due to slab techniques, see, e.g., Klawonn et al. (2015), the condition number is expected to stay bounded independently of the coefficients.

Reducing the Number of Edge Constraints The second approach uses the strategy discussed before and discards additionally edge constraints from face eigenvalue problems, if there are no coefficient jumps in the neighborhood of the edge.

5 Numerical Results

In this section, we will give numerical results for five different algorithms. First, we will present results for our new algorithm that is covered by theory (denoted by '*Alg. Ia*') and two modifications thereof; see also Klawonn et al. (2016a) where these algorithms were introduced for elasticity. By '*Alg. Ib*' we will denote the modification using only the first strategy presented in Sect. 4. We will also test a variant using both heuristics of Sect. 4. This algorithm will be denoted '*Alg. Ic*'.

The performance of these algorithms will be compared to the approaches of Mandel et al. (2012). By '*Alg. III*' we denote the 'classic' approach which discards all edge constraints from face eigenvalue problems. The coarse space enriched by those edge constraints but without edge constraints from edge eigenvalue problems will be denoted by '*Alg. II*'.

For all algorithms we will start with an extended first coarse space. Given the coarse space consisting of primal vertices, we will add some additional edge nodes. We will set those edge nodes primal that belong to an edge eigenvalue problem on a short edge, i.e., an edge with only one dual node. Then, the corresponding edge eigenvalue problem will become superfluous.

We use a singular value decomposition with a drop tolerance of $1e - 6$ to orthogonalize all adaptively computed constraints. We use the balancing preconditioner to enforce the resulting constraints. For simplicity, we assume $\rho(x)$ to be constant on each finite element and we use ρ-scaling in the form of patch-ρ-scaling. The coefficient at a node will be set as the maximum coefficient on the support of the corresponding nodal basis function; cf. (Klawonn and Rheinbach 2007). In the experiments, we use an irregular partitioning of the domain using the METIS graph partitioner with options -ncommon=3 and -contig. Let us note that Alg. III might be sufficient if regular decompositions are chosen and jumps only appear at subdomain faces; see Mandel et al. (2012). We will therefore just test irregular decompositions.

In all tables, "κ" denotes the condition number of the preconditioned FETI-DP operator, "*its*" is the number of iterations of the pcg algorithm and "$|U|$" denotes the size of the corresponding second coarse space. By "N" we denote the number of subdomains. For our modified coarse space, we also give the number of edge eigenvalue problem as "$\#\mathcal{E}_{ewp}$" and in parentheses the percentage of these in the total number of eigenvalue problems. Our stopping criterion for the pcg algorithm is a relative reduction of the starting residual by 10^{-10}, and the maximum number of iterations is set to 500. The condition numbers κ, which we report in the tables, are estimates from the Krylov process. We will consider $\Omega = [0, 1]^3$, discretized by a structured fine mesh of cubes, each containing five tetrahedra. We apply Dirichlet boundary conditions for the face with $x = 0$ and zero Neumann boundary conditions elsewhere. Moreover, let $f = 0.1$ and $\rho(x) \in \{1, 1e + 6\}$.

A Composite Material We consider a soft matrix material with $\rho_1 = 1$ and stiff inclusions in the form of $4N^{2/3}$ beams with $\rho_2 = 1e + 06$; see Fig. 1. In Table 1, we see that Alg. III always leads to high condition numbers and even to nonconvergence ($its = 500$) in three of four cases. The use of edge constraints from face eigenvalue problems (cf. Alg. II) can neither guarantee small condition numbers but results

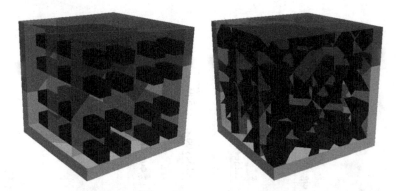

Fig. 1 Composite material (*left*) and randomly distributed coefficients (*right*) with irregular decomposition. High coefficients $\rho_2 = 1e + 06$ are shown in *dark purple* in the picture; low coefficients $\rho_1 = 1$ are not shown. Subdomains are shown in *different colors* in the background and by half-transparent slices. Visualization for $N = 8$ and $H/h = 5$

in convergence within a maximum of about 90 iterations. Although only Alg. Ia is covered by our theoretical bound, Alg. Ia, Ib, and Ic can guarantee condition numbers around the size of the prescribed tolerance and convergence within 30–40 iterations. Here, Alg. Ic gives the best performance: it uses the smallest coarse space and leads to convergence in a small number of iterations.

Let us note that the number of edge eigenvalue problems here is larger than in the case of linear elasticity (cf. Klawonn et al. 2016a). This is due to the fact that, in case of elasticity, we have to select additional primal vertices to remove hinge modes on curved edges. Then, edge eigenvalue problems on certain short edges become superfluous. Since this is not necessary for the diffusion equation, and since it also enlarges the primal coarse space, we do not carry this out here and accept a higher number of eigenvalue problems.

Random Coefficients We now perform 100 runs using randomly generated coefficients (20% high and 80% low) for different numbers of subdomains; see Table 2. For $N \in \{4^3, 5^3\}$, we see that the classical Alg. III does not converge in any single run and always leads to a condition number of at least $1e + 05$. Although Alg. II converges in all cases it exhibits a condition number of $1e + 05$ or higher in 71 ($N = 4^3$) and 73 ($N = 5^3$) runs. The performance of Alg. Ia, Ib, and Ic is almost identical. For these algorithms, the condition number is always lower than 15, and convergence is reached within 35 iterations.

Table 1 Compressible linear elasticity with $E_1 = 1$, $E_2 = 1e + 06$. Coarse spaces for TOL $= 10$ for all generalized eigenvalue problems

Composite material, irregular partitioning and $H/h = 5$

N		Algs. Ia, Ib, and Ic				Alg. II			Alg. III		
		κ	Its	$|U|$	ε_{exp} (%)	κ	Its	$|U|$	κ	Its	$|U|$
4^3	(a)	9.54	36	1784	41(14.9)	9.78	37	1765	2.23e+06	500	609
	(b)	9.78	36	1783	30(11.3)						
	(c)	10.68	39	1475	30(11.3)						
6^3	(a)	11.72	38	6455	166(15.1)	5.13e+05	98	6364	3.13e+06	500	2057
	(b)	11.72	38	6455	134(12.6)						
	(c)	11.72	39	5701	134(12.6)						
8^3	(a)	12.34	39	15292	390(14.1)	2.27e+05	62	15120	2.99e+06	500	4921
	(b)	12.34	39	15292	334(12.4)						
	(c)	12.34	40	13682	334(12.4)						

Table 2 Compressible linear elasticity with $E_1 = 1$, $E_2 = 1e + 06$. Coarse spaces for TOL $= 10$ for all generalized eigenvalue problems. Randomly distributed coefficients, irregular partitioning, and $H/h = 5$

N			Algs. Ia, Ib, and Ic				Alg. II			Alg. III		
			κ	Its	$|U|$	$\#\mathcal{E}_{evp}$	κ	Its	$|U|$	κ	Its	$|U|$
4^3	$\bar{x}$	(a)	8.81	30.64	1913.92	41 (14.9%)	3.92e+05	43.61	1889.83	2.62e+06	500	675.53
		(b)	8.81	30.64	1913.92	41 (14.9%)						
		(c)	8.81	30.64	1913.72	41 (14.9%)						
	$\tilde{x}$	(a)	8.76	31	1918	41 (14.9%)	2.31e+05	42.5	1893.5	2.57e+06	500	676
		(b)	8.76	31	1918	41 (14.9%)						
		(c)	8.76	31	1918	41 (14.9%)						
	σ	(a)	0.88	1.32	43.57	–	5.12e+05	10.41	43.25	7.42e+05	0	22.05
		(b)	0.88	1.32	43.57	–						
		(c)	0.88	1.32	43.67	–						
5^3	$\bar{x}$	(a)	9.26	32.19	3992.86	61 (10.3%)	2.29e+05	55.35	3954.5	2.96e+06	500	1357.53
		(b)	9.26	32.19	3992.86	61 (10.3%)						
		(c)	9.26	32.19	3992.55	61 (10.3%)						
	$\tilde{x}$	(a)	9.20	32	3997.5	61 (10.3%)	2.01e+05	52.5	3955.5	2.79e+06	500	1359.5
		(b)	9.20	32	3997.5	61 (10.3%)						
		(c)	9.20	32	3996	61 (10.3%)						
	σ	(a)	0.86	0.88	69.31	–	2.09e+05	15.05	68.58	7.52e+05	0	33.67
		(b)	0.86	0.88	69.31	–						
		(c)	0.86	0.90	69.38	–						

References

C. Dohrmann, C. Pechstein, Modern domain decomposition solvers - BDDC, deluxe scaling, and an algebraic approach, in *Slides to a talk at NuMa Seminar, JKU Linz, Dec 10th, 2013*, ed. by C. Pechstein. http://people.ricam.oeaw.ac.at/c.pechstein/pechstein-bddc2013.pdf

V. Dolean, F. Nataf, R. Scheichl, N. Spillane, Analysis of a two-level Schwarz method with coarse spaces based on local Dirichlet-to-Neumann maps. Comput. Methods Appl. Math. **12**(4), 391–414 (2012)

C. Farhat, M. Lesoinne, K. Pierson, A scalable dual-primal domain decomposition method. Numer. Linear Alg. Appl. **7**(7–8), 687–714 (2000)

J. Galvis, Y. Efendiev, Domain decomposition preconditioners for multiscale flows in high-contrast media. Multiscale Model. Simul. **8**(4), 1461–1483 (2010)

H.H. Kim, E.T. Chung, A BDDC algorithm with enriched coarse spaces for two-dimensional elliptic problems with oscillatory and high contrast coefficients. Multiscale Model. Simul. **13**(2), 571–593 (2015)

A. Klawonn, O. Rheinbach, Robust FETI-DP methods for heterogeneous three dimensional elasticity problems. Comput. Methods Appl. Mech. Eng. **196**(8), 1400–1414 (2007)

A. Klawonn, O. Rheinbach, Deflation, projector preconditioning, and balancing in iterative substructuring methods: connections and new results. SIAM J. Sci. Comput. **34**(1) A459–A484, (2012)

A. Klawonn, P. Radtke, O. Rheinbach, FETI-DP methods with an adaptive coarse space. SIAM J. Numer. Anal. **53**, 297–320 (2015)

A. Klawonn, M. Kühn, O. Rheinbach, Adaptive coarse spaces for FETI-DP in three dimensions. SIAM J. Sci. Comput. **38**(5), A2880–A2911 (2016a). doi:10.1137/15M1049610

A. Klawonn, P. Radtke, O. Rheinbach, A comparison of adaptive coarse spaces for iterative substructuring in two dimensions. Electron. Trans. Numer. Anal. 45, 75–106 (2016b)

J. Mandel, B. Sousedík, Adaptive selection of face coarse degrees of freedom in the BDDC and the FETI-DP iterative substructuring methods. Comput. Methods Appl. Mech. Eng. **196**(8), 1389–1399 (2007)

J. Mandel, B. Sousedík, J. Šístek, Adaptive BDDC in three dimensions. Math. Comput. Simul. **82**(10), 1812–1831 (2012)

N. Spillane, D.J. Rixen, Automatic spectral coarse spaces for robust finite element tearing and interconnecting and balanced domain decomposition algorithms. Int. J. Numer. Methods Eng. **95**(11), 953–990 (2013)

A. Toselli, O.B. Widlund, *Domain Decomposition Methods - Algorithms and Theory*, vol. 34. Springer Series in Computational Mathematics (Springer, Berlin, 2005)

Newton-Krylov-FETI-DP with Adaptive Coarse Spaces

Axel Klawonn, Martin Lanser, Balthasar Niehoff, Patrick Radtke, and Oliver Rheinbach

1 Introduction

Newton-Krylov domain decomposition methods are approaches for solving nonlinear problems arising from the discretization of nonlinear partial differential equations. These methods are based on an iterative solution of linearized systems using a domain decomposition preconditioner. In this paper, we use FETI-DP as an iterative method and compute an adaptive coarse space, first introduced in Mandel and Sousedík (2007), to improve the condition number and thus the convergence of the iterative method. A theory has been developed in Klawonn et al. (2016c) for this coarse space in two dimensions and later, in Klawonn et al. (2016a), for three dimensions. In this paper, several heuristic strategies are introduced to reduce the computational effort for nonlinear problems, where a sequence of related linear problems have to be solved. These approaches show the potential of reducing the number of eigenvalue problems necessary for the construction of adaptive coarse spaces. A different but related approach was presented in Gosselet et al. (2013).

A. Klawonn (✉) • M. Lanser • B. Niehoff • P. Radtke
Mathematisches Institut, Universität zu Köln, Weyertal 86-90, 50931 Köln, Germany
e-mail: axel.klawonn@uni-koeln.de; martin.lanser@uni-koeln.de; patrick.radtke@uni-koeln.de

O. Rheinbach
Institut für Numerische Mathematik und Optimierung, Fakultät für Mathematik und Informatik, Technische Universität Bergakademie Freiberg, Akademiestr. 6, 09596 Freiberg, Germany
e-mail: oliver.rheinbach@math.tu-freiberg.de

© Springer International Publishing AG 2017   197
C.-O. Lee et al. (eds.), *Domain Decomposition Methods in Science and Engineering XXIII*, Lecture Notes in Computational Science and Engineering 116, DOI 10.1007/978-3-319-52389-7_19

2 Newton-Krylov-FETI-DP

In order to solve a discrete nonlinear equation

$$\widehat{K}(\hat{u}) - \hat{f} = 0, \tag{1}$$

associated with a computational domain Ω, we perform a Newton linearization of (1) and compute an update $\delta\hat{u}$ by solving the linearized system

$$D\widehat{K}(\hat{u})\,\delta\hat{u} = \widehat{K}(\hat{u}) - \hat{f}. \tag{2}$$

We always consider an iterative Krylov method such as CG to solve (2) using a domain decomposition preconditioner. In this paper, we always consider a FETI-DP (Finite Element Tearing and Interconnecting - Dual-Primal) preconditioner although a BDDC method could also be used. Therefore, we decompose Ω into nonoverlapping subdomains Ω_i, $i = 1, \ldots, N$, and assume the subdomains to be unions of finite elements. We denote the finite element space associated with Ω by $\widehat{W}$ and the local finite element spaces associated with the subdomains by W_i, $i = 1, \ldots, N$. Let us define local nonlinear problems in W_i, $i = 1, \ldots, N$, by

$$K^{(i)}(u_i) = f_i. \tag{3}$$

These local problems arise from a finite element discretization on subdomains Ω_i, $i = 1, \ldots, N$. The corresponding tangential matrices are defined as $DK^{(i)}(u_i)$. We introduce the block vectors

$$K(u) := \begin{pmatrix} K^{(1)}(u_1) \\ \vdots \\ K^{(N)}(u_N) \end{pmatrix}, \quad u := \begin{pmatrix} u_1 \\ \vdots \\ u_N \end{pmatrix}, \quad f := \begin{pmatrix} f_1 \\ \vdots \\ f_N \end{pmatrix}, \tag{4}$$

and the block tangential matrix

$$DK(u) = \begin{bmatrix} DK^{(1)}(u_1) & & \\ & \ddots & \\ & & DK^{(N)}(u_N) \end{bmatrix}. \tag{5}$$

In FETI-DP type methods, we divide all degrees of freedom into variables inside subdomains (I), dual interface variables (Δ), and primal variables (Π). Using the partial assembly operator R^T, well-known from the standard (linear) FETI-DP literature (Farhat et al. 2001; Klawonn and Rheinbach 2007, 2010; Klawonn and Widlund 2006), we can define the partially assembled operator $\widetilde{K}(\tilde{u}) := R^T K(R\tilde{u})$. Here, we perform a global assembly in all primal variables Π, but not in the remaining part of the interface. Equivalently, we partially assemble the right hand side $\tilde{f} := R^T f$ and the tangential matrix $D\widetilde{K}(\tilde{u}) := R^T DK(R\tilde{u})R$. We define the space of partially assembled functions by $\widetilde{W} \subset W := W_1 \times \cdots \times W_N$. Introducing the standard FETI-DP jump operator B and Lagrange multipliers to enforce the constraint $B\tilde{u} = 0$, the FETI-DP master system reads

$$\begin{pmatrix} D\widetilde{K}(\tilde{u}) & B^T \\ B & 0 \end{pmatrix} \begin{pmatrix} \delta\tilde{u} \\ \lambda \end{pmatrix} = \begin{pmatrix} \widetilde{K}(\tilde{u}) - \tilde{f} \\ 0 \end{pmatrix}. \tag{6}$$

At convergence, the solution $\delta\tilde{u}$ of (6) is continuous on the interface and thus can be assembled to the solution $\delta\hat{u}$ in (2). We finally obtain a solution of system (6) by eliminating all variables of $\delta\tilde{u}$ and using a preconditioned Krylov subspace method and solve

$$M_{BP}^{-1} F \lambda := M_{BP}^{-1} B \, (D\widetilde{K}(\tilde{u}))^{-1} \, B^T \, \lambda = M_{BP}^{-1} B \, (D\widetilde{K}(\tilde{u}))^{-1} \, (\widetilde{K}(\tilde{u}) - \tilde{f}). \tag{7}$$

In this paper, we use the balancing preconditioner M_{BP}^{-1}, see, e.g., Klawonn and Rheinbach (2012), for the Lagrange multipliers, implementing a second, adaptive coarse space computed from eigenvalue problems based on localized tangential matrices; see Sect. 3. The preconditioner M_{BP}^{-1} is defined by $M_{BP}^{-1} = (I - P)M^{-1}(I - P) + U(U^T F U)^{-1} U^T$, where $P = U(U^T F U)^{-1} U^T F$ is an F-orthogonal projection onto range U. The columns of U represent additional constraints of the form $U^T B\tilde{u} = 0$. For more details on the balancing preconditioner applied to FETI-DP methods, we refer to Klawonn and Rheinbach (2012). We denote the resulting algorithm by Adaptive-Newton-Krylov-FETI-DP; see Fig. 1 for the algorithm.

ADAPTIVE-NEWTON-KRYLOV-FETI-DP

Init: $\tilde{u}^{(0)} \in W$, continuous
for $k = 0, \ldots, convergence$

    build: $\widetilde{K}(\tilde{u}^{(k)})$ and $D\widetilde{K}(\tilde{u}^{(k)})$
    if cond_func$(k, r^{(0)}, \ldots, r^{(k)}, its(0), \ldots, its(k-1))$
        compute adaptive coarse space using tangent $D\widetilde{K}(\tilde{u}^{(k)})$
    else
        recycle adaptive coarse space from step $k-1$
    end if
    solve with preconditioned CG:
    $M_{BP}^{-1} B \, (D\widetilde{K}(\tilde{u}^{(k)}))^{-1} B^T \lambda = M_{BP}^{-1} B \, (D\widetilde{K}(\tilde{u}^{(k)}))^{-1} (\widetilde{K}(\tilde{u}^{(k)}) - \tilde{f})$
    compute:
    $\delta\tilde{u}^{(k)} = D\widetilde{K}(\tilde{u}^{(k)})^{-1} (\widetilde{K}(\tilde{u}^{(k)}) - \tilde{f} - B^T \lambda)$ // Compute $\delta\tilde{u}$ from λ.
    compute: steplength $\alpha^{(k)}$
    update: $\tilde{u}^{(k+1)} := \tilde{u}^{(k)} - \alpha^{(k)} \delta\tilde{u}^{(k)}$

end

Fig. 1 Algorithmic description of Adaptive-Newton-Krylov-FETI-DP

3   Adaptive Coarse Space

In the following, we briefly describe an adaptive approach first introduced in Mandel and Sousedík (2007). For other uses of this coarse space and modifications, see, e.g., Jan Mandel et al. (2012), Sousedík et al. (2013), Klawonn et al. (2016c). A theory is provided in Klawonn et al. (2016a,c). Due to space limitations, for further references on other adaptive coarse spaces, see, e.g., Spillane and Rixen (2013), Spillane (2015), Kim and Chung (2015), and the references therein. Let the Schur complements S_l be obtained by eliminating the interior degrees of freedom in $DK^{(l)}(u_l)$, $l = i, j$. We define $B_{D,ij}$ as the matrix with rows of $[B_D^{(i)} B_D^{(j)}]$ which correspond to Lagrange multipliers connecting degrees of freedom on $\partial \Omega_i \cap \partial \Omega_j$ and by B_{ij} the corresponding rows in $[B^{(i)} B^{(j)}]$. We then build a local operator $P_{D,ij} = B_{D,ij}^T B_{ij}$. Let $\widetilde{W}_{ij}$ be the subspace of functions in $W_i \times W_j$ which are continuous at those primal vertices that the two substructures Ω_i and Ω_j have in common. Let Π_{ij} be the l_2-orthogonal projection from $W_i \times W_j$ onto $\widetilde{W}_{ij}$. Let $\sigma > 0$ and $\overline{\Pi}_{ij}$ be the l_2-orthogonal projection that projects orthogonally the elements of $\ker(\Pi_{ij} S_{ij} \Pi_{ij} + \sigma(I - \Pi_{ij}))$ onto constants. In our computations we use $\sigma = \max(\mathrm{diag}(S_{ij}))$. To compute adaptive constraints, for each pair of substructures (Ω_i, Ω_j) having an edge in common, we solve the eigenvalue problem

$$\overline{\Pi}_{ij} \Pi_{ij} P_{D_{ij}}^T S_{ij} P_{D_{ij}} \Pi_{ij} \overline{\Pi}_{ij} w_{ij,m}$$

$$= \mu_{ij,m} (\overline{\Pi}_{ij} (\Pi_{ij} S_{ij} \Pi_{ij} + \sigma(I - \Pi_{ij})) \overline{\Pi}_{ij} + \sigma(I - \overline{\Pi}_{ij})) w_{ij,m}, \tag{8}$$

for eigenpairs where $\mu_{ij,m} \geq \mathrm{TOL}$, $m = k, \dots, n$. We implement the constraints $w_{ij,m}^T P_{D_{ij}}^T S_{ij} P_{D_{ij}} w_{ij} = 0$ for $w_{ij} \in W_i \times W_j$ and $m = k, \dots, n$. The adaptive constraint vectors are then given by $u_{ij,m} = B_{D_{ij}} S_{ij} P_{D_{ij}} w_{ij,m}$. They are extended by zero on the remaining interface and aggregated in the matrix U.

In our Adaptive-Newton-Krylov-FETI-DP method, we also use heuristic strategies to decide if the adaptive coarse space can be recycled in a certain Newton step. Only if some condition $cond\_func(k, r^{(0)}, \dots, r^{(k)}, its(0), \dots, its(k-1))$ is fulfilled in the k-th Newton step, we do compute a new adaptive coarse space. Otherwise, we recycle the coarse space already used in the previous Newton step. We suppose, that conditions can be provided that depend on the nonlinear residuals $r^{(l)} := \widetilde{K}(u^{(l)}) - \tilde{f}$, $l = 0, \dots, k$, the current iteration k, or the number of Krylov iterations $its(l)$ in the previous Newton steps $l = 0, \dots k - 1$. In the present paper, we propose three different strategies. **Strategy (a)**: $cond\_func := true$, **Strategy (b)**: $cond\_func := (k == 0)$, or **Strategy (c)**: $cond\_func := ((its(k-1)/its(c) < 0.75) \vee (its(c)/its(k-1) < 0.75))$. For **Strategy (a)**, we can prove a theoretical condition number bound for each linearization; see Klawonn et al. (2016c). **Strategy (b)** is based on the assumption that the optimal coarse space mainly depends on a coefficient function ρ. Therefore, the coarse space computed in the first Newton iteration can be recycled, since the coefficient function ρ does not change during the iteration. In **Strategy (c)** we compute an adaptive coarse space in the first Newton step. In the following steps we consider the number of Krylov iterations in the previous Newton step ($its(k-1)$) and the last Newton step in which an adaptive coarse space has been computed ($its(c)$). We always compute a new coarse space if $its(k-1)$ and $its(c)$ differ strongly. This strategy is based on the assumption that the quality of the coarse space in the c-th Newton step is verified by theoretical results and thus we can recycle our current coarse space as long as we have similar iteration counts as in step c. Let us remark that Strategy (b) will not succeed for elastoplasticity problems, see Klawonn et al. (2016b), for which we suggest the use of **Strategy (a)**. Alternatively, the knowledge of the plastic zones could be included into the heuristic function *cond_ func*. This is ongoing research and will be published elsewhere.

4 Numerical Results

As a model problem, we consider the p-Laplace equation with $p = 4$

$$
\begin{aligned}
-\mathrm{div}(\rho\,|\nabla u|^{p-2}\nabla u) &= 1 && \text{in } \Omega \\
u &= 0 && \text{on } \partial\Omega,
\end{aligned}
\tag{9}
$$

where $\rho : \Omega \to \mathbb{R}$ is a coefficient function given by

$$
\rho(x) = \begin{cases} 1e6 & \text{if } x \in \Omega_C \\ 1 & \text{elsewhere;} \end{cases}
\tag{10}
$$

see Fig. 2 for a definition of Ω_C. Let us remark that, given a finite element basis $\{\varphi_1, \ldots, \varphi_{N_i}\}$ on a subdomain Ω_i, we have

$$
K^{(i)}(u_i) := \left(\int_{\Omega_i} \rho\,|\nabla u_i|^{p-2}\,\nabla u_i^T \nabla \varphi_1 dx\,, \;\ldots\,, \int_{\Omega_i} \rho\,|\nabla u_i|^{p-2}\nabla u_i^T \nabla \varphi_{N_i} dx \right)^T .
$$

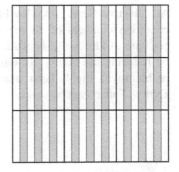

Fig. 2 Decomposition of $\Omega = [0, 1] \times [0, 1]$ into 3×3 subdomains. Each subdomain is intersected by 3 channels (*gray color*). All channels are unions of finite elements and the union of all channels is denoted by Ω_C

For the tangential matrices $DK^{(i)}(u_i)$, we obtain

$$(DK^{(i)}(u_i))_{j,k} := \int_{\Omega_i} \rho \, |\nabla u_i|^{p-2} \nabla \varphi_j^T \nabla \varphi_k dx$$

$$+ (p-2) \int_{\Omega_i} \rho \, |\nabla u_i|^{p-4} \, (\nabla u_i^T \nabla \varphi_j) \, (\nabla u_i^T \nabla \varphi_k) dx.$$

This tangential matrix is symmetric positive definite for all nonconstant functions u. We present numerical results for model problem (9) in Table 1 comparing Newton-Krylov-FETI-DP with Adaptive-Newton-Krylov-FETI-DP. We always make all subdomain vertex values primal initially. In all computations we use a moderate tolerance TOL$=$ 1000 to keep our adaptive coarse spaces small. All three adaptive strategies reduce the number of CG iterations drastically in comparison to classical Newton-Krylov-FETI-DP. Using **Strategy (a)** and computing a new coarse space in each Newton step, the condition number stays below the theoretical bound $C \cdot$ TOL. The coarse spaces generated are sufficiently small with a size of less than 5% of the size of the interface. Using **Strategy (b)**, the number of CG iterations is even lower. This is caused by a comparably large coarse space computed in the first Newton step. In this approach, the adaptive coarse space has only to be computed once, which results in a large reduction of local computational work compared to **Strategy (a)**. Unfortunately, the number of CG iterations in the different Newton steps and the average size of the coarse space strongly differs from the theoretically verified **Strategy (a)** and thus a control using tolerance TOL is no longer possible. In contrast, **Strategy (c)** can nearly reproduce the average size of the coarse space and the number of CG iterations of **Strategy (a)**. Additionally, the number of adaptive coarse space computations and thus the number of local eigenvalue problems is reduced by a factor of 5.0 to 6.0. For a graphical comparison of all methods see also Fig. 3. Especially the similar behavior of **Strategies (a)** and **(c)** can be observed.

5   Conclusion

An adaptive Newton-Krylov-FETI-DP approach has been presented, where the condition numbers of all preconditioned tangential matrices are bounded by a constant. Additionally, heuristic strategies have been introduced saving local work by reducing the number of eigenvalue problems. Results for a p-Laplace model problem with highly heterogeneous coefficient have been presented, showing the ability of adaptive coarse spaces to save CG iterations.

Table 1 Numerical results for model problem (9); each subdomain is a union of $2 \times 28 \times 28$ linear triangular finite elements; tolerance TOL$= 1000$ for adaptive coarse space; **N**: number of subdomains; **Strategy**: strategy chosen for cond_func; "—" denotes the case without an adaptive coarse space; **Max./Min. Krylov It.**: maximal/minimal number of Krylov subspace iterations during the Newton iteration; **Total Krylov It.**: total number of Krylov subspace iterations during the Newton iteration; **Max./Min. cond.**: maximal/minimal condition number during the Newton iteration; **Interface d.o.f.**: degrees of freedom on the interface; **Avg. size U**: average size of the adaptive coarse spaces during Newton iteration; **EP Solves**: Number of Newton steps in which a new adaptive coarse space is computed

TOL=1000

N	Strategy	Newton It.	Max. Krylov It.	Min. Krylov It.	Total Krylov It.	Max. cond.	Min. cond.	Interface d.o.f.	Avg. size U	EP solves
4	—	20	7	5	132	1.2	1.0	113	—	0
	(a)	20	7	5	132	1.2	1.0	113	0	20
	(b)	20	7	5	132	1.2	1.0	113	0	1
	(c)	20	7	5	132	1.2	1.0	113	0	2
16	—	22	129	25	890	363,714.3	653.7	675	—	0
	(a)	22	77	8	557	216.6	1.3	675	10.4	22
	(b)	22	108	6	335	569.0	1.1	675	36.0	1
	(c)	22	108	8	541	569.0	1.3	675	13.4	4
64	—	24	1148	111	5908	674,804.3	2000.9	3143	—	0
	(a)	24	109	8	1465	243.1	1.3	3143	65.9	24
	(b)	24	163	6	777	2740.5	1.1	3143	168.0	1
	(c)	24	113	8	1483	433.0	1.3	3143	68.7	5
256	—	26	3417	352	18,764	696,950.1	5083.7	13,455	—	0
	(a)	26	136	8	2406	247.8	1.3	13,455	325.9	26
	(b)	26	141	7	1086	5413.9	1.3	13,455	720.0	1
	(c)	26	206	8	2397	5413.9	1.3	13,455	389.0	4

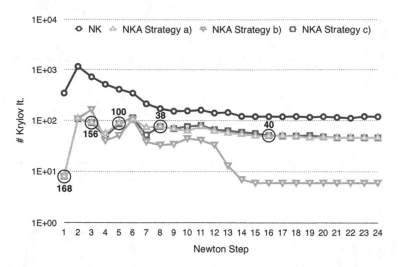

Fig. 3 Results for 64 subdomains from Table 1 showing the number of Krylov subspace iterations in each Newton step; **NK** (*blue curve*) denotes Newton-Krylov-FETI-DP without adaptive coarse spaces; **NKA Strategy (a)/(b)/(c)** (*green/yellow/red curve*) denotes Strategy (a)/(b)/(c); the *five black circles* mark the Newton steps in which **Strategy (c)** decided to compute a new coarse space and the numbers give the sizes of the coarse spaces

Acknowledgements This work was supported in part by the German Research Foundation (DFG) through the Priority Programme 1648 "Software for Exascale Computing" (**SPPEXA**) under KL 2094/4-1, KL 2094/4-2, RH 122/2-1, RH 122/3-2 .

References

C. Farhat, M. Lesoinne, P. LeTallec, K. Pierson, D. Rixen, FETI-DP: A dual-primal unified FETI method - part I: a faster alternative to the two-level FETI method. Int. J. Numer. Methods Eng. **50**, 1523–1544 (2001)

P. Gosselet, C. Rey, J. Pebrel, Total and selective reuse of Krylov subspaces for the resolution of sequences of nonlinear structural problems. Int. J. Numer. Methods Eng. **94**(1), 60–83 (2013)

H.H. Kim, E.T. Chung, A BDDC algorithm with enriched coarse spaces for two-dimensional elliptic problems with oscillatory and high contrast coefficients. Multiscale Model. Simul. **13**(2), 571–593 (2015)

A. Klawonn, M. Kühn, O. Rheinbach, Adaptive coarse spaces for FETI-DP in three dimensions. SIAM J. Sci. Comput. **38**(5), A2880–A2911 (2016a). Preprint 2015-11 available at http://tu-freiberg.de/sites/default/files/media/fakultaet-fuer-mathematik-und-informatik-fakultaet-1-9277/prep/2015-11_fertig.pdf

A. Klawonn, P. Radtke, O. Rheinbach, A Newton-Krylov-FETI-DP method with an adaptive coarse space applied to elastoplasticity, in *Domain Decomposition Methods in Science and Engineering XXII*, vol. 104. Lecture Notes in Computational Science and Engineering (Springer, Berlin, 2016b), pp. 293–300.

A. Klawonn, P. Radtke, O. Rheinbach, A comparison of adaptive coarse spaces for iterative substructuring in two dimensions. Electron. Trans. Numer. Anal. **45**, 75–106 (2016c)

A. Klawonn, O. Rheinbach, Robust FETI-DP methods for heterogeneous three dimensional elasticity problems. Comput. Methods Appl. Mech. Eng. **196**(8), 1400–1414 (2007)

A. Klawonn, O. Rheinbach, Highly scalable parallel domain decomposition methods with an application to biomechanics. ZAMM Z. Angew. Math. Mech. **90**(1), 5–32 (2010)

A. Klawonn, O. Rheinbach, Deflation, projector preconditioning, and balancing in iterative substructuring methods: connections and new results. SIAM J. Sci. Comput. **34**(1), A459–A484 (2012)

A. Klawonn, O.B. Widlund, Dual-primal FETI methods for linear elasticity. Commun. Pure Appl. Math. **59**(11), 1523–1572 (2006)

J. Mandel, B. Sousedík, Adaptive selection of face coarse degrees of freedom in the BDDC and the FETI-DP iterative substructuring methods. Comput. Methods Appl. Mech. Eng. **196**(8), 1389–1399 (2007)

J. Mandel, B. Sousedík, J. Šístek, Adaptive BDDC in three dimensions. Math. Comput. Simul. **82**(10), 1812–1831 (2012)

B. Sousedík, J. Šístek, J. Mandel, Adaptive-multilevel BDDC and its parallel implementation. Computing **95**, 1087–1119 (2013)

N. Spillane, Adaptive multi preconditioned conjugate gradient: algorithm, theory and an application to domain decomposition (2015). https://hal.archives-ouvertes.fr/hal-01170059

N. Spillane, D.J. Rixen, Automatic spectral coarse spaces for robust finite element tearing and interconnecting and balanced domain decomposition algorithms. Int. J. Numer. Methods Eng. **95**(11), 953–990 (2013)

New Nonlinear FETI-DP Methods Based on a Partial Nonlinear Elimination of Variables

Axel Klawonn, Martin Lanser, Oliver Rheinbach, and Matthias Uran

1 Introduction

We introduce two new nonlinear FETI-DP (Finite Element Tearing and Interconnecting—Dual-Primal) methods based on a partial nonlinear elimination and provide a comparison to Newton-Krylov-FETI-DP, Nonlinear-FETI-DP-1, and -2 methods (Klawonn et al. 2014a,b). In contrast to classical Newton-Krylov-FETI-DP methods, where a geometrical decomposition after linearization is performed, in nonlinear FETI-DP methods, the nonlinear problem is decomposed before linearization. The approaches help to localize work and thus are well suited for modern computer architectures. Recently, an inexact nonlinear FETI-DP implementation using PETSc and BoomerAMG has scaled, for nonlinear hyperelasticity, to the largest supercomputers currently available, i.e., to more than half a million MPI ranks (Klawonn et al. 2015) on the JUQUEEN supercomputer (Julich Supercomputing Centre), more than half a million cores (Klawonn et al. 2015) on the Mira supercomputer (Argonne National Laboratory), and later (Klawonn et al. 2015) the complete Mira ($786K$ cores). To the best of our knowledge, this is the largest range of parallel scalability reported for any domain decomposition method. Here, we now describe new variants of nonlinear FETI-DP methods.

A. Klawonn • M. Lanser • M. Uran
Mathematisches Institut, Universität zu Köln, Weyertal 86-90, 50931 Köln, Germany
e-mail: axel.klawonn@uni-koeln.de; martin.lanser@uni-koeln.de

O. Rheinbach (✉)
Institut für Numerische Mathematik und Optimierung, Fakultät für Mathematik und Informatik, Technische Universität Bergakademie Freiberg, Akademiestr. 6, 09596 Freiberg, Germany
e-mail: oliver.rheinbach@math.tu-freiberg.de

© Springer International Publishing AG 2017
C.-O. Lee et al. (eds.), *Domain Decomposition Methods in Science and Engineering XXIII*, Lecture Notes in Computational Science and Engineering 116, DOI 10.1007/978-3-319-52389-7_20

2 Nonlinear FETI-DP Methods

In all nonlinear FETI-DP methods, a geometrical decomposition of the computational domain Ω into nonoverlapping subdomains $\Omega_i, i = 1, \ldots, N$ is performed
before linearizing the nonlinear problem. In the more traditional Newton-Krylov-
FETI-DP approach a discrete nonlinear problem $A(u) = 0$ associated with Ω is
linearized first. Let $K_i(u_i) = f_i$, $i = 1, \ldots, N$, be the local finite element problem on
subdomain Ω_i and let W_i be the associated finite element space; see Klawonn et al.
(2014b), for a detailed definition. We define the nonlinear, discrete block operator
$K(u)$ and the corresponding vectors u and f by

$$
K(u) := \begin{pmatrix} K_1(u_1) \\ \vdots \\ K_N(u_N) \end{pmatrix}, \; u := \begin{pmatrix} u_1 \\ \vdots \\ u_N \end{pmatrix}, \text{ and } f := \begin{pmatrix} f_1 \\ \vdots \\ f_N \end{pmatrix}. \tag{1}
$$

As in linear FETI-DP, we decompose the degrees of freedom into variables interior
to subdomains (I), dual interface variables (Δ), and primal variables (Π), e.g.,
on vertices. Using the standard partial assembly operator R_Π^T, (Farhat et al. 2001;
Klawonn and Rheinbach 2010) we define the nonlinear, partially assembled operator
$\widetilde{K}(\tilde{u}) := R_\Pi^T K(R_\Pi \tilde{u})$ and the right hand side $\tilde{f} := R_\Pi^T f$. We define the usual space
of partially continuous discrete functions by $\widetilde{W} \subset W := W_1 \times \cdots \times W_N$. Using
the standard FETI-DP jump operator B, we can formulate the nonlinear FETI-DP
master system, first introduced in Klawonn et al. (2014a)

$$
\begin{aligned}
\widetilde{K}(\tilde{u}) + B^T \lambda - \tilde{f} &= 0 \\
B\tilde{u} \quad\quad\quad &= 0.
\end{aligned} \tag{2}
$$

In Klawonn et al. (2014b), two approaches have been suggested to solve the
nonlinear system (2): linearize first (Nonlinear-FETI-DP-1 or NL-1) and eliminate
first (Nonlinear-FETI-DP-2 or NL-2). The first variant is based on a Newton
linearization of the saddle point system and a solution of the resulting linear system.
The second variant is based on a nonlinear elimination of the variable $\tilde{u}$ in (2) before
linearization. While in NL-1 nonlinear problems in $\widetilde{W}$ are solved as an initial guess,
in NL-2 the solution of nonlinear problems in $\widetilde{W}$ is included into each Newton step,
often resulting into faster convergence. In both methods the quality of the coarse
space directly influences the Newton convergence. Thus, for problems where a good
coarse space is known, NL-2 is often the best choice. However, if a good coarse
space is not available, current nonlinear FETI-DP methods might fail to converge
without spending effort in globalization. Here, we introduce new nonlinear FETI-
DP methods based on a *partial* nonlinear elimination. In these methods, all primal
variables are linearized before elimination, which also allows the definition of
inexact FETI-DP variants; see also Klawonn et al. (2015), Klawonn and Rheinbach
(2010). In the new methods, the choice of primal variables has a weaker influence
on the Newton convergence and local nonlinear problems are also computationally
cheaper.

3 Nonlinear FETI-DP Based on Partial Elimination

Derivation of the Method We partition $\tilde{u} := (\tilde{u}_E^T, \tilde{u}_L^T)^T$ and $\tilde{f} := (\tilde{f}_E^T, \tilde{f}_L^T)^T$ into a set of variables $E \subseteq B := [I\ \Delta]$, and the remaining variables $L := (B \setminus E) \cup \Pi$. The variables $\tilde{u}_E$ will be eliminated from the nonlinear saddle point system (2) while the variables $\tilde{u}_L$ will be linearized. Accordingly, we partition

$$\widetilde{K}(\tilde{u}) = (\widetilde{K}_E(\tilde{u}_E, \tilde{u}_L)^T, \widetilde{K}_L(\tilde{u}_E, \tilde{u}_L)^T)^T, \text{ and}$$

$$D\widetilde{K}(\tilde{u}) = \begin{bmatrix} D_{\tilde{u}_E}\widetilde{K}_E(\tilde{u}_E, \tilde{u}_L) & D_{\tilde{u}_L}\widetilde{K}_E(\tilde{u}_E, \tilde{u}_L) \\ D_{\tilde{u}_E}\widetilde{K}_L(\tilde{u}_E, \tilde{u}_L) & D_{\tilde{u}_L}\widetilde{K}_L(\tilde{u}_E, \tilde{u}_L) \end{bmatrix} =: \begin{bmatrix} D\widetilde{K}_{EE} & D\widetilde{K}_{EL} \\ D\widetilde{K}_{LE} & D\widetilde{K}_{LL} \end{bmatrix}. \tag{3}$$

We can reformulate the nonlinear FETI-DP saddle point system (2) as

$$\begin{aligned} \widetilde{K}_E(\tilde{u}_E, \tilde{u}_L) + B_E^T\lambda - \tilde{f}_E &= 0 \\ \widetilde{K}_L(\tilde{u}_E, \tilde{u}_L) + B_L^T\lambda - \tilde{f}_L &= 0 \\ B_E\tilde{u}_E + B_L\tilde{u}_L &= 0, \end{aligned} \tag{4}$$

with $B = [B_E\ B_L]$. We perform a (local) nonlinear elimination of $\tilde{u}_E$. To construct our new nonlinear FETI-DP methods, we first derive a nonlinear Schur complement in $(\tilde{u}_L, \lambda)$. Let $(\tilde{u}_E^*, \tilde{u}_L^*, \lambda^*)$ be a solution of (4). We assume there is an implicit function h with the following property in a neighborhood of $(\tilde{u}_E^*, \tilde{u}_L^*, \lambda^*)$:

$$\widetilde{K}_E(h(\tilde{u}_L^*, \lambda^*), \tilde{u}_L^*) + B_E^T\lambda^* - \tilde{f}_E = 0. \tag{5}$$

Here, we consider the first equation from (4). The derivative of the implicit function is

$$Dh(\tilde{u}_L, \lambda) = (D_{\tilde{u}_L}h(\tilde{u}_L, \lambda), D_\lambda h(\tilde{u}_L, \lambda)), \tag{6}$$

where $D_{\tilde{u}_L}h(\tilde{u}_L, \lambda) = -(D_{\tilde{u}_E}\widetilde{K}_E(h(\tilde{u}_L, \lambda), \tilde{u}_L))^{-1}D_{\tilde{u}_L}\widetilde{K}_E(h(\tilde{u}_L, \lambda), \tilde{u}_L) \tag{7}$

and $D_\lambda h(\tilde{u}_L, \lambda) = -(D_{\tilde{u}_E}\widetilde{K}_E(h(\tilde{u}_L, \lambda), \tilde{u}_L))^{-1}B_E^T. \tag{8}$

Inserting the implicit function into equations two and three from (4) we can define a nonlinear Schur complement by

$$S_L(\tilde{u}_L, \lambda) := \begin{bmatrix} \widetilde{K}_L(h(\tilde{u}_L, \lambda), \tilde{u}_L) + B_L^T\lambda - \tilde{f}_L \\ B_E h(\tilde{u}_L, \lambda) + B_L\tilde{u}_L \end{bmatrix}. \tag{9}$$

We finally solve the nonlinear problem $S_L(\tilde{u}_L^*, \lambda^*) = 0$ with Newton's method and obtain the iteration

$$\begin{pmatrix} \tilde{u}_L^{(k+1)} \\ \lambda^{(k+1)} \end{pmatrix} = \begin{pmatrix} \tilde{u}_L^{(k)} \\ \lambda^{(k)} \end{pmatrix} - (DS_L(\tilde{u}_L^{(k)}, \lambda^{(k)}))^{-1} S_L(\tilde{u}_L^{(k)}, \lambda^{(k)}). \tag{10}$$

Using (7) and (8), the short hand notation introduced in (3), and, for simplicity, omitting the variables and indices, we obtain

$$DS_L(\tilde{u}_L, \lambda) = \begin{bmatrix} D\widetilde{K}_{LL} - D\widetilde{K}_{LE}D\widetilde{K}_{EE}^{-1}D\widetilde{K}_{EL} & -D\widetilde{K}_{LE}D\widetilde{K}_{EE}^{-1}B_E^T + B_L^T \\ -B_E D\widetilde{K}_{EE}^{-1}D\widetilde{K}_{EL} + B_L & -B_E D\widetilde{K}_{EE}^{-1}B_E^T \end{bmatrix}. \tag{11}$$

It is easy to verify that the derivative of the nonlinear Schur complement in (11) is equal to the Schur complement of the derivative of the nonlinear saddle point system in (4). Therefore, we can use any FETI-DP type method and solve a linear system equivalent to the linear system in (10). In order to assemble and solve (10) we need to compute $h(\tilde{u}_{\Pi}^{(k)}, \lambda^{(k)})$ first. We consider local nonlinear problems in each global Newton step, arising from the first equation in (4)

$$\widetilde{K}_E(h(\tilde{u}_L^{(k)}, \lambda^{(k)}), \tilde{u}_L^{(k)}) + B_E \lambda^{(k)} - \tilde{f}_E = 0. \tag{12}$$

Since $\tilde{u}_L^{(k)}$ and $\lambda^{(k)}$ are given as results of the k-th step of the global Newton iteration (10), we can simply perform a local Newton iteration to find $\tilde{u}_E^{(k)} = h(\tilde{u}_L^{(k)}, \lambda^{(k)})$. The local iteration writes

$$\tilde{u}_E^{(l+1)} = \tilde{u}_E^{(l)} - (D\widetilde{K}(\tilde{u}_E^{(l)}, \tilde{u}_L^{(k)}))_{EE}^{-1} (\widetilde{K}_E(\tilde{u}_E^{(l)}, \tilde{u}_L^{(k)}) + B_E \lambda^{(k)} - \tilde{f}_E). \tag{13}$$

Let us finally remark that, since $E \cap \Pi = \emptyset$, $D\widetilde{K}(\tilde{u}_E^{(l)}, \tilde{u}_L^{(k)}))_{EE}$ is block diagonal and thus all computations in (13) are local to the subdomains.

Two Different Variants We suggest two different choices of E. First, we define $E := B = [I \; \Delta]$ as the set of interior and dual variables. Consequently, we have $L = \Pi$, $B_E = B_B$, and $B_L = 0$. This defines the Nonlinear-FETI-DP-3 (NL-3) method, where local nonlinear problems in u_B are solved in each global Newton step; see Fig. 1. In this method, the coarse space can slightly influence the convergence of Newton's method, since primal constraints on edges, or faces in three dimensions, influence the variables u_B. As a second choice, we use $E := I$ and thus we have $L = \Delta \cup \Pi =: \Gamma$, $B_E = 0$, and $B_L = B_\Gamma$. This leads to the Nonlinear-FETI-DP-4 (NL-4) method, where local nonlinear problems in u_I are solved in each global Newton step; see Fig. 2. In this method, the coarse space cannot influence Newton's method, since the local problems are independent of the variables on the interface.

Init: $(u_B^{(0)}, \tilde{u}_\Pi^{(0)}) = \tilde{u}^{(0)} \in \widetilde{W}$, $\lambda^{(0)} = 0$
for $k = 0, ..., convergence$

    for $l = 0, ..., convergence$
        build: $\tilde{K}(\tilde{u}^{(l)})$ and $D\tilde{K}(\tilde{u}^{(l)})$
        solve: $(D\tilde{K}(\tilde{u}^{(l)}))_{BB} \delta u_B^{(l)} = K_B(\tilde{u}^{(l)}) + B_B^T \lambda^{(k)} - f_B$ //local problems
        compute steplength $\alpha^{(l)}$
        update: $\tilde{u}^{(l+1)} := \tilde{u}^{(l)} - \alpha^{(l)} \left(\delta u_B^{(l)T}, 0 \right)^T$ //update only on B
    end
    $\tilde{u}^{(k)} := \tilde{u}^{(l+1)}$
    build: $\tilde{K}(\tilde{u}^{(k)})$ and $D\tilde{K}(\tilde{u}^{(k)})$
    solve: $DS_\Pi(\tilde{u}_\Pi^{(k)}, \lambda^{(k)}) \begin{pmatrix} \delta \tilde{u}_\Pi^{(k)} \\ \delta \lambda^{(k)} \end{pmatrix} = \begin{pmatrix} \tilde{K}_\Pi(\tilde{u}^{(k)}) - \tilde{f}_\Pi \\ B_B u_B^{(k)} \end{pmatrix}$ //solve equivalent
    FETI-DP system
    compute steplength $\alpha^{(k)}$
    update: $\lambda^{(k+1)} := \lambda^{(k)} - \alpha^{(k)} \delta \lambda^{(k)}$
    update: $\tilde{u}_\Pi^{(k+1)} := \tilde{u}_\Pi^{(k)} - \alpha^{(k)} \delta \tilde{u}_\Pi^{(k)}$
    $\tilde{u}^{(0)} := \left(u_B^{(l+1)T}, \tilde{u}_\Pi^{(k+1)T} \right)^T$
    $\lambda^{(0)} := \lambda^{(k+1)}$

end

Fig. 1 Pseudocode of Nonlinear-FETI-DP-3

Init: $(u_I^{(0)}, \tilde{u}_\Gamma^{(0)}) = \tilde{u}^{(0)} \in \widetilde{W}$, $\lambda^{(0)} = 0$
for $k = 0, ..., convergence$

    for $l = 0, ..., convergence$
        build: $\tilde{K}(\tilde{u}^{(l)})$ and $D\tilde{K}(\tilde{u}^{(l)})$
        solve: $(D\tilde{K}(\tilde{u}^{(l)}))_{II} \delta u_I^{(l)} = K_I(\tilde{u}^{(l)}) - f_I$ //local problems
        compute steplength $\alpha^{(l)}$
        update: $\tilde{u}^{(l+1)} := \tilde{u}^{(l)} - \alpha^{(l)} \left(\delta u_I^{(l)T}, 0 \right)^T$ //update only on I
    end
    $\tilde{u}^{(k)} := \tilde{u}^{(l+1)}$
    build: $\tilde{K}(\tilde{u}^{(k)})$ and $D\tilde{K}(\tilde{u}^{(k)})$
    solve: $DS_\Gamma(\tilde{u}_\Gamma^{(k)}, \lambda^{(k)}) \begin{pmatrix} \delta \tilde{u}_\Gamma^{(k)} \\ \delta \lambda^{(k)} \end{pmatrix} = \begin{pmatrix} \tilde{K}_\Gamma(\tilde{u}^{(k)}) + B_\Gamma^T \lambda^{(k)} - \tilde{f}_\Gamma \\ B_\Gamma u_\Gamma^{(k)} \end{pmatrix}$ //solve equiv-
    alent FETI-DP system
    compute steplength $\alpha^{(k)}$
    update: $\lambda^{(k+1)} := \lambda^{(k)} - \alpha^{(k)} \delta \lambda^{(k)}$
    update: $\tilde{u}_\Gamma^{(k+1)} := \tilde{u}_\Gamma^{(k)} - \alpha^{(k)} \delta \tilde{u}_\Gamma^{(k)}$
    $\tilde{u}^{(0)} := \left(u_I^{(l+1)T}, \tilde{u}_\Gamma^{(k+1)T} \right)^T$
    $\lambda^{(0)} := \lambda^{(k+1)}$

end

Fig. 2 Pseudocode of Nonlinear-FETI-DP-4

4 Numerical Results

As a first model problem, we consider a scaled p-Laplace equation

$$-\text{div}(\alpha|\nabla u|^2\nabla u - \beta\nabla u) = 1 \ \text{ in } \Omega, \ \ u = 0 \ \text{ on } \partial\Omega, \tag{14}$$

where $\alpha, \beta : \Omega \to \mathbb{R}$ are coefficient functions given by

$$\alpha(x) = \begin{cases} 10^6 & \text{if } x \in \Omega_C \\ 0 & \text{elsewhere} \end{cases} \qquad \beta(x) = \begin{cases} 0 & \text{if } x \in \Omega_C \\ 1 & \text{elsewhere;} \end{cases} \tag{15}$$

see Fig. 3 for a definition of Ω_C.

In Table 1, we present results for the p-Laplace problem (14). Here, NL-4 and Newton-Krylov-FETI-DP both require many Krylov iterations. The local nonlinear problems on the interior part of the subdomains solved in NL-4 cannot resolve the strongly global nonlinearity of the channels. Comparable good results in terms of Krylov space iterations are obtained using NL-2 and NL-3. The new NL-3 method additionally reduces the number of FETI-DP coarse solves drastically and thus is potentially faster in a parallel setup. In contrast to NL-2, where in each global Newton step nonlinear problems in W including the FETI-DP coarse problem have to be solved, in NL-3 and NL-4 the coarse solves are only necessary in the global Newton iteration.

Our second model problem is a nonlinear hyperelasticity problem. We consider a Neo-Hooke material ($\nu = 0.3$) with a soft matrix material ($E = 210$) and stiff inclusions ($E = 210,000$); see Fig. 4 (left) for the geometry. The strain energy density function W (Holzapfel 2000) is given by $W(u) = \frac{\mu}{2}\left(\text{tr}(F^T F) - 3\right) - \mu\ln(\det(F)) + \frac{\lambda}{2}\ln^2(\det(F))$ with the Lamé constants $\lambda = \frac{\nu E}{(1+\nu)(1-2\nu)}$, $\mu = \frac{E}{2(1+\nu)}$ and the deformation gradient $F(x) := \nabla\varphi(x)$. Here, $\varphi(x) = x + u(x)$ denotes the deformation and $u(x)$ the displacement of x. The energy functional of which

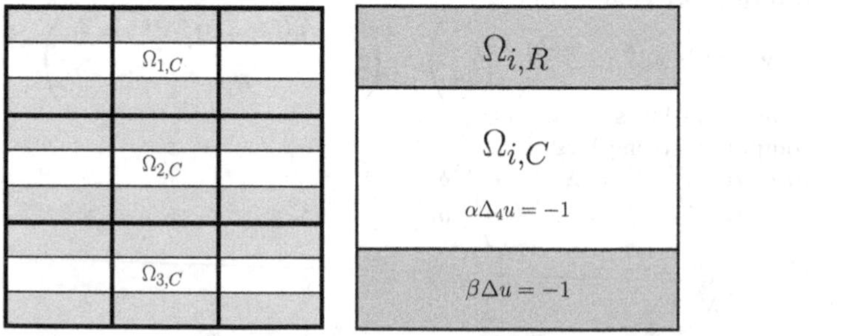

Fig. 3 *Left*: example for a decomposition of Ω in $N = 9$ subdomains, intersected by 3 channels $\Omega_{i,C}, i = 1, 2, 3$. We define $\Omega_C = \bigcup_i \Omega_{i,C}$. *Right*: subdomain Ω_i with channel $\Omega_{i,C}$ of width $\frac{H}{2}$, where H is the size of a subdomain

Table 1 p-Laplace problem; channels of p-Laplace ($p = 4$) with high coefficient $1e6$ in standard linear Laplacian matrix

N	Solver	# Krylov It.	# local solves	# coarse solves	Min. cond.	Max. cond.
	NK-FETI-DP	864	19	19	95.9	31,265.6
	Nonlinear-FETI-DP-1	537	26	26	39.5	151.5
64	Nonlinear-FETI-DP-2	225	34	34	39.6	95.9
	Nonlinear-FETI-DP-3	264	36	6	30.4	95.9
	Nonlinear-FETI-DP-4	1343	56	17	95.8	32,520.7
	NK-FETI-DP	2341	19	19	158.1	59,730.5
	Nonlinear-FETI-DP-1	1128	26	26	60.5	255.2
256	Nonlinear-FETI-DP-2	**481**	**34**	34	60.6	158.4
	Nonlinear-FETI-DP-3	529	38	**6**	39.6	158.9
	Nonlinear-FETI-DP-4	2766	54	18	158.0	60,415.5

N: number of subdomains; **Krylov It.**: sum of CG iterations over all Newton steps; **local solves**: number of local factorizations on subdomains; **coarse solves**: number of FETI-DP coarse problem factorizations. Best results are marked in **bold face**

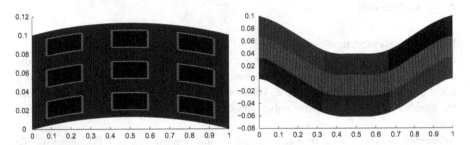

Fig. 4 *Left*: initial value (reference configuration) and two different materials with $v = 0.3$ everywhere, $E_1 = 210,000$ in the red inclusions, and $E_2 = 210$ in the blue matrix material. *Right*: solution when a volume force $f_v = [0, -10]^T$ is applied

stationary points are computed, is given by

$$J(u) = \int_\Omega W(u) - V(u)dx - \int_\Gamma G(u)ds,$$

where $V(u)$ and $G(u)$ are functionals related to the volume and traction forces. The nonlinear elasticity problem is discretized with piecewise linear finite elements. In Table 2 we present the results for our Neo-Hooke model problem described in Fig. 4. We only considered continuity in vertices as primal constraints, which is not an optimal coarse space for highly heterogeneous elasticity problems. This leads to divergence of NL-1 and NL-2 when using no further globalization strategy. Since the coarse space does not influence the convergence behavior of Newton-Krylov-FETI-DP and NL-4, both methods converge. Due to the local nonlinear problems

Table 2 Heterogeneous Neo-Hooke problem; see Fig. 4

D.o.f.	N	Solver	#Krylov-It.	# local solves	# coarse solves
		NK-FETI-DP	595	10	10
51,842	64	NL-FETI-DP-4	356	12	6
		NK-FETI-DP	939	**10**	10
206,082	256	NL-FETI-DP-4	**491**	12	**6**

Using GMRES as Krylov solver and primal vertex constraints; **d.o.f.**: problem size; **N**: number of subdomains; **Krylov It.**: sum of GMRES iterations over all Newton steps; **local solves**: number of local factorizations on subdomains; **coarse solves**: number of FETI-DP coarse problem factorizations. Best results are marked in **bold face**

solved in NL-4, the number of GMRES iterations is reduced up to 47% compared to Newton-Krylov-FETI-DP. Also the number of necessary coarse solves is reduced in NL-4. Of coarse, in the nonlinear variant, the local work is increased slightly.

5 Conclusion

We have presented new nonlinear FETI-DP variants based on a partial nonlinear elimination of interior and interface variables. These methods can remove the influence of the coarse space to the Newton convergence and can be superior if a good coarse space is not available. We have seen that the new methods can reduce the number of FETI-DP coarse solves drastically.

Acknowledgements This work was supported by the German Research Foundation (DFG) through the Priority Programme 1648 "Software for Exascale Computing" (**SPPEXA**) under KL 2094/4-1, KL 2094/4-2, RH 122/2-1 and RH 122/3-2.

References

C. Farhat, M. Lesoinne, P. LeTallec, K. Pierson, D. Rixen, FETI-DP: a dual-primal unified FETI method - part I: a faster alternative to the two-level FETI method. Int. J. Numer. Methods Eng. **50**, 1523–1544 (2001)

G.A. Holzapfel, *Nonlinear Solid Mechanics. A Continuum Approach for Engineering* (Wiley, Chichester, 2000)

A. Klawonn, O. Rheinbach, Highly scalable parallel domain decomposition methods with an application to biomechanics. ZAMM Z. Angew. Math. Mech. **90**(1), 5–32 (2010)

A. Klawonn, M. Lanser, P. Radtke, O. Rheinbach, On an adaptive coarse space and on nonlinear domain decomposition, in *Domain Decomposition Methods in Science and Engineering XXI*, ed. by J. Erhel, M.J. Gander, L. Halpern, G. Pichot, T. Sassi, O.B. Widlund. Lecture Notes in Computational Science and Engineering, vol. 98 (Springer, Berlin, 2014a), pp. 71–83

A. Klawonn, M. Lanser, O. Rheinbach, Nonlinear FETI-DP and BDDC methods. SIAM J. Sci. Comput. **36**(2), A737–A765 (2014b)

A. Klawonn, M. Lanser, O. Rheinbach, FE^{2}TI: computational scale bridging for dual-phase steels. in *Parallel Computing: On the Road to Exascale*. IOS Series Advances in Parallel Computing, vol. 27, pp. 797–806 (2015). Proceedings of ParCo 2015. http://dx.doi.org/10.3233/978-1-61499-621-7-797 . Also TUBAF Preprint 2015-12. http://tu-freiberg.de/fakult1/forschung/preprints

A. Klawonn, M. Lanser, O. Rheinbach, Toward extremely scalable nonlinear domain decomposition methods for elliptic partial differential equations. SIAM J. Sci. Comput. **37**(6), C667–C696 (2015)

Direct and Iterative Methods for Numerical Homogenization

Ralf Kornhuber, Joscha Podlesny, and Harry Yserentant

1 Introduction

Numerical approximation usually aims at modifications of standard finite element approximations of partial differential equations with highly oscillatory coefficients that preserve the accuracy known in the smooth case. Using classical homogenization as a guideline, these modifications are obtained from local auxiliary problems (Abdulle et al., 2012; Efendiev and Hou, 2009; Hughes et al., 1998). The error analysis for these kinds of methods is typically restricted to coefficients with separated scales and often requires periodicity (Abdulle, 2011; Abdulle et al., 2012; Hou et al., 1999). These restrictions were overcome in a recent paper by Målqvist and Peterseim (2014) that provides quasioptimal energy and L^2 error estimates without any additional assumptions on periodicity and scale separation (Henning et al., 2014; Målqvist and Peterseim, 2014). While their approach relies on (approximate) orthogonal subspace decomposition, alternative decompositions into a coarse space and local fine-grid spaces associated with low and high frequencies has been recently considered by Kornhuber and Yserentant (2015). Here, we review these two decomposition techniques providing direct (Målqvist and Peterseim, 2014) and iterative methods (Kornhuber and Yserentant, 2015) for numerical homogenization in order to better understand conceptual similarities and differences. We also illustrate the performance of the iterative variant by first numerical experiments in $d = 3$ space dimensions.

R. Kornhuber (✉) • J. Podlesny
FU Berlin, Berlin, Germany
e-mail: ralf.kornhuber@fu-berlin.de; joscha.podlesny@fu-berlin.de

H. Yserentant
TU Berlin, Berlin, Germany
e-mail: yserentant@math.tu-berlin.de

© Springer International Publishing AG 2017
C.-O. Lee et al. (eds.), *Domain Decomposition Methods in Science and Engineering XXIII*, Lecture Notes in Computational Science and Engineering 116, DOI 10.1007/978-3-319-52389-7_21

Both approaches rely on subspace decomposition in function space while practical, discrete variants aim at approximating a sufficiently accurate, computationally unfeasible fine-grid solution up to discretization accuracy. This approximation is either obtained directly from a linear system as derived from local fine-grid problems (Målqvist and Peterseim, 2014) or iteratively by repeated solution of coarse- and local fine-grid problems (Kornhuber and Yserentant, 2015). Comparing the computational effort, the direct method requires assembly of the multiscale stiffness matrix and usually leads to larger local fine-grid problems than the iterative approach. In addition, the local fine-grid problems involve a saddle point structure (Målqvist and Peterseim, 2014, Remark 4.5) rather than positive-definite stiffness matrices (Kornhuber and Yserentant, 2015). However, in contrast to iterative homogenization the direct approach provides a reduced multiscale basis that incorporates all relevant features and has various advantages, e.g., in case of many different right-hand sides.

2 Elliptic Problems with Oscillating Coefficients

Let $\Omega \subset \mathbb{R}^d$, $d = 2$ or $d = 3$, be a bounded convex domain with polygonal or polyhedral boundary $\partial\Omega$. We consider the variational problem

$$u \in V: \qquad a(u, v) = (f, v) \qquad \forall v \in V, \tag{1}$$

where $V = H_0^1(\Omega)$ is a closed subspace of $H^1(\Omega)$, $(\cdot, \cdot)$ is the canonical scalar product in $L^2(\Omega)$, and $f \in L^2(\Omega)$. The bilinear form $a(\cdot, \cdot)$ takes the form $a(v, w) = \int_\Omega \nabla v(x) \cdot A(x) \nabla w(x)\, dx$, $v, w \in V$, where $A(x) \in \mathbb{R}^{d \times d}$ is a symmetric matrix with sufficiently smooth, but intentionally highly oscillating entries and

$$\delta |\eta|^2 \leq \eta \cdot A(x) \eta \leq M |\eta|^2 \tag{2}$$

holds for all $\eta \in \mathbb{R}^d$ and almost all $x \in \Omega$ with positive constants δ, M independent of x and η. It is well-known that (1) admits a unique solution and, for ease of presentation, we assume $u \in V \cap H^2(\Omega)$. As a model problem, one might think of two separate scales

$$A(x) = \alpha \left(x, \frac{x}{\varepsilon} \right) I, \qquad x \in \Omega, \tag{3}$$

with the identity matrix I and a fine-scale parameter $\varepsilon > 0$. For periodic coefficients α, the oscillatory problem (1) can be treated by classical homogenization via the solution of certain continuous cell problems. However, no scale separation, periodicity, or exact solvability of continuous cell problems will be assumed throughout the rest of the presentation.

Let $\mathcal{T}_H$ denote a regular partition of Ω into simplices with maximal diameter $H > 0$. The corresponding space of piecewise affine finite elements

$$\mathcal{S}_H = \{v \in C(\overline{\Omega}) \mid v|_{\partial\Omega} = 0 \text{ and } v|_t \text{ affine } \forall t \in \mathcal{T}_H\}$$

is spanned by the nodal basis $\lambda_p \in \mathcal{S}_H$, $p \in \mathcal{N}_H$, where $\mathcal{N}_H$ stands for the set of interior vertices of $\mathcal{T}_H$. The usual finite element approximation is given by $u_H = P_{\mathcal{S}_H} u$ with $P_{\mathcal{S}_H} : V \to \mathcal{S}_H$ denoting the Ritz projection defined by

$$P_{\mathcal{S}_H} w \in \mathcal{S}_H : \qquad a(P_{\mathcal{S}_H} w, v) = a(w, v) \qquad \forall v \in \mathcal{S}_H.$$

We have the well-known error estimate $\|u - u_H\| \lesssim H\|u\|_{H^2(\Omega)}$, where $\|\cdot\| = a(\cdot, \cdot)^{1/2}$ signifies the energy norm. Here and throughout this paper, we write $a \lesssim b$, if $a \leq cb$ holds with a constant c only depending on the contrast M/δ and on the shape regularity of $\mathcal{T}_H$. Unfortunately, $\|u\|_{H^2(\Omega)}$ depends on the oscillatory behavior of A. For example, we have $\|u\|_{H^2(\Omega)} = \mathcal{O}(\varepsilon^{-1})$ and thus $\|u - u_H\| \lesssim \varepsilon^{-1} H$ in the model case (3). Numerical homogenization is aiming at a modified finite element space $\mathcal{S}_H^{ms}$ with $\dim \mathcal{S}_H^{ms} = \dim \mathcal{S}_H$ such that $u_H^{ms} = P_{\mathcal{S}_H^{ms}}$ satisfies $\|u - u_H^{ms}\| \lesssim H$.

3 Direct Homogenization by Localized Orthogonal Decomposition

Let $\Pi : V \to \mathcal{S}_H$ denote a quasi-interpolation with the property

$$\|v - \Pi v\|_{0,t} \leq C_\Pi H \|\nabla v\|_{0,\omega_t} \qquad \forall t \in \mathcal{T}_H, \quad \forall v \in V, \tag{4}$$

with local L^2-norms $\|\cdot\|_{0,t}$, $\|\cdot\|_{0,\omega_t}$ on t, ω_t, respectively, and let ω_t be the union of $t' \in \mathcal{T}_H$ with $t \cap t' \neq \emptyset$. A possible choice is the Clément-type operator (Clément, 1975)

$$\Pi v = \sum_{p \in \mathcal{N}_H} v_p \lambda_p, \qquad v_p = \frac{1}{\omega_p} \int_{\omega_p} v \, dx, \qquad \omega_p = \text{int supp } \lambda_p. \tag{5}$$

The main idea taken from Målqvist and Peterseim (2014) is to consider the a-orthogonal decomposition

$$V = \mathcal{S}_H^{ms} + V^f \tag{6}$$

into the kernel V^f of Π and its a-orthogonal complement $\mathcal{S}_H^{ms} = (I - P_{V^f})V$.

Proposition 1 *The Ritz projection $u_H^{ms} \in \mathcal{S}_H^{ms}$ of u on $\mathcal{S}_H^{ms}$ satisfies*

$$\|u - u_H^{ms}\| \lesssim H. \tag{7}$$

Proof Orthogonality of the splitting (6) implies that $w = u - u_H^{ms} \in V^f$ fulfills $\|w\|^2 = (f, w)$. Utilizing the local L^2 scalar product $(\cdot, \cdot)_t$, (4), (2), the local energy norm $\| \cdot \|_t$, the binomial formula, and the L^2 norm $\| \cdot \|_0$, we get

$$(f, w) = \sum_{t \in \mathcal{T}_H} (f, w)_t = \sum_{t \in \mathcal{T}_H} (f, w - \Pi w)_t \lesssim \sum_{t \in \mathcal{T}_H} \|f\|_{0,t} H \|\nabla w\|_{0,\omega_t}$$
$$\lesssim \sum_{t \in \mathcal{T}_H} s^{-1} H \|f\|_{0,t} s \|w\|_{\omega_t} \lesssim \tfrac{1}{2} s^{-2} H^2 \|f\|_0^2 + \tfrac{1}{2} c s^2 \|w\|^2$$

with positive $s \in \mathbb{R}$. The assertion follows by choosing s sufficiently small.   □

Note that different choices of Π give rise to different multiscale methods. We refer to Henning et al. (2014) and Målqvist and Peterseim (2014) for a detailed discussion.

A basis $\lambda_p^{ms} = (I - P_{V^f})\lambda_p$ of $\mathcal{S}_H^{ms}$ is obtained from the local problems

$$\mu_p^{ms} \in V^f : \qquad a(\mu_p^{ms}, v) = a(\lambda_p, v) \qquad \forall v \in V^f \tag{8}$$

for the multiscale corrections $\mu_p^{ms} = P_{V^f} \lambda_p$. Unfortunately, the resulting multiscale basis functions λ_p^{ms} have global support so that sparsity of the corresponding stiffness matrix is lost. As a way out, Målqvist and Peterseim (2014) consider the localized orthogonal projection

$$\mu_p^k \in V^f(\omega_{p,k}) : \qquad a(\mu_p^k, v) = a(\lambda_p, v) \qquad \forall v \in V^f(\omega_{p,k}) \tag{9}$$

with local patches $\omega_{p,k}$ of order $k \in \mathbb{N}$ defined by

$$\omega_{p,1} = \omega_p, \qquad \omega_{p,k} = \text{int}\,\{t \in \mathcal{T}_H \mid t \cap \omega_{p,k-1} \neq \emptyset\}, \quad k > 1, \tag{10}$$

and $V^f(\omega_{p,k}) = \{v \in V^f \mid \text{int supp } v \in \omega_{p,k}\}$. The resulting multiscale finite element space now reads $\mathcal{S}_H^k = \text{span}\,\{\lambda_p^k = \lambda_p - \mu_p^k \mid p \in \mathcal{N}_H\}$. Exploiting the decay properties of Green's functions (Målqvist and Peterseim, 2014) [see Henning et al. (2014) for a later, more elegant proof] were able to show that the desired error estimate (7) is preserved under localization (9).

Theorem 1 *The Ritz projection u_H^k of the solution u of (1) to $\mathcal{S}_H^k$ admits the error estimate $\|u - u_H^k\| \lesssim H$ for sufficiently large $k \gtrsim H^{-1}$.*

The solution of the localized problems (9) is computationally unfeasible, because $\dim V^f = \infty$. As a way out, the continuous solution space V is replaced by a possibly unfeasibly fine finite element space $\mathcal{S}_h$ providing an approximation $u_h = P_{\mathcal{S}_h} u$ with accuracy $\|u - u_h\| \lesssim H$. In the model case (3), we might choose $\mathcal{S}_h$ associated with a uniform partition $\mathcal{T}_h$ with mesh size $h = H\varepsilon^{-1}$. Repeating the above reasoning with V^f replaced by $V_h^f = \ker \Pi|_{\mathcal{S}_h}$, $V^f(\omega_{p,k})$ replaced by $V_h^f(\omega_{p,k}) = V^f(\omega_{p,k}) \cap V_h^f$, etc., we obtain the multiscale finite element space $\mathcal{S}_{H,h}^k = \text{span}\,\{\lambda_{p,h}^k = \lambda_p - \mu_{p,h}^k \mid p \in \mathcal{N}_H\}$ with discrete multiscale corrections

$\mu_{p,h}^k$ obtained from

$$\mu_{p,h}^k \in V_h^f(\omega_{p,k}) : \qquad a(\mu_{p,h}^k, v) = a(\lambda_p, v) \qquad \forall v \in V_h^f(\omega_{p,k}). \tag{11}$$

For quasi-interpolations Π like the one defined in (5), there is no local basis of the linearly constrained subspaces $V_h^f = \ker \Pi|_{S_h}$. Hence, the constraint $\Pi v = 0$ is usually enforced by a Lagrange multiplier so that the algebraic solution of (11) amounts to solving a saddle point problem. Utilizing essentially the same arguments as before, the error estimates in Proposition 1 and Theorem 1 directly carry over to the discrete case.

Theorem 2 *The Ritz projection $u_{H,h}^k$ of the solution u of (1) to $S_{H,h}^k$ admits the error estimate $\|u - u_{H,h}^k\| \lesssim H$ for sufficiently large $k \gtrsim H^{-1}$.*

Note that localized orthogonal decomposition can be regarded as a direct method to approximate u_h up to the discretization error by the solution $u_{H,h}^k$ of a much smaller problem. From such a perspective, multiscale finite element methods appears to be a kind of model reduction.

4 Iterative Homogenization by Subspace Correction

The main idea of iterative homogenization is to derive an iterative scheme that allows for solving the given boundary value problem (1) up to a prescribed accuracy with a number of steps that depends only on the contrast M/δ from (2) and on the shape regularity of $\mathcal{T}_H$. To this end, we consider the splitting

$$V = S_H + \sum_{p \in \mathcal{N}_H} V_p, \qquad V_p = H_0^1(\omega_p), \tag{12}$$

with ω_p defined in (5) and $\mathcal{N}_H$ consisting of all vertices of $\mathcal{T}_H$. This splitting induces a parallel subspace correction method providing the preconditioner

$$T = P_{S_H} + \sum_{p \in \mathcal{N}_H} P_{V_p}. \tag{13}$$

Utilizing basic results from subspace correction (Xu, 1992; Yserentant, 1993), spectral equivalence

$$K_1^{-1} a(v, v) \le a(Tv, v) \le K_2 a(v, v) \qquad \forall v \in V, \tag{14}$$

follows from the stability of the splitting (12). This means that for any $v \in V$ there is a decomposition $v = v_H + \sum_{p \in \overline{\mathcal{N}}_H} v_p$ into $v_H \in \mathcal{S}_H$ and $v_p \in V_p, p \in \overline{\mathcal{N}}_H$, such that

$$\|v_H\|^2 + \sum_{p \in \overline{\mathcal{N}}_H} \|v_p\|^2 \leq K_1 \|v\|^2 \tag{15}$$

is satisfied with a constant $K_1 > 0$ and such that

$$\|v\|^2 \leq K_2(\|v_H\|^2 + \sum_{p \in \overline{\mathcal{N}}_H} \|v_p\|^2) \tag{16}$$

holds with a constant $K_2 > 0$ for any such decomposition. The following proposition taken from Kornhuber and Yserentant (2015) is crucial for the rest of this exposition.

Proposition 2 *The splitting* (12) *is stable with positive constants* K_1, K_2 *depending only on the contrast* M/δ *and on the shape regularity of* $\mathcal{T}_H$.

It is not difficult to realize that (16) with $K_2 = d + 2$ follows from the Cauchy-Schwarz inequality. Exploiting the quasi-interpolation Π defined in (5) and that the functions $\lambda_p, p \in \overline{\mathcal{N}}_H$, form a partition of unity, it turns out that (15) holds for the decomposition $v_H = \Pi v$, $v_p = \lambda_p(v - \Pi v), p \in \overline{\mathcal{N}}_H$. We refer to Kornhuber and Yserentant (2015) for details.

Note that, in contrast to direct numerical homogenization as explained above, the quasi-interpolation Π now only enters the proof of the condition number estimate, but not the algorithm itself.

Employing spectral equivalence (14), we can use the spectral mapping theorem to obtain usual error bounds for preconditioned cg iterations in function space.

Theorem 3 *The convergence rate* ρ *of the preconditioned cg iteration with preconditioner* T *satisfies* $\rho \leq \frac{\sqrt{\kappa}-1}{\sqrt{\kappa}+1}, \kappa \leq K_1 K_2$, *so that the error estimate* $\|u - u^\nu\| \lesssim Tol$ *holds for* $\nu \gtrsim \log(Tol^{-1})$ *and any given tolerance* $Tol > 0$.

Note that, in contrast to direct numerical homogenization, the achievable accuracy is independent of the choice of $\mathcal{S}_H$.

Of course, the preconditioner (13) is computationally unfeasible, because the evaluation of the local Ritz projections $P_{V_p}, p \in \overline{\mathcal{N}}_H$, amounts to the solution of continuous variational problems. As in the previous section, the continuous solution space V is therefore replaced by a, possibly unfeasibly large, finite element space $\mathcal{S}_h \subset V$ that provides an approximation $u_h = P_{\mathcal{S}_H} u$ with accuracy of order H. We then consider the discrete splitting

$$\mathcal{S}_h = \mathcal{S}_H + \sum_{p \in \overline{\mathcal{N}}_H} V_{p,h}, \qquad V_{p,h} = \mathcal{S}_h \cap H_0^1(\omega_p), \tag{17}$$

and the associated preconditioner

$$T_h = P_{\mathcal{S}_H} + \sum_{p \in \overline{\mathcal{N}}_H} P_{V_{p,h}}. \tag{18}$$

Similar arguments as in the continuous case provide the stability of the discrete splitting (17) with constants K_1, K_2 depending only on the contrast M/δ from (2) and on the shape regularity of $\mathcal{T}_H$. Hence, spectral equivalence

$$K_1^{-1} \, a(v, v) \leq a(T_h v, v) \leq K_2 \, a(v, v) \qquad \forall v \in \mathcal{S}_h \tag{19}$$

follows from well-known results, e.g., in Xu (1992) and Yserentant (1993). As a consequence, the preconditioned cg iteration in $\mathcal{S}_h$ with preconditioner T_h exhibits mesh-independent convergence rates.

Theorem 4 *The preconditioned cg iteration with preconditioner T_h provides the error estimate $\|u - u_h^v\| \lesssim H$ for $v \gtrsim \log(H^{-1})$ iteration steps as applied to a fixed initial iterate $u_h^0 \in \mathcal{S}_h$.*

Note that the achievable accuracy is limited only by the selection of the space $\mathcal{S}_h$ but not by the space $\mathcal{S}_H$ as opposed to the direct approach.

Each evaluation of the preconditioner T_h requires the evaluation of the Ritz projections to $\mathcal{S}_H$ and $V_{p,h}$, $p \in \overline{\mathcal{N}}_H$, respectively. As local bases of these subspaces are readily available, this amounts to the solution of symmetric, positive-definite, linear systems associated with the coarse grid $\mathcal{T}_H$ and with the local fine grids $\omega_p \cap \mathcal{T}_h$, $p \in \overline{\mathcal{N}}_H$, and not to saddle point problems (11) as in direct numerical homogenization.

Similar results can be achieved for successive subspace corrections based on the splitting (17). We refer to Kornhuber and Yserentant (2015) for further information.

5 Numerical Experiments

We consider the unit cube $\Omega = (0, 1)^3$ and its uniform partition into cubes of edge length $H = 1/8$ which are further subdivided into cubes of edge length $h = 1/32$ (one more uniform refinement step would lead to computations with more than 2×10^6 unknowns). The simplical partitions $\mathcal{T}_H$ and $\mathcal{T}_h$ are obtained by subdividing each cube into six tetrahedra by the Coxeter-Freudenthal-Kuhn triangulation. We consider (1) with $f \equiv 1$ in the model case (3) with a scalar coefficient $\alpha(x)$ which is piecewise constant on a $32 \times 32 \times 32$ cube grid, with values that are uniformly distributed random numbers in an interval with lower bound $\delta = 1$ and upper bound M.

The reduction factors for the energy error $\|u_h - u_h^v\|$ of the preconditioned cg iteration with preconditioner T_h given in (18) and initial iterate $u_h^0 = u_H$ is listed

Table 1 Error reduction factors of preconditioned cg iteration with preconditioner T_h

Step	$M/\delta = 10^0$	$M/\delta = 10^1$	$M/\delta = 10^2$	$M/\delta = 10^4$	$M/\delta = 10^6$
1	0.42289	0.43180	0.43730	0.43673	0.43747
2	0.40494	0.43488	0.44331	0.44399	0.44364
3	0.29253	0.34578	0.34930	0.34953	0.35052
4	0.32946	0.30560	0.30561	0.30714	0.30635
5	0.38972	0.39920	0.40461	0.39976	0.39907
6	0.38917	0.37999	0.38262	0.37489	0.37601
7	0.30847	0.34791	0.35729	0.35498	0.35238
8	0.33201	0.36407	0.38412	0.38667	0.37269
9	0.40475	0.45993	0.47379	0.47412	0.46402
10	0.34971	0.41312	0.41947	0.42260	0.41620

in Table 1 for the ratios $M/\delta = 1, 10, 10^2, 10^4$, and 10^6. The convergence speed does not decrease significantly from $M/\delta = 10^0$, i.e., the simple Laplace equation, to larger and larger contrast, less and less covered by theory. The stopping criterion $\|u_h - u_h^\nu\| \leq \|u_{h/2} - u_h\| \leq \|u - u_h\|$ was reached with at most $\nu = 2$ iteration steps for all considered values of M/δ. Replacing ω_p in (12) by $\omega_{p,k}$, $k > 1$, thus introducing larger overlap, leads to a further improvement of reduction factors. Though error reduction will probably saturate at slightly larger values for mesh sizes $h < 1/32$, we found a similar convergence behavior for $h = 1/512$ in 2D and these computations confirm the potential of iterative methods for numerical homogenization.

Acknowledgements This research has been funded by Deutsche Forschungsgemeinschaft (DFG) through grant CRC 1114.

References

A. Abdulle, A priori and a posteriori error analysis for numerical homogenization: a unified framework. Ser. Contemp. Appl. Math. **16**, 280–305 (2011)

A. Abdulle, E. Weinan, B. Engquist, E. Vanden-Eijnden, The hetereogeneous multiscale method. Acta Numer. **21**, 1–87 (2012)

Ph. Clément, Approximation by finite element functions using local regularization. Rev. Franc. Automat. Inform. Rech. Operat. **9**, 77–84 (1975)

Y. Efendiev, T.Y. Hou, *Multiscale Finite Element Methods: Theory and Applications* (Springer, New York, 2009)

P. Henning, A. Målqvist, D. Peterseim, A localized orthogonal decomposition method for semilinear elliptic problems. ESAIM: Math. Model. Numer. Anal. **48**, 1331–1349 (2014)

T.Y. Hou, X.-H. Wu, Z. Cai, Convergence of a multiscale finite element method for elliptic problems with rapidly oscillating coefficients. Math. Comput. **68**, 913–943 (1999)

T.J.R. Hughes, G.R. Feijó, L.M. Mazzei, J.-B. Quincy, The variational multiscale method - a paradigm for computational mechanics. Comput. Methods Appl. Mech. Eng. **166**, 3–24 (1998)

R. Kornhuber, H. Yserentant, Numerical homogenization of elliptic multiscale problems by subspace decomposition. Multiscale Model. Simul. **14**, 1017–1036 (2016)

A. Målqvist, D. Peterseim, Localization of elliptic multiscale problems. Math. Comput. **83**, 2583–2603 (2014)

J. Xu, Iterative methods by space decomposition and subspace correction. SIAM Rev. **34**, 581–613 (1992)

H. Yserentant, Old and new convergence proofs for multigrid methods. Acta Numer. **2**, 285–326 (1993)

Nonlinear Multiplicative Schwarz Preconditioning in Natural Convection Cavity Flow

Lulu Liu, Wei Zhang, and David E. Keyes

1 Introduction

The multiplicative Schwarz preconditioned inexact Newton (MSPIN) algorithm, as a complement to additive Schwarz preconditioned inexact Newton (ASPIN), provides a Gauss-Seidel-like way to improve the global convergence of systems with unbalanced nonlinearities. To demonstrate, a natural convection cavity flow PDE system is solved using nonlinear multiplicative Schwarz preconditioners resulting from different groupings and orderings of the PDEs and their associated fields, and convergence results are reported over a range of Rayleigh number, a dimensionless parameter representing the ratio of convection to diffusion, and in this case, of the magnitude of nonlinear to the linear terms in the transport PDEs. The robustness of nonlinear convergence with respect to Rayleigh number is sensitive to the grouping strategy.

Globally nonlinearly implicit methods, such as Newton-Krylov-Schwarz, work well for many problems, but they may be frustrated by "nonlinear stiffness," which

L. Liu (✉)
Institute of Computational Science, Università della Svizzera italiana (USI), 6900, Lugano, Switzerland
e-mail: lulu.liu@usi.ch

W. Zhang
Program in Mechanical Engineering, King Abdullah University of Science and Technology (KAUST), 23955-6900 Thuwal, Saudi Arabia
e-mail: wei.zhang@kaust.edu.sa

D.E. Keyes
Program in Applied Mathematics and Computational Science and Extreme Computing Research Center, King Abdullah University of Science and Technology (KAUST), 23955-6900 Thuwal, Saudi Arabia
e-mail: david.keyes@kaust.edu.sa

© Springer International Publishing AG 2017
C.-O. Lee et al. (eds.), *Domain Decomposition Methods in Science and Engineering XXIII*, Lecture Notes in Computational Science and Engineering 116, DOI 10.1007/978-3-319-52389-7_22

results in stagnation of residual norms or even failure of global Newton iterations. Nonlinear preconditioning may improve global convergence of nonlinearly stiff problems by changing coordinates and solving a different system possessing the same root by an outer Jacobian-free (Knoll and Keyes, 2004) Newton method.

Though algebraically related, ASPIN and MSPIN arise from different motivations. Additive Schwarz preconditioned inexact Newton (Cai and Keyes, 2002), was based on domain decomposition when proposed in 2002. It is shown in, e.g., Cai and Keyes (2002), Cai et al. (2002a), Cai et al. (2002b), Hwang and Cai (2005) and Skogestad et al. (2013) that ASPIN is effective in reducing the number of globally synchronizing outer Newton iterations, at the price of solving in parallel many smaller subdomain-scale nonlinear systems. Motivated instead by splitting physical fields, multiplicative Schwarz preconditioned inexact Newton algorithm (Liu and Keyes, 2015) was introduced in 2015. MSPIN solves physical submodels sequentially, and different groupings and different orderings result in different preconditioned functions. These two types of preconditioning can be nested.

2 MSPIN

Given the discrete nonlinear function $F : R^n \to R^n$, we want to find $x^* \in R^n$ such that

$$F(x^*) = 0, \tag{1}$$

where $F(x) = [F_1(x), F_2(x), \ldots, F_n(x)]^T$ and $x = [x_1, x_2, \ldots, x_n]^T$. We assume that $F(x)$ in (1) is continuously differentiable. The function $F(x)$ is split into $2 \leq N \leq n$ nonoverlapping components representing distinct physical features as

$$F(x) = F(u_1, \ldots, u_N) = \begin{bmatrix} \hat{F}_1(u_1, \ldots, u_N) \\ \vdots \\ \hat{F}_N(u_1, \ldots, u_N) \end{bmatrix} = 0, \tag{2}$$

where $x = [x_1, \ldots, x_n]^T = [u_1, \ldots, u_N]^T \in R^n$. u_i and $\hat{F}_i$ denote conformal subpartitions of x and F, respectively, $i = 1, \ldots, N$.

The inexact Newton method with backtracking (INB) (Dennis and Schnabel, 1996; Eisenstat and Walker, 1994; Pernice and Walker, 1998) serves as the basic component of MSPIN, so we first review the framework of INB.

Algorithm 1 (INB) An initial guess $x^{(0)}$ is given. For $k = 0, 1, 2, \ldots$ until convergence:

1. Choose η_k and find an approximate Newton step $d^{(k)}$ such that

$$\|F(x^{(k)}) - F'(x^{(k)})d^{(k)}\| \leq \eta_k \|F(x^{(k)})\|. \tag{3}$$

2. Determine $\lambda^{(k)}$ using a backtracking linesearch technique (Dennis and Schnabel, 1996).
3. Update $x^{(k+1)} = x^{(k)} - \lambda^{(k)}d^{(k)}$.

$\eta_k \in [0, 1)$ is a "forcing term," and determines how accurately we solve $F'(x^{(k)})d^{(k)} = F(x^{(k)})$. As η_k approaches 0, INB becomes ordinary Newton with backtracking (NB).

In the MSPIN algorithm, the submodels are solved sequentially for the physical variable corrections, and the preconditioned system consists of the sum of these corrections. The multiplicative Schwarz preconditioned function

$$\mathcal{F}(x) = \begin{bmatrix} T_1(u_1, \ldots, u_N) \\ \vdots \\ T_N(u_1, \ldots, u_N) \end{bmatrix} \tag{4}$$

is obtained by solving the following equations:

$$\begin{aligned} &\hat{F}_1(u_1 - T_1(x), u_2, u_3, \ldots, u_N) = 0, \\ &\hat{F}_2(u_1 - T_1(x), u_2 - T_2(x), u_3, \ldots, u_N) = 0, \\ &\qquad\qquad \vdots \\ &\hat{F}_N(u_1 - T_1(x), u_2 - T_2(x), u_3 - T_3(x), \ldots, u_N - T_N(x)) = 0. \end{aligned} \tag{5}$$

As with ASPIN, MSPIN solves the global preconditioned problem in (4) using INB in Algorithm 1, which requires only Jacobian-vector multiplication.

In general, the Jacobian $\mathcal{F}'(x) = \mathcal{J}(x)$ is dense. Fortunately, as shown in Liu and Keyes (2015), the Jacobian of preconditioned function $\mathcal{F}(x)$ can be written as follows:

$$\mathcal{J}(x) = \begin{bmatrix} \frac{\partial \hat{F}_1}{\partial \delta_1} & & & \\ \frac{\partial \hat{F}_2}{\partial \delta_1} & \frac{\partial \hat{F}_2}{\partial \delta_2} & & \\ \vdots & \vdots & \ddots & \\ \frac{\partial \hat{F}_N}{\partial \delta_1} & \frac{\partial \hat{F}_N}{\partial \delta_2} & \cdots & \frac{\partial \hat{F}_N}{\partial \delta_N} \end{bmatrix}^{-1} \begin{bmatrix} \frac{\partial \hat{F}_1}{\partial \delta_1} & \frac{\partial \hat{F}_1}{\partial u_2} & \frac{\partial \hat{F}_1}{\partial u_3} & \cdots & \frac{\partial \hat{F}_1}{\partial u_N} \\ \frac{\partial \hat{F}_2}{\partial \delta_1} & \frac{\partial \hat{F}_2}{\partial \delta_2} & \frac{\partial \hat{F}_2}{\partial u_3} & \cdots & \frac{\partial \hat{F}_2}{\partial u_N} \\ \vdots & \vdots & \vdots & & \vdots \\ \frac{\partial \hat{F}_N}{\partial \delta_1} & \frac{\partial \hat{F}_N}{\partial \delta_2} & \frac{\partial \hat{F}_N}{\partial \delta_3} & \cdots & \frac{\partial \hat{F}_N}{\partial \delta_N} \end{bmatrix}, \tag{6}$$

where $\delta_i = u_i - T_i(x)$. Due to the continuity of $F(x)$, we know that $T_i(x) \to 0$ and $\delta_i \to x$ when x approaches the exact solution x^*. In practical implementations, it is more convenient to use the following approximate Jacobian

$$\hat{\mathcal{J}}(x) = L(x)^{-1} J(x)|_{x=[\delta_1, \ldots, \delta_N]^T}, \tag{7}$$

where $J(x) = F'(x) = \left(\frac{\hat{F}_i}{u_j}\right)_{N \times N}$ and $L(x)$ is the lower triangular part of $J(x)$. Functions from the original code may be used to compute $\hat{J}(y)z$ for any given vectors y, z, matrix-free, rather than forming Jacobian $J(x)$ explicitly.

3   Natural Convection Cavity Flow Problem

We consider a benchmark problem (De Vahl Davis, 1983) that describes the two-dimensional natural convection cavity flow of a Boussinesq fluid with Prandtl number 0.71 in an upright square cavity $\Omega = (0, 1) \times (0, 1)$. Following Zhang et al. (2010), the nondimensional steady-state Navier-Stokes equations in vorticity-velocity form and energy equation are formulated as:

$$\begin{cases} -\Delta u - \frac{\partial \omega}{\partial y} = 0, \\ -\Delta v + \frac{\partial \omega}{\partial x} = 0, \\ -(\frac{Pr}{Ra})^{0.5} \Delta \omega + u\frac{\partial \omega}{\partial x} + v\frac{\partial \omega}{\partial y} - \frac{\partial T}{\partial x} = 0, \\ -(\frac{1}{PrRa})^{0.5} \Delta T + u\frac{\partial T}{\partial x} + v\frac{\partial T}{\partial y} = 0, \end{cases} \tag{8}$$

where Pr and Ra denote the Prandtl number and the Rayleigh number, respectively. There are four unknowns: the velocities u, v, the vorticity ω, and the temperature T.

The upright square cavity is filled with air ($Pr = 0.71$). Boundary conditions are described as follows. On the solid walls, both velocity components u, v are zero, and the vorticity is determined from its definition:

$$\omega(x, y) = -\frac{\partial u}{\partial y} + \frac{\partial v}{\partial x}. \tag{9}$$

The horizontal (top and bottom) walls are insulated, $\frac{\partial T}{\partial y} = 0$, and the vertical walls are maintained at temperatures $T = 0.5$ (left) and $T = -0.5$ (right). The temperature difference drives circulation in the cavity through the Boussinesq buoyancy term in the vorticity equation. In Fig. 1, we compare contours of temperature T at different Rayleigh numbers, where higher Ra boosts the buoyant convection relative to diffusion.

Considering the partition with respect to velocity unknowns, the vorticity unknown, and the temperature unknown, we split the system (8) into three sub-models:

$$F_T: \quad -(\frac{1}{PrRa})^{0.5} \Delta T + u\frac{\partial T}{\partial x} + v\frac{\partial T}{\partial y} = 0, \tag{10}$$

$$F_\omega: \quad -(\frac{Pr}{Ra})^{0.5} \Delta \omega + u\frac{\partial \omega}{\partial x} + v\frac{\partial \omega}{\partial y} - \frac{\partial T}{\partial x} = 0, \tag{11}$$

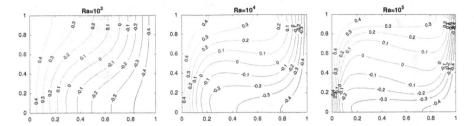

Fig. 1 Contours of temperature T at Rayleigh numbers over 2 orders of magnitude

$$F_{u,v} : \quad \begin{cases} -\Delta u - \frac{\partial \omega}{\partial y} = 0, \\ -\Delta v + \frac{\partial \omega}{\partial x} = 0. \end{cases} \tag{12}$$

A finite difference scheme with the 5-point stencil is used to discretize the PDEs, and the first order upwinding is used in both the vorticity equation and the temperature equation.

3.1 Effect of Ordering

In the framework of MSPIN, even when the partition of unknowns and equations is determined, different orderings for solving subproblems result in different nonlinear preconditioners.

We consider two different orderings in the MSPIN algorithm for the natural convection cavity flow problem:

- Ordering A:

$$\hat{F}_1(x) = \begin{bmatrix} F_T \\ F_\omega \end{bmatrix}, \qquad \hat{F}_2(x) = F_{u,v}. \tag{13}$$

- Ordering B:

$$\hat{F}_1(x) = F_{u,v}, \qquad \hat{F}_2(x) = \begin{bmatrix} F_T \\ F_\omega \end{bmatrix}. \tag{14}$$

Independent of ordering, $\hat{F}_1(x)$ and $\hat{F}_2(x)$ are both linear among their own unknowns, and are thus solved by GMRES alone with the tolerance $\epsilon_{sub-lin-rtol}$ ($\equiv \epsilon_{sub-nonlin-rtol}$) $= 10^{-5}$. The nonlinear system (8) is discretized on 100×100 mesh. We set the tolerances for outer Newton iterations as $\epsilon_{global-lin-rtol} = 10^{-6}$ and $\epsilon_{global-nonlin-rtol} = 10^{-10}$. The initial guess is zero for u, v, and ω, and linear interpolation in x for T. Figure 2 compares the convergence history of nonlinear

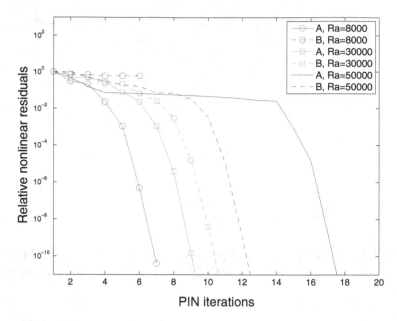

Fig. 2 Convergence history of nonlinear preconditioners using Ordering A (*solid lines*) and Ordering B (*dashed lines*)

preconditioners corresponding to Ordering A and Ordering B at different Rayleigh numbers. Using Ordering A MSPIN converges for all tests, while using Ordering B it fails at $Ra = 8000$ due to failure of backtracking. However, performance is inconsistent; compared with B, A requires fewer global Newton iterations at $Ra = 30,000$, but more iterations at $Ra = 50,000$. As shown in Table 1, for high Rayleigh numbers on fine grids, with a "cold" initial iterate as above, unpreconditioned globalized Newton stagnates outside of the zone of quadratic convergence.

3.2 Effect of Grouping

For the natural convection cavity flow problem, we can obtain different nonlinear preconditioners by grouping different PDEs and their corresponding unknowns. We consider four grouping-ordering schemes:

- Grouping A with two subsystems, $\hat{F}_1 : F_T \mid \hat{F}_2 : F_\omega, F_{u,v}$
- Grouping B with two subsystems, $\hat{F}_1 : F_T, F_\omega \mid \hat{F}_2 : F_{u,v}$
- Grouping C with two subsystems, $\hat{F}_1 : F_T, F_{u,v} \mid \hat{F}_2 : F_\omega$
- Grouping D with three subsystems, $\hat{F}_1 : F_T \mid \hat{F}_2 : F_\omega \mid \hat{F}_3 : F_{u,v}$

Table 1 Global nonlinear iterations for NB and MSPIN (plus global linear iterations for MSPIN) at 3 mesh resolutions for each Rayleigh number corresponding to Fig. 1

Ra	No MSPIN	Grouping A		Grouping B		Grouping C		Grouping D	
	NB	$F_T|F_\omega, F_{u,v}$		$F_T, F_\omega|F_{u,v}$		$F_T, F_{u,v}|F_\omega$		$F_T|F_\omega|F_{u,v}$	
	Newton iter.	Newton	GMRES	Newton	GMRES	Newton	GMRES	Newton	GMRES
64×64 *mesh, 4 subdomains*									
10^3	5	4	5	5	17	4	15	5	17
10^4	*	*		7	27	8	23	6	27
10^5	*	*		18	61		–	17	65
128×128 *mesh, 16 subdomains*									
10^3	5	4	5	5	18	4	16	5	18
10^4	*	*		7	28	10	30	7	28
10^5	*	*		18	110		–	16	83
256×256 *mesh, 64 subdomains*									
10^3	5	4	5	5	18	4	16	4	18
10^4	*	*		7	31	9	32	7	31
10^5	*	*			–		–	19	97

The initial guess is zero for u, v, and ω, and linear interpolation in x for T. $\epsilon_{global-nonlin-rtol} = 10^{-10}$, $\epsilon_{global-lin-rtol} = 10^{-6}$, $\epsilon_{sub-nonlin-rtol} = 10^{-4}$, and $\epsilon_{sub-lin-rtol} = 10^{-6}$
"*" indicates that one or more subproblems fail to converge or outer backtracking fails
"–" indicates that linear iterations fail to converge within allowed limits

The subproblems corresponding to Groupings B and D are linear, and are solved here by GMRES with BoomerAMG preconditioning. With Groupings A and C, one subproblem is linear and the other one is still nonlinear, which is solved by an internal invocation of INB. The elements of the global MSPIN Jacobians $\hat{\mathcal{J}}$ are not explicitly available, so the global linear problems inherit a conditioning from the subproblem solutions that is hard to improve further; hence, we tabulate the total number of linear iterations required in all of the Newton steps.

Table 1 compares a global Newton method with backtracking (NB), in which the Newton correction is always solved for accurately, with MSPIN algorithms corresponding to different grouping-ordering schemes. When MSPIN algorithms with Groupings B and D converge on a given mesh at a given Rayleigh number, they have similar numbers of Newton iterations and GMRES iterations. In Table 1, MSPIN algorithms with Grouping A, B or C fail to converge in some cases. Sometimes, GMRES on $\hat{\mathcal{J}}$ does not converge within the allowed number of iterations. Sometimes, the outer INB still cannot converge due to failure of the global line search, even though residuals decrease in the early iterations. However, the most decomposed MSPIN algorithm, Grouping D, works in all cases. Experimentally, the groupings play an essential role in determining the quality of nonlinear preconditioning.

Checking corresponding entries for nonlinear iteration count across different mesh densities at the same Rayleigh number in Table 1, we observe that Newton is asymptotically insensitive to the mesh resolution, as expected by theory.

As shown in Liu and Keyes (2015) on a related forced convection problem, additive field-split nonlinear preconditioning can be much less robust than multiplicative. However, classical ASPIN based on domain decomposition can be effective for such problems at high Reynolds or Raleigh numbers, when properly tuned. ASPIN for system (8) with $Ra = 10^5$ on a 128×128 mesh with 16 subdomains and overlap=3, with the same tolerance parameters used in Table 1, converges in 8 Newton iterations. However, this case fails with smaller overlap.

4 Conclusions

MSPIN is used to solve a nonlinear flow problem, with backtracking linesearch as the only globalization technique, in the absence of any other physically based globalization strategy normally employed in Newton's method on such problems, such as mesh sequencing or parameter continuation. We experiment with different groups and orderings, since there is not yet a theory for their selection in nonlinear Schwarz preconditioning. Groupings are exhibited that robustify Newton's method even on a fine mesh at high Rayleigh number from a "cold start" initial guess—a regime in which a traditional global Newton method with backtracking alone is completely ineffective.

Acknowledgements The authors acknowledge support from KAUST's Extreme Computing Research Center and the PETSc group of Argonne National Laboratory.

References

X.-C. Cai, D.E. Keyes, Nonlinearly preconditioned inexact Newton algorithms. SIAM J. Sci. Comput. **24**(1), 183–200 (2002)

X.-C. Cai, D.E. Keyes, L. Marcinkowski, Nonlinear additive Schwarz preconditioners and application in computational fluid dynamics. Int. J. Numer. Methods Fluids **40**(12), 1463–1470 (2002a)

X.-C. Cai, D.E. Keyes, D.P. Young, A nonlinear additive Schwarz preconditioned inexact Newton method for shocked duct flows, in *Domain Decomposition Methods in Science and Engineering (Lyon, 2000)* (CIMNE, Barcelona, 2002b), pp. 345–352

G. De Vahl Davis, Natural convection of air in a square cavity: a bench mark numerical solution. Int. J. Numer. Methods Fluids **3**(3), 249–264 (1983)

J.E. Dennis, Jr., R.B. Schnabel. Numerical methods for unconstrained optimization and nonlinear equations, *Classics in Applied Mathematics*, vol. 16 (SIAM, Philadelphia, 1996)

S.C. Eisenstat, H.F. Walker, Globally convergent inexact Newton methods. SIAM J. Optim. **4**(2), 393–422 (1994)

F.-N. Hwang, X.-C. Cai, A parallel nonlinear additive Schwarz preconditioned inexact Newton algorithm for incompressible Navier-Stokes equations. J. Comput. Phys. **204**(2), 666–691 (2005)

D.A. Knoll, D.E. Keyes, Jacobian-free Newton-Krylov methods: a survey of approaches and applications. J. Comput. Phys. **193**(2), 357–397 (2004)

L. Liu, D.E. Keyes, Field-split preconditioned inexact Newton algorithms. SIAM J. Sci. Comput. **37**(3), A1388–A1409 (2015)

M. Pernice, H.F. Walker, NITSOL: a Newton iterative solver for nonlinear systems. SIAM J. Sci. Comput. **19**(1), 302–318 (1998)

J.O. Skogestad, E. Keilegavlen, J.M. Nordbotten, Domain decomposition strategies for nonlinear flow problems in porous media. J. Comput. Phys. **234**, 439–451 (2013)

C.-H. Zhang, W. Zhang, G. Xi, A pseudospectral multidomain method for conjugate conduction-convection in enclosures. Numer. Heat. Tr. B-FUND **57**(4), 260–282 (2010)

Treatment of Singular Matrices in the Hybrid Total FETI Method

A. Markopoulos, L. Říha, T. Brzobohatý, P. Jirůtková, R. Kučera, O. Meca, and T. Kozubek

1 From FETI to HTFETI Method

The FETI (Finite Element Tearing and Interconnecting) method is based on eliminating primal unknowns so that dual linear systems in terms of Lagrange multipliers are solvable by the projected conjugate gradient method (see Farhat and Roux 1994). The projections on the kernel of $\mathbf{G}^\top$ are computed by the orthogonal projector

$$\mathbf{P} = \mathbf{I} - \mathbf{G} \left(\mathbf{G}^\top \mathbf{G} \right)^{-1} \mathbf{G}^\top. \tag{1}$$

The H(ybrid) FETI method (see Klawonn and Rheinbach 2010) combines the classical FETI method and the FETI-DP method (see Farhat et al. 2001) with the aim to adapt a code to parallel computer architectures. In this paper, we use another variant of the Hybrid FETI method (see Brzobohatý et al. 2013) that starts from the T(otal) FETI method (see Dostál et al. 2006). Its implementation (HTFETI) does not differ significantly from the original approach (TFETI). In some sense, having both algorithms in one library requires just a few additions across the code of the TFETI method. Note that TFETI approach also enforces the boundary conditions by Lagrange multipliers so that stiffness matrices on all subdomains exhibit the same defect and kernel matrices may be easily assembled.

We will shortly introduce our HTFETI method for the 2-dimensional problem given by cantilever beam, see Fig. 1a. After discretization, domain decomposition,

A. Markopoulos (✉) • L. Říha • T. Brzobohatý • P. Jirůtková • R. Kučera • O. Meca • T. Kozubek
IT4Innovations National Supercomputing Centre, 17. listopadu, 15/2172, Ostrava, Czech Republic
e-mail: alexandros.markopoulos@vsb.cz; lubomir.riha@vsb.cz; tomas.brzobohaty@vsb.cz; pavla.jirutkova@vsb.cz; radek.kucera@vsb.cz; ondrej.meca@vsb.cz; tomas.kozubek@vsb.cz

© Springer International Publishing AG 2017   237
C.-O. Lee et al. (eds.), *Domain Decomposition Methods in Science and Engineering XXIII*, Lecture Notes in Computational Science and Engineering 116, DOI 10.1007/978-3-319-52389-7_23

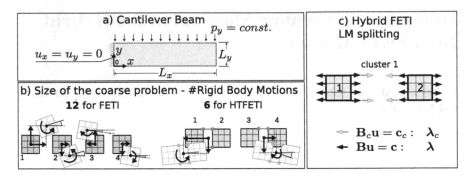

Fig. 1 Cantilever beam in 2D

and linear algebra object assembly, the linear system reads as follows:

$$
\begin{pmatrix}
\mathbf{K}_1 & \mathbf{O} & \mathbf{O} & \mathbf{O} & \mathbf{B}_{c,1}^\top & \mathbf{O} & \mathbf{B}_1^\top \\
\mathbf{O} & \mathbf{K}_2 & \mathbf{O} & \mathbf{O} & \mathbf{B}_{c,2}^\top & \mathbf{O} & \mathbf{B}_2^\top \\
\mathbf{O} & \mathbf{O} & \mathbf{K}_3 & \mathbf{O} & \mathbf{O} & \mathbf{B}_{c,3}^\top & \mathbf{B}_3^\top \\
\mathbf{O} & \mathbf{O} & \mathbf{O} & \mathbf{K}_4 & \mathbf{O} & \mathbf{B}_{c,4}^\top & \mathbf{B}_4^\top \\
\mathbf{B}_{c,1} & \mathbf{B}_{c,2} & \mathbf{O} & \mathbf{O} & \mathbf{O} & \mathbf{O} & \mathbf{O} \\
\mathbf{O} & \mathbf{O} & \mathbf{B}_{c,3} & \mathbf{B}_{c,4} & \mathbf{O} & \mathbf{O} & \mathbf{O} \\
\mathbf{B}_1 & \mathbf{B}_2 & \mathbf{B}_3 & \mathbf{B}_4 & \mathbf{O} & \mathbf{O} & \mathbf{O}
\end{pmatrix}
\begin{pmatrix}
\mathbf{u}_1 \\ \mathbf{u}_2 \\ \mathbf{u}_3 \\ \mathbf{u}_4 \\ \boldsymbol{\lambda}_{c,1} \\ \boldsymbol{\lambda}_{c,2} \\ \boldsymbol{\lambda}
\end{pmatrix}
=
\begin{pmatrix}
\mathbf{f}_1 \\ \mathbf{f}_2 \\ \mathbf{f}_3 \\ \mathbf{f}_4 \\ \mathbf{o} \\ \mathbf{o} \\ \mathbf{c}
\end{pmatrix} .
\tag{2}
$$

We denote:

$$
\mathbf{B}_c = \begin{pmatrix} \mathbf{B}_{c,1} & \mathbf{B}_{c,2} & \mathbf{O} & \mathbf{O} \\ \mathbf{O} & \mathbf{O} & \mathbf{B}_{c,3} & \mathbf{B}_{c,4} \end{pmatrix}, \quad \mathbf{B} = \begin{pmatrix} \mathbf{B}_1 & \mathbf{B}_2 & \mathbf{B}_3 & \mathbf{B}_4 \end{pmatrix}.
$$

The matrix $\mathbf{B}_c$ is a copy of specific rows from the matrix $\mathbf{B}$ corresponding to components of $\boldsymbol{\lambda}$ acting on the corners between subdomains 1,2, and 3,4, respectively (see Fig. 1c). Although the whole matrix in (2) is singular, it beneficially affects convergence of the iterative process (Farhat and Roux 1994). If the redundant rows of $\mathbf{B}_c$ are omitted, the primal solution components remain the same. To simplify our presentation, we permute (2) as

$$
\begin{pmatrix}
\mathbf{K}_1 & \mathbf{O} & \mathbf{B}_{c,1}^\top & \mathbf{O} & \mathbf{O} & \mathbf{O} & \mathbf{B}_1^\top \\
\mathbf{O} & \mathbf{K}_2 & \mathbf{B}_{c,2}^\top & \mathbf{O} & \mathbf{O} & \mathbf{O} & \mathbf{B}_2^\top \\
\mathbf{B}_{c,1} & \mathbf{B}_{c,2} & \mathbf{O} & \mathbf{O} & \mathbf{O} & \mathbf{O} & \mathbf{O} \\
\mathbf{O} & \mathbf{O} & \mathbf{O} & \mathbf{K}_3 & \mathbf{O} & \mathbf{B}_{c,3}^\top & \mathbf{B}_3^\top \\
\mathbf{O} & \mathbf{O} & \mathbf{O} & \mathbf{O} & \mathbf{K}_4 & \mathbf{B}_{c,4}^\top & \mathbf{B}_4^\top \\
\mathbf{O} & \mathbf{O} & \mathbf{O} & \mathbf{B}_{c,3} & \mathbf{B}_{c,4} & \mathbf{O} & \mathbf{O} \\
\mathbf{B}_1 & \mathbf{B}_2 & \mathbf{O} & \mathbf{B}_3 & \mathbf{B}_4 & \mathbf{O} & \mathbf{O}
\end{pmatrix}
\begin{pmatrix}
\mathbf{u}_1 \\ \mathbf{u}_2 \\ \boldsymbol{\lambda}_{c,1} \\ \mathbf{u}_3 \\ \mathbf{u}_4 \\ \boldsymbol{\lambda}_{c,2} \\ \boldsymbol{\lambda}
\end{pmatrix}
=
\begin{pmatrix}
\mathbf{f}_1 \\ \mathbf{f}_2 \\ \mathbf{o} \\ \mathbf{f}_3 \\ \mathbf{f}_4 \\ \mathbf{o} \\ \mathbf{c}
\end{pmatrix} ,
\tag{3}
$$

and then we introduce a new notation consistently with the line partition in (3):

$$
\left(\begin{array}{cc|c}
\tilde{\mathbf{K}}_1 & \mathbf{O} & \tilde{\mathbf{B}}_1^\top \\
\hline
\mathbf{O} & \tilde{\mathbf{K}}_2 & \tilde{\mathbf{B}}_2^\top \\
\hline
\tilde{\mathbf{B}}_1 & \tilde{\mathbf{B}}_2 & \mathbf{O}
\end{array}\right)
\left(\begin{array}{c}
\tilde{\mathbf{u}}_1 \\
\tilde{\mathbf{u}}_2 \\
\tilde{\lambda}
\end{array}\right)
=
\left(\begin{array}{c}
\tilde{\mathbf{f}}_1 \\
\tilde{\mathbf{f}}_2 \\
\tilde{\mathbf{c}}
\end{array}\right).
\tag{4}
$$

Eliminating $\tilde{\mathbf{u}}_i$, $i = 1, 2$, we also eliminate the subset of dual variables $\lambda_{c,j}, j = 1, 2$ related to the matrix $\mathbf{B}_c$. Therefore, the structure behaves like a problem decomposed into two clusters: the first and second subdomains belong to the first cluster, the third and fourth subdomains belong to the second cluster, see Fig. 1b. Here, $\tilde{\mathbf{K}}_1, \tilde{\mathbf{K}}_2$ can be interpreted as the cluster stiffness matrices with the kernels $\tilde{\mathbf{R}}_1, \tilde{\mathbf{R}}_2$, respectively. Denoting $\tilde{\mathbf{K}} = \mathrm{diag}(\tilde{\mathbf{K}}_1, \tilde{\mathbf{K}}_2)$, $\tilde{\mathbf{B}} = (\tilde{\mathbf{B}}_1, \tilde{\mathbf{B}}_2)$, $\tilde{\mathbf{R}} = \mathrm{diag}(\tilde{\mathbf{R}}_1, \tilde{\mathbf{R}}_2)$, $\tilde{\mathbf{F}} = \tilde{\mathbf{B}}\tilde{\mathbf{K}}^+\tilde{\mathbf{B}}^\top$, $\tilde{\mathbf{G}} = -\tilde{\mathbf{B}}\tilde{\mathbf{R}}$, $\tilde{\mathbf{d}} = \tilde{\mathbf{B}}\tilde{\mathbf{K}}^+\tilde{\mathbf{f}} - \tilde{\mathbf{c}}$, and $\tilde{\mathbf{e}} = -\tilde{\mathbf{R}}^\top\tilde{\mathbf{f}}$, we arrive at the Schur complement system

$$
\begin{pmatrix}
\tilde{\mathbf{F}} & \tilde{\mathbf{G}} \\
\tilde{\mathbf{G}}^\top & \mathbf{O}
\end{pmatrix}
\begin{pmatrix}
\tilde{\lambda} \\
\tilde{\alpha}
\end{pmatrix}
=
\begin{pmatrix}
\tilde{\mathbf{d}} \\
\tilde{\mathbf{e}}
\end{pmatrix}
\tag{5}
$$

that can be solved by the same iterative method as in the classical FETI method. The dimension of the new coarse problem $\tilde{\mathbf{G}}^\top\tilde{\mathbf{G}}$ is smaller (size = 6) compared to the FETI case. To keep optimality of the HTFETI approach, the matrix $\tilde{\mathbf{K}}$ can not be factorized directly. The implicit factorization will be demonstrated by its first block (cluster). It is obtained by solving the linear system $\tilde{\mathbf{K}}_1\tilde{\mathbf{x}}_1 = \tilde{\mathbf{b}}_1$, i.e.,

$$
\begin{pmatrix}
\mathbf{K}_{1:2} & \mathbf{B}_{c,1:2}^\top \\
\mathbf{B}_{c,1:2} & \mathbf{O}
\end{pmatrix}
\begin{pmatrix}
\mathbf{x}_1 \\
\mu
\end{pmatrix}
=
\begin{pmatrix}
\mathbf{b} \\
\mathbf{z}
\end{pmatrix},
\tag{6}
$$

where $\mathbf{K}_{1:2} = \mathrm{diag}(\mathbf{K}_1, \mathbf{K}_2)$ and $\mathbf{B}_{c,1:2} = (\mathbf{B}_{c,1}, \mathbf{B}_{c,2})$. The subindex $1:2$ adverts to the first and the last ordinal number of the subdomains in the cluster. Although (6) can be interpreted as a FETI problem, we solve it by a direct solver. The respective Schur complement system reads as:

$$
\begin{pmatrix}
\mathbf{F}_{c,1:2} & \mathbf{G}_{c,1:2} \\
\mathbf{G}_{c,1:2}^\top & \mathbf{O}
\end{pmatrix}
\begin{pmatrix}
\mu \\
\beta
\end{pmatrix}
=
\begin{pmatrix}
\mathbf{d}_{c,1:2} \\
\mathbf{e}_{c,1:2}
\end{pmatrix},
\tag{7}
$$

where $\mathbf{F}_{c,1:2} = \mathbf{B}_{c,1:2}\mathbf{K}_{1:2}^+\mathbf{B}_{c,1:2}^\top$, $\mathbf{G}_{c,1:2} = -\mathbf{B}_{c,1:2}\mathbf{R}_{1:2}$, $\mathbf{d}_{c,1:2} = \mathbf{B}_{c,1:2}\mathbf{K}_{1:2}^+\mathbf{b} - \mathbf{z}$, $\mathbf{e}_{c,1:2} = -\mathbf{R}_{1:2}^\top\mathbf{b}$, and $\mathbf{R}_{1:2} = \mathrm{diag}(\mathbf{R}_1, \mathbf{R}_2)$. To obtain the vector $\tilde{\mathbf{x}}_1$, both systems (6), (7) are subsequently solved in three steps:

$$
\begin{aligned}
\beta &= \mathbf{S}_{c,1:2}^+ \left(\mathbf{G}_{c,1:2}^\top \mathbf{F}_{c,1:2}^{-1} \mathbf{d}_{c,1:2} - \mathbf{e}_{c,1:2} \right), \\
\mu &= \mathbf{F}_{c,1:2}^{-1} \left(\mathbf{d}_{c,1:2} - \mathbf{G}_{c,1:2}\beta \right), \\
\mathbf{x} &= \mathbf{K}_{1:2}^+ \left(\mathbf{b} - \mathbf{B}_{c,1:2}^\top \mu \right) + \mathbf{R}_{1:2}\beta,
\end{aligned}
\tag{8}
$$

where $\mathbf{S}_{c,1:2} = \mathbf{G}_{c,1:2}^\top\mathbf{F}_{c,1:2}^{-1}\mathbf{G}_{c,1:2}$ is the singular Shur complement matrix.

The kernel $\tilde{\mathbf{R}}_1$ of $\tilde{\mathbf{K}}_1$ is the last object going to be effectively evaluated. The orthogonality condition $\tilde{\mathbf{K}}_1\tilde{\mathbf{R}}_1 = \mathbf{O}$ can be written by

$$\left(\begin{array}{c|c} \mathbf{K}_{1:2} & \mathbf{B}_{c,1:2}^{\mathsf{T}} \\ \hline \mathbf{B}_{c,1:2} & \mathbf{O} \end{array}\right)\left(\begin{array}{c} \mathbf{R}_{1:2} \\ \mathbf{O} \end{array}\right)\mathbf{H}_{1:2} = \left(\begin{array}{c} \mathbf{O} \\ \mathbf{O} \end{array}\right), \tag{9}$$

where $\tilde{\mathbf{R}}_1 = (\mathbf{R}_{1:2}^{\mathsf{T}}, \mathbf{O}^{\mathsf{T}})^{\mathsf{T}}\mathbf{H}_{1:2}$. Assuming that the subdomain kernels $\mathbf{R}_1$ and $\mathbf{R}_2$ are known, it remains to determine $\mathbf{H}_{1:2}$. The first equation in (9) does not impose any condition onto $\mathbf{H}_{1:2}$. The second equation gives

$$\mathbf{B}_{c,1:2}\mathbf{R}_{1:2}\mathbf{H}_{1:2} = -\mathbf{G}_{c,1:2}\mathbf{H}_{1:2} = \mathbf{O}, \tag{10}$$

implying that $\mathbf{H}_{1:2}$ is the kernel of $\mathbf{G}_{c,1:2}$, which is not full-column rank matrix due to the absence of the Dirichlet boundary condition in $\mathbf{B}_{c,1:2}$.

Preprocessing in the HTFETI method starts in the same way as in the FETI approach preparing factors $\mathbf{K}_i$ and kernels $\mathbf{R}_i$ for each subdomain. Then, only one pair consisting of $\mathbf{F}_{c,j:k}$ and $\mathbf{S}_{c,j:k}$ is assembled and factorized on each cluster. The dimension of $\mathbf{F}_{c,1:2}$ is controlled by the number of Lagrange multipliers $\lambda_{c,1}$ glueing the cluster subdomains. The dimension of $\mathbf{S}_{c,1:2}$ is given by the sum of defects of all matrices $\mathbf{K}_i$ belonging to a particular cluster.

2 Solving a Singular System via Kernel Detection

This work continues with the results of Dostál et al. (2011), Brzobohatý et al. (2011), Kučera et al. (2012), Kučera et al. (2013), and it queries from work published by Suzuki and Roux (2014).

If a problem with large jumps in the material coefficients and/or with an irregular decomposition is solved by the FETI method, direct factorizations of singular symmetric stiffness matrices $\mathbf{K}_i$ can be very unstable due to unclear criteria for distinguishing null pivots. We propose a heuristic technique for detecting kernels $\mathbf{R}_i$ of symmetric positive semi-definite (SPSD) matrices utilizing direct solvers designed primarily for non-singular cases. The mesh of the subdomain, the stiffness matrix of which is assembled above, must be given by the specific graph decomposition. In the three-dimensional case, e.g., deleting any two nodes of the relevant graph does not yield two components (the resulting graph will remain connected). The analyzed matrix should be also diagonally scaled. Via fixing nodes (FNs) the goal is to find (see Dostál et al. 2011) an appropriate set of indices s (size(s) $\geq$ defect($\mathbf{K}_i$)) and a complementary set of indices r characterizing the singular and non-singular part of $\mathbf{K}_i$, respectively. The original stiffness matrix $\mathbf{K}_i$ (the subindex will be omitted in the rest of this section) can be permuted by the matrix $\mathbf{Q}$ so that

$$\mathbf{Q}\mathbf{K}\mathbf{Q}^T = \begin{pmatrix} \mathbf{K}_{rr} & \mathbf{K}_{rs} \\ \mathbf{K}_{sr} & \mathbf{K}_{ss} \end{pmatrix},$$

where $\mathbf{K}_{rr}$ is the well-conditioned matrix. It is sufficient to find at least three noncollinear nodes from the finite element mesh in the case of 3-dimensional linear elasticity. The DOFs corresponding to these nodes determine the set s. Our choice of the FNs is based on a random number generator. From mechanical point of view, the structure is sufficiently supported by those FNs against any rigid movement. As the Schur complement $\mathbf{S} = \mathbf{K}_{ss} - \mathbf{K}_{sr}\mathbf{K}_{rr}^{-1}\mathbf{K}_{rs}$ is a relatively small matrix, it can be analysed by robust algorithms for dense matrices.

Once the Schur complement is correctly defined, it is spectrally decomposed using, e.g., LAPACK to $\mathbf{U}\mathbf{\Sigma}\mathbf{U}^T$. Its eigenvalues are stored in $\mathbf{\Sigma} = \mathrm{diag}(\sigma_1, \sigma_2, \cdots, \sigma_n)$ in the descending order. The kth eigenvalue is considered to be zero, if

$$\sigma_k/\sigma_{k-1} < 10^{-4}.$$

Such information determines splitting $\mathbf{U} = (\hat{\mathbf{U}}, \mathbf{R}_s)$ where $\mathbf{R}_s$ consists of last columns of $\mathbf{U}$ starting with the column index k, and it is already a part of the searched kernel of $\mathbf{K}$. If $\mathbf{R}_s$ is known, its supplement $\mathbf{R}_r = -\mathbf{K}_{rr}^{-1}\mathbf{K}_{rs}\mathbf{R}_s$ is obtained from

$$\begin{pmatrix} \mathbf{K}_{rr} & \mathbf{K}_{rs} \\ \mathbf{K}_{sr} & \mathbf{K}_{ss} \end{pmatrix} \begin{pmatrix} \mathbf{R}_r \\ \mathbf{R}_s \end{pmatrix} = \begin{pmatrix} \mathbf{0} \\ \mathbf{0} \end{pmatrix}. \tag{11}$$

As an example, a uniformly meshed cube ($L = 30\,\mathrm{mm}$, $E = 2.1 \cdot 10^5\,\mathrm{MPa}$, $\mu = 0.3$, $\rho = 7850\,\mathrm{kg/m^3}$, $g = 9.81\,\mathrm{m/s^2}$) is used with a variable number of nodes controlled by n (number of nodes in x, y, and z direction). The singular set s is selected via several DOFs belonging to randomly chosen FNs. The quality of a selection (see Fig. 2) is measured by the ratio of bad choices (collinear nodes) to all possible combinations for a given number of FNs and the size of mesh n. Probability curves for 3, 4, and 5 FNs depending on the mesh parameter n are shown in Fig. 2. Increasing FNs for fixed mesh (constant n) intuitively helps to ensure noncollinear nodes. For instance, for $n = 10$ with 3 FNs the probability of a bad choice is $9.068 \cdot 10^{-2}$, with 4 FNs it decreases to $3.272 \cdot 10^{-4}$, and with 5 FNs to $1.545 \cdot 10^{-6}$. Surprisingly enough, for a fixed number of FNs and a simultaneously enlarging parameter n (mesh refinement), the probability of collinear FNs decreases as well.

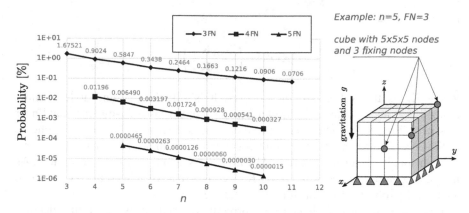

Fig. 2 Probability of collinear fixing nodes (FN)

3   ExaScale PaRallel FETI SOlver: ESPRESO

ESPRESO is a highly efficient parallel solver which contains several FETI method based algorithms including the HTFETI method suitable for parallel machines with tens or hundreds of thousands of cores. The solver is based on a highly efficient communication layer based on MPI, and it is able to run on massively parallel machines with thousands of compute nodes and hundreds of thousands of CPU cores. ESPRESO is also being developed to support modern many-core accelerators. We are currently developing four major versions of the solver:

- **ESPRESO CPU** is a CPU version using sparse representation of system matrices;
- **ESPRESO MIC** is an Intel Xeon Phi accelerated version working with dense representation of system matrices in the form of Schur complement;
- **ESPRESO GPU** is a GPU accelerated version working with dense structures. Support for sparse structures using cuSolver is under development;
- **ESPRESO GREEN** is a power efficient version developed under the H2020 READEX project. This version is in the very early development stage.

In order to solve real engineering problems, we are developing a FEM/BEM library that enables database files from ANSYS simulation software to be imported and all inputs required by the FETI or HTFETI solver generated. In addition, we are developing an interface to ELMER that allows ESPRESO to be used as its linear solver. This integration is done through API that can be used as an interface to many other applications.

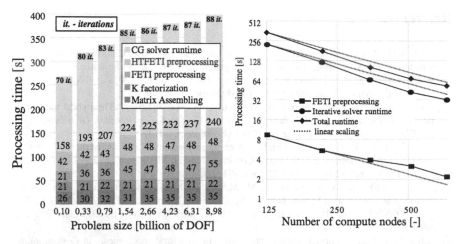

Fig. 3 Weak and strong scalability

4 Numerical Experiments

Efficiency of the HTFETI method is presented in the ESPRESO library on the cube benchmark described in Sect. 2. Weak scalability of the solver, see Fig. 3 left, includes matrix assembly, linear solver preprocessing (preprocessing of the TFETI and HTFETI method), and iterative solver runtime measured on 1–729 compute nodes of IT4Innovations Salomon supercomputer. Benchmark configuration: subdomain size 14,739 DOFs ($n = 17$); 1000 subdomains per cluster; Lumped preconditioner, stopping criteria 10^{-3}. Strong scalability on 126, 216, 343, 512, and 729 compute nodes of Salomon supercomputer is seen in Fig. 3 right. The problem size is 1.5 billions of unknowns.

5 Conclusions

This paper presents the HTFETI method, an extension of FETI algorithm for problems with the larger number of subdomains to handle the coarse problem more effectively. The basic principles are explained and demonstrated on linear elasticity problem. In the second part, the methodology for factorizing SPSD matrix using robust applications, e.g., PARDISO, is shown. Efficiency is proved by the numerical test performed in ESPRESO library for almost nine billions of unknowns.

Acknowledgements This work was supported by The Ministry of Education, Youth and Sports from the National Programme of Sustainability (NPU II) project "IT4Innovations excellence in science—LQ1602" and from the Large Infrastructures for Research, Experimental Development and Innovations project "IT4Innovations National Supercomputing Center—LM2015070".

References

T. Brzobohatý, Z. Dostál, T. Kozubek, P. Kovář, A. Markopoulos, Cholesky decomposition with fixing nodes to stable computation of a generalized inverse of the stiffness matrix of a floating structure. Int. J. Numer. Methods Eng. **88**(5), 493–509 (2011). ISSN 0029-5981. doi:10.1002/nme.3187. http://dx.doi.org/10.1002/nme.3187

T. Brzobohatý, M. Jarošová, T. Kozubek, M. Menšík, A. Markopoulos, The hybrid total FETI method, in *Proceedings of the Third International Conference on Parallel, Distributed, Grid and Cloud Computing for Engineering* (Civil-Comp, Pécs, 2013)

Z. Dostál, D. Horák, R. Kučera, Total FETI—an easier implementable variant of the FETI method for numerical solution of elliptic PDE. Commun. Numer. Methods Eng. **22**(12), 1155–1162 (2006). ISSN 1069-8299. doi:10.1002/cnm.881. http://dx.doi.org/10.1002/cnm.881

Z. Dostál, T. Kozubek, A. Markopoulos, M. Menšík, Cholesky decomposition of a positive semidefinite matrix with known kernel. Appl. Math. Comput. **217**(13), 6067–6077 (2011). ISSN 0096-3003. doi:10.1016/j.amc.2010.12.069. http://dx.doi.org/10.1016/j.amc.2010.12.069

C. Farhat, F.-X. Roux, Implicit parallel processing in structural mechanics, in *Computational Mechanics Advances*, ed. by J.T. Oden, vol. 2 (1) (North-Holland, Amsterdam, 1994), pp. 1–124

C. Farhat, M. Lesoinne, P. LeTallec, K. Pierson, D. Rixen, FETI-DP: a dual-primal unified FETI method. I. A faster alternative to the two-level FETI method. Int. J. Numer. Methods Eng. **50**(7), 1523–1544 (2001). ISSN 0029-5981. doi:10.1002/nme.76. http://dx.doi.org/10.1002/nme.76

A. Klawonn, O. Rheinbach, Highly scalable parallel domain decomposition methods with an application to biomechanics. Z. Angew. Math. Mech. **90**(1), 5–32 (2010). ISSN 0044-2267. doi:10.1002/zamm.200900329. http://dx.doi.org/10.1002/zamm.200900329

R. Kučera, T. Kozubek, A. Markopoulos, J. Machalová, On the Moore-Penrose inverse in solving saddle-point systems with singular diagonal blocks. Numer. Linear Algebra Appl. **19**(4), 677–699 (2012). ISSN 1070-5325. doi:10.1002/nla.798. http://dx.doi.org/10.1002/nla.798

R. Kučera, T. Kozubek, A. Markopoulos, On large-scale generalized inverses in solving two-by-two block linear systems. Linear Algebra Appl. **438**(7), 3011–3029 (2013). ISSN 0024-3795. doi:10.1016/j.laa.2012.09.027. http://dx.doi.org/10.1016/j.laa.2012.09.027

A. Suzuki, F.-X. Roux, A dissection solver with kernel detection for symmetric finite element matrices on shared memory computers. Int. J. Numer. Methods Eng. **100**(2), 136–164 (2014). ISSN 0029-5981. doi:10.1002/nme.4729. http://dx.doi.org/10.1002/nme.4729

From Surface Equivalence Principle to Modular Domain Decomposition

Florian Muth, Hermann Schneider, and Timo Euler

1 Introduction

Real-world electromagnetic problems such as mounted antennas often involve multiple electromagnetic scales and properties: These kinds of problems may contain antenna models with extremely detailed structures and complex materials besides electrically very large platforms of hundreds of wavelengths. Potentially, even complete systems, e.g. additionally including the feeding circuits of the antennas, need to be simulated. There are existing methods suitable to solve the full-wave MAXWELL's equations for each part of the described complex problem. E.g. the Finite Integration Technique (Weiland, 1977) or the finite element method (Monk, 1992) could be used for the comparatively small and complex antennas, while a boundary element method (Chew et al., 2001) or an asymptotic approach (McNamara et al., 1990) would be more appropriate for the electrically large platform. All these methods have their strengths regarding particular types of electromagnetic problems, but their capabilities are limited, especially if a combination of the mentioned problem types occur.

Here, domain decomposition methods come into play. The goal is to spatially decompose the original model into smaller subdomains and to apply the most suitable method in each subdomain. To obtain the overall solution, a global iterative solver is needed. An example for this approach is depicted in Fig. 1.

The presented project pursues a modular domain decomposition approach to enable the simple integration of existing electromagnetic solvers. Here, the subdomains are coupled via surface currents. This allows for adding arbitrary methods

F. Muth (✉) • H. Schneider • T. Euler
CST – Computer Simulation Technology AG, Darmstadt, Germany
e-mail: florian.muth@cst.com; hermann.schneider@cst.com; timo.euler@cst.com

© Springer International Publishing AG 2017
C.-O. Lee et al. (eds.), *Domain Decomposition Methods in Science and Engineering XXIII*, Lecture Notes in Computational Science and Engineering 116, DOI 10.1007/978-3-319-52389-7_24

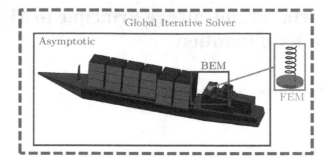

Fig. 1 Complex models, e.g. involving multiple scales, can be decomposed into smaller subdomains to apply the most suitable solver to each subdomain

to the developed black box framework, to make use of the full potential of available electromagnetic solvers.

2 LOVE's Equivalence Principle

The method described in this paper is based on the surface equivalence principle as developed by A.E.H. LOVE and described in Schelkunoff (1936). The coupling of the subdomains is realized by exchanging boundary data in terms of surface currents. LOVE's equivalence principle is illustrated in Fig. 2.

Let's assume an original model domain Ω is decomposed into two subdomains Ω_1 and Ω_2 by introducing a closed surface S, see Fig. 2a. $\mathbf{E}_i$ and $\mathbf{H}_i$ are the solutions of the original model for the electric and magnetic fields in subdomain Ω_i. ε_i and μ_i are the permittivity and the permeability of the material in the respective subdomain. The field solution on the surface S is denoted by $\mathbf{E}_S$ and $\mathbf{H}_S$.

According to LOVE's equivalence principle, the sources and material distributions enclosed by surface S can be replaced by equivalent electric and magnetic surface currents $\mathbf{J}_S = \mathbf{n}_S \times \mathbf{H}_S$ and $\mathbf{M}_S = \mathbf{E}_S \times \mathbf{n}_S$. Here, $\mathbf{n}_S$ is the unit normal vector of S pointing outwards. The resulting equivalent model for the outer domain Ω_1 as shown in Fig. 2b reproduces the solution of the original model in Ω_1, i.e. $\mathbf{E}_1^{(e)} = \mathbf{E}_1$ and $\mathbf{H}_1^{(e)} = \mathbf{H}_1$, and null fields in Ω_2. In the equivalent model, it is irrelevant what is modelled inside of the surface S, since the fields of the solution are forced to zero anyway.

The same applies for the corresponding inner equivalent model. Equivalent surface currents are defined in the same way on S, but the unit normal vector $\mathbf{n}_S$ is inverted pointing inwards. As for the outer equivalent model this results in null fields in Ω_1 and reproduces the solution of the original model in Ω_2.

Figure 3 illustrates again the above described principle with the help of a reflector antenna setup simulated with CST MICROWAVE STUDIO®. Additionally, the inner equivalent model (Fig. 3c) is shown besides the original and the outer equivalent models.

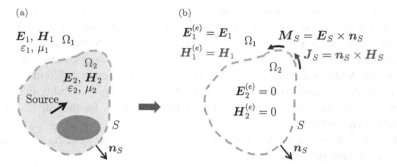

Fig. 2 According to LOVE's equivalence principle, sources and material distributions enclosed by a surface S in an original model (**a**) can be replaced by equivalent electric and magnetic surface currents $\mathbf{J}_S$ and $\mathbf{M}_S$ on S to obtain an equivalent model with the same solution outside of S (**b**)

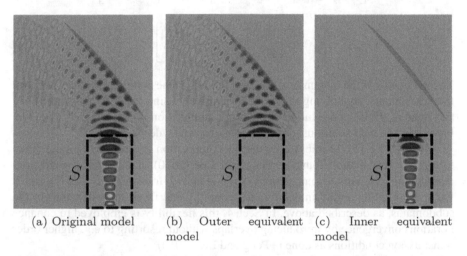

(a) Original model   (b) Outer equivalent model   (c) Inner equivalent model

Fig. 3 LOVE's equivalence principle is demonstrated by means of a reflector antenna setup using CST MICROWAVE STUDIO®: By monitoring the tangential fields on S in the original model (**a**), either the inside (**b**) or the outside (**c**) of the closed surface S can be replaced by equivalent surface currents

3 Iteration Scheme for Modular Domain Decomposition

The principle described in the previous section will be utilized for the black box domain decomposition approach. In this way, the subdomains need only provide surface currents to realize the coupling to the other subdomains. In the end, this will result in an iterative domain decomposition method, which will be explained in the following section.

The reflector antenna model from Sect. 2 is again considered. After decomposing it into the two subdomains Ω_1 and Ω_2, we obtain a typical coupled system. Now, the idea is to solve this coupled system by making use of LOVE's surface equivalence principle. But, instead of knowing the solution of the original model $\mathbf{E}_S$ and $\mathbf{H}_S$ beforehand, only approximations $\tilde{\mathbf{E}}_S$ and $\tilde{\mathbf{H}}_S$ are available, since the subdomains have to be solved separately. Here, the subdomains can basically be truncated by arbitrary boundary conditions, even transparent boundary conditions can be considered. Additionally, the exchange surfaces can be chosen in different locations. This gives the resulting domain decomposition method a high flexibility in defining the coupling interfaces between the subdomains and allows for the introduction of overlaps between them.

The above approach finally results in the following linear system, whose terms will be explained subsequently:

$$
\begin{bmatrix} I & \overline{R}_1 A_1^{-1} C_{12} \overline{R}_2^T \\ \overline{R}_2 A_2^{-1} C_{21} \overline{R}_1^T & I \end{bmatrix} \begin{bmatrix} \overline{x}_1 \\ \overline{x}_2 \end{bmatrix} = \begin{bmatrix} \overline{R}_1 A_1^{-1} b_1 \\ \overline{R}_2 A_2^{-1} b_2 \end{bmatrix} \tag{1}
$$

$$
\overline{x}_1 = \begin{bmatrix} \tilde{\mathbf{H}}_{S+}^{(1)} \\ \tilde{\mathbf{E}}_{S+}^{(1)} \end{bmatrix} ; \quad \overline{x}_2 = \begin{bmatrix} \tilde{\mathbf{H}}_S^{(2)} \\ \tilde{\mathbf{E}}_S^{(2)} \end{bmatrix} \tag{2}
$$

The unknowns of the system $\overline{x}_1$ and $\overline{x}_2$ are defined on the coupling surfaces between the subdomains. By solving this system iteratively using a GMRES solver (Saad and Schultz, 1986), the solution of the original model on the surface S is obtained. From this, the field solutions in the subdomains can be derived.

Although Eq. (1) describes a domain decomposition formally very much alike to e.g. the formulation found in Peng and Lee (2010), it goes far beyond non-overlapping domain decompositions with standard transmission conditions: It features a high flexibility in defining the coupling interfaces and extensions of the subdomains, as described above. In Sect. 4, this flexibility is employed to enhance iteration convergence by introducing overlaps without resorting to e.g. higher order transmission conditions as done in Peng and Lee (2010).

The iteration scheme of the presented method is illustrated in Fig. 4. Boundary data in terms of surface fields is iteratively exchanged between the subdomains, where each iteration mainly consists of three parts. First, the monitored surface fields $\overline{x}_j$ from subdomain Ω_j are transformed into a current source for subdomain Ω_i, represented by the operator $C_{ji} \overline{R}_i^T$. Afterwards, subdomain Ω_i is solved by applying its inverted system operator A_i^{-1}. In the last step, the operator $\overline{R}_i$ restricts the obtained solution to the corresponding coupling surface S. In practice, the last step is realized by monitoring the fields on the coupling surface. After each iteration, the feedback is exchanged between the subdomains to take into account the influence of the other parts of the model.

As shown in Fig. 4, the surfaces where the fields are monitored and the currents are imprinted do not coincide. This follows from the jumping fields due to the imprinted electric and magnetic currents.

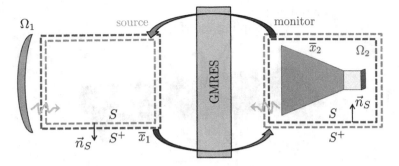

Fig. 4 Boundary data is iteratively exchanged by monitoring surface fields, which are then imprinted as current sources in the other domain. By solving the corresponding linear system using e.g. a GMRES solver, the solution of the original model is obtained

4 Investigations

The area of application of the presented black box framework mainly comprises models with a small number of user-defined, coupled subdomains as is the case for antenna placement scenarios. Here, the priority is not on scalability regarding the number of subdomains, but on the flexibility of the overall domain decomposition framework.

For first investigations, an "array" of two patch antennas is considered. The setup of this model and how it is decomposed into two subdomains is illustrated in Fig. 5. Each of the patch elements is simulated with CST's finite element frequency domain solver using an absorbing boundary condition (ABC). By shifting the coupling interfaces, non-overlapping ($d = 0$) as well as overlapping ($d > 0$) setups can be realized. In the latter case, each subdomain is extended towards the other one by modelling the structure of the original model in the overlap region. The discretizations of the subdomains can be chosen independently of each other and don't need to match in the overlap region nor at the coupling interfaces.

For the validation of the results of the presented domain decomposition method, the absolute value of the electric field is evaluated along the array axis and slightly above the surface of the patch elements for $d = 0$. In Fig. 6, the corresponding curves are depicted showing the smooth transition from one subdomain to the other at $x = -3$ cm. Furthermore, the results precisely match the solution of the original model.

An interesting aspect for future investigations is the relationship between the relative residual of the global iterative solver and the error of the quantities of interest. For the investigated model ($d = 0$), the absolute error of the S-parameter as the quantity of interest is already smaller than 10^{-3} after the first iteration, which is sufficient for typical engineering applications (Fig. 7).

As pointed out in Sect. 3, overlaps can be used to accelerate the convergence of the global iterative solver. Figure 8 compares the convergence of the relative residual of the global iterative solver for different overlap sizes d. The larger the

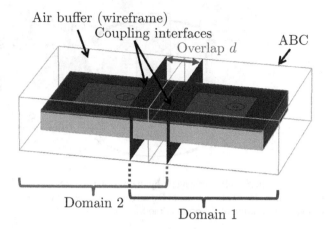

Fig. 5 The 1×2 patch antenna array is decomposed into two subdomains, each calculated by CST's finite element frequency domain solver. The subdomains are truncated by an absorbing boundary condition (ABC) and can partly overlap by a size d

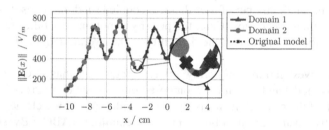

Fig. 6 The results of the presented method precisely match the solution of the original model for the non-overlapping setup ($d = 0$). Especially, a smooth transition between the subdomains at $x = -3$ cm can be observed

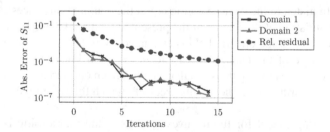

Fig. 7 Comparison of the relative residual of the global iterative solver with the absolute error of the S-parameter for the non-overlapping setup ($d = 0$): For typical engineering applications, an absolute error smaller than 10^{-3} is already sufficient. The value from the fifteenth iteration was taken as reference

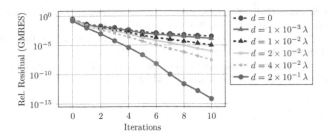

Fig. 8 The convergence of the presented method can be accelerated by introducing an overlap $d > 0$. There is no significant performance drawback, since the overlap size is in the range of a fraction of the wavelength λ

overlap size the faster the presented method converges. At the same time, there is no significant performance drawback, since the overlaps are still quite small in terms of the wavelength λ. E.g. $d = 4 \times 10^{-2} \lambda$ corresponds to an overlap size of approximately one mesh cell layer and reduces the number of iterations from 7 to 4 to reach a relative residual smaller than 10^{-3}.

5 Discussion and Conclusion

This paper has presented a domain decomposition approach, which is suitable for electrically large and complex setups. The main advantage is its modularity due to the coupling of the subdomains via surface currents motivated by the equivalence principle. The resulting black box framework allows for any numerical method in each subdomain. Another feature is the high flexibility in defining the coupling interfaces between the subdomains. In this way, overlapping setups can easily be introduced.

Promising results regarding the coupling of finite element subdomains were shown. The presented method was proven to converge for both the non-overlapping and the overlapping setup. By introducing a small overlap of a fraction of a wavelength, the convergence of the method could be accelerated drastically.

References

W.C. Chew, J.-M. Jin, E. Michielssen, J. Song, *Fast and Efficient Algorithms in Computational Electromagnetics* (Artech House, Norwood, 2001)

D.A. McNamara, C.W.I. Pistorius, J.A.G. Malherbe, *Introduction to the Uniform Geometrical Theory of Diffraction.* Antennas and Propagation Library (Artech House, Norwood, 1990)

P. Monk, A finite element method for approximating the time-harmonic Maxwell equations. Numer. Math. **63**(1), 243–261 (1992)

Z. Peng, J.-F. Lee, Non-conformal domain decomposition method with second-order transmission conditions for time-harmonic electromagnetics. J. Comput. Phys. **229**(16), 5615–5629 (2010)

Y. Saad, M.H. Schultz, GMRES: a generalized minimal residual algorithm for solving nonsymmetric linear systems. SIAM J. Sci. Stat. Comput. **7**(3), 856–869 (1986)

S.A. Schelkunoff, Some equivalence theorems of electromagnetics and their application to radiation problems. Bell Syst. Tech. J. **15**(1), 92–112 (1936)

T. Weiland, A discretization model for the solution of Maxwell's equations for six-component fields. Archiv fuer Elektronik Uebertragungstechnik **31**, 116–120 (1977)

Space-Time CFOSLS Methods with AMGe Upscaling

Martin Neumüller, Panayot S. Vassilevski, and Umberto E. Villa

1 Introduction

In this paper we explore a robust approach to derive combined space-time discretization methods for two classes (parabolic and hyperbolic) of time-dependent PDEs. We use the popular FOSLS (first order systems least-squares) approach (cf., e.g., Cai et al. 1994 or Carey et al. 1995) treating time as an additional space variable and, in addition, we prescribe a space-time divergence equation as a constraint in order to maintain certain space-time mass conservation (following, e.g., Adler and Vassilevski 2014).

More specifically, our approach is applied to the following model problem

$$\frac{\partial S}{\partial t} + \text{div}(\mathcal{L}(S)) = q_0(\mathbf{x}, t), \qquad \mathbf{x} \in \Omega \subset \mathbb{R}^d, t \in (0, T), \tag{1}$$

M. Neumüller (✉)
Institute of Computational Mathematics, Johannes Kepler University Linz, Altenberger Straße 69, 4040 Linz, Austria
e-mail: neumueller@numa.uni-linz.ac.at

P.S. Vassilevski
Lawrence Livermore National Laboratory, Center for Applied Scientific Computing, P.O. Box 808, L-561, Livermore, CA 94551, USA

Fariborz Maseeh Department of Mathematics and Statistics, Portland State University, OR 97201, Portland
e-mail: panayot@llnl.gov; panayot@pdx.edu

U.E. Villa
Institute for Computational Engineering and Sciences (ICES), The University of Texas at Austin, 201 E. 24th Street, Stop C0200, Austin, TX 78712-0027, USA
e-mail: uvilla@ices.utexas.edu

© Springer International Publishing AG 2017
C.-O. Lee et al. (eds.), *Domain Decomposition Methods in Science and Engineering XXIII*, Lecture Notes in Computational Science and Engineering 116, DOI 10.1007/978-3-319-52389-7_25

where $\mathcal{L}$ is at most a first-order differential operator with respect to the space variable $\mathbf{x}$ only. At $t = 0$ we impose an initial condition $S = S_0$ and on $\partial\Omega$ for all $t \in (0, T)$ we apply some appropriate boundary conditions (if any). More specifically we consider differential operators of the form

$$\mathcal{L}(S) := -k\nabla_{\mathbf{x}}S \quad \text{and} \quad \mathcal{L}(S) := f(S)\,\mathbf{u}(\cdot)$$

for respectively parabolic and hyperbolic problems, as explained in more details in Sects. 4 and 5.

2 Space-Time Constrained First Order System Least Squares

Problem (1) can be rewritten as a first order system by introducing the "flux" variable $\sigma := [\mathcal{L}(S); S]^\top$ as

$$\sigma - \begin{bmatrix} \mathcal{L}(S) \\ S \end{bmatrix} = 0,$$
$$\mathrm{div}_{\mathbf{x},t}\sigma = q_0, \tag{2}$$

where $\mathrm{div}_{\mathbf{x},t}$ is the $d + 1$-dimensional space-time divergence operator. We then introduce the FOSLS functional as

$$J(\sigma, S) = \left\| \sigma - \begin{bmatrix} \mathcal{L}(S) \\ S \end{bmatrix} \right\|_{0,\,K^{-1}}^2 + \|q_0 - \mathrm{div}_{\mathbf{x},t}\sigma\|_0^2,$$

where $K = K(\mathbf{x}) \in \mathbb{R}^{(d+1)\times(d+1)}$ is a symmetric and positive definite coefficient matrix and $\|\cdot\|_0$ ($\|\cdot\|_{0,K^{-1}}$) denotes the (weighted) $L_2(\Omega_T)$-norm with respect to the space-time domain $\Omega_T := \Omega \times (0, T)$. A constrained least-square version of (2) is given by minimizing the functional $J(\sigma, S)$ under the constraint which is given by the conservation equation

$$(\mathrm{div}_{\mathbf{x},t}\sigma, w) = (q_0, w) \quad \text{for all} \quad w \in L_2(\Omega_T).$$

Here we denote with $(\cdot, \cdot)$ the inner product with respect to $L_2(\Omega_T)$. First order optimality conditions for the constrained minimization problem lead to the system of variational equations: Find $\sigma \in H(\mathrm{div}_{\mathbf{x},t}; \Omega_T)$, $S \in V$ and $\mu \in L_2(\Omega_T)$, such that

$$(\sigma, \boldsymbol{\psi})_{K^{-1}} + (\mathrm{div}_{\mathbf{x},t}\sigma, \ \mathrm{div}_{\mathbf{x},t}\boldsymbol{\psi}) - \left(\begin{bmatrix} \mathcal{L}(S) \\ S \end{bmatrix}, \boldsymbol{\psi} \right)_{K^{-1}} +(\mu, \ \mathrm{div}_{\mathbf{x},t}\boldsymbol{\psi}) = (q_0, \mathrm{div}_{\mathbf{x},t}\boldsymbol{\psi}),$$

$$-\left(\sigma, \begin{bmatrix} \mathcal{L}(\phi) \\ \phi \end{bmatrix} \right)_{K^{-1}} + \left(\begin{bmatrix} \mathcal{L}(S) \\ S \end{bmatrix}, \begin{bmatrix} \mathcal{L}(\phi) \\ \phi \end{bmatrix} \right)_{K^{-1}} = 0,$$

$$(\mathrm{div}_{\mathbf{x},t}\sigma, \ w) = (q_0, w) \tag{3}$$

holds for all $\psi \in H(\mathrm{div}_{\mathbf{x},t}; \Omega_T)$, all $\phi \in V$ and all $w \in L_2(\Omega_T)$. Here V denotes an appropriate function space for the unknown S, such that $\mathcal{L} : V \to L^2$ is a bounded operator. In a straightforward manner we obtain the finite element discretization of the CFOSLS system (3) by using appropriate finite dimensional spaces, i.e. we use $\sigma_h \in \mathbf{R}_h \subset H(\mathrm{div}_{\mathbf{x},t}; \Omega_T), S_h \in V_h \subset V$ and $\mu_H \in W_H \subset L_2(\Omega_T)$. Note that the Lagrangian multiplier μ_H belongs to the space W_H of discontinuous piecewise polynomials defined on a coarser mesh $\mathcal{T}_H$ (the lowest order being piecewise constants). The fine mesh $\mathcal{T}_h$ is constructed by performing one uniform refinement of $\mathcal{T}_H$. This choice leads to a *relaxed* Petrov-Galerkin discretization of the mass conservation equation and prevents overconstraining the resulting system. A relevant error analysis of the above discretization has been presented in Adler and Vassilevski (2014). Finally, using appropriate basis functions for the discrete function spaces, we obtain the system of linear equations for the saddle point problem

$$
\begin{bmatrix} A & B^\top & D^\top \\ B & C & 0 \\ D & 0 & 0 \end{bmatrix} \begin{bmatrix} \sigma_h \\ S_h \\ \mu_H \end{bmatrix} = \begin{bmatrix} f_h \\ 0 \\ g_H \end{bmatrix}. \tag{4}
$$

3   AMGe Upscaling

The AMGe (element agglomeration) coarsening has been developed at LLNL, originally to derive hierarchies of finite element spaces for designing multigrid solvers for bilinear forms corresponding to an entire de Rham sequence of spaces (H^1-conforming, $H(\mathrm{curl})$-conforming, and $H(\mathrm{div})$-conforming), (Pasciak and Vassilevski, 2008), and more recently (Lashuk and Vassilevski, 2012, 2014) to ensure that these hierarchies of spaces have guaranteed approximation properties. Such spaces are hence suitable to construct accurate coarse discretizations and can be used as a tool for dimension reduction, also refereed to as numerical upscaling.

The CFOSLS space-time discretization approach leads to saddle-point systems involving function spaces in the divergence constraint that are $H(\mathrm{div})$-conforming. This allows to solve combined space-time problems up to 2 space dimensions using the existing AMGe upscaling framework for 3D Raviart-Thomas elements. The goal in the near future is to extend this framework to 4D Raviart-Thomas analogs. This paper, as a first step, demonstrates the feasibility of the AMGe upscaling approach applied to combined space-time discretization that is both accurate, mass-conservative and achieving reasonable dimension reduction, which makes the expensive direct space-time approach (applied on the fine grid) feasible at coarser upscaled levels.

In the next sections we study the presented approach in detail for the two differential operators introduced in the beginning of this work. The finite element library MFEM (MFEM) is used to assemble the discretized systems which are then solved using the algebraic multigrid solvers (AMG) in hypre (HYPRE).

4 Parabolic Problem

Here we choose the differential operator $\mathcal{L}(S) := -k\nabla_{\mathbf{x}}S$, where $k = k(\mathbf{x})$ is a given positive coefficient. For simplicity, we use homogeneous Dirichlet boundary conditions on $\partial\Omega$ for all $t \in (0, T)$. For the variational problem (3) we then introduce the weight

$$K = \begin{bmatrix} kI_d & 0 \\ 0 & 1 \end{bmatrix}.$$

A natural space for the unknown S is then given by $V = L_2(0, T, H_0^1(\Omega))$. For the discretization, we use a standard conforming subspace $V_h \subset V$ consisting of piecewise Lagrangian polynomials which are globally continuous. We then solve the discretized saddle-point problem (4) by using the MINRES method with the block diagonal preconditioner

$$\hat{P} = \begin{bmatrix} \hat{A} & 0 & 0 \\ 0 & \hat{C} & 0 \\ 0 & 0 & \hat{W} \end{bmatrix},$$

where $\hat{A}$ denotes the auxiliary space AMG solver for $H(\text{div})$-problem applied to the matrix A (HypreADS, Kolev and Vassilevski 2012), $\hat{C}$ is a standard AMG preconditioner for C (BoomerAMG, HYPRE), and $\hat{W}$ represents the diagonal of the $L_2(\Omega_T)$ mass matrix W.

Example 1 In this example we let $\Omega = (0, 1)^2$, $T = 1$ and $k \equiv 1$. The exact solution is given by $u(x_1, x_2, t) = e^{-t}\sin(\pi x_1)\sin(\pi x_2)$.

The initial—fine—space-time mesh (level 0) is an unstructured tetrahedral mesh with $490, 200$ elements. We use graph partitioning algorithms (Karypis and Kumar, 1998) to construct the agglomerated space-time meshes shown in Fig. 1. For the discretization, we use lowest order finite element spaces on the fine grid and then we construct the hierarchy of coarse spaces as explained in Sect. 3. Table 1 reports the errors with respect to the exact solution. We observe that the upscaling procedure allows to dramatically reduce the number of unknowns maintaining reasonable good approximations, see also Fig. 1.

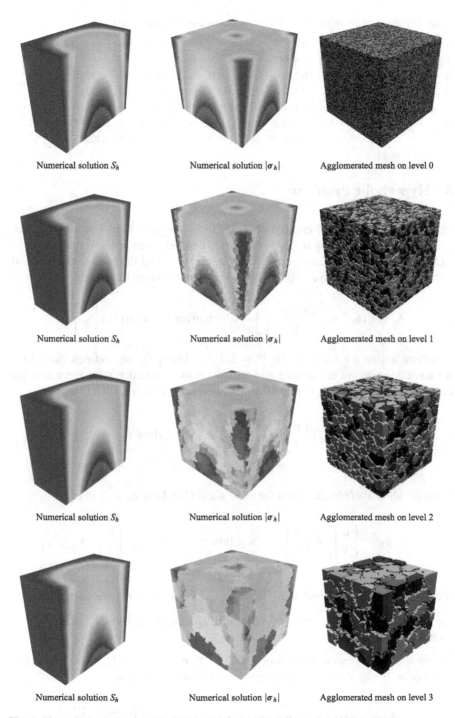

Numerical solution S_h Numerical solution $|\sigma_h|$ Agglomerated mesh on level 0

Numerical solution S_h Numerical solution $|\sigma_h|$ Agglomerated mesh on level 1

Numerical solution S_h Numerical solution $|\sigma_h|$ Agglomerated mesh on level 2

Numerical solution S_h Numerical solution $|\sigma_h|$ Agglomerated mesh on level 3

Fig. 1 Numerical solutions and agglomerated meshes for different levels (Example 1)

Table 1 Numerical errors for different agglomeration levels for Example 1

Level	Elements	Dof	$\|S - S_H\|_0$	$\|\sigma - \sigma_H\|_0$	$\|u_h - u_H\|_0$	$\|\sigma_h - \sigma_H\|_0$	Iter
0	490, 200	1, 579, 808	3.4360E−03	2.4217E−02	–	–	107
1	7, 700	218, 089	6.2509E−03	3.2351E−02	2.0985E−03	3.5408E−02	80
2	1, 043	59, 085	2.5489E−02	7.5482E−02	8.3829E−03	1.0854E−01	102
3	179	12, 366	8.1318E−02	1.7308E−01	2.6544E−02	2.5752E−01	60
4	39	3, 127	2.3470E−01	3.7018E−01	7.6846E−02	5.5365E−01	34
5	8	635	3.0685E−01	5.1457E−01	1.0064E−01	7.7024E−01	27

5 Hyperbolic Problem

Here we consider the differential operator $\mathcal{L}(S) := f_0(S_*)S\mathbf{u}(\cdot)$, with the given velocity field $\mathbf{u}$ (satisfying $\mathbf{u} \cdot \mathbf{n_x} = 0$ on $\partial\Omega$) and the given positive function $f_0 = f_0(S_*)$. Such equations can be used, for example, to model the evolution in time of water or gas saturation in an oil reservoir. We then introduce the weight

$$K = K(S_*) = \begin{bmatrix} f_0(S_*)I_d & 0 \\ 0 & 1 \end{bmatrix} \quad \text{which gives} \quad \sigma = K(S_*)\begin{bmatrix} \mathbf{u} \\ 1 \end{bmatrix}S.$$

A natural setting for S is given by $V = L_2(\Omega_T)$. Using the second equation of (3) we can eliminate the unknown S and we obtain the reduced system for σ and the Lagrange multiplier μ: Find $\sigma \in H(\mathrm{div}; \Omega_T)$ and $\mu \in L_2(\Omega_T)$, such that

$$\left(\left(K^{-1} - \delta_K^{-1} \begin{bmatrix} \mathbf{u} \\ 1 \end{bmatrix}\begin{bmatrix} \mathbf{u} \\ 1 \end{bmatrix}^\top \right) \sigma, \, \psi \right) + (\mu, \mathrm{div}\psi) = 0,$$

$$(\mathrm{div}\sigma, w) = (q, w) \tag{5}$$

holds for all $\psi \in H(\mathrm{div}; \Omega_T)$ and for all $w \in L_2(\Omega_T)$. Here $\delta_K \in \mathbb{R}$ is given by

$$\delta_K = \begin{bmatrix} \mathbf{u} \\ 1 \end{bmatrix}^\top K \begin{bmatrix} \mathbf{u} \\ 1 \end{bmatrix} \quad \text{and further} \quad S = \delta_K^{-1} \begin{bmatrix} \mathbf{u} \\ 1 \end{bmatrix}^\top \sigma.$$

It can be shown that the matrix $K^{-1} - \delta_K^{-1} \begin{bmatrix} \mathbf{u} \\ 1 \end{bmatrix}\begin{bmatrix} \mathbf{u} \\ 1 \end{bmatrix}^\top$ in (5) is positive definite on the nullspace of the divergence operator, if $\mathrm{div}_x(f_0(S_*)\mathbf{u}) \geq 0$ in Ω and $\mathbf{u} \cdot \mathbf{n_x} = 0$ on $\partial\Omega$.

Example 2 In this example we consider $\Omega = \{\mathbf{x} \in \mathbb{R}^2 : |\mathbf{x}| < 1\}$, $T = 2$, $f_0(S_*) \equiv 1$ and $q_0 \equiv 0$ with the velocity function and the initial condition

$$\mathbf{u}(x_1, x_2, t) = \begin{bmatrix} -x_2 \\ x_1 \end{bmatrix} \quad \text{and} \quad S_0(x_1, x_2) = e^{-100\left[(x_1-0.5)^2 + x_2^2\right]}.$$

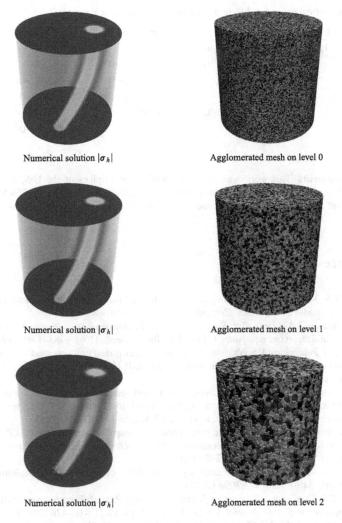

Fig. 2 Numerical solution and agglomerated meshes for different levels (Example 2)

For the discretization we use Raviart-Thomas pairs $\mathbf{R}_h$, W_h for σ and the Lagrange multiplier μ. The initial fine mesh (an unstructured tetrahedral mesh with 1,315,708 elements) and the agglomerated meshes are shown in Fig. 2. Table 2 shows (similarly to what already observed for the parabolic example) that upscaling allows to achieve both effective dimension reduction and good approximation of the fine grid solution (level 0). The divergence free solver Christensen et al. (2015) allows for the robust solution of the discretized saddle point problem at each level as shown by the number of iterations reported in Table 2.

Table 2 Numerical errors for different agglomeration levels for Example 2

Level	Elements	Dof	$\|\sigma_h - \sigma_H\|_0$	$\|\mu_h - \mu_H\|_0$	Iter
0	1, 315, 708	3, 970, 948	–	–	39
1	164, 495	1, 636, 016	1.1665E−03	1.2176E−09	39
2	21, 009	495, 815	5.0647E−03	2.2788E−04	33
3	3, 215	99, 004	9.1879E−03	4.6800E−04	24
4	684	22, 324	1.0483E−02	5.6677E−04	19
5	200	8, 041	1.2115E−02	7.1052E−04	16

Acknowledgements This work was performed under the auspices of the U.S. Department of Energy by Lawrence Livermore National Laboratory under Contract DE-AC52-07NA27344. The work was partially supported by ARO under US Army Federal Grant # W911NF-15-1-0590.

References

J.H. Adler, P.S. Vassilevski, Error analysis for constrained first-order system least-squares finite-element methods. SIAM J. Sci. Comput. **36**(3), A1071–A1088 (2014)

Z. Cai, R. Lazarov, T.A. Manteuffel, S.F. McCormick, First-order system least squares for second-order partial differential equations. I. SIAM J. Numer. Anal. **31**(6), 1785–1799 (1994)

G.F. Carey, A.I. Pehlivanov, P.S. Vassilevski, Least-squares mixed finite element methods for non-selfadjoint elliptic problems. II. Performance of block-ILU factorization methods. SIAM J. Sci. Comput. **16**(5), 1126–1136 (1995)

M. Christensen, U. Villa, P.S. Vassilevski, Multilevel techniques lead to accurate numerical Upscaling and Scalable Robust Solvers for Reservoir Simulation, in *SPE Reservoir Simulation Symposium, 23–25 February, Houston, Texas, USA*, SPE-173257-MS, 2015

HYPRE, A library of high performance preconditioners, http://www.llnl.gov/CASC/hypre/

G. Karypis, V. Kumar, A fast and high quality multilevel scheme for partitioning irregular graphs. SIAM J. Sci. Comput. **20**(1), 359–392 (1998)

T.V. Kolev, P.S. Vassilevski, Parallel auxiliary space AMG solver for H(div) problems. SIAM J. Sci. Comput. **34**(6), A3079–A3098 (2012)

I.V. Lashuk, P.S. Vassilevski, Element agglomeration coarse Raviart-Thomas spaces with improved approximation properties. Numer. Linear Algebra Appl. **19**(2), 414–426 (2012)

I.V. Lashuk, P.S. Vassilevski, The construction of the coarse de Rham complexes with improved approximation properties. Comput. Methods Appl. Math. **14**(2), 257–303 (2014)

MFEM, Modular finite element methods, mfem.org

J.E. Pasciak, P.S. Vassilevski, Exact de Rham sequences of spaces defined on macro-elements in two and three spatial dimensions. SIAM J. Sci. Comput. **30**(5), 2427–2446 (2008)

Scalable BDDC Algorithms for Cardiac Electromechanical Coupling

L.F. Pavarino, S. Scacchi, C. Verdi, E. Zampieri, and S. Zampini

1 Introduction

The spread of electrical excitation in the cardiac muscle and the subsequent contraction-relaxation process is quantitatively described by the cardiac electromechanical coupling model. The electrical model consists of the Bidomain system, which is a degenerate parabolic system of two nonlinear partial differential equations (PDEs) of reaction-diffusion type, describing the evolution in space and time of the intra- and extracellular electric potentials. The PDEs are coupled through the reaction term with a stiff system of ordinary differential equations (ODEs), the *membrane model*, which describes the flow of the ionic currents through the cellular membrane and the dynamics of the associated gating variables. The mechanical model consists of the quasi-static finite elasticity system, modeling the cardiac tissue as a nearly-incompressible transversely isotropic hyperelastic material, and coupled with a system of ODEs accounting for the development of biochemically generated active force.

L.F. Pavarino (✉)
Dipartimento di Matematica, Università di Pavia, Via Ferrata 5, 27100 Pavia, Italy
e-mail: luca.pavarino@unipv.it

S. Scacchi • C. Verdi • E. Zampieri
Dipartimento di Matematica, Università degli Studi di Milano, Via Saldini 50, 20133 Milano, Italy
e-mail: simone.scacchi@unimi.it; claudio.verdi@unimi.it; elena.zampieri@unimi.it

S. Zampini
Extreme Computing Research Center, Computer Electrical and Mathematical Sciences &
Engineering Division, King Abdullah University of Science and Technology, Thuwal,
Saudi Arabia
e-mail: stefano.zampini@kaust.edu.sa

© Springer International Publishing AG 2017   261
C.-O. Lee et al. (eds.), *Domain Decomposition Methods in Science
and Engineering XXIII*, Lecture Notes in Computational Science
and Engineering 116, DOI 10.1007/978-3-319-52389-7_26

The numerical approximation of the cardiac electromechanical coupling is a challenging multiphysics problem, because the space and time scales associated with the electrical and mechanical models are very different, see e.g. Chapelle et al. (2012), Sundnes et al. (2014). Moreover, the discretization of the model leads to the solution of a large nonlinear system at each time step, which is often decoupled by an operator splitting techniques into the solution of a large linear system for the electrical part and a nonlinear system for the mechanical part.

While several studies in the last decade have been devoted to the development of efficient solvers and preconditioners for the Bidomain model, see e.g. Plank et al. (2007), Pavarino and Scacchi (2008), Zampini (2014) and the recent monograph by Colli Franzone et al. (2014), a few studies have focused on the development of efficient solvers for the quasi-static cardiac mechanical model, see Vetter and McCulloch (2000), Rossi et al. (2012), Gurev et al. (2011).

In this paper, we present new numerical results for a Balancing Domain Decomposition by Constraints (BDDC) preconditioner, first introduced in Dohrmann (2003), here embedded in a Newton-Krylov (NKBDDC) method, introduced in Pavarino et al. (2015) for the nonlinear system arising from the discretization of the finite elasticity equations. The Jacobian system arising at each Newton step is solved iteratively by a BDDC preconditioned GMRES method. We report here the results of three-dimensional numerical tests on a BlueGene/Q machine, showing the scalability of the NKBDDC mechanical solver.

2   Cardiac Electromechanical Models

(a) **Mechanical model of cardiac tissue.** We denote by $\mathbf{X} = (X_1, X_2, X_3)^T$ the material coordinates of the undeformed cardiac domain $\widehat{\Omega}$, by $\mathbf{x} = (x_1, x_2, x_3)^T$ the spatial coordinates of the deformed cardiac domain $\Omega(t)$ at time t, and by $\mathbf{F}(\mathbf{X}, t) = \frac{\partial \mathbf{x}}{\partial \mathbf{X}}$ the deformation gradient. The cardiac tissue is modeled as a nonlinear hyperelastic material satisfying the steady-state force equilibrium equation

$$\mathrm{Div}(\mathbf{FS}) = \mathbf{0}, \qquad \mathbf{X} \in \widehat{\Omega}. \tag{1}$$

The second Piola-Kirchoff stress tensor $\mathbf{S} = \mathbf{S}^{pas} + \mathbf{S}^{vol} + \mathbf{S}^{act}$ is the sum of passive, volumetric and active components. The passive and volumetric components are defined as $S_{ij}^{pas,vol} = \frac{1}{2} \left(\frac{\partial W^{pas,vol}}{\partial E_{ij}} + \frac{\partial W^{pas,vol}}{\partial E_{ji}} \right)$ $i, j = 1, 2, 3$, where $\mathbf{E} = \frac{1}{2}(\mathbf{C} - \mathbf{I})$ and $\mathbf{C} = \mathbf{F}^T \mathbf{F}$ are the Green-Lagrange and Cauchy strain tensors, W^{pas} is an exponential strain energy function (derived from Eriksson et al. 2013) modeling the myocardium as a transversely isotropic hyperelastic material, and $W^{vol} = K(J - 1)^2$ is a volume change penalization term accounting for the almost incompressibility of the myocardium, with K a positive bulk modulus and $J = det(\mathbf{F})$.

(b) **Mechanical model of active tension.** The active component $\mathbf{S}^{act}$ develops along the myofiber direction, $\mathbf{S}^{act} = T_a \frac{\widehat{\mathbf{a}_l} \otimes \widehat{\mathbf{a}_l}}{\mathbf{a}_l^T \mathbf{C} \mathbf{a}_l}$, where $\widehat{\mathbf{a}}_l$ is the fiber direction and $T_a = T_a\left(Ca_i, \lambda, \frac{d\lambda}{dt}\right)$ is the biochemically generated active tension, which depends on intracellular calcium concentrations, and the myofiber stretch $\lambda = \sqrt{\widehat{\mathbf{a}}_l^T \mathbf{C} \widehat{\mathbf{a}}_l}$ and stretch-rate $\frac{d\lambda}{dt}$ (see Land et al. 2012).

(c) **Electrical model of cardiac tissue: the Bidomain model.** We will use the following parabolic-elliptic formulation of the modified Bidomain model on the reference configuration $\widehat{\Omega} \times (0, T)$,

$$\begin{cases} c_m J \dfrac{\partial \widehat{v}}{\partial t} - \text{Div}(J\,\mathbf{F}^{-1} D_i \mathbf{F}^{-T}\,\text{Grad}(\widehat{v} + \widehat{u}_e)) + J\,i_{ion}(\widehat{v}, \widehat{\mathbf{w}}, \widehat{\mathbf{c}}) = 0 \\ -\text{Div}(J\,\mathbf{F}^{-1} D_i \mathbf{F}^{-T}\,\text{Grad}\,\widehat{v}) - \text{Div}(J\,\mathbf{F}^{-1}(D_i + D_e)\mathbf{F}^{-T}\,\text{Grad}\,\widehat{u}_e) = J\,\widehat{i}_{app}^{e} \\ \dfrac{\partial \widehat{\mathbf{w}}}{\partial t} - \mathbf{R}_w(\widehat{v}, \widehat{\mathbf{w}}) = 0, \qquad \dfrac{\partial \widehat{\mathbf{c}}}{\partial t} - \mathbf{R}_c(\widehat{v}, \widehat{\mathbf{w}}, \widehat{\mathbf{c}}) = 0. \end{cases} \tag{2}$$

for the transmembrane potential $\widehat{v}$, the extracellular potential $\widehat{u}_e$, and the gating and ionic concentrations variables $(\widehat{\mathbf{w}}, \widehat{\mathbf{c}})$. This system is completed by prescribing initial conditions, insulating boundary conditions, and the applied current $\widehat{i}_{app}^{e}$; see Colli Franzone et al. (2016) for further details. The axisymmetric conductivity tensors are given by $D_{i,e}(\mathbf{x}) = \sigma_l^{i,e}\,\mathbf{a}_l(\mathbf{x})\mathbf{a}_l^T(\mathbf{x}) + \sigma_t^{i,e}\,\mathbf{a}_t(\mathbf{x})\mathbf{a}_t^T(\mathbf{x})$, where $\sigma_l^{i,e}$, $\sigma_t^{i,e}$ are the conductivity coefficients in the intra- and extracellular media measured along and across the fiber direction $\mathbf{a}_l$, $\mathbf{a}_t$.

(d) **Ionic membrane model and stretch-activated channel current.** The functions $I_{ion}(v, \mathbf{w}, \mathbf{c})$ ($i_{ion} = \chi I_{ion}$), $R_w(v, \mathbf{w})$ and $R_c(v, \mathbf{w}, \mathbf{c})$ in the Bidomain model (2) are given by the ionic membrane model introduced by ten Tusscher et al. (2004), available from the cellML depository (models.cellml.org/cellml). χ denotes the cellular surface to volume ratio.

3 Methods

Space and Time Discretization We discretize the cardiac domain with a hexahedral structured grid T_{h_m} for the mechanical model (1) and T_{h_e} for the electrical Bidomain model (2), where T_{h_e} is a refinement of T_{h_m}. We then discretize all scalar and vector fields of both mechanical and electrical models by isoparametric Q_1 finite elements in space. The time discretization is performed by a semi-implicit splitting method; see Colli Franzone et al. (2016) for further details.

Computational Kernels Due to the discretization strategies described above, the main computational kernels of our solver at each time step are the following:

1. solve the nonlinear system deriving from the discretization of the mechanical problem (1) using an inexact Newton method. At each Newton step, a

nonsymmetric Jacobian system $Kx = f$ is solved inexactly by the GMRES iterative method preconditioned by a BDDC preconditioner, described in the next section.

2. solve the symmetric positive semidefinite linear system deriving from the discretization of the Bidomain model by using the Conjugate Gradient method preconditioned by the Multilevel Additive Schwarz preconditioner developed in Pavarino and Scacchi (2008).

3.1 Iterative Substructuring, Schur Complement System and BDDC Preconditioner

To keep the notation simple, in the remainder of this section and the next, we denote the reference domain by Ω instead of $\widehat{\Omega}$. Let us consider a decomposition of Ω into N nonoverlapping subdomains Ω_i of diameter H_i (see e.g. Toselli and Widlund 2004, Ch. 4) $\Omega = \bigcup_{i=1}^{N} \Omega_i$, and set $H = \max H_i$. As in classical iterative substructuring, we reduce the problem to the interface $\Gamma := \left(\bigcup_{i=1}^{N} \partial\Omega_i \right) \backslash \partial\Omega$ by eliminating the interior degrees of freedom associated to basis functions with support in the interior of each subdomain, hence obtaining the Schur complement system

$$S_\Gamma x_\Gamma = g_\Gamma, \tag{3}$$

where $S_\Gamma = K_{\Gamma\Gamma} - K_{\Gamma I} K_{II}^{-1} K_{I\Gamma}$ and $g = f_\Gamma - K_{\Gamma I} K_{II}^{-1} f_I$ are obtained from the original discrete problem $Kx = f$ by reordering the finite element basis functions in interior (subscript I) and interface (subscript Γ) basis functions. The Schur complement system (3) is solved iteratively by the GMRES method using a BDDC preconditioner M_{BDDC}^{-1}

$$M_{\text{BDDC}}^{-1} S_\Gamma x_\Gamma = M_{\text{BDDC}}^{-1} f_\Gamma. \tag{4}$$

Once the interface solution x_Γ is computed, the internal values x_I can be recovered by solving local problems on each subdomain Ω_i.

BDDC preconditioners represent an evolution of balancing Neumann-Neumann methods where all local and coarse problems are treated additively due to a choice of so-called primal continuity constraints across the interface of the subdomains. These primal constraints can be point constraints and/or averages or moments over edges or faces of the subdomains. BDDC preconditioners were introduced in Dohrmann (2003) and first analyzed in Mandel and Dohrmann (2003). For the construction of BDDC preconditioners applied to the nonlinear elasticity system constituting the cardiac electromechanical coupling problem, we refer to Pavarino et al. (2015).

4 Numerical Results

We present here the results of parallel numerical experiments run on the IBM-BlueGene/Q machine of Cineca (www.cineca.it). Our FORTRAN90 code is based on the open source PETSc library, see Balay et al. (2016). At each Newton iteration of the mechanical solver, the Jacobian system is solved by GMRES preconditioned by the BDDC preconditioner, using as stopping criterion a 10^{-8} reduction of the relative residual l_2-norm. The BDDC method is available as a preconditioner in PETSc and it has been contributed to the library by Zampini (2016).

The values of the Bidomain electrical conductivity coefficients used in all the numerical tests are $\sigma_l^i = 3.0$, $\sigma_l^e = 2.0$, $\sigma_t^i = 0.315$, $\sigma_t^e = 1.35$, all in mΩ^{-1}cm^{-1}. The parameter values in the transversely isotropic strain energy function are chosen as in the original work of Eriksson et al. (2013). The domains used in the simulations model are wedges of the ventricular wall. They are either slabs or truncated ellipsoidal domains; for details on the dimensions, see Pavarino et al. (2015). The myocardial fibers are modeled to rotate intramurally linearly with the depth of the ventricular wall for a total amount of 120°.

Test 1: Weak Scaling We first consider a weak scaling test on slab and truncated ellipsoidal domains of increasing size. The number of subdomains (processors) is increased from 256 to 8192, with the largest domain being a slab or a truncated half ellipsoid. The physical dimensions of the domains are chosen so that the electrical mesh size h is kept fixed to the value of about $h = 0.01$ cm and so that the local mesh on each subdomain is fixed ($20 \cdot 20 \cdot 20$). The mechanical mesh size is four times smaller than the electrical one in each direction, thus on each subdomain the local mechanical mesh is $5 \cdot 5 \cdot 5$. The discrete nonlinear elasticity system increases from about 100 thousand degrees of freedom for the case with 256 subdomains to 3 million degrees of freedom for the case with 8192 subdomains. Motivated by the results of our previous study (Pavarino et al., 2010) of BDDC methods for almost incompressible linear elasticity, we have considered several choices of primal constraints in our BDDC preconditioner: subdomain vertices (V), vertices + edges (VE), vertices + edges + faces (VEF), vertices + edges + edge moments (VEm), vertices + edges + edge moments + faces (VEmF). The simulation is run for 10 electrical time steps of size $\tau_e = 0.05$ ms during the excitation phase and for 2 mechanical time steps of size $\tau_m = 0.25$ ms.

The results regarding the mechanical solver reported in Table 1 show that the linear GMRES iteration (lit) are completely scalable due to the use of the BDDC preconditioner, as well as the nonlinear Newton iterations (not shown), while the cpu times increase with the number of processors. This is due to the superlinear cost of the coarse problem and will require further research with a three-level BDDC preconditioner. For slab domains, even if the number of GMRES iterations is the largest, the best choice of primal space in terms of CPU times is the minimal one (V), using only the vertices. For truncated ellipsoidal domains, instead, the GMRES iterations with only vertices as primal space grow considerably, and the best primal choice in terms of timings is vertices + edges (VE).

Table 1 Weak scaling test on slab and ellipsoidal domains

		V		VE		VEF		VEm		VEmF	
Procs.	Dof	Lit	Time	Lit	Time	Lit	Time	Lit	Time	Lit	Time
Slab domains											
256	105,903	94	1.0	42	0.9	38	1.1	32	1.2	26	1.2
512	209,223	90	1.1	40	1.1	37	1.3	32	1.5	26	1.5
1042	413,343	86	1.4	38	1.6	36	1.9	30	2.1	24	2.2
2048	807,003	85	2.2	38	2.9	36	3.5	30	3.9	24	4.1
4096	1,604,043	84	5.2	39	6.6	–	–	–	–	–	–
8192	3,188,283	88	16.7	–	–	–	–	–	–	–	–
Ellipsoidal domains											
256	105,903	475	3.3	180	2.3	168	2.6	119	2.5	106	2.4
512	209,223	533	4.2	191	2.8	174	3.3	126	3.0	109	3.0
1042	413,343	558	5.8	173	4.0	158	4.6	125	4.7	106	4.9
2048	807,003	674	9.4	179	6.3	169	7.5	130	7.2	107	7.5
4096	1,604,043	686	15.9	176	12.3	–	–	–	–	–	–

Mechanical solver with GMRES-BDDC and different choices of primal constraints: vertices (V), vertices + edges (VE), vertices + edges + faces (VEF), vertices + edges + edge moments (VEm), vertices + edges + edge moments + faces (VEmF). Fixed local mechanical mesh: $5 \times 5 \times 5$ elements. Local mechanical problem size $= 648$. The table reports the number of processors (procs., that equals the number of subdomains), the total number of degrees of freedom (dof), the average GMRES-BDDC iterations per Newton iteration (lit) and the average CPU time in seconds per Newton iteration (time). The missing results (denoted by –) correspond to out-of-memory runs

Test 2: Whole Heartbeat Simulation We then present the results of a whole heart beat simulation (500 ms, 10,000 time steps) on 256 processors. The domain is a truncated ellipsoid discretized with a $96 \times 32 \times 8$ mechanical mesh (86,427 dof) nested in a $768 \times 256 \times 64$ electrical mesh (25,692,290 dof). Figure 1, top panels, reports the transmembrane potential distributions on the deforming epicardial surface and selected transmural sections of the cardiac domain at six selected time instants during the heartbeat.

We compare our BDDC solver (with only subdomain vertices primal constraints) vs. the widely used parallel AMG preconditioner BoomerAMG provided within the Hypre library (Henson and Yang, 2002); we used the default BoomerAMG parameters without any specific tuning. The table in Fig. 1, bottom, shows the average GMRES iterations per time step are 821 and 138 for the AMG and the BDDC solver, respectively. The average CPU times per time step are 32 and 3 s for the AMG and the BDDC solver, respectively. Thus the BDDC solver yields a reduction of computational costs and cpu times of about a factor 10 with respect to the default AMG preconditioner considered (this gain would probably be reduced by a proper AMG parameter tuning).

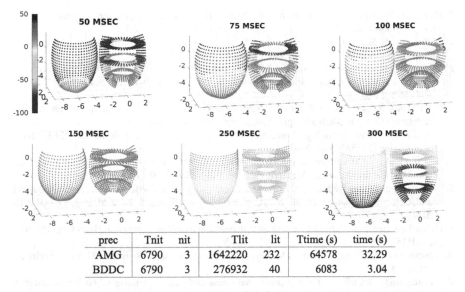

prec	Tnit	nit	Tlit	lit	Ttime (s)	time (s)
AMG	6790	3	1642220	232	64578	32.29
BDDC	6790	3	276932	40	6083	3.04

Fig. 1 Whole heartbeat simulation. Top: mechanical deformation of the cardiac domain at six time instants, from 50 to 300 msec. At each instant, the plot shows the transmembrane potential v at each point, ranging from resting (blue, -85 mV) to excited (red, 45 mV) values, on the epicardial surface and on selected transmural sections. The values on the axis are expressed in centimeters. Bottom: table reporting the comparison between the AMG and BDDC preconditioners: total Newton iterations (Tnit), average Newton iterations per time step (nit), total GMRES iterations (Tlit), average GMRES iterations per Newton iteration (lit), total CPU time (Ttime) in seconds, average CPU time per time step (time) in seconds

References

S. Balay et al., PETSc users manual. Tech. Rep. ANL-95/11 - Revision 3.7, Argonne National Laboratory, 2016

D. Chapelle et al., An energy-preserving muscle tissue model: formulation and compatible discretizations. J. Multiscale Comput. Eng. **10**, 189–211 (2012)

P. Colli Franzone, L.F. Pavarino, S. Scacchi, *Mathematical Cardiac Electrophysiology*. MSA, vol. 13 (Springer, New York, 2014)

P. Colli Franzone, L.F. Pavarino, S. Scacchi, Bioelectrical effects of mechanical feedbacks in a strongly coupled cardiac electro-mechanical model. Math. Models Methods Appl. Sci. **26**, 27–57 (2016)

C.R. Dohrmann, A preconditioner for substructuring based on constrained energy minimization. SIAM J. Sci. Comput. **25**, 246–258 (2003)

T.S.E. Eriksson et al., Influence of myocardial fiber/sheet orientations on left ventricular mechanical contraction. Math. Mech. Solids **18**, 592–606 (2013)

V. Gurev et al., Models of cardiac electromechanics based on individual hearts imaging data: Image-based electromechanical models of the heart. Biomech. Model Mechanobiol. **10**, 295–306 (2011)

V.E. Henson, U.M. Yang, BoomerAMG: a parallel algebraic multigrid solver and preconditioner. Appl. Numer. Math. **41**, 155–177 (2002)

S. Land et al., An analysis of deformation-dependent electromechanical coupling in the mouse heart. J. Physiol. 590, 4553–4569 (2012)

J. Mandel, C.R. Dohrmann, Convergence of a balancing domain decomposition by constraints and energy minimization. Numer. Linear Algebra Appl. **10**, 639–659 (2003)

L.F. Pavarino, S. Scacchi, Multilevel additive Schwarz preconditioners for the Bidomain reaction-diffusion system. SIAM J. Sci. Comput. **31**, 420–443 (2008)

L.F. Pavarino, S. Zampini, O.B. Widlund, BDDC preconditioners for spectral element discretizations of almost incompressible elasticity in three dimensions. SIAM J. Sci. Comput. **32**(6), 3604–3626 (2010)

L.F. Pavarino, S. Scacchi, S. Zampini, Newton-Krylov-BDDC solvers for non-linear cardiac mechanics. Comput. Methods Appl. Mech. Eng. **295**, 562–580 (2015)

G. Plank et al., Algebraic multigrid preconditioner for the cardiac bidomain model. IEEE Trans. Biomed. Eng. **54**, 585–596 (2007)

S. Rossi et al., Orthotropic active strain models for the numerical simulation of cardiac biomechanics. Int. J. Numer. Methods Biomed. Eng. **28**, 761–788 (2012)

J. Sundnes et al., Improved discretisation and linearisation of active tension in strongly coupled cardiac electro-mechanics simulations. Comput. Methods Biomech. Biomed. Eng. **17**, 604–615 (2014)

K.H. ten Tusscher et al., A model for human ventricular tissue. Am. J. Physiol. Heart Circ. Physiol. **286**, H1573–H1589 (2004)

A. Toselli, O.B. Widlund, *Domain Decomposition Methods: Algorithms and Theory* (Springer, Berlin, 2004)

F.J. Vetter, A.D. McCulloch, Three-dimensional stress and strain in passive rabbit left ventricle: a model study. Ann. Biomed. Eng. **28**, 781–792 (2000)

S. Zampini, Dual-primal methods for the cardiac bidomain model. Math. Models Methods Appl. Sci. **24**, 667–696 (2014)

S. Zampini, PCBDDC: a class of robust dual-primal preconditioners in PETSc. SIAM J. Sci. Comput. **38**(5), S282–S306 (2016)

A BDDC Algorithm for Weak Galerkin Discretizations

Xuemin Tu and Bin Wang

1 Introduction

The weak Galerkin (WG) methods are a class of nonconforming finite element methods, which were first introduced for a second order elliptic problem in Wang and Ye (2014). The idea of the WG is to introduce weak functions and their weak derivatives as distributions, which can be approximated by polynomials of different degrees. For second elliptic problems, weak functions have the form of $v = \{v_0, v_b\}$, where v_0 is defined inside each element and v_b is defined on the boundary of the element. v_0 and v_b can both be approximated by polynomials. The gradient operator is approximated by a *weak gradient* operator, which is further approximated by polynomials. These weakly defined functions and derivatives make the WG methods highly flexible and these WG methods have been extended to different applications such as Darcy in Lin et al. (2014), Stokes in Wang and Ye (2016), bi-harmonic in Mu et al. (2014), Maxwell in Mu et al. (2015c), Helmholtz in Mu et al. (2015b), and Brinkman equations in Mu et al. (2014). In Mu et al. (2015a), the optimal order of polynomial spaces is studied to minimize the number of degrees of freedom in the computation.

The WG methods are closely related to the hybridizable discontinuous Galerkin (HDG) methods, which were introduced by Cockburn and his collaborators in Cockburn et al. (2009). As most DG methods, the WG methods result in a large number of degrees of freedom and therefore require solving large linear systems with condition number deteriorating with the refinement of the mesh. Efficient fast solvers for the resulting linear system are necessary. However, so far there

X. Tu (✉) • B. Wang
Department of Mathematics, University of Kansas, 1460 Jayhawk Blvd, Lawrence, KS 66045-7594, USA
e-mail: xtu@math.ku.edu; binwang@math.ku.edu

© Springer International Publishing AG 2017
C.-O. Lee et al. (eds.), *Domain Decomposition Methods in Science and Engineering XXIII*, Lecture Notes in Computational Science and Engineering 116, DOI 10.1007/978-3-319-52389-7_27

are relatively few fast solvers for the WG methods. Some multigrid methods, based on conforming finite element discretization, are studied in Chen et al. (2015).

The BDDC algorithms, introduced by Dohrmann for second order elliptic problem in Dohrmann (2003), see also Mandel and Dohrmann (2003), Mandel et al. (2005), are non-overlapping domain decomposition methods, which are similar to the balancing Neumann-Neumann (BNN) algorithms. In the BDDC algorithm, the coarse problems are given in terms of a set of primal constraints. An important advantage with such a coarse problem is that the Schur complements that arise in the computation will all be invertible. The BDDC algorithms have been extended to the second order elliptic problem with mixed and hybrid formulations in Tu (2005, 2007) and the Stokes problem in Li and Widlund (2006b).

In this paper, we apply the BDDC preconditioner directly to the system arising from the WG discretization and estimate the condition number of the resulting preconditioned operator using its spectral equivalence with that of a hybridized RT method, which have been studied in Tu (2007).

The rest of the paper is organized as follows. An elliptic problem and its WG discretization are described in Sect. 2. We introduce the BDDC algorithms in Sect. 3 and analyze the condition number of the resulting preconditioned operator in Sect. 4. Finally, some computational results are given in Sect. 5.

2 An Elliptic Problem and Its WG Discretization

We consider the following elliptic problem on a bounded polygonal domain Ω, in two dimensions, with a Dirichlet boundary condition:

$$
\begin{cases}
-\nabla \cdot (\rho \nabla u) = f \text{ in } \quad \Omega, \\
u = g \qquad \text{ on } \quad \partial \Omega,
\end{cases}
\tag{1}
$$

where ρ is a positive definite matrix function with entries in $L^\infty(\Omega)$ satisfying

$$
\xi^T \rho(\mathbf{x}) \xi \geq \alpha \|\xi\|^2, \quad \text{for a.e. } \mathbf{x} \in \Omega,
$$

for some positive constant α, $f \in L^2(\Omega)$, and $g \in H^{1/2}(\partial \Omega)$. Without loss of generality, we assume that $g = 0$. If Ω is convex or has a C^2 boundary, the Eq. (1), with sufficiently smooth coefficient ρ, has a unique solution $u \in H^2(\Omega)$.

We will approximate u by introducing discontinuous finite element spaces. Let $\mathcal{T}_h$ be a shape-regular and quasi-uniform triangulation of Ω and denote an the element in $\mathcal{T}_h$ by κ. Let h_κ be the diameter of κ and the mesh size be $h = \max_{\kappa \in \mathcal{T}_h} h_\kappa$. Define $\mathcal{E}$ to be the union of edges of elements κ. $\mathcal{E}_i$ and $\mathcal{E}_\partial$ are the sets of the edges which are in interior of the domain and on its boundary, respectively.

Let $P_k(D)$ be the space of polynomials of order at most k on D and $\mathbf{P}_k(D) = [P_k(D)]^2$. Define the weak Galerkin finite element spaces associated with $\mathcal{T}_h$ as:

$$V_k = \{v = \{v_0, v_b\} : v_0|_\kappa \in P_k(\kappa),\ v_b|_e \in P_{k-1}(e),\quad \forall \kappa \in \mathcal{T}_h, e \in \partial \kappa\}$$

$$= \{v = \{v_0, v_b\} : v_0 \in W_k,\ v_b \in M_{k-1}\},$$

where

$$W_k = \{w_h \in L^2(\Omega) : w_h|_\kappa \in P_k(\kappa),\quad \forall \kappa \in \mathcal{T}_h\},$$

$$M_k = \{\mu_h \in L^2(\mathcal{E}) : \mu_h|_e \in P_k(e),\quad \forall e \in \mathcal{E}\}.$$

A function $v \in V_k$ has a single value v_b on each $e \in \mathcal{E}$.

Let

$$V_k^0 = \{v \in V_k\ v_b = 0 \text{ on } \partial \Omega\}.$$

Denoted by $\nabla_{w,k-1}$, the discrete weak gradient operator on the finite element space V_k is defined as follows: for $v = \{v_0, v_b\} \in V_k$, on each element $\kappa \in \mathcal{T}_h$, $\nabla_{w,k-1} v|_\kappa \in \mathbf{P}_{k-1}(\kappa)$ is the unique solution of the following equation

$$(\nabla_{w,k-1} v|_\kappa, \mathbf{q})_\kappa = -(v_{0,\kappa}, \nabla \cdot \mathbf{q}) + <v_{b,\kappa}, \mathbf{q} \cdot \mathbf{n}>_{\partial \kappa},\quad \forall \mathbf{q} \in \mathbf{P}_{k-1}(\kappa),$$

where $v_{0,\kappa}$ and $v_{b,\kappa}$ are the restrictions of v_0 and v_b to κ, respectively, $(u, w)_\kappa = \int_\kappa uw dx$, and $<u, w>_{\partial \kappa} = \int_{\partial \kappa} uw ds$. To simplify the notation, we will drop the subscript $k - 1$ in the discrete weak gradient operator $\nabla_{w,k-1}$.

The discrete problem resulting from the WG discretization of (1) can be written as: find $u_h = \{u_0, u_b\} \in V_k$ such that

$$a_s(u_h, v_h) = a(u_h, v_h) + s(u_h, v_h) = (f, v_h),\quad \forall v_h = \{v_0, v_b\} \in V_k,\qquad (2)$$

where

$$a(u_h, v_h) = \sum_{\kappa \in \mathcal{T}_h} (\rho \nabla_w u_h, \nabla_w v_h)_\kappa,$$

$$s(u_h, v_h) = \sum_{\kappa \in \mathcal{T}_h} h_\kappa^{-1} <Q_b u_0 - v_b, Q_b v_0 - v_b>_{\partial \kappa},$$

and where Q_b is the L^2-projection from $L^2(e)$ to $P_{k-1}(e)$, for $e \in \partial \kappa$. In Mu et al. (2015a), (2) is proved to have a unique solution and the approximation properties of the WG methods are also studied.

Given a $u_h \in V_k$, let $\mathbf{q}|_\kappa = \nabla_w u_h|_\kappa$ and write (2) as a system of $\mathbf{q}, u_0, u_b$, which is similar to the linear system resulting from the HDG discretization with the local stabilization parameter h_κ^{-1}. Given the value of u_b on $\partial \kappa$, $\mathbf{q}_\kappa$ and u_0 can be uniquely determined, see Cockburn et al. (2009). Therefore, by eliminating $\nabla_w u|_\kappa$ and u_0 locally in each element, (2) can be reduced to a system in u_b only

$$Au_b = b, \tag{3}$$

where b is the corresponding right-hand-side function.

In next section, we will develop a BDDC algorithm to solve the system in (3) for the u_b. To make the notation simple, we will denote u_b by λ and the finite element space for u_b by $\Lambda = \{\mu \in M_{k-1} : \mu|_e = 0 \; \forall e \in \partial\Omega\}$.

3 The BDDC Algorithms and Condition Number Bound

We decompose Ω into N non-overlapping subdomains Ω_i with diameters H_i, $i = 1, \cdots, N$, and set $H = \max_i H_i$. We assume that each subdomain is a union of shape-regular coarse triangles and that the number of such triangles forming an individual subdomain is uniformly bounded. We also assume $\rho(\mathbf{x})$, the coefficient of (1), is constant in each subdomain. We reduce the global problem (3) to a subdomain interface problem. Let Γ be the interface between subdomains. The set of the interface nodes Γ_h is defined as $\Gamma_h = \left(\cup_{i \neq j}\partial\Omega_{i,h} \cap \partial\Omega_{j,h}\right) \setminus \partial\Omega_h$, where $\partial\Omega_{i,h}$ is the set of nodes on $\partial\Omega_i$ and $\partial\Omega_h$ is the set of nodes on $\partial\Omega$.

We can decompose Λ into the subdomain interior and interface parts as

$$\Lambda = \bigoplus_{i=1}^{N} \Lambda_I^{(i)} \bigoplus \widehat{\Lambda}_\Gamma.$$

We denote the subdomain interface space of Ω_i by $\Lambda_\Gamma^{(i)}$, and the associate product space by $\Lambda_\Gamma = \prod_{i=1}^{N} \Lambda_\Gamma^{(i)}$. $R_\Gamma^{(i)}$ is the operator which maps functions in the continuous interface numerical trace space $\widehat{\Lambda}_\Gamma$ to their subdomain components in the space $\Lambda_\Gamma^{(i)}$. The direct sum of the $R_\Gamma^{(i)}$ is denoted by R_Γ. We can eliminate the subdomain interior variables $\lambda_I^{(i)}$ in each subdomain independently and define the subdomain Schur complement $S_\Gamma^{(i)}$ by: given $\lambda_\Gamma^{(i)} \in \Lambda_\Gamma^{(i)}$, $S_\Gamma^{(i)}\lambda_\Gamma^{(i)}$ is determined by such that

$$\begin{bmatrix} A_{II}^{(i)} & A_{I\Gamma}^{(i)} \\ A_{I\Gamma}^{(i)^T} & A_{\Gamma\Gamma}^{(i)} \end{bmatrix} \begin{bmatrix} \lambda_I^{(i)} \\ \lambda_\Gamma^{(i)} \end{bmatrix} = \begin{bmatrix} 0 \\ S_\Gamma^{(i)}\lambda_\Gamma \end{bmatrix}. \tag{4}$$

The global interface problem is assembled from the subdomain interface problems, and can be written as: find $\lambda_\Gamma \in \widehat{\Lambda}_\Gamma$, such that

$$\widehat{S}_\Gamma \lambda_\Gamma = b_\Gamma, \tag{5}$$

where $b_\Gamma = \sum_{i=1}^{N} R_\Gamma^{(i)^T} b_\Gamma^{(i)}$, and $\widehat{S}_\Gamma = \sum_{i=1}^{N} R_\Gamma^{(i)^T} S_\Gamma^{(i)} R_\Gamma^{(i)}$. Thus, $\widehat{S}_\Gamma$ is a symmetric, positive definite operator defined on the interface space $\widehat{\Lambda}_\Gamma$. We will propose a BDDC preconditioner for solving (5) with a preconditioned conjugate gradient method.

In order to introduce the BDDC precondition, we first introduce a partially assembled interface space $\widetilde{\Lambda}_\Gamma$ by

$$\widetilde{\Lambda}_\Gamma = \widehat{\Lambda}_\Pi \bigoplus \Lambda_\Delta = \widehat{\Lambda}_\Pi \bigoplus \left(\prod_{i=1}^{N} \Lambda_\Delta^{(i)} \right).$$

Here, $\widehat{\Lambda}_\Pi$ is the coarse level, primal interface space which is spanned by subdomain interface edge basis functions with constant values at the nodes of the edge for two dimensions. We change the variables so that the degree of freedom of each primal constraint is explicit, see Li and Widlund (2006a) and Klawonn and Widlund (2006). The new variables are called the primal unknowns. The space Λ_Δ is the direct sum of the $\Lambda_\Delta^{(i)}$, which are spanned by the remaining interface degrees of freedom with a zero average over each edge/face. In the space $\widetilde{\Lambda}_\Gamma$, we relax most continuity constraints across the interface but retain the continuity at the primal unknowns, which makes all the linear systems nonsingular.

We need to introduce several restriction, extension, and scaling operators between different spaces. $\overline{R}_\Gamma^{(i)}$ restricts functions in the space $\widetilde{\Lambda}_\Gamma$ to the components $\Lambda_\Gamma^{(i)}$ of the subdomain Ω_i. $R_\Delta^{(i)}$ maps the functions from $\widehat{\Lambda}_\Gamma$ to $\Lambda_\Delta^{(i)}$, its dual subdomain components. $R_{\Gamma\Pi}$ is a restriction operator from $\widehat{\Lambda}_\Gamma$ to its subspace $\widehat{\Lambda}_\Pi$. $\overline{R}_\Gamma : \widetilde{\Lambda}_\Gamma \to \Lambda_\Gamma$ is the direct sum of the $\overline{R}_\Gamma^{(i)}$ and $\widetilde{R}_\Gamma : \widehat{\Lambda}_\Gamma \to \widetilde{\Lambda}_\Gamma$ is the direct sum of $R_{\Gamma\Pi}$ and the $R_\Delta^{(i)}$. We define a positive scaling factor $\delta_i^\dagger(x)$ as follows: for $\gamma \in [1/2, \infty)$,

$$\delta_i^\dagger(x) = \frac{\rho_i^\gamma(x)}{\sum_{j \in \mathcal{N}_x} \rho_j^\gamma(x)}, \quad x \in \partial\Omega_{i,h} \cap \Gamma_h,$$

where $\mathcal{N}_x$ is the set of indices j of the subdomains such that $x \in \partial\Omega_j$. We note that $\delta_i^\dagger(x)$ is constant on each edge/face, since we assume that the $\rho_i(x)$ is constant in each subdomain. Multiplying each row of $R_\Delta^{(i)}$, with the scaling factor $\delta_i^\dagger(x)$, gives us $R_{D,\Delta}^{(i)}$. The scaled operators $\widetilde{R}_{D,\Gamma}$ is the direct sum of $R_{\Gamma\Pi}$ and the $R_{D,\Delta}^{(i)}$.

The partially assembled interface Schur complement is defined by $\widetilde{S}_\Gamma = \overline{R}_\Gamma^T \text{diag}(S_\Gamma^{(i)}) \overline{R}_\Gamma$ and the preconditioned BDDC operator is then of the form: find $\lambda_\Gamma \in \widehat{\Lambda}_\Gamma$, such that

$$\widetilde{R}_{D,\Gamma}^T \widetilde{S}_\Gamma^{-1} \widetilde{R}_{D,\Gamma} \widehat{S}_\Gamma \lambda_\Gamma = \widetilde{R}_{D,\Gamma}^T \widetilde{S}_\Gamma^{-1} \widetilde{R}_{D,\Gamma} b_\Gamma. \tag{6}$$

This preconditioned problem is the product of two symmetric, positive definite operators and we can use the preconditioned conjugate gradient method to solve it.

4 Condition Number Bound

We first introduce one useful norm, which is defined in Gopalakrishnan (2003) and Cockburn et al. (2014). For any domain D, we denote the L^2 norm by $\| \cdot \|_D$. For any $\lambda \in \Lambda(D)$, define

$$\| \lambda \|_D^2 = \left(\frac{1}{h} \sum_{\kappa \in \mathcal{T}_h, \kappa \subseteq \bar{D}} \| \lambda - m_\kappa(\lambda) \|_{L^2(\partial\kappa)}^2 \right)^{1/2}, \tag{7}$$

where $m_\kappa = \frac{1}{|\partial\kappa|} \int_{\partial\kappa} \lambda \, ds$, and $|\partial\kappa|$ is the length of the boundary of κ.

We define the interface averaging operator E_D, by

$$E_D = \widetilde{R}_\Gamma \widetilde{R}_{D,\Gamma}^T, \tag{8}$$

which computes a weighted average across the subdomain interface Γ and then distributes the averages to the degrees of freedom on the boundary of the subdomains.

Similarly to the proof of Tu and Wang (2016, Lemma 5), using the spectral equivalence of A, defined in (3), the linear system from the hybridized RT method, and the norm defined in (7), we obtain that the interface averaging operator E_D satisfies the following bound:

Lemma 1 *For any* $\lambda_\Gamma \in \widetilde{\Lambda}_\Gamma$,

$$|E_D \lambda_\Gamma|_{\widehat{S}_\Gamma}^2 \leq C \left(1 + \log \frac{H}{h} \right)^2 |\lambda_\Gamma|_{\widehat{S}_\Gamma}^2,$$

where C is a positive constant independent of H, h, and the coefficient of (1).

As in the proof of Li and Widlund (2006b, Theorem 1) and Tu and Wang (2016, Theorem 1), using Lemma 1, we can obtain

Theorem 1 *The condition number of the preconditioned operator $M^{-1}\widehat{S}_\Gamma$ is bounded by $C(1 + \log \frac{H}{h})^2$, where C is a constant which is independent of h, H, and the coefficients ρ of (1).*

5 Numerical Experiments

We have applied our BDDC algorithms to the model problem (1), where $\Omega = [0, 1]^2$. We decompose the unit square into $N \times N$ subdomains with the side-length $H = 1/N$. Equation (1) is discretized, in each subdomain, by the kth-order WG

Table 1 Performance with $H/h = 8$, # sub=64

| | | $\rho = 1$ | | | | ρ Checkboard pattern | | | |
| | | $k = 1$ | | $k = 1$ | | $k = 1$ | | $k = 2$ | |
H/h	#sub	Cond.	Iter.	Cond.	Iter.	Cond.	Iter.	Cond.	Iter.
8	4×4	2.22	6	3.50	7	1.80	5	2.37	5
	8×8	2.45	13	3.85	16	2.08	9	2.76	10
	16×16	2.45	14	3.86	17	2.16	14	2.87	15
	24×24	2.46	14	3.87	17	2.17	15	2.89	15
	32×32	2.46	14	3.87	17	2.18	15	2.90	16
4	8×8	1.78	11	2.90	14	1.67	9	2.33	10
8		2.45	13	3.86	16	2.08	9	2.76	10
16		3.29	15	4.95	18	2.49	10	3.18	10
24		3.85	17	5.67	18	2.74	10	3.43	11
32		4.28	17	6.21	19	2.91	10	3.60	11

method with a element diameter h. The preconditioned conjugate gradient iteration is stopped when the relative l_2-norm of the residual has been reduced by a factor of 10^6.

We have carried out two different sets of experiments to obtain iteration counts and condition number estimates. The results are listed in Table 1. In the first set of experiments, we take the coefficient $\rho \equiv 1$. In the second set of experiments, we take the coefficient $\rho = 1$ in half the subdomains and $\rho = 1000$ in the neighboring subdomains, in a checkerboard pattern. All the experimental results are fully consistent with our theory.

Acknowledgements This work was supported in part by National Science Foundation Contract No. DMS-1419069.

References

L. Chen, J. Wang, Y. Wang, X. Ye, An auxiliary space multigrid preconditioner for the weak Galerkin method. Comput. Math. Appl. **70**(4), 330–344 (2015)

B. Cockburn, J. Gopalakrishnan, R. Lazarov, Unified hybridization of discontinuous Galerkin, mixed, and continuous Galerkin methods for second order elliptic problems. SIAM J. Numer. Anal. **47**(2), 1319–1365 (2009)

B. Cockburn, O. Dubois, J. Gopalakrishnan, S. Tan, Multigrid for an HDG method. IMA J. Numer. Anal. **34**(4), 1386–1425 (2014)

C.R. Dohrmann, A preconditioner for substructuring based on constrained energy minimization. SIAM J. Sci Comput. **25**(1), 246–258 (2003)

J. Gopalakrishnan, A Schwarz preconditioner for a hybridized mixed method. Comput. Methods Appl. Math. **3**(1), 116–134 (electronic) (2003); Dedicated to Raytcho Lazarov

A. Klawonn, O.B. Widlund, Dual-primal FETI methods for linear elasticity. Commun. Pure Appl. Math. **59**(11), 1523–1572 (2006)

J. Li, O.B. Widlund, FETI–DP, BDDC, and block Cholesky methods. Int. J. Numer. Methods Eng. **66**, 250–271 (2006a)

J. Li, O.B. Widlund, BDDC algorithms for incompressible Stokes equations. SIAM J. Numer. Anal. **44**(6), 2432–2455 (2006b)

G. Lin, J. Liu, L. Mu, X. Ye, Weak Galerkin finite element methods for Darcy flow: anisotropy and heterogeneity. J. Comput. Phys. **276**, 422–437 (2014)

J. Mandel, C.R. Dohrmann, Convergence of a balancing domain decomposition by constraints and energy minimization. Numer. Linear Algebra Appl. **10**(7), 639–659 (2003)

J. Mandel, C.R. Dohrmann, R. Tezaur, An algebraic theory for primal and dual substructuring methods by constraints. Appl. Numer. Math. **54**(2), 167–193 (2005)

L. Mu, J. Wang, X. Ye, A stable numerical algorithm for the Brinkman equations by weak Galerkin finite element methods. J. Comput. Phys. **273**, 327–342 (2014)

L. Mu, J. Wang, X. Ye, A weak Galerkin finite element method with polynomial reduction. J. Comput. Appl. Math. **285**, 45–58 (2015a)

L. Mu, J. Wang, X. Ye, A new weak Galerkin finite element method for the Helmholtz equation. IMA J. Numer. Anal. **35**(3), 1228–1255 (2015b)

L. Mu, J. Wang, X. Ye, S. Zhang, A weak Galerkin finite element method for the Maxwell equations. J. Sci. Comput. **65**(1), 363–386 (2015c)

X. Tu, A BDDC algorithm for a mixed formulation of flows in porous media. Electron. Trans. Numer. Anal. **20**, 164–179 (2005)

X. Tu, A BDDC algorithm for flow in porous media with a hybrid finite element discretization. Electron. Trans. Numer. Anal. **26**, 146–160 (2007)

X. Tu, B. Wang, A BDDC algorithm for second order elliptic problems with hybridizable discontinuous Galerkin discretizations. Electron. Trans. Numer. Anal. **45**, 354–370 (2016)

J. Wang, X. Ye, A weak Galerkin mixed finite element method for second order elliptic problems. Math. Comput. **83**(289), 2101–2126 (2014)

J. Wang, X. Ye, A weak Galerkin finite element method for the Stokes equations. Adv. Comput. Math. **42**(1), 155–174 (2016)

Parallel Sums and Adaptive BDDC Deluxe

Olof B. Widlund and Juan G. Calvo

1 Introduction

There has recently been a considerable activity in developing adaptive methods for the selection of primal constraints for BDDC algorithms and, in particular, for BDDC deluxe variants. The primal constraints of a BDDC or FETI-DP algorithm provide the global, coarse part of such a preconditioner and are of crucial importance for obtaining rapid convergence of these preconditioned conjugate gradient methods for the case of many subdomains. When the primal constraints are chosen adaptively, we aim at selecting a primal space, which for a certain dimension of the coarse space, provides the fastest rate of the convergence for the iterative method. In the alternative, we can try to develop criteria which will guarantee that the condition number of the iteration stays below a given tolerance.

A particular inspiration for our own work has been a talk, see Dohrmann and Pechstein (2012), by Clark Dohrmann at DD21, held in Rennes, France, in June 2012. Dohrmann had then started joint work with Clemens Pechstein, see also Pechstein and Dohrmann (2016).

Much of this work for BDDC and FETI-DP iterative substructuring algorithms, which has been supported by theory, has been confined to developing primal constraints for equivalence classes with two elements such as those related to subdomain edges for problems defined on domains in the plane; see a recent survey paper, Klawonn et al. (2016b). In our context, the equivalence classes are sets of

O.B. Widlund (✉)
Courant Institute, 251 Mercer Street, New York, NY 10012, USA
e-mail: widlund@cims.nyu.edu

J.G. Calvo
CIMPA, Universidad de Costa Rica, San Jose 11501, Costa Rica
e-mail: juan.calvo@ucr.ac.cr

© Springer International Publishing AG 2017
C.-O. Lee et al. (eds.), *Domain Decomposition Methods in Science and Engineering XXIII*, Lecture Notes in Computational Science and Engineering 116, DOI 10.1007/978-3-319-52389-7_28

finite element nodes which belong to the boundaries of more than one subdomain with the equivalence relation defined by the sets of subdomain boundaries to which the nodes belong. While it is important to further study the best way of handling all cases, the basic issues appear to be well settled when the equivalence classes all have just two elements.

We note that this work is relevant for problems posed in $H(\text{div})$ even in three dimensions (3D) since the degrees of freedom on the interface between subdomains for Raviart-Thomas and Brezzi-Douglas-Marini elements are associated only with faces of the elements, see Oh et al. (2015), Zampini (2016). (These papers also concern BDDC three–level algorithms choosing two levels of primal constraints adaptively.) But for other elliptic problems in 3D, there is a need to develop algorithms and results for equivalence classes with three or more elements.

There is early work by Mandel, Šístek, and Sousedík, who developed condition number indicators, cf. Mandel and Sousedík (2007), Mandel et al. (2012). Talks by Clark Dohrmann and Axel Klawonn at DD23, held on Jeju Island, the Republic of Korea in July 2015, see Klawonn et al. (2016a), reported on recent progress to give similar algorithms a firm theoretical basis. A talk by Hyea Hyun Kim in the same mini-symposium also reported considerable progress for a different kind of algorithm. Her main new algorithm for problems in three dimensions is similar but not the same as ours; see further Kim et al. (2015). Our main result, developed independently, was reported on by the first author in the same mini-symposium; see further Calvo and Widlund (2016) and, for applications to isogeometric analysis, Beirão da Veiga et al. (2015).

This paper will focus on using parallel sums for general equivalence classes. Such an approach for equivalence classes with two elements has proven very successful in simplifying the formulas and arguments; see in particular Pechstein and Dohrmann (2013) and Sect. 2. Parallel sums for equivalence classes with more than two elements have also been quite successfully in numerical experiments by Simone Scacchi and Stefano Zampini, reported in Beirão da Veiga et al. (2015), for problems arising in isogeometric analysis and also by Zampini in a study of 3D problems formulated in $H(\mathbf{curl})$, based in part on Dohrmann and Widlund (2016), and reported on in this mini-symposium.

In this paper, we will focus on low order, nodal finite element approximations for scalar elliptic problems in three dimensions,

$$- \nabla \cdot (\rho(x)\nabla u) = f(x), \quad x \in \Omega, \quad \rho(x) > 0, \tag{1}$$

resulting in a linear system of equations to be solved using BDDC domain decomposition algorithms, especially its deluxe variant. We will always assume that the choice of boundary conditions results in a positive definite, symmetric stiffness matrix.

2 Equivalence Classes and BDDC Algorithms

BDDC algorithms, see, e.g., Li and Widlund (2006), are domain decomposition algorithms based on the decomposition of the domain Ω of an elliptic operator into non-overlapping subdomains Ω_i, each often associated with tens of thousands of degrees of freedom. The subdomain interface Γ_i of Ω_i does not cut through any elements and is defined by $\Gamma_i := \partial\Omega_i \setminus \partial\Omega$. The equivalence classes are associated with the subdomain faces, edges, and vertices of $\Gamma := \cup_i \Gamma_i$, the interface of the entire decomposition. Thus, for a problem in three dimensions, a subdomain face is associated with the degrees of freedom of the nodes belonging to the interior of the intersection of two boundaries of two neighboring subdomains Ω_i and Ω_j. Those of a subdomain edge are typically associated with a set of nodes common to three or more subdomain boundaries, while the endpoints of the subdomain edges are the subdomain vertices which are associated with even more subdomains.

Given the stiffness matrix $A^{(i)}$ of the subdomain Ω_i, we obtain a subdomain Schur complement $S^{(i)}$ by eliminating the interior variables, i.e., all those that do not belong to Γ_i. We will also work with principal minors of these Schur complements associated with faces, F, and edges, E, denoting them by $S_{FF}^{(i)}$ and $S_{EE}^{(i)}$, respectively.

The interface space is divided into a *primal* subspace of functions which are continuous across Γ and a complementary, *dual* subspace for which we will allow multiple values across the interface during part of the iteration. In our study, all the subdomain vertex variables will always belong to the primal set. We have three product spaces of finite element functions/vectors defined by their interface nodal values:

$$\widehat{W}_\Gamma \subset \widetilde{W}_\Gamma \subset W_\Gamma.$$

W_Γ is a product space without any continuity constraints across the interface. Elements of $\widetilde{W}_\Gamma$ have common values of the primal variables but allow multiple values of the dual variables while the elements of $\widehat{W}_\Gamma$ are continuous at all nodes on Γ. We will change variables, explicitly introducing the primal variables and a complementary sets of dual variables in order to simplify the presentations. We note that the change of basis will not in any way change the results of the computation. After eliminating the interior variables, we can then write the subdomain Schur complements as

$$S^{(i)} = \begin{pmatrix} S_{\Delta\Delta}^{(i)} & S_{\Delta\Pi}^{(i)} \\ S_{\Pi\Delta}^{(i)} & S_{\Pi\Pi}^{(i)} \end{pmatrix}.$$

We will partially subassemble the $S^{(i)}$, obtaining $\widetilde{S}$, enforcing the continuity of the primal variables only. Thus, we then work in $\widetilde{W}_\Gamma$. In each step of the iteration, we solve a linear system with the coefficient matrix $\widetilde{S}$. Solving these linear systems will be considerably much faster than if we work with the fully assembled system if the dimension of the primal space is modest. At the end of each iteration, the approximate solution is made continuous at all nodal points of the interface by applying a weighted averaging operator E_D. We always accelerate the iteration with the preconditioned conjugate gradient algorithm.

2.1 BDDC Deluxe

When designing a BDDC algorithm, we have to choose an effective set of primal constraints and also a good recipe for the averaging across the interface. Our paper concerns the choice of the primal constraints while we will always use the deluxe recipe in the construction of the averaging operator E_D.

We note that in work on three-dimensional problems formulated in $H(\textbf{curl})$, it was found that traditional averaging recipes did not always work well; cf. Dohrmann and Widlund (2016). The same is true for problems in $H(\text{div})$; see Oh et al. (2015). This occasional failure has its roots in the fact that there are two sets of material parameters in these applications. The deluxe scaling that was then introduced has also proven quite successful for a variety of other applications.

A face component of the average operator E_D across a subdomain face $F \subset \Gamma$, common to two subdomains Ω_i and Ω_j, is defined in terms of principal minors $S_{FF}^{(k)}$ of the $S^{(k)}, k = i, j$:

$$\bar{w}_F := (E_D w)_F := (S_{FF}^{(i)} + S_{FF}^{(j)})^{-1}(S_{FF}^{(i)} w_F^{(i)} + S_{FF}^{(j)} w_F^{(j)}).$$

Here $w_F^{(i)}$ is the restriction of $w^{(i)}$ to the face F, etc.

Deluxe averaging operators are also developed for subdomain edges and the operator E_D is assembled from all these components; see further Sect. 3. Our bound for this operator will be obtained from bounds for certain eigenvalues for the individual equivalence sets and will include factors that depend quadratically on the number of equivalence classes associated with the faces and edges of the individual subdomains. We have found that the performance consistently is far better than these bounds.

The core of any estimate for a BDDC algorithm is the norm of the averaging operator E_D. By an algebraic argument known, for FETI-DP since 2002, we know that the condition number of the iteration satisfies

$$\kappa(M_{BDDC}^{-1}\widehat{S}) \le \|E_D\|_{\widehat{S}}; \tag{2}$$

see Klawonn et al. (2002). Here M_{BDDC}^{-1} denotes the BDDC preconditioner and $\widehat{S}$ the fully assembled Schur complement of the problem. Instead of developing an estimate for E_D, we will work with $P_D := I - E_D$ and estimate $(R_F^T(w_F^{(i)} - \bar{w}_F))^T S^{(i)} R_F^T(w_F^{(i)} - \bar{w}_F)$. Here R_F denotes the restriction to the face F. We find, following Pechstein, that the sum of this quadratic form and a similar contribution from the neighboring subdomain Ω_j equals

$$(w_F^{(i)} - w_F^{(j)})^T (S_{FF}^{(i)} : S_{FF}^{(j)})(w_F^{(i)} - w_F^{(j)})$$

where

$$A : B := A(A + B)^{-1}B$$

is the parallel sum of A and B; cf. Anderson Jr. and Duffin (1969). We note that if A and B are positive definite, then $A : B = (A^{-1} + B^{-1})^{-1}$. If $A + B$ is only positive semi-definite, we can replace $(A + B)^{-1}$ by $(A + B)^\dagger$, any generalized inverse. The quadratic form can be estimated from above by

$$2(w_F^{(i)} - w_{F\Pi})^T (S_{FF}^{(i)} : S_{FF}^{(j)})(w_F^{(i)} - w_{F\Pi}) + 2(w_F^{(j)} - w_{F\Pi})^T (S_{FF}^{(i)} : S_{FF}^{(j)})(w_F^{(j)} - w_{F\Pi})$$

where $w_{F\Pi}$ is the restriction of an arbitrary element of the primal space to the face. We note that each of these terms can be estimated by an expression which is local to only one subdomain.

With $w_{F\Delta}^{(i)} := w_F^{(i)} - w_{F\Pi}$, we now estimate $w_{F\Delta}^{(i)T}(S_{FF}^{(i)} : S_{FF}^{(j)})w_{F\Delta}^{(i)}$ by the energy of $w^{(i)}$. We then need the minimum norm extension of any finite element function defined on F, which will provide a uniform bound for any extension of the values on F to the rest of Γ_i. We find that the relevant matrix is

$$\widetilde{S}_{FF}^{(i)} := S_{FF}^{(i)} - S_{F'F}^{(i)T}S_{F'F'}^{(i)-1}S_{F'F}^{(i)}.$$

Here $S_{F'F'}^{(i)}$ is the principal minor of $S^{(i)}$ with respect to $\Gamma_i \backslash F$ and $S_{F'F}^{(i)}$ an off-diagonal block of $S^{(i)}$. By appropriate choices of the primal space and of $w_{F\Pi}$, we are able to show that

$$w_{F\Delta}^{(i)T}(\widetilde{S}_{FF}^{(i)} : \widetilde{S}_{FF}^{(j)})w_{F\Delta}^{(i)} \leq w^{(i)T}S^{(i)}w^{(i)},$$

where $w^{(i)}$ is an arbitrary extension of the values of $w_F^{(i)}$.

For an adaptive algorithm, we can complete the estimate by using a generalized eigenvalue problem:

$$\widetilde{S}_{FF}^{(i)} : \widetilde{S}_{FF}^{(j)}\phi = \lambda S_{FF}^{(i)} : S_{FF}^{(j)}\phi. \qquad (3)$$

Primal constraints are then generated by using the eigenvectors of a few of the smallest eigenvalues of (3) and making $(\widetilde{S}_{FF}^{(i)} : \widetilde{S}_{FF}^{(j)})(w_F^{(i)} - w_F^{(j)})$ orthogonal to these eigenvectors.

A bound can now be obtained in terms of the smallest eigenvalue associated with the eigenvectors not used in deriving the primal constraints. Numerical studies show a very rapid decay of the eigenvalues of $S_{FF}^{(i)-1}(S_{FF}^{(i)} - \widetilde{S}_{FF}^{(i)})$; this property can also be proven assuming that Ω_i is Lipschitz and the coefficient $\rho(x)$ a constant. Therefore only a few primal constraints will greatly improve the bound on the norm of $(E_D w)_F$.

3 Equivalence Classes with More than Two Elements

We begin this section by considering parallel sums of more than two operators. We will work with symmetric matrices which all are at least positive semi-definite. For three positive definite matrices, we can define their parallel sum by

$$A : B : C := (A^{-1} + B^{-1} + C^{-1})^{-1},$$

with similar formulas for four or more matrices. A quite complicated formula for $A : B : C$ is given in Tian (2002) for the general case when some or all of the matrices might be only positive semi-definite. It is also shown, in Tian (2002, Theorem 3), that $A : B : C = (A^\dagger + B^\dagger + C^\dagger)^\dagger$ if and only if the three operators A, B, and C have the same range. In our context, this is not always the case since the matrix $\widetilde{S}_{EE}^{(i)}$, defined below, will be singular if Ω_i is an interior subdomain while it will be non-singular if $\partial\Omega_i$ intersects a part of $\partial\Omega$ where a Dirichlet condition is imposed. This issue can be avoided by making all operators non-singular by adding a small positive multiple of the identity to the singular operators.

We will first focus on a case of an equivalence class common to three subdomains as arising for most subdomain edges in a three-dimensional finite element context if the subdomains are generated using a mesh partitioner. We will use the notation $S_{EE}^{(i)}$, $S_{EE}^{(j)}$, and $S_{EE}^{(k)}$ for the principal minors, of the degrees of freedom of an edge E, of the subdomain Schur complements of the three subdomains that have this subdomain edge in common. The Schur complements of the Schur complements representing the minimal energy extensions to individual subdomains, of given values on the subdomain edge E, will be denoted by $\widetilde{S}_{EE}^{(i)}$, $\widetilde{S}_{EE}^{(j)}$, etc., and are defined by

$$\widetilde{S}_{EE}^{(i)} := S_{EE}^{(i)} - S_{E'E}^{(i)T} S_{E'E'}^{(i)-1} S_{E'E}^{(i)}. \tag{4}$$

Here $S_{E'E'}^{(i)}$ is the principal minor of $S^{(i)}$ of $\Gamma_i \setminus E$ and $S_{E'E}^{(i)}$ an off-diagonal block.

We can now introduce the deluxe average over the edge E by

$$\bar{w}_E := (S_{EE}^{(i)} + S_{EE}^{(j)} + S_{EE}^{(k)})^{-1}(S_{EE}^{(i)} w_E^{(i)} + S_{EE}^{(j)} w_E^{(j)} + S_{EE}^{(k)} w_E^{(k)}).$$

By using elementary inequalities, we can now obtain a bound of the square of the norm of an edge component of $P_D w$ by

$$3 w_{E\Delta}^{(i)T} S_{EE}^{(i)} : (S_{EE}^{(j)} + S_{EE}^{(k)}) w_{E\Delta}^{(i)}$$

and two similar terms obtained by changing the superscripts appropriately.

Returning to the search for adaptive primal spaces, we note that ideally, we would now like to prove that the three operators $T_E^{(i)} := S_{EE}^{(i)} : (S_{EE}^{(j)} + S_{EE}^{(k)})$, $T_E^{(j)} := S_{EE}^{(j)} : (S_{EE}^{(i)} + S_{EE}^{(k)})$, and $T_E^{(k)} := S_{EE}^{(k)} : (S_{EE}^{(i)} + S_{EE}^{(j)})$ all can be bounded uniformly from above by

$$S_{EE}^{(i)} : S_{EE}^{(j)} : S_{EE}^{(k)} := (S_{EE}^{(i)-1} + S_{EE}^{(j)-1} + S_{EE}^{(k)-1})^{-1}. \tag{5}$$

If this were possible, we could use that same matrix for estimates for $w_{E\Delta}^{(i)}, w_{E\Delta}^{(j)}$, and $w_{E\Delta}^{(k)}$; we could use arguments very similar to those of the previous section. But we are not that lucky; good bounds are only possible if $S_{EE}^{(i)}, S_{EE}^{(j)}$, and $S_{EE}^{(k)}$ are spectrally equivalent with good bounds. However, it is easy to find interesting examples where this does not hold. We therefore have to find a different common upper bound for $T_E^{(i)}, T_E^{(j)}$, and $T_E^{(k)}$ and accomplish this by using the trivial inequality

$$T_E^{(i)} \le T_E^{(i)} + T_E^{(j)} + T_E^{(k)},$$

and define our generalized eigenvalue problem as

$$(\widetilde{S}_{EE}^{(i)} : \widetilde{S}_{EE}^{(j)} : \widetilde{S}_{EE}^{(k)})\phi = \lambda(T_E^{(i)} + T_E^{(j)} + T_E^{(k)})\phi. \tag{6}$$

We note that these arguments extend directly to equivalence classes with more than three elements.

This is the recipe that we have used in most of our numerical experiments, which have proven quite successful; cf. Calvo and Widlund (2016) for many more details. However, it deserves to be noted that the distribution of the eigenvalues associated with the subdomain edges, in our experience, is less favorable than those of the subdomain faces but that we can benefit from the fact that the number of degrees of freedom of an edge typically is much smaller than that of a face.

Given the success, by others, with using parallel sums of each of the two sets of three Schur complements, we have also carried out experiments with that alternative generalized eigenvalue problem. The performance is very similar to that of our algorithm.

In our experiments, we have compared the performance of our adaptive algorithms with standard choices of the primal spaces. In choosing our primal constraints, we have, in some of our experiments, used tolerances introduced in Kim et al. (2015). We have found that our adaptive algorithm also works quite well for irregular subdomains generated by the METIS mesh partitioner.

References

W.N. Anderson Jr., R.J. Duffin, Series and parallel addition of matrices. J. Math. Anal. Appl. **26**, 576–594 (1969)

L. Beirão da Veiga, L.F. Pavarino, S. Scacchi, O.B. Widlund, S. Zampini, Adaptive selection of primal constraints for isogeometric BDDC deluxe preconditioners. SIAM J. Sci. Comput. (2017) (to appear)

J.G. Calvo, O.B. Widlund,   An adaptive choice of primal constraints for BDDC domain decomposition algorithms. TR2015-979, Courant Institute, New York University, 2016

C.R. Dohrmann, C. Pechstein, Constraint and weight selection algorithms for BDDC. Slides for a talk by Dohrmann at DD21 in Rennes, France, June 2012 (2012), http://www.numa.uni-linz. ac.at/~clemens/dohrmann-pechstein-dd21-talk.pdf

C.R. Dohrmann, O.B. Widlund, A BDDC algorithm with deluxe scaling for three-dimensional H(curl) problems. Commun. Pure Appl. Math. **69**(4), 745–770 (2016)

H.H. Kim, E.T. Chung, J. Wang, BDDC and FETI-DP algorithms with adaptive coarse spaces for three-dimensional elliptic problems with oscillatory and high contrast coefficients (2015), http://arxiv.org/abs/1606.07560

A. Klawonn, O.B. Widlund, M. Dryja, Dual-primal FETI methods for three-dimensional elliptic problems with heterogeneous coefficients. SIAM J. Numer. Anal. **40**(1), 159–179 (2002)

A. Klawonn, M. Kühn, O. Rheinbach, Adaptive coarse spaces for FETI–DP in three dimensions. SIAM J. Sci. Comput. (2016a, to appear)

A. Klawonn, P. Radtke, O. Rheinbach, A comparison of adaptive coarse spaces for iterative substructuring in two dimensions. Electron. Trans. Numer. Anal. **46**, 75–106 (2016b)

J. Li, O.B. Widlund, FETI–DP, BDDC, and block Cholesky methods. Int. J. Numer. Methods Eng. **66**(2), 250–271 (2006)

J. Mandel, B. Sousedík, Adaptive selection of face coarse degrees of freedom in the BDDC and the FETI-DP iterative substructuring methods. Comput. Methods Appl. Mech. Eng. **196**(8), 1389–1399 (2007)

J. Mandel, B. Sousedík, J. Šístek, Adaptive BDDC in three dimensions. Math. Comput. Simul. **82**(10), 1812–1831 (2012)

D.-S. Oh, O.B. Widlund, S. Zampini, C.R. Dohrmann, BDDC algorithms with deluxe scaling and adaptive selection of primal constraints for Raviart-Thomas vector fields. Math. Comput. (2017) (to appear)

C. Pechstein, C.R. Dohrmann, Modern domain decomposition methods, BDDC, deluxe scaling, and an algebraic approach. Talk by Pechstein in Linz, Austria, December 2013, http://people. ricam.oeaw.ac.at/c.pechstein/pechstein-bddc2013.pdf

C. Pechstein, C.R. Dohrmann, A unified framework for adaptive BDDC. Technical Report 2016-20, Johann Radon Institute for Computational and Applied Mathematics (RICAM) (2016), http://www.ricam.oeaw.ac.at/files/reports/16/rep16-20.pdf

Y. Tian, How to express a parallel sum of k matrices. J. Math. Anal. Appl. **266**(2), 333–341 (2002)

S. Zampini, PCBDDC: a class of robust dual-primal preconditioners in PETSc. SIAM J. Sci. Comput. **38**(5), S282–S306 (2016)

Adaptive BDDC Deluxe Methods for H(curl)

Stefano Zampini

1 Introduction

We present two- and three-dimensional numerical results obtained using BDDC *deluxe* preconditioners, cf. Dohrmann and Widlund (2013), for the linear systems arising from finite element discretizations of

$$\int_{\Omega} \alpha \, \nabla \times \mathbf{u} \cdot \nabla \times \mathbf{v} + \beta \, \mathbf{u} \cdot \mathbf{v} \, dx. \tag{1}$$

This bilinear form originates from implicit time-stepping schemes of the quasi-static approximation of the Maxwell's equations in the time domain, cf. Rieben and White (2006). The coefficient α is the reciprocal of the magnetic permeability, whereas β is proportional to the ratio between the conductivity of the medium and the time step. Anisotropic, tensor-valued, conductivities can be handled as well. We only present results for essential boundary conditions, but the generalization of the algorithms to natural boundary conditions is straightforward.

The operator $\nabla \times$ is the *curl* operator, defined, e.g., in Boffi et al. (2013); the vector fields belong to the space $H_0(\mathrm{curl})$, which is the subspace of $H(\mathrm{curl})$ of functions with vanishing tangential traces over $\partial \Omega$. The space $H(\mathrm{curl})$ is often discretized using Nédélec elements; those of lowest order use polynomials with continuous tangential components along the edges of the elements. While most existing finite element codes for electromagnetics use lowest order elements, those

---

S. Zampini (✉)
Extreme Computing Research Center, Computer, Electrical and Mathematical Sciences and Engineering Division, King Abdullah University of Science and Technology, Thuwal, Saudi Arabia
e-mail: stefano.zampini@kaust.edu.sa

© Springer International Publishing AG 2017
C.-O. Lee et al. (eds.), *Domain Decomposition Methods in Science and Engineering XXIII*, Lecture Notes in Computational Science and Engineering 116, DOI 10.1007/978-3-319-52389-7_29

of higher order have shown to require fewer degrees of freedom (dofs) for a fixed accuracy; see, e.g., Schwarzbach et al. (2011) and Grayver and Kolev (2015). We note that higher order elements have been neglected in the domain decomposition (DD) literature with the exception of spectral elements. Section 3 contains novel results for two-dimensional discretizations of (1) using arbitrary order Nédélec elements of first and second kind on triangles.

The design of solvers for edge-element approximations of (1) poses significant difficulties, since the kernel of the curl operator is non-trivial. Moreover, finding logarithmically stable decompositions for edge-element approximations in three dimensions is challenging, due to the strong coupling that exists between dofs located on the subdomain edges and on the subdomain faces. Among non-overlapping DD solvers, it is worth citing the wirebasket algorithms developed by Dohrmann and Widlund (2012) and by Hu et al. (2013). To save space, we omit citing some of the related DD literature; references can be found in Dohrmann and Widlund (2012), and Dohrmann and Widlund (2016).

The edge-element approximations of (1) have also received a lot of attention from the multigrid community; for Algebraic Multigrid (AMG) methods see Hu et al. (2006) and the references therein. Robust and efficient multigrid solvers can be obtained combining AMG and auxiliary space techniques, that require some extra information on the mesh connectivity and on the dofs, cf. Hiptmair and Xu (2007), Kolev and Vassilevski (2009). This approach has recently proven to be quite successful in 3D even with higher order elements, cf. Grayver and Kolev (2015).

An analysis for 3D FETI-DP algorithms with the lowest order Nédélec elements of the first kind was given in Toselli (2006), a paper which also highlighted the importance of changing the basis on the subdomain edges. Recently, Toselli's results have been significantly improved by Dohrmann and Widlund (2016), who were able to obtain sharp and quasi-optimal condition number bounds, with a mild dependence on the material parameters through the factor $1 + \beta H^2/\alpha$. Deluxe scaling proved to be critical to obtain bounds independent on the jumps of the material coefficients in 3D.

While BDDC algorithms are often robust with respect to jumps in the material parameters, their convergence rates drastically deteriorate when these jumps are not aligned with the interface of the subdomains. After the pioneering study of Mandel and Sousedík (2007), primal space enrichment techniques have been the focus of much recent work on BDDC and FETI-DP algorithms; cf. Mandel et al. (2012), Pechstein and Dohrmann (2013), Kim et al. (2015), Klawonn et al. (2015), Calvo and Widlund (2016) and the references therein. Section 3 contains numerical results using heterogeneous material coefficient distributions, for triangular elements of both kinds, and for the lowest order tetrahedral elements of the first kind. All the results of this paper have been obtained using the BDDC implementation developed by the author, and which is available in the current version of the PETSc library (Balay et al., 2015). For details on the implementation, see Zampini (2016).

2 Adaptive BDDC Deluxe Methods

Non-overlapping DD algorithms are often designed using the stiffness matrix $A^{(i)}$ assembled on each subdomain Ω_i. We note that for the problem of interest, these matrices are always symmetric and positive definite. The recipe for the construction of a BDDC preconditioner consists in the design of a partially continuous space $\widetilde{\mathbf{W}}$, the direct sum of a continuous *primal* space $\mathbf{W}_\Pi$ and a discontinuous *dual* space $\mathbf{W}_\Delta$, and in the choice of an averaging operator E_D for the partially continuous dofs, cf. Mandel et al. (2005). A remarkably simple formula, related to the stability of the average operator with respect to the energy norm, provides an upper bound for the condition number (κ) of the BDDC preconditioned operator

$$\kappa \le \max_{\mathbf{w} \in \widetilde{\mathbf{W}}} \frac{\mathbf{w}^T E_D^T S E_D \mathbf{w}}{\mathbf{w}^T S \mathbf{w}},$$

where S is the direct sum of the subdomain Schur complements $S^{(i)}$, obtained by condensing out from $A^{(i)}$ the dofs in the interior of the subdomains. We can then control the convergence rate of the methods by enriching the primal space $\mathbf{W}_\Pi$, and this can be accomplished by solving a few local generalized eigenvalue problems, associated to the equivalence classes of the interface.

For the BDDC deluxe algorithms, a local generalized eigenvalue problem for each equivalence class C, shared by two subdomains, is given by

$$(\widetilde{S}_{CC}^{(i)-1} + \widetilde{S}_{CC}^{(j)-1})\Phi = \lambda (S_{CC}^{(i)-1} + S_{CC}^{(j)-1})\Phi, \tag{2}$$

with $S_{CC}^{(i)}$ a principal minor of $S^{(i)}$ relative to C. The $\widetilde{S}_{CC}^{(i)}$ matrices are obtained by energy-minimization as $\widetilde{S}_{CC}^{(i)} = S_{CC}^{(i)} - S_{C'C}^{(i)T} S_{C'C'}^{(i)-1} S_{C'C}^{(i)}$, with C' the set of complementary interface dofs of C, cf. Pechstein and Dohrmann (2013). Elements in the dual space are then made orthogonal, in the inner product $(S_{CC}^{(i)-1} + S_{CC}^{(j)-1})^{-1}$, to a few selected eigenvectors of (2), with eigenvalues greater than a given tolerance μ.

More complicated generalized eigenvalue problems arise when controlling the energies contributed by interface classes shared by three or more subdomains; even if they lead to fully controllable condition number bounds, they could potentially generate unnecessary primal constraints, cf. Kim et al. (2015), Calvo and Widlund (2016). In our algorithm, we instead consider the eigenvectors associated to the largest eigenvalues of

$$(\widetilde{S}_{CC}^{(i)-1} + \widetilde{S}_{CC}^{(j)-1} + \widetilde{S}_{CC}^{(k)-1})\Phi = \lambda (S_{CC}^{(i)-1} + S_{CC}^{(j)-1} + S_{CC}^{(k)-1})\Phi, \tag{3}$$

that is a generalization of (2), so far without a theoretical validation. With tetrahedral meshes, classes shared by more than three subdomains are rarely encountered. Therefore, we impose full continuity on the partially assembled space for the few dofs that belong to these classes.

We also provide results for adaptive algorithms working with the economic variant of the deluxe approach (e-deluxe), where the $S^{(i)}$ are obtained by eliminating the interior dofs in two layers of elements next to the subdomain part of the interface.

3 Numerical Experiments

The triangulation of Ω and the assembly of the subdomain matrices have been performed with the DOLFIN library, cf. Logg and Wells (2010). ParMETIS (Karypis, 2011) is used to decompose the meshes, and each subdomain is assigned to a different MPI process. MUMPS (Amestoy et al., 2001) is used for the subdomain interior solvers and for the explicit computation of the $S^{(i)}$. A relative residual reduction of 10^{-8} is used as the stopping criterion of the conjugate gradients; random right-hand sides are always considered.

Results will be given sometimes as a function of the ratio H/h, where $H = \max_i\{\max_{P_1,P_2 \in \partial\Omega_{i,h}} d(P_1,P_2)\}$, with P_1 and P_2 two vertices of the boundary mesh $\partial\Omega_{i,h}$ of Ω_i, and $d(P_1,P_2)$ their Euclidean distance. N_p^1 and N_p^2 denote Nédélec first and second kind elements on simplices, respectively, with p the polynomial order.

For the numerical results, we always consider decompositions of the unit domain into 40 irregular subdomains; large scale numerical results for adaptive BDDC algorithms with N_1^1 tetrahedral elements can be found in Zampini and Keyes (2016).

3.1 2D Results

We first report on the quasi-optimality and on the dependence of p. The material coefficients are subdomain-wise constant, but they have jumps between subdomains, which are subdivided in even and odd groups according to their MPI rank. $\alpha = \beta = 100$ for odd subdomains, $\alpha = \beta = 0.01$ for even subdomains. The primal space is characterized in terms of the continuity of the tangential traces along the subdomain edges, cf. Toselli and Vasseur (2005). The quadrature weights for such constraints can easily be obtained by exploiting the Stokes theorem, i.e.,

$$\int_{\Omega_i} \nabla \times \mathbf{u}\, dx = \int_{\partial\Omega_i} \mathbf{u} \cdot \mathbf{t}\, ds.$$

Figure 1 shows the quasi-optimality of the deluxe methods with N_p^1 (left) and N_p^2 (center) elements. The results in the right panel, obtained with a fixed mesh and by increasing p, seem to indicate a polylogarithmic bound.

We then analyze adaptive BDDC deluxe algorithms with the heterogeneous coefficients distribution given in Fig. 2. The mesh is fixed (H/h=140.7), as well as the number of dofs, which varies from 800K for N_1^1 to 11M for N_4^2. Figure 3 shows the condition number, the iteration count, and the relative size, in terms of the number interface dofs, of the adaptively generated coarse spaces, all given as a

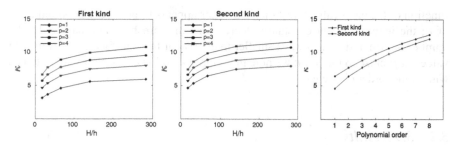

Fig. 1 2D results. κ as a function of H/h. *Left*: N_p^1. *Center*: N_p^2. *Right*: κ as a function of p ($H/h = 66$)

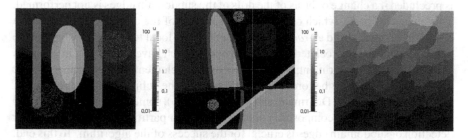

Fig. 2 2D distributions of α (*left*) and β (*center*). *Right*: decomposition in 40 subdomains

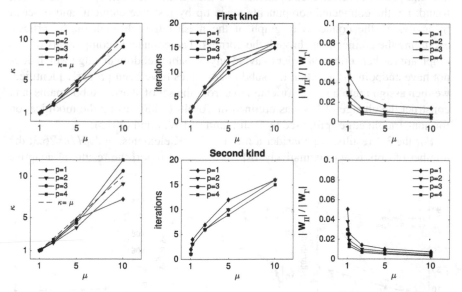

Fig. 3 2D results. κ (*left*), iterations (*center*), and relative size of $\mathbf{W}_{\Pi}$ (*right*) as a function of μ. *Top*: N_p^1. *Bottom*: N_p^2. α, β as in Fig. 2

function of the eigenvalue threshold. The latter appears to be a very good indicator of κ; the iteration count constantly decrease as the threshold approaches 1. The number of primal dofs is always smaller than 10% of the interface dofs, even with values of μ close to the limit; we note that more favorable coarsenings are obtained with higher order elements.

3.2   3D Results

As first highlighted by Toselli (2006), the existence of a stable decomposition in 3D is precluded if a change of basis of the dofs of the subdomain edges is not performed. This change of basis, which consists in the splitting of the dofs of each subdomain edge E in a *constant* and a *gradient* component, is not local to E, as it involves all the other interface dofs associated to those elements which have a fine edge in common with E. In our 3D experiments, we consider only N_1^1 elements; constructing suitable changes of basis for higher order elements could be the subject of future research.

As already noted by Dohrmann and Widlund (2016), some care must be exercised when considering a decomposition obtained by mesh partitioners, since the proper detection of subdomain edges is crucial for the success of the algorithm. To this end, we first construct the connectivity graph of the mesh vertices through mesh edges, and analyze its connected components. We then mark the *corners* that have been found, i.e. the connected components made up by just one element, and proceed by analyzing the connectivity graph of the mesh edges through mesh vertices, excluding the connections through the corners. The connected components of this graph are further refined in order to avoid any possible subdomain edge which does not have endpoints. Once that the subdomain edges have been properly identified, we then assign them a unique orientation across the set of sharing subdomains, and construct the change of basis as outlined in Toselli (2006), using the modification for non-straight edges proposed by Dohrmann and Widlund (2016).

For the 3D results, we consider a mesh of 750K elements, with $H/h=26.3$; the number of dofs is approximatively 1M. In Fig. 4 we report the results of adaptive

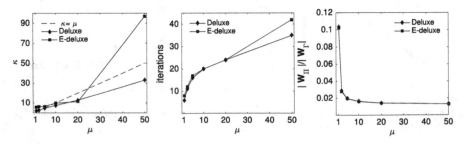

Fig. 4   3D results. κ (*left*), iterations (*center*) and relative size of $\mathbf{W}_\Pi$ (*right*) as a function of μ. (x, y) distributions of α and β as in Fig. 2 (extruded in the z-direction)

Table 1 3D results

	Deluxe				E-deluxe			
	$q_\alpha = 0$	$q_\alpha = 1$	$q_\alpha = 2$	$q_\alpha = 3$	$q_\alpha = 0$	$q_\alpha = 1$	$q_\alpha = 2$	$q_\alpha = 3$
$q_\beta = 0$	3.82 (15)	4.17 (15)	4.26 (16)	7.61 (20)	4.62 (15)	4.14 (15)	4.48 (16)	7.43 (19)
$q_\beta = 1$	9.34 (24)	9.34 (24)	9.33 (24)	8.66 (22)	9.15 (24)	9.15 (23)	8.98 (23)	8.29 (23)
$q_\beta = 2$	8.08 (22)	8.09 (22)	8.14 (22)	7.82 (22)	8.22 (22)	8.22 (22)	8.25 (22)	7.88 (22)
$q_\beta = 3$	8.19 (20)	8.21 (20)	8.28 (20)	8.39 (20)	8.06 (20)	8.07 (20)	8.16 (20)	8.30 (20)

κ and iterations (in parentheses) for adaptive BDDC algorithms. Randomly distributed $\alpha \in [10^{-q_\alpha}, 10^{q_\alpha}]$, $\beta \in [10^{-q_\beta}, 10^{q_\beta}]$; $\mu = 10$

algorithms using an extrusion in the z-direction of the coefficients distributions in Fig. 2, and compare the deluxe and e-deluxe generated primal spaces. Notably, e-deluxe gives very similar results to the deluxe case. The eigenvalue threshold results in a very good indicator of κ even in 3D, despite the lack of a theoretical validation for the eigenvalue problem (3). The iterations constantly decrease as the threshold approaches one in both cases. The relative size of the primal problem is larger than in the 2D case, but it still shows interesting coarsening factors.

We close with a test case where α and β are exponentially and randomly chosen in $[10^{-q_\alpha}, 10^{q_\alpha}]$ and $[10^{-q_\beta}, 10^{q_\beta}]$, and using $\mu = 10$. The results, provided in Table 1 as a function of q_α and q_β, provide a clear evidence that the condition number is fully controllable.

References

P.R. Amestoy, I.S. Duff, J.-Y. L'Excellent, J. Koster, A fully asynchronous multifrontal solver using distributed dynamic scheduling. SIAM J. Matrix Anal. Appl. **23**(1), 15–41 (2001)

S. Balay et al., PETSc users manual. Technical Report ANL-95/11 - Revision 3.6, Argonne National Lab, 2015

D. Boffi, F. Brezzi, M. Fortin, *Mixed Finite Element Methods and Applications*. Springer Series in Computational Mathematics, vol. 44 (Springer, Heidelberg, 2013)

J.G. Calvo, O.B. Widlund, An adaptive choice of primal constraints for BDDC domain decomposition algorithms. Technical Report TR2015-979, Courant Institute of Mathematical Sciences, 2016

C.R. Dohrmann, O.B. Widlund, An iterative substructuring algorithm for two-dimensional problems in H(curl). SIAM J. Numer. Anal. **50**(3), 1004–1028 (2012)

C.R. Dohrmann, O.B. Widlund, Some recent tools and a BDDC algorithm for 3D problems in H(curl), in *Domain Decomposition Methods in Science and Engineering XX*. Lecture Notes in Computational Science and Engineering, vol. 91 (Springer, Heidelberg, 2013), pp. 15–25

C.R. Dohrmann, O.B. Widlund, A BDDC algorithm with deluxe scaling for three-dimensional H(curl) problems. Commun. Pure Appl. Math. **69**(4), 745–770 (2016)

A.V. Grayver, T.V. Kolev, Large-scale 3D geoelectromagnetic modeling using parallel adaptive high-order finite element method. Geophysics **80**(6), E277–E291 (2015)

R. Hiptmair, J. Xu, Nodal auxiliary space preconditioning in H(curl) and H(div) spaces. SIAM J. Numer. Anal. **45**(6), 2483–2509 (2007)

J.J. Hu, R.S. Tuminaro, P.B. Bochev, C.J. Garasi, A.C. Robinson, Toward an h-independent algebraic multigrid method for Maxwell's equations. SIAM J. Sci. Comput. **27**(5), 1669–1688 (2006)

Q. Hu, S. Shu, J. Zou, A substructuring preconditioner for three-dimensional Maxwell's equations, in *Domain Decomposition Methods in Science and Engineering XX*. Lecture Notes in Computational Science and Engineering, vol. 91 (Springer, Heidelberg, 2013), pp. 73–84

G. Karypis, METIS and ParMETIS, in *Encyclopedia of Parallel Computing*, ed. by D. Padua (Springer, New York, 2011), pp. 1117–1124

H.H. Kim, E.T. Chung, J. Wang, BDDC and FETI-DP algorithms with adaptive coarse spaces for three-dimensional elliptic problems with oscillatory and high contrast coefficients. (2015, submitted). https://arxiv.org/abs/1606.07560

A. Klawonn, M. Kühn, O. Rheinbach, Adaptive coarse spaces for FETI-DP in three dimensions. Technical Report 2015-11, Mathematik und Informatik, Bergakademie Freiberg, 2015

T.V. Kolev, P.S. Vassilevski, Parallel auxiliary space AMG for H(curl) problems. J. Comput. Math. **27**(5), 604–623 (2009)

A. Logg, G.N. Wells, Dolfin: automated finite element computing. ACM Trans. Math. Softw. **37**(2), 20:1–20:28 (2010)

J. Mandel, B. Sousedík, Adaptive selection of face coarse degrees of freedom in the BDDC and the FETI-DP iterative substructuring methods. Comput. Methods Appl. Mech. Eng. **196**(8), 1389–1399 (2007)

J. Mandel, C.R. Dohrmann, R. Tezaur, An algebraic theory for primal and dual substructuring methods by constraints. Appl. Numer. Math. **54**(2), 167–193 (2005)

J. Mandel, B. Sousedík, J. Šístek, Adaptive BDDC in three dimensions. Math. Comput. Simul. **82**(10), 1812–1831 (2012)

C. Pechstein, C.R. Dohrmann, Modern domain decomposition methods, BDDC, deluxe scaling, and an algebraic approach (2013), http://people.ricam.oeaw.ac.at/c.pechstein/pechstein-bddc2013.pdf

R.N. Rieben, D.A. White, Verification of high-order mixed finite-element solution of transient magnetic diffusion problems. IEEE Trans. Magn. **42**(1), 25–39 (2006)

C. Schwarzbach, R.-U. Börner, K. Spitzer, Three-dimensional adaptive higher order finite element simulation for geo-electromagnetics: a marine CSEM example. Geophys. J. Int. **187**(1), 63–74 (2011)

A. Toselli, Dual-primal FETI algorithms for edge finite-element approximations in 3D. IMA J. Numer. Anal. **26**(1), 96–130 (2006)

A. Toselli, X. Vasseur, Dual-primal FETI algorithms for edge element approximations: two-dimensional h and p finite elements on shape-regular meshes. SIAM J. Numer. Anal. **42**(6), 2590–2611 (2005)

S. Zampini, PCBDDC: a class of robust dual-primal methods in PETSc. SIAM J. Sci. Comput. **38**(5), S282–S306 (2016)

S. Zampini, D.E. Keyes, On the robustness and prospects of adaptive BDDC methods for finite element discretizations of elliptic PDEs with high-contrast coefficients, in *Proceedings of the Platform for Advanced Scientific Computing Conference*, PASC'16 (ACM, New York, 2016)

Part III
Contributed Talks (CT) and Posters

A Study of the Effects of Irregular Subdomain Boundaries on Some Domain Decomposition Algorithms

Erik Eikeland, Leszek Marcinkowski, and Talal Rahman

1 Introduction

In the standard domain decomposition theory the resulting subdomains are often assumed to have a certain regularity, as in Toselli and Widlund (2005, Assumption 4.3), where each subdomain is a finite union of coarse scale elements and the number of coarse elements forming the subdomain are uniformly bounded. This assumption does not always hold. Subdomains might be generated from a mesh partitioner, or be the result of a decomposition scheme with slight or systematic alterations of the subdomain following refinement, e.g. see the type 3 domain in Dohrmann et al. (2008a, figure 5.1) and the snowflake domain in Fig. 1. In this paper we will assume that each subdomain is a connected union of fine scale elements.

Several papers, Dohrmann et al. (2008a,b), Klawonn et al. (2008), Widlund (2009), have developed theory for such less regular or irregular subdomains. In these studies the subdomains are assumed to be uniform or John domains; see Dohrmann et al. (2008a), Klawonn et al. (2008) for definitions of these families of domains. While these domains are not necessarily Lipschitz, a number of the tools important to the development of theory of domain decomposition algorithms have been developed for such domains in the plane. We note that the Poincaré inequality is particularly important; see Dohrmann et al. (2008a).

In this paper we primarily consider the Additive Average method, introduced in Bjørstad et al. (1997). We note that Toselli and Widlund (2005, Assumption 4.3)

E. Eikeland (✉) • T. Rahman
Bergen Engineering College, Inndalsveien 28, 5063 Bergen, Norway
e-mail: erik.eikeland@hib.no; Talal.Rahman@hib.no

L. Marcinkowski
University of Warsaw, Banacha 2, 02-097 Warszawa, Poland
e-mail: lmarcin@mimuw.edu.pl

© Springer International Publishing AG 2017
C.-O. Lee et al. (eds.), *Domain Decomposition Methods in Science and Engineering XXIII*, Lecture Notes in Computational Science and Engineering 116, DOI 10.1007/978-3-319-52389-7_30

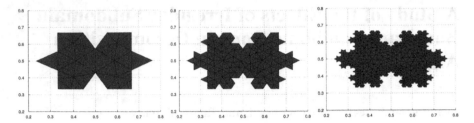

Fig. 1 Here we have three different levels of refinement of a snowflake domain. This domain has constant area but its boundary is growing by a factor $4/3$ with each refinement

was not needed in the original proof. The original proof uses the trace theorem, and to our knowledge this theorem is not available if the subdomains are only John domains. In Dryja and Sarkis (2010), the authors proved a condition number estimate of the Additive Average method for the scalar elliptic equation in $\mathbb{R}^2$ without the use of a trace theorem. Following the setup of Dryja and Sarkis (2010), we have extended the result to $\mathbb{R}^3$ and can show that this convergence estimate, with some modification, holds also when subdomain are John domains. To our knowledge convergence estimates for methods where the subdomains are John or uniform domains have previously only been available for methods in $\mathbb{R}^2$. We have obtained an estimate valid for both $\mathbb{R}^2$ and $\mathbb{R}^3$. In addition, when restricted to $\mathbb{R}^2$ our result may be improved so that it is comparable with the results of Dohrmann and Widlund (2012a). In this paper we must leave out the proof due to page restrictions.

In certain cases of domain decomposition, the length of the subdomain boundaries can grow with refinement. One example is the snowflake domain shown in Fig. 1. In Dohrmann and Widlund (2012a,b) it was pointed out that such domains introduce a factor into the condition number bound which depends on the Hausdorff dimension of the resulting boundary as h goes to zero. For the snowflake domain in Fig. 1, we have a bound of this factor. Numerical results in Sect. 4 are presented to indicate that this factor need to be present in the condition number bound.

This paper has the following layout. In Sect. 2 we present the test problem, assumptions and definitions. In Sect. 3 we introduce the additive average Schwarz preconditioner with convergence estimate as our main result. Finally we present some numerical results in Sect. 4, mainly to illustrate effects of various subdomains on the condition number.

2 The Differential Problem

Find $u \in H_0^1(\Omega)$ such that

$$a(u, v) = f(v), \quad v \in H_0^1(\Omega), \tag{1}$$

where

$$a(u, v) := (\alpha(\cdot)\nabla u, \nabla v)_{L_2(\Omega)}, \quad f(v) := \int_\Omega f v \, dx \tag{2}$$

We assume that $\alpha \in L_\infty(\Omega)$, with $\alpha(x) \geq \alpha_0 > 0$ and that $f \in L_2(\Omega)$. Here Ω is a polygonal or polyhedral region in $\mathbb{R}^n$ where $n \in \{2, 3\}$. Let $\mathcal{T}^h(\Omega)$ be the shape regular triangulation of Ω into triangular or tetrahedral elements. Let V_h be a space of piecewise linear continuous functions.

$$V_h(\Omega) := \left\{ v \in C_0(\Omega); v_{|e_k} \in P_1(x) \right\},$$

where e_k are elements of $\mathcal{T}^h(\Omega)$ and $P_1(x)$ is the set of linear polynomials.

The finite element problem is then defined as: Find $u_h \in V_h(\Omega)$ such that

$$a(u_h, v) = f(v), \quad v \in V_h(\Omega). \tag{3}$$

2.1   Assumptions

Let Ω be divided into disjoint subdomains Ω_i, $\overline{\Omega} = \cup_i \overline{\Omega}_i$, $i \in \{1, \cdots, N\}$, where each Ω_i is a John domain, as defined in Dohrmann et al. (2008a), with a uniformly bounded John constant. Let the boundary $\partial \Omega_i$ be aligned with the triangulation of $\mathcal{T}^h(\Omega)$ such that the inherited triangulation of Ω_i is shape regular with a mesh parameter h_i and $H_i := \text{diam}(\Omega_i)$. According to Dohrmann et al. (2008a), $\text{diam}(\Omega_i)$ can be estimated above and below by $|\Omega_i|^{\frac{1}{n}}$ with one of the constants depending on the John constant C_J. Denote by Ω_i^h the layer around $\partial \Omega_i$ which is a union the of of $e_k^{(i)}$ the element of $\mathcal{T}^h(\Omega_i)$ which touch $\partial \Omega_i$, the boundary of Ω_i. We assume that all elements in Ω_i^h are quasi uniform. We also, as in Dryja and Sarkis (2010), introduce

$$\overline{\alpha}_i := \sup_{x \in \overline{\Omega}_i^h} \alpha(x), \quad \underline{\alpha}_i := \inf_{x \in \overline{\Omega}_i^h} \alpha(x). \tag{4}$$

2.2   The Snowflake Domain

When proving the condition number estimate in Theorem 1, we needed to estimate the number of elements in the internal boundary layer given by $\Omega_i^h \cap \Omega_i$. Usually such an estimate is given by $c(H_i/h_i)^{n-1}$ where c is a constants not depending on the mesh parameter. This is not correct for all types of subdomains.

The snowflake domain follows a rule of refinement. It starts with a square with a boundary node in each corner. With each refinement all boundary edges are divided into three equal parts, and the middle part is replaced with an equilateral triangle. In Fig. 1, we see the first 3 refinements of the a snowflake domain. For the particular domain in the figure, we see that the triangles at the top and at the bottom always point into the domain, subtracting from its area, while the triangles at the left and the right side, always point outwards, adding to its area. The net change of the domains area is zero. With each refinement, the length of the boundary of the subdomain increases by a factor $4/3$. It is possible to show that the asymptotic boundary of the snowflake domain is a von Koch curve with a Hausdorff dimension greater then 1. In Dohrmann and Widlund (2012a,b), it is pointed out that such a domain introduce a factor into the condition number which depends on the Hausdorff dimension, and particularly for the snowflake domain a bound for this factor is given by $c(4/3)^{\log(H_i/h_i)}$, with c independent of mesh parameters. This bound can be rewritten as $C(H_i/h_i)^{0.262}$ with C independent of mesh parameters.

3   Additive Average Schwarz Method

Let us decompose $V_h(\Omega) = V_0(\Omega) + V_1(\Omega) + \ldots + V_N(\Omega)$, and define $V_i(\Omega) = V_h(\Omega) \cap H_0^1(\Omega_i)$ on Ω_i and extend by zero outside Ω_i for $i \in \{1, \cdots, N\}$. The coarse space $V_0(\Omega)$ is defined as the range of the following interpolation operator I_A. For $u \in V_h(\Omega)$, let $I_A u \in V_h(\Omega)$ be defined so that on Ω_i

$$I_A u = \begin{cases} u_j, & \text{if } x_j \in \partial\Omega_{ih} \\ \bar{u}_j, & \text{if } x_j \in \Omega_{ih} \setminus \partial\Omega_{ih} \end{cases} \tag{5}$$

where

$$\bar{u}_j := \frac{1}{n_i} \sum_{x_j \in \partial\Omega_{ih}} u_j. \tag{6}$$

Here Ω_{ih} and $\partial\Omega_{ih}$ are the sets of nodal points x_j on Ω_i and $\partial\Omega_i$, respectively, and n_i is the number of nodes on $\partial\Omega_{ih}$. u_j is the value of u at a nodal point.

For $i \in \{1, \cdots, N\}$, let us introduce

$$b_i(u, v) := a_i(u, v), \quad u, v \in V_i(\Omega), \tag{7}$$

where $a_i(\cdot, \cdot)$ is the restriction of $a(\cdot, \cdot)$ to Ω_i.

For $i = 0$ we introduce

$$b_0(u, v) := \sum_{i=1}^{N} \bar{\alpha} h_i^{n-2} \sum_{x_j \in \partial\Omega_{ih}} (u_j - \bar{u}_j) v_j. \tag{8}$$

3.1   The Preconditioner

For $i \in \{0, \cdots, N\}$, we define the operator $T_i^{(A)} : V_h(\Omega) \to V_i(\Omega)$ by $b_i(T_i^{(A)}u, v) = a(u, v)$, with $v \in V_i(\Omega)$. Of course, each of these problems have a unique solution. Let us introduce $T_A := T_0^{(A)} + T_1^{(A)} + \cdots + T_N^{(A)}$. We replace (3) by the operator equation

$$T_A u_h = g_h \tag{9}$$

where $g_h = \sum_{i=0}^{N} g_i$, and $g_i = T_i^{(A)} u_h$ and u_h is the solution of 3.

The main result

Theorem 1 *For any $u \in V_h(\Omega)$ the following holds:*

$$C_1 \beta_1^{-1} a(u, u) \leq a(Tu, u) \leq C_2 a(u, u), \tag{10}$$

where $\beta_1 = (\overline{\alpha}/\underline{\alpha}) \max_i \chi_i^2 (H_i/h_i)^2$, and C_1 and C_2 depend on the parameter of an isoperimetric inequality, and the John constant, but not on the mesh parameter, and χ_i is a factor related to the Hausdorff dimension of the subdomain boundary. This factor χ_i might be mesh dependent, and can be estimated from the condition that $C\chi_i(H/h)^{n-1}$ are the number of patches needed to cover Ω_i^h, where C is a mesh independent constant and n is the dimension of the problem.

Due to page restrictions, we leave out the proof. It is similar to that in Dryja and Sarkis (2010) but extended to $\mathbb{R}^3$, and valid for subdomains being John domains using some results from Dohrmann et al. (2008a).

Remark 1 When restricted to $\mathbb{R}^2$ with α constant in Ω, we can show that β_1 in Theorem 1 can be reduced to $\beta_1 = \max_i \chi_i ((1 + log(H_i/h_i))(H_i/h_i))$.

4   Numerical Results

Here we present numerical results, for the simple Poisson equation in $\mathbb{R}^2$, for a variety of more or less irregular subdomains. The purpose of these results is to illustrate how the geometrical features of the subdomains impact the condition number. All tests have been done with the Additive Average method, and with the method in Dohrmann et al. (2008a). In all the tests the two methods have shown similar performance. All methods are implemented in MATLAB using pcgeig with a default tolerance of 10^{-6}.

In Table 1, we present results from solving the Poisson equation on the unit square with 16 subdomain of various shapes. We mainly look for effects on the condition number from boundary deformations, and from the use of subdomains with mesh dependent John constants. We use the results from the square subdomains with constant boundaries and a mesh independent John constant as a reference.

Based on the definition of a John domain in Dohrmann et al. (2008a), the subdomains with fingers, see Fig. 2, are designed to have a mesh dependent John constant that is doubling with each refinement of h. This does not cause an increase the condition number in the range of refinement tested as shown in Table 1. Similar results where observed with the method in Dohrmann et al. (2008a). Subdomains from the partitioner METIS result in an increase in the condition number, but it is

Table 1 This table shows iteration and condition numbers when solving the Poisson equation on different subdomains using additive average Schwarz method

		Square subdomains		Square subdomains with fingers		METIS subdomains		Type 2 subdomains	
N	h	itr	cond	itr	cond	itr	cond	itr	cond
16	1/16	17	6.22	13	4.20	21	9.18	13	4.20
16	1/32	26	16.61	28	18.91	38	25.26	22	14.00
16	1/64	46	38.34	43	44.47	62	66.85	35	36.41
16	1/128	68	82.32	65	97.42	91	126.29	53	82.86
16	1/256	84	170.58	94	205.13	135	282.69	81	171.98

The number of subdomains is fixed at $N = 16$ and $h = \left\{ \frac{1}{16}, \frac{1}{32}, \frac{1}{64}, \frac{1}{128}, \frac{1}{256} \right\}$

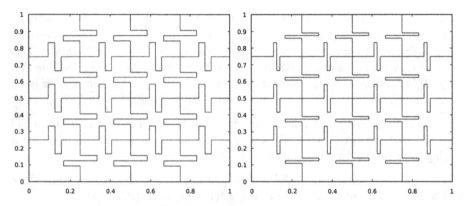

Fig. 2 Figures showing square subdomains with fingers on the edges. These fingers have length $1/3H$ and width h thus growing thiner with refinement of h. In the left figure $h = 1/32$, and in the right figure $h = 1/64$. This should give a growing John constant with refinement of h

hard to estimate what geometrical feature causes this increase. It is surprising that the type 2 subdomains of Dohrmann et al. (2008a) does not increase the condition number compared to the reference domain. The type 2 subdomain boundary is growing with refinement, however we see that the number of elements along the boundary is given by $C(H/h)$ with C independent of mesh parameters. This might explain why we do not see any increase in the condition number from this choice of subdomain geometry.

The deliberately poor choice of rectangular subdomains, as shown in Fig. 3, illustrate a type of domain where the John constant increases as the number of subdomains increases. Theory establishes that for the domains given in Fig. 3, we can estimate $H_i = C_J |\Omega_i|^{\frac{1}{2}}$ with a constant which depends on the John constant. In Table 2, we observe an increase in the condition number even though the method in principle should be scalable and H/h is kept fixed.

Finally in Table 3 the results for snowflake domains are listed. Looking at the ratio of the condition number with different proposed estimates it seems clear that the original estimate for the additive average Schwarz method given in Bjørstad et al. (1997) does not hold. If we take into account the Hausdorff dimension of the subdomain boundary, and adjust the classical convergence estimate by the bound of

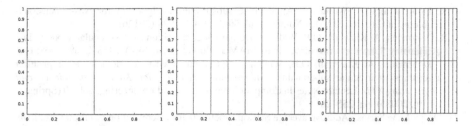

Fig. 3 Figures showing rectangular subdomains of length $1/2$ and width $2H^2$. Here theory for irregular subdomains estimates that $H_i = C_J |\Omega_i|^{\frac{1}{2}}$

Table 2 This table shows iteration and condition numbers when solving the Poisson equation on both square and rectangular subdomains

		Square subdomains		Rectangle subdomains	
N	H/h	itr	cond	itr	cond
4	16	13	20.57	13	20.57
16	16	27	20.66	36	55.94
64	16	32	20.69	84	350.61

The numerics is done with fixed $\frac{H}{h} = 16$ for $N = \{4, 16, 64\}$ subdomains. Using the method presented in Dohrmann et al. (2008a)

Table 3 This table shows iteration and condition numbers when solving the Poisson equation on snowflake subdomains using additive average Schwarz method

Snowflake subdomains						
N	H/h	itr	cond	$\frac{cond}{(H/h)}$	$\frac{cond}{(H/h)^\beta}$	$\frac{cond}{\log(H/h)(H/h)^\beta}$
9	3	15	6.94	2.31	1.73	1.58
9	9	35	28.53	3.17	1.78	0.81
9	27	75	121.62	4.50	1.90	0.58
9	81	154	488.57	6.03	1.91	0.43

Here $\beta = 1.262$

the factor χ, then this would result in an estimate $C(H/h)^\beta$ with $\beta = 1.262$. This estimate fits well with the numerical results. The condition number is well within the bounds established for irregular domains. Similar results were obtained when using the method of Dohrmann et al. (2008a) on snowflake subdomains.

Acknowledgements Leszek Marcinkowski was partially supported by Polish Scientific Project 2016/21/B/ST1/00350.

References

P.E. Bjørstad, M. Dryja, E. Vainikko, Additive Schwarz methods without subdomain overlap and with new coarse spaces, in *Domain Decomposition Methods in Sciences and Engineering (Beijing, 1995)* (Wiley, Chichester, 1997), pp. 141–157

C.R. Dohrmann, O.B. Widlund, An alternative coarse space for irregular subdomains and an overlapping Schwarz algorithm for scalar elliptic problems in the plane. SIAM J. Numer. Anal. **50**(5), 2522–2537 (2012a)

C.R. Dohrmann, O.B. Widlund, An iterative substructuring algorithm for two-dimensional problems in H (curl). SIAM J. Numer. Anal. **50**(3), 1004–1028 (2012b)

C.R. Dohrmann, A. Klawonn, O.B. Widlund, Domain decomposition for less regular subdomains: overlapping Schwarz in two dimensions. SIAM J. Numer. Anal. **46**(4), 2153–2168 (2008a)

C.R. Dohrmann, A. Klawonn, O.B. Widlund, Extending theory for domain decomposition algorithms to irregular subdomains, in *Domain Decomposition Methods in Science and Engineering XVII*. Lecture Notes in Computational Science and Engineering, vol. 60 (Springer, Berlin, 2008b), pp. 255–261

M. Dryja, M.V. Sarkis, Additive average Schwarz methods for discretization of elliptic problems with highly discontinuous coefficients. Comput. Methods Appl. Math. **10**(2), 164–176 (2010)

A. Klawonn, O. Rheinbach, O.B. Widlund, An analysis of a FETI-DP algorithm on irregular subdomains in the plane. SIAM J. Numer. Anal. **46**(5), 2484–2504 (2008)

A. Toselli, O. Widlund, *Domain Decomposition Methods—Algorithms and Theory*. Springer Series in Computational Mathematics, vol. 34 (Springer, Berlin, 2005). ISBN 3-540-20696-5

O.B. Widlund, Accomodating irregular subdomains in domain decomposition theory, in *Domain Decomposition Methods in Science and Engineering XVIII*. Lecture Notes in Computational Science and Engineering, vol. 70 (Springer, Berlin, 2009), pp. 87–98

On the Definition of Dirichlet and Neumann Conditions for the Biharmonic Equation and Its Impact on Associated Schwarz Methods

Martin J. Gander and Yongxiang Liu

1 Introduction

We are interested in formulating and analyzing Schwarz methods for the biharmonic equation

$$\Delta^2 u = f \quad \text{in } \Omega, \tag{1}$$

where Δ denotes the Laplacian, f is a source term and Ω is a domain in $\mathbb{R}^2$. The biharmonic equation is quite different from the Laplace equation, since it requires two boundary conditions, and not just one.

A classical clamped boundary condition would impose the value and normal derivative at the boundary,

$$\mathcal{D}_1(u) := \begin{bmatrix} u \\ \frac{\partial u}{\partial n} \end{bmatrix}, \tag{2}$$

and a two level additive Schwarz method with this "Dirichlet" boundary condition at the interfaces between subdomains was studied in Brenner (1996), where a condition number estimate of order $1 + (\frac{H}{\delta})^4$ was proved for large overlap and

M.J. Gander
Section of Mathematics, University of Geneva, 2-4 rue du Lièvre, CP 64, CH-1211 Genève, Switzerland
e-mail: Martin.Gander@unige.ch

Y. Liu (✉)
Microsystems and Terahertz Research Center, China Academy of Engineering Physics, Chengdu, Sichuan 610200, China
e-mail: liuyongxiang@mtrc.ac.cn

© Springer International Publishing AG 2017
C.-O. Lee et al. (eds.), *Domain Decomposition Methods in Science and Engineering XXIII*, Lecture Notes in Computational Science and Engineering 116, DOI 10.1007/978-3-319-52389-7_31

order $1 + (\frac{H}{\delta})^3$ for small overlap. A non-overlapping Schwarz preconditioner for a discontinuous Galerkin discretization was introduced in Feng and Karakashian (2005), with a condition number estimate of order $(1 + \frac{H}{h})^3$. The convergence rate for the classical Schwarz method with "Dirichlet" condition (2) was also studied in Shang and He (2009).

Considering (2) as "Dirichlet" condition, there are two corresponding possibilities for the associated "Neumann" conditions, depending on which functional minimization led to the necessary optimality condition in (1). If the problem comes from a Stokes formulation (Ciarlet, 1978), the variational derivative leads for the "Neumann" conditions to

$$\mathcal{N}_1(u) := \begin{bmatrix} \Delta u \\ -\partial_n \Delta u \end{bmatrix}. \tag{3}$$

If one however uses the energy functional of a thin plate, see Gander and Kwok (2012) and references therein, the "Neumann" condition associated with (2) is

$$\mathcal{N}_2(u) := \begin{bmatrix} \Delta u - (1 - \sigma)\partial_{\tau\tau}u \\ -\partial_n \Delta u - (1 - \sigma)\partial_\tau(\partial_{n\tau}u) \end{bmatrix}, \tag{4}$$

where ∂_τ is the tangential derivative along the boundary and $\sigma \in (0, 1)$ is a material constant. While condition (3) does not always lead to a well posed problem for the biharmonic equation, condition (4), which can be interpreted as the freely supported boundary condition for the plate problem, is always well posed up to a linear function, analogously to the Neumann condition for the Laplace equation. A FETI method using (2) and (4) was proposed and studied in Farhat and Mandel (1998), and later in Mandel et al. (1999), where continuity of the transverse displacements is enforced at substructure cross points, and a condition number estimate of order $(1 + \log \frac{H}{h})^3$ was obtained. An optimized Schwarz waveform relaxation method based on combining the "Dirichlet" condition (2) with the "Neumann" condition (3) was introduced in Nourtier-Mazauric and Blayo (2010) for the corresponding time dependent problem, and an optimized choice of the combining parameters in the transmission conditions was illustrated by numerical experiments.

The clamped condition (2) is however not the only possible choice for a "Dirichlet" condition. Instead of (2) and (3), one could also consider

$$\mathcal{D}_3(u) := \begin{bmatrix} u \\ \Delta u \end{bmatrix} \tag{5}$$

as the "Dirichlet" condition, and then naturally the corresponding "Neumann" condition would be

$$\mathcal{N}_3(u) := \begin{bmatrix} \partial_n u \\ -\partial_n \Delta u \end{bmatrix}, \tag{6}$$

see for example Dolean et al. (2008), which also corresponds to our personal communication with Yingxiang Xu (2016). Similarly, in the thin plate case, instead of (2) and (4), another choice for the "Dirichlet" condition would be

$$\mathcal{D}_4(u) := \begin{bmatrix} u \\ \Delta u - (1 - \sigma)\partial_{\tau\tau}u \end{bmatrix}, \tag{7}$$

and then the corresponding "Neumann" condition would be

$$\mathcal{N}_4(u) := \begin{bmatrix} \partial_n u \\ -\partial_n \Delta u - (1 - \sigma)\partial_\tau(\partial_{n\tau}u) \end{bmatrix}. \tag{8}$$

When the boundary is flat, conditions (5) and (7) are essentially equivalent, since imposing u also imposes $\partial_{\tau\tau}$. Similarly also conditions (6) and (8) are equivalent for flat boundaries. For curved boundaries however, and as transmission conditions, these conditions are different.

Because of these different choices for the "Dirichlet" conditions, the classical Schwarz methods studied in Brenner (1996) and Shang and He (2009) are not the only possible ones for the biharmonic equation, and similarly there are also more possibilities for optimized Schwarz methods than the one in Nourtier-Mazauric and Blayo (2010). We will show that a different choice of "Dirichlet" conditions in the classical Schwarz method permits the removal of the typical power of 3 in the convergence estimates, and leads to faster methods, while optimized Schwarz methods are robust with respect to which condition is chosen to be the "Dirichlet" one.

2 Classical Schwarz Methods

Because of the three different possibilities for the "Dirichlet" conditions in (2), (5) and (7), we get three classical Schwarz methods which we index by $j \in \{1, 3, 4\}$. To simplify the description and analysis, we consider an unbounded domain $\Omega = \mathbb{R}^2$ and solutions u decaying at infinity. We assume that Ω is divided into two subdomains $\Omega_1 = (-\infty, L) \times \mathbb{R}$ and $\Omega_2 = (0, +\infty) \times \mathbb{R}$, where $L \geq 0$ denotes the overlap.

Given an initial approximation u_2^0, the three classical alternating Schwarz methods indexed by $j \in \{1, 3, 4\}$ compute for $n = 1, 2, \ldots$

$$\begin{array}{llll} \Delta^2 u_1^n = f_1 & \text{in } \Omega_1, & \Delta^2 u_2^n = f_2 & \text{in } \Omega_2, \\ \mathcal{D}_j(u_1^n) = \mathcal{D}_j(u_2^{n-1}) & \text{at } x = L, & \mathcal{D}_j(u_2^n) = \mathcal{D}_j(u_1^n) & \text{at } x = 0. \end{array} \tag{9}$$

Taking a Fourier transform in the y direction with Fourier symbol k, and assuming that the relevant numerical Fourier frequencies $|k|$ lie in the interval $[k_{min}, k_{max}]$ with

$k_{min}, k_{max} > 0$, we obtain by a direct computation (see also Shang and He 2009 for $j = 1$):

Theorem 1 *If $L > 0$, the convergence factors ρ_j for the Algorithm (9) are*

$$\rho_1(L) = (k_{min}L + \sqrt{k_{min}^2 L^2 + 1})^2 e^{-2k_{min}L} \sim 1 - \frac{1}{3}k_{min}^3 L^3,$$

$$\rho_{3,4}(L) = e^{-2k_{min}L} \sim 1 - 2k_{min}L.$$

We see that the classical clamped "Dirichlet" transmission condition (2) leads to a convergence factor depending on the overlap L cubed, whereas using the other two possible "Dirichlet" conditions (5) or (7), the convergence factor only depends linearly on L. This substantially improved convergence factor, which is now like for Laplace's equation (Gander, 2006), is illustrated for an example in Fig. 1 on the left.

3 Optimal and Optimized Schwarz Methods

Optimized Schwarz methods (Gander, 2006) use a combination of Dirichlet and Neumann conditions as transmission conditions, and allowing a non-local operator for this combination can lead to optimal Schwarz methods which converge in a finite number of steps (two in the case of two subdomains, see Gander (2006) and references therein). Letting $\mathcal{D}_2 := \mathcal{D}_1$, such a method, again indexed by $j \in \{1, 2, 3, 4\}$, computes for an initial approximation u_2^0 and $n = 1, 2, \ldots$

$$
\begin{aligned}
\Delta^2 u_1^n &= f_1 & \text{in } \Omega_1, \\
(\mathcal{N}_j + P_j \mathcal{D}_j)(u_1^n) &= (\mathcal{N}_j + P_j \mathcal{D}_j)(u_2^{n-1}) & \text{at } x = L, \\
\Delta^2 u_2^n &= f_2 & \text{in } \Omega_2, \\
(\mathcal{N}_j + P_j \mathcal{D}_j)(u_2^n) &= (\mathcal{N}_j + P_j \mathcal{D}_j)(u_1^n) & \text{at } x = 0,
\end{aligned}
\tag{10}
$$

where P_j is a two by two matrix to be chosen for best performance of the method, depending on the choice of "Dirichlet" and "Neumann" conditions $\mathcal{D}_j$ and $\mathcal{N}_j$ we made. The following result can be obtain by a direct but lengthy calculation using Fourier analysis.

Theorem 2 *If the symbols of the elements in the matrix P_j for variant j of Algorithm (10) are chosen in the Fourier domain as*

$$
\hat{P}_1 = \begin{bmatrix} 2|k|^2 & 2|k| \\ 2|k|^3 & 2|k|^2 \end{bmatrix}, \quad
\hat{P}_2 = \begin{bmatrix} (1+\sigma)|k|^2 & 2|k| \\ 2|k|^3 & (1+\sigma)|k|^2 \end{bmatrix},
$$

$$
\hat{P}_3 = \begin{bmatrix} |k| & \frac{1}{2|k|} \\ 0 & -|k| \end{bmatrix}, \quad
\hat{P}_4 = \begin{bmatrix} \frac{1}{2}(1+\sigma)|k| & \frac{1}{2|k|} \\ \frac{1}{2}(1-\sigma)(\sigma+3)|k|^3 & -\frac{1}{2}(1+\sigma)|k| \end{bmatrix},
\tag{11}
$$

then the resulting optimal Schwarz method converges in two iterations.

Remark 1 The choice of the matrix P_j, $j \in \{1, 2, 3, 4\}$ in Theorem 2 leads in each case to the transparent boundary condition, and the associated algorithm can be interpreted as an exact factorization independently of the PDE one considers, see Gander and Nataf (2000) and references therein, and also the more recent variants (Chen and Xiang, 2013; Chen et al., 2016; Engquist and Ying, 2011; Stolk, 2013). Such factorizations are theoretically still possible in the presence of cross points, see Gander and Kwok (2011).

The optimal choice of $\hat{P}_j$ in Theorem 2 corresponds to a non-local operator once back-transformed using the inverse Fourier transform, and thus is often approximated using an absorbing boundary condition or perfectly matched layers to obtain a more practical algorithm. Theorem 2 also indicates a very simple, structurally consistent local approximation: replacing $|k|$ by a constant $p \geq 0$ will make the approximation exact for precisely this frequency $|k|$, and leads to the following results.

Theorem 3 *With the structural consistent approximations for $p \geq 0$,*

$$P_1^a = \begin{bmatrix} 2p^2 & 2p \\ 2p^3 & 2p^2 \end{bmatrix}, \quad P_3^a = \begin{bmatrix} p & \frac{1}{2p} \\ 0 & -p \end{bmatrix}, \tag{12}$$

the convergence factor of the optimized Schwarz algorithm (10) *is*

$$\rho(L) = \left(\frac{p - |k|}{p + |k|} \right)^2 e^{-2|k|L} < 1. \tag{13}$$

With overlap, $L > 0$, the optimal choice for p for best performance, and the associated contraction factor are for L small

$$p \sim \left(\frac{k_{\min}^2}{2L} \right)^{1/3}, \quad \rho(L) \sim 1 - 4(2k_{\min})^{1/3} L^{1/3}, \tag{14}$$

where $k_{\min}$ is an estimate for the lowest frequency along the interface. Without overlap, $L = 0$, and with $k_{\max}$ an estimate for the largest frequency along the interface, one obtains

$$p = \sqrt{k_{\min} k_{\max}}, \quad \rho(0) = \left(\frac{\sqrt{k_{\max}} - \sqrt{k_{\min}}}{\sqrt{k_{\max}} + \sqrt{k_{\min}}} \right)^2 \sim 1 - 4\sqrt{\frac{k_{\min}}{k_{\max}}}, \quad k_{\max} \text{ large.} \tag{15}$$

Proof The convergence factor (13) can be obtained by a direct computation, and noticing that it is identical to the case of the Laplace equation, the results from Gander (2006) can then be used to obtain (14) and (15).

Theorem 4 *With the structural consistent approximations for $p \geq 0$,*

$$P_2^a = \begin{bmatrix} (1+\sigma)p^2 & 2p \\ 2p^3 & (1+\sigma)p^2 \end{bmatrix}, \; P_4^a = \begin{bmatrix} \frac{1}{2}(1+\sigma)p & \frac{1}{2p} \\ \frac{1}{2}(1-\sigma)(\sigma+3)p^3 & -\frac{1}{2}(1+\sigma)p \end{bmatrix},$$
(16)

the convergence factor of the optimized Schwarz algorithm (10) for $j = 2$ and $j = 4$ coincide. With overlap, $L > 0$, the optimal choice of p for best performance, and the associated contraction factor are for L small

$$p \sim \frac{1}{2^{1/3}} \left(\frac{6k_{\min}^4}{(1-\sigma^2)L} \right)^{1/5}, \quad \rho(L) \sim 1 - \frac{16}{3} \frac{(6^2 k_{\min}^3 (1-\sigma^2))^{1/5}}{3 - 2\sigma - \sigma^2} L^{3/5}.$$
(17)

Without overlap, one obtains for $k_{\max}$ large

$$p \sim \sqrt{k_{\min} k_{\max}}, \quad \rho(0) \sim 1 - \frac{16 k_{\min}^{3/2}}{3 - 2\sigma - \sigma^2} \frac{1}{k_{\max}^{3/2}}.$$
(18)

The proof of Theorem 4 requires a detailed asymptotic analysis and is too long for this short manuscript. We see however that the constant σ from the plate problem enters the convergence factor, and the convergence of algorithm (10) for $j \in \{2, 4\}$ is worse than in the case $j \in \{1, 3\}$. Theorems 3 and 4 also show that the optimized Schwarz algorithms have the same performance, independently of the choice of "Dirichlet" condition, in contrast to the classical Schwarz method.

One might be wondering what the importance is of the structural consistent choice of the approximate transmission condition in Theorems 3 and 4. Our next result answers this question for one particular case.

Theorem 5 *For algorithm (10) in the case $j = 1$ without overlap, if we permit the general matrix*

$$P_1^g = \begin{bmatrix} p_{11} & p_{12} \\ p_{21} & p_{22} \end{bmatrix},$$
(19)

then the optimal choice of the parameters is

$$p_{11} = p_{22} \geq 0, \quad p_{12}p_{21} = p_{11}^2, \quad \frac{p_{21}}{p_{12}} = k_{min}k_{max}.$$
(20)

Therefore, the structural choice in Theorem 3 is optimal.

The proof of Theorem 5 is technical and too long for this short paper.

4 Numerical Results

We solve the biharmonic equation (1) numerically on the unit square domain $\Omega = (0, 1) \times (0, 1)$ with the homogeneous "Dirichlet" conditions $\mathcal{D}_1(u) = 0$ on $\partial\Omega$, and choose for the right hand side $f := 24y^2(1 - y)^2 + 24x^2(1 - x)^2 + 8[(1 - 2x)^2 - 2(x - x^2)][(1 - 2y)^2 - 2(y - y^2)]$, so that the exact solution is $u = x^2(1 - x)^2 y^2(1 - y)^2$. We discretize (1) using a standard 13-point finite difference scheme obtained by taking the square of the standard five point Laplacian, see Gander and Kwok (2012). We divide the domain into two equal overlapping subdomains Ω_1 and Ω_2. We stop the Schwarz iteration when $\frac{\|u^n - u\|_{l2}}{\|u\|_{l2}} \leq 10^{-6}$, where u^n denotes the discrete approximation at iteration n, and u is the discrete solution obtained by a direct method.

We compare for $j = 1, 3$ the classical Schwarz algorithm (9) to the optimized Schwarz algorithm (10). The results in Table 1 clearly show how the good choice of "Dirichlet" greatly improves the performance, and also the superiority of the optimized Schwarz method, as one would expect from the contraction factor plot in Fig. 1 on the left. In Fig. 1 on the right we show the plot corresponding to Table 1, and we can clearly see the asymptotic difference in behavior as predicted by Theorems 1 and 3.

Table 1 Iteration numbers for classical Schwarz (9) and optimized Schwarz (10)

	Classical Schwarz $j = 1$				Classical Schwarz $j = 3$				Optimized Schwarz $j = 1, 3$			
$L \setminus h$	1/16	1/32	1/64	1/128	1/16	1/32	1/64	1/128	1/16	1/32	1/64	1/128
h	853	6469	50,906	>200,000	34	68	134	267	6	9	12	14
$2h$	235	1655	12,819	101,157	18	35	67	135	5	8	11	14
$4h$	53	305	2189	16,971	9	17	34	67	4	7	9	13

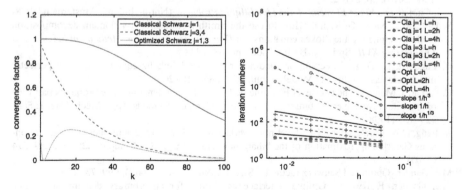

Fig. 1 *Left*: convergence factors corresponding to an overlap $L = 1/50$ for the biharmonic equation and various Schwarz algorithms. *Right*: graphical representation of the results from Table 1, and theoretical prediction from Theorems 1 and 3

5 Conclusions

We showed that using the classical clamped boundary conditions as "Dirichlet" transmission conditions for a Schwarz algorithm applied to the biharmonic equation leads to a convergence that depends on the overlap cubed, see also Brenner (1996) and Shang and He (2009). A better choice of "Dirichlet" conditions involving a Laplacian leads to a convergence that only depends linearly on the overlap, like in the case of Laplace's equation, without additional computational cost, since the Laplacian appearing in this new "Dirichlet" condition is naturally available, for example in a mixed formulation. We then proved that optimized Schwarz methods do not depend on the choice of what the "Dirichlet" condition is, and they all lead to a still substantially better convergence behavior than the classical Schwarz method with the best "Dirichlet" condition. We also found that transmission conditions based on the thin plate model ($\mathcal{D}_j$ and $\mathcal{N}_j$ for $j = 2, 4$) are inferior in performance compared to the ones coming from the Stokes model ($\mathcal{D}_j$ and $\mathcal{N}_j$ for $j = 1, 3$).

Acknowledgements The work of Yongxiang Liu was supported by the Science Challenge Project No. JCKY2016212A503.

References

S.C. Brenner, A two-level additive Schwarz preconditioner for nonconforming plate elements. Numer. Math. **72**(4), 419–447 (1996)

Z. Chen, X. Xiang, A source transfer domain decomposition method for Helmholtz equations in unbounded domain. SIAM J. Numer. Anal. **51**(4), 2331–2356 (2013)

Z. Chen, M.J. Gander, H. Zhang, On the relation between optimized Schwarz methods and source transfer, in *Domain Decomposition Methods in Science and Engineering XXII* (Springer, New York, 2016), pp. 217–225

P.G. Ciarlet, *The Finite Element Method for Elliptic Problems* (North-Holland, Amsterdam, 1978)

V. Dolean, F. Nataf, G. Rapin, How to use the Smith factorization for domain decomposition methods applied to the Stokes equations, in *Domain Decomposition Methods in Science and Engineering XVII* (Springer, Berlin, 2008), pp. 331–338

B. Engquist, L. Ying, Sweeping preconditioner for the Helmholtz equation: moving perfectly matched layers. Multiscale Model. Simul. **9**(2), 686–710 (2011)

C. Farhat, J. Mandel, The two-level FETI method for static and dynamic plate problems part I: an optimal iterative solver for biharmonic systems. Comput. Methods Appl. Mech. Eng. **155**(1), 129–151 (1998)

X. Feng, O.A. Karakashian, Two-level non-overlapping Schwarz preconditioners for a discontinuous Galerkin approximation of the biharmonic equation. J. Sci. Comput. **22**(1–3), 289–314 (2005)

M.J. Gander, Optimized Schwarz methods. SIAM J. Numer. Anal. **44**(2), 699–731 (2006)

M.J. Gander, F. Kwok, Optimal interface conditions for an arbitrary decomposition into subdomains, in *Domain Decomposition Methods in Science and Engineering XIX* (Springer, Berlin, 2011), pp. 101–108

M.J. Gander, F. Kwok, Chladni figures and the Tacoma bridge: motivating PDE eigenvalue problems via vibrating plates. SIAM Rev. **54**(3), 573–596 (2012)

M.J. Gander, F. Nataf, AILU: A preconditioner based on the analytic factorization of the elliptic operator. Numer. Linear Algebra Appl. **7**, 505–526 (2000)

J. Mandel, R. Tezaur, C. Farhat, A scalable substructuring method by Lagrange multipliers for plate bending problems. SIAM J. Numer. Anal. **36**(5), 1370–1391 (1999)

E. Nourtier-Mazauric, E. Blayo, Towards efficient interface conditions for a Schwarz domain decomposition algorithm for an advection equation with biharmonic diffusion. Appl. Numer. Math. **60**(1), 83–93 (2010)

Y. Shang, Y. He, Fourier analysis of Schwarz domain decomposition methods for the biharmonic equation. Appl. Math. Mech. **30**, 1177–1182 (2009)

C. Stolk, A rapidly converging domain decomposition method for the Helmholtz equation. J. Comput. Phys. **241**, 240–252 (2013)

SHEM: An Optimal Coarse Space for RAS and Its Multiscale Approximation

Martin J. Gander and Atle Loneland

1 Introduction and Model Problem

Domain decomposition methods for elliptic problems need a coarse space component in order to be scalable, and there are many now classical results in the literature on such two level Schwarz, balancing Neumann-Neumann and FETI methods, see Toselli and Widlund (2005) and references therein. Coarse spaces can however do much more for a subdomain iteration than just make it scalable. For each domain decomposition method, there exists an optimal coarse space which will make it converge in only one iteration, i.e. makes the method into a direct solver. A first such coarse space component was discovered within transmission conditions in Gander and Kwok (2011). A separate optimal coarse space was developed in Gander and Halpern (2012), and also introduced in Gander et al. (2014b), with easy to use approximations to get practical coarse spaces, see also Gander et al. (2014a) where the case of discontinuous subdomain iterates was treated. The full potential of these new coarse spaces for additive Schwarz methods (AS) applied to multiscale problems was realized in Gander et al. (2015), where also a convergence analysis can be found.

We explain here what this optimal coarse space is for Restricted Additive Schwarz (RAS). RAS was discovered in Cai and Sarkis (1999), and it represents a consistent discretization of the parallel Schwarz method that was introduced by Lions in the first DD conference (Lions, 1988), see Efstathiou and Gander (2003)

M.J. Gander
Section of Mathematics, University of Geneva, 1211 Geneva 4, Switzerland
e-mail: Martin.Gander@unige.ch

A. Loneland (✉)
Department of Informatics, University of Bergen, 5020 Bergen, Norway
e-mail: atle.loneland@ii.uib.no

© Springer International Publishing AG 2017
C.-O. Lee et al. (eds.), *Domain Decomposition Methods in Science and Engineering XXIII*, Lecture Notes in Computational Science and Engineering 116, DOI 10.1007/978-3-319-52389-7_32

and Gander (2008) for more explanations. There is no general convergence theory for RAS, but the results of Lions apply in the discrete setting. The optimal coarse space and its approximation also differ from the case of AS, since RAS iterates are in general discontinuous.

Our approximations of the optimal coarse space are related to more recent developments of robust coarse spaces for high contrast problems, see Aarnes and Hou (2002) and the analysis in Graham et al. (2007), where multiscale finite elements were proposed for the coarse space. The idea to enrich the coarse space goes back to Galvis and Efendiev (2010a) and Galvis and Efendiev (2010b), where subdomain eigenfunctions are combined with partition of unity functions, see also Efendiev et al. (2012). A different approach is using eigenfunctions of the Dirichlet to Neumann map of each subdomain, see Dolean et al. (2012), the improved variant based on a generalized eigenvalue problem in the overlaps in Spillane et al. (2014), and also the recent adaptive coarse spaces for BDD(C) and FETI(-DP) methods (Klawonn et al., 2015; Mandel and Sousedík, 2007). A good overview of the most recent approaches can be found in Scheichl (2013). The main difference in our approach is that we start with an optimal coarse space depending on the method for which we want to construct the coarse space, and that we do not need volume eigenproblems in our construction.

Our model problem is the elliptic boundary value problem

$$- \nabla \cdot (\alpha(x) \nabla u) = f \text{ in } \Omega, \quad u = 0 \text{ on } \partial\Omega, \tag{1}$$

where Ω is a bounded convex domain in $\mathbb{R}^2$, $f \in L^2(\Omega)$ and $\alpha \in L^\infty(\Omega)$ such that $\alpha \geq \alpha_0$ for some positive constant α_0. Discretizing this problem using a P1 finite element method leads to the linear system

$$A\mathbf{u} = \mathbf{f}. \tag{2}$$

Based on a decomposition of the domain Ω into J non-overlapping subdomains $\widetilde{\Omega}_j$, which are enlarged to create overlapping subdomains Ω_j, one can construct non-overlapping restriction matrices $\widetilde{R}_j$, associated overlapping restriction matrices R_j, and local subdomain matrices $A_j := R_j A R_j^T$ to define RAS,

$$\mathbf{u}^{n+1} = \mathbf{u}^n + \sum_{j=1}^{J} \widetilde{R}_j^T A_j^{-1} R_j (\mathbf{f} - A\mathbf{u}^n), \tag{3}$$

see Cai and Sarkis (1999), and Efstathiou and Gander (2003), Gander (2008) for more details.

2   Optimal Coarse Space

To discover the optimal coarse space for RAS, we define the error $\mathbf{e}^n := \mathbf{u} - \mathbf{u}^n$ and look at properties of the error after one iteration. First note that the solution satisfies (3) at the fixed point, i.e.

$$\mathbf{u} = \mathbf{u} + \sum_{j=1}^{J} \widetilde{R}_j^T A_j^{-1} R_j (\mathbf{f} - A\mathbf{u}). \tag{4}$$

Taking the difference between (4) and (3), and using that for any vector $\mathbf{e}^0$ we have $\mathbf{e}^0 = \sum_{j=1}^{J} \widetilde{R}_j R_j \mathbf{e}^0$ by the definition of R_j and $\widetilde{R}_j$, we obtain

$$\mathbf{e}^1 = \mathbf{e}^0 - \sum_{j=1}^{J} \widetilde{R}_j^T A_j^{-1} R_j A \mathbf{e}^0 = \sum_{j=1}^{J} \widetilde{R}_j^T A_j^{-1} A_j R_j \mathbf{e}^0 - \sum_{j=1}^{J} \widetilde{R}_j^T A_j^{-1} R_j A \mathbf{e}^0$$

$$= \sum_{j=1}^{J} \widetilde{R}_j^T A_j^{-1} (A_j R_j - R_j A) \mathbf{e}^0 = \sum_{j=1}^{J} \widetilde{R}_j^T A_j^{-1} (R_j A R_j^T R_j - R_j A) \mathbf{e}^0$$

$$= \sum_{j=1}^{J} \widetilde{R}_j^T A_j^{-1} R_j A (R_j^T R_j - I) \mathbf{e}^0.$$

Now since $(R_j^T R_j - I)\mathbf{e}^0$ contains only non-zero elements outside subdomain Ω_j, $A(R_j^T R_j - I)\mathbf{e}^0$ represents precisely boundary conditions for Ω_j, and thus

$$\widetilde{R}_j \mathbf{e}^1 = \widetilde{R}_j \widetilde{R}_j^T A_j^{-1} R_j A (R_j^T R_j - I) \mathbf{e}^0$$

is a discrete harmonic function on each $\widetilde{\Omega}_j$. This is illustrated in Fig. 1 for the case of the Poisson equation in the top row, where we see that the error is harmonic in the $\widetilde{\Omega}_j$ on the left and on the right we show the associated residual, which is zero in each Ω_j, since the error is harmonic there. In the bottom row we show the corresponding results for the high contrast problem from Fig. 2 on the left, and we see that even though the error looks very different, it is still the solution of the homogeneous equation, i.e. "harmonic", in each non-overlapping subdomain, the residual is zero there.

   If the coarse space should remove all of $\mathbf{e}^1$ for RAS, it needs to contain all discrete harmonic functions on each non-overlapping subdomain $\widetilde{\Omega}_j$. Putting these functions into the columns of the coarse restriction matrix R_0, the coarse correction step with $A_0 := R_0 A R_0^T$ leads to the exact solution,

$$\mathbf{u} = \mathbf{u}^1 + R_0^T A_0^{-1} R_0 (\mathbf{f} - A\mathbf{u}^1).$$

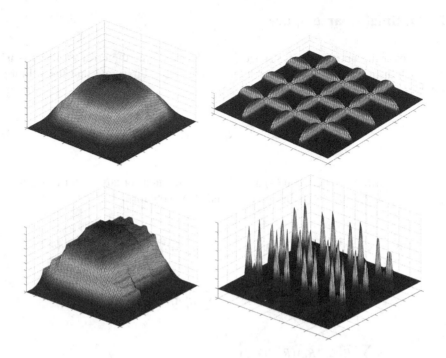

Fig. 1 Error (*left*) and residual (*right*) of the 1-level method with minimal overlap h after one iteration for the Poisson problem in the *top row*, and for the high contrast problem from Fig. 2 on the left in the *bottom row*

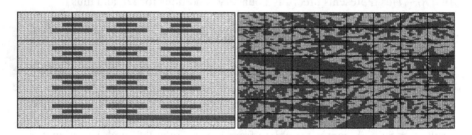

Fig. 2 *Left*: channel distributions of α for a geometry with $h = \frac{1}{64}$, $H = 16h$. *Right*: irregular distribution of α for a geometry with $h = \frac{1}{128}$, $H = 16h$

A simple basis for the optimal coarse space is to choose the functions whose value equals 1 at one node of the interface of the non-overlapping subdomains, zero at all the others, and then to harmonically extend this data inside the non-overlapping subdomain. The dimension of this optimal coarse space is thus twice the number of interface nodes of the non-overlapping decomposition, and would be infinite dimensional at the continuous level.

3   Approximation of the Optimal Coarse Space

Since the full discrete harmonic space is very large, we propose to approximate it, and it is best to explain this using as example the decomposition of the square into four sub-squares which represent the non-overlapping subdomains $\widetilde{\Omega}_j$. The first four basis functions which we put into the coarse space are shown in Fig. 3 on the left. In the constant coefficient case, i.e. the Poisson equation, this would just correspond to Q1 finite elements in these square subdomains, as we see in the top row, but in the more general case of a specific distribution α as shown in Fig. 2, we solve a one dimensional boundary value problem along the edges where the function is non-zero, see Gander et al. (2015). To get a better coarse space, we enrich the former one by adding harmonically extended eigenfunctions on each non-overlapping subdomain from an interface eigenvalue problem along each edge of the non-overlapping decomposition (Gander et al., 2015), which leads to the Spectral Harmonically Enriched Multiscale coarse space we call SHEM$_j$, where j indicates how many functions were added for the enrichment. An example of two such spectral coarse functions based on the first eigenfunction is shown in Fig. 3 on

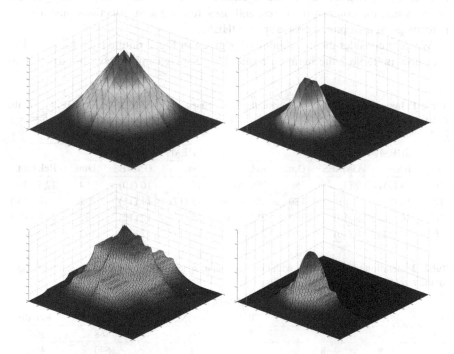

Fig. 3 Discontinuous multiscale finite element basis functions (*left*) and first spectral enrichment functions (*right*) corresponding to the Poisson case for $h = 1/32$ and $H = 16h$ in the *top row*, and a multiscale problem with distribution α given in Fig. 2 on the left for $h = 1/64$ and $H = 32h$ in the *bottom row*

the right for the Poisson equation on top, and below for the multiscale problem with distribution α given in Fig. 2 on the left. If we add all spectral enrichment functions, we obtain again the optimal coarse space OHEM (Optimal Harmonically Enriched Multiscale coarse space).

4   Numerical Results

The first numerical experiment is for the distribution α shown in Fig. 2 on the left. The iteration counts and the size of the coarse space compared to the optimal coarse space are shown in Table 1, where we run RAS or GMRES preconditioned with RAS until the l_2 norm of the initial residual is reduced by a factor of 10^6. For the solution of the generalized 1D eigenvalue problems we used eig in Matlab. We see that SHEM$_3$ is a robust method, independently of h, which is related to the fact that in the distribution α given in Fig. 2 on the left, there are at most three channels crossing any one given interface. This motivates to use an adaptive variant we call SHEM$_a$, where we include an adaptive number of enrichment functions on each interface, based on the size of the eigenvalues. Table 1 shows that SHEM$_a$ is also robust when the contrast increases, and uses fewer coarse functions, just a small percentage of the optimal coarse space OHEM.

We next consider the distribution of α given in Fig. 2 on the right for $\hat{\alpha} = 10^4$. We show in Table 2 the iteration counts for an increasing number of coarse basis

Table 1 Iteration count for RAS with the new coarse space SHEM$_3$ and SHEM$_a$ for the distribution in Fig. 2 on the left, with $h = \frac{1}{64}$, $H = 16h$ and overlap $2h$ (in parentheses $h = \frac{1}{256}$, $H = 64h$ and overlap $8h$)

	SHEM$_3$				SHEM$_a$			
$\hat{\alpha}$	Iter.	GMRES	Dim.	Rel. dim.	Iter.	GMRES	Dim.	Rel. dim.
10^0	8 (8)	7 (7)	180	25% (6%)	15 (17)	10 (10)	84	12% (3%)
10^2	10 (11)	9 (9)	180	25% (6%)	15 (17)	11 (11)	132	18% (4%)
10^4	10 (11)	9 (10)	180	25% (6%)	15 (17)	12 (12)	132	18% (4%)
10^6	10 (11)	9 (10)	180	25% (6%)	15 (17)	12 (12)	132	18% (4%)

Table 2 Iteration count for RAS with the new coarse space SHEM$_j$ for the distribution in Fig. 2 on the right with $h = \frac{1}{128}$, $H = 16h$

	SHEM$_j$ $\delta = 2h$		SHEM$_j$ $\delta = H$			
j	Iter.	GMRES	Iter.	GMRES	Dim.	Rel. dim.
3	34	13	7	6	868	26%
6	9	8	5	4	1540	46%
9	7	7	4	4	2212	66%
12	6	6	4	4	2884	86%
15	1	1	1	1	3360	100%

Table 3 Iteration count for RAS with SHEM$_a$ for the distribution in Fig. 2 on the right with $h = \frac{1}{128}$, $H = 16h$ and overlap $2h$ (in parentheses $h = \frac{1}{256}$, $H = 32h$ and overlap $4h$)

| | SHEM$_a$ $\delta = 2h$ $(4h)$ | | SHEM$_a$ $\delta = H$ | | | |
Min.	Iter.	GMRES	Iter.	GMRES	Dim.	Rel. dim.
1	39 (43)	20 (20)	10 (12)	7 (8)	532 (551)	16% (8%)
2	17 (21)	12 (13)	7 (7)	6 (6)	747 (782)	22% (11%)
3	13 (14)	10 (11)	6 (6)	5 (5)	980 (988)	29% (14%)

functions on each edge. For this example we consider both small overlap $\delta = 2h$ and large overlap $\delta = H$.

These results show that SHEM for RAS performs very well for the fairly hard distribution of α in Fig. 2 on the right. We see also that by systematically increasing the number of spectral enrichment functions on each edge we eventually reach a maximal degree where OHEM turns RAS into a direct solver, as predicted. We also note that RAS without Krylov acceleration performs about as well as RAS with GMRES when SHEM$_j$ is used with $j \geq 6$, which shows that the iterative solver is now so good that Krylov acceleration is not needed any more, a bit like multigrid for the Poisson equation.

In Table 3 we give the iteration count for the same distribution of α in Fig. 2 on the right, except that we now consider an adaptive variant of the coarse space. For both small overlap $\delta = 2h$ and large overlap $\delta = H$ we consider three experiments: For the first experiment we choose the threshold for including eigenfunctions into the coarse space such that we are guaranteed that at least one spectral function is included on each subdomain edge segment. For the second experiment, the threshold is chosen such that we are guaranteed at least two spectral functions on each of the subdomain edge segments and for the last experiment, the threshold is chosen so that at least three spectral functions are guaranteed. The numerical results in Table 3 show that a comparable performance as the one given in Table 2 can be achieved with a considerably smaller coarse space as long as all the bad eigenmodes that are due to the discontinuities in the coefficients are included in the coarse space, and the results are similar when the mesh is refined.

We finally show a numerical experiment where we use an irregular decomposition of the domain into subdomains, as shown in Fig. 4 on the left. As in the case of a regular decomposition in Fig. 3, we can compute the corresponding multiscale coarse basis functions and spectral enrichment functions for each subdomain, and obtain the iteration counts in Fig. 4 on the right. We clearly see that SHEM also works very well for an irregular domain decomposition, and just enriching the coarse space with the adaptively chosen number of spectral enrichment functions leads to a robust solver.

Fig. 4 *Left*: Irregular decomposition of Ω into 16 subdomains with $h = 1/64$. *Right*: Iteration count for RAS with $SHEM_0$ and $SHEM_a$ for the distribution in Fig. 2 on the left with $h = \frac{1}{64}$ and Ω subdivided as on the left, with overlap $3h$

5 Conclusions

We presented an optimal coarse space for RAS called OHEM, which leads to convergence of RAS in one iteration, both when used as an iterative solver and as a preconditioner for GMRES. We then proposed an approximation called SHEM based on multiscale finite elements in each subdomain, enriched with spectral harmonic functions. We showed numerically that SHEM is robust for problems with high contrast, and also derived an adaptive variant.

References

J. Aarnes, T.Y. Hou, Multiscale domain decomposition methods for elliptic problems with high aspect ratios. Acta Math. Appl. Sin. Engl. Ser. **18**(1), 63–76 (2002)

X.-C. Cai, M. Sarkis, A restricted additive Schwarz preconditioner for general sparse linear systems. SIAM J. Sci. Comput. **21**(2), 792–797 (1999)

V. Dolean, F. Nataf, R. Scheichl, N. Spillane, Analysis of a two-level Schwarz method with coarse spaces based on local Dirichlet-to-Neumann maps. Comput. Methods Appl. Math. **12**(4), 391–414 (2012)

Y. Efendiev, J. Galvis, R. Lazarov, J. Willems, Robust domain decomposition preconditioners for abstract symmetric positive definite bilinear forms. ESAIM Math. Model. Numer. Anal. **46**(5), 1175–1199 (2012)

E. Efstathiou, M.J. Gander, Why restricted additive Schwarz converges faster than additive Schwarz. BIT **43**, 945–959 (2003)

J. Galvis, Y. Efendiev, Domain decomposition preconditioners for multiscale flows in high-contrast media. Multiscale Model. Simul. **8**(4), 1461–1483 (2010a)

J. Galvis, Y. Efendiev, Domain decomposition preconditioners for multiscale flows in high contrast media: reduced dimension coarse spaces. Multiscale Model. Simul. **8**(5), 1621–1644 (2010b)

M.J. Gander, Schwarz methods over the course of time. Electron. Trans. Numer. Anal. **31**, 228–255 (2008)

M.J. Gander, L. Halpern, Méthodes de décomposition de domaine. Encyclopédie électronique pour les ingénieurs, (2012)

M.J. Gander, L. Halpern, K. Santugini, Discontinuous coarse spaces for DD-methods with discontinuous iterates, in *Domain Decomposition Methods in Science and Engineering XXI*. Lecture Notes in Computational Science and Engineering (Springer, Cham, 2014a), pp. 607–616

M.J. Gander, L. Halpern, K. Santugini, A new coarse grid correction for RAS/AS, in *Domain Decomposition Methods in Science and Engineering XXI*. Lecture Notes in Computational Science and Engineering (Springer, Cham, 2014b), pp. 275–284

M.J. Gander, F. Kwok, Optimal interface conditions for an arbitrary decomposition into subdomains, in *Domain Decomposition Methods in Science and Engineering XIX* (Springer, Berlin, 2011), pp. 101–108

M.J. Gander, A. Loneland, T. Rahman, Analysis of a new harmonically enriched multiscale coarse space for domain decomposition methods. arXiv preprint arXiv:1512.05285 (2015)

I.G. Graham, P.O. Lechner, R. Scheichl, Domain decomposition for multiscale PDEs. Numer. Math. **106**(4), 589–626 (2007)

A. Klawonn, P. Radtke, O. Rheinbach, FETI-DP methods with an adaptive coarse space. SIAM J. Numer. Anal. **53**(1), 297–320 (2015)

P.-L. Lions, On the Schwarz alternating method. I, in *First International Symposium on Domain Decomposition Methods for Partial Differential Equations*, Paris, France, 1988, pp. 1–42

J. Mandel, B. Sousedík, Adaptive coarse space selection in the BDDC and the FETI-DP iterative substructuring methods: optimal face degrees of freedom, in *Domain Decomposition Methods in Science and Engineering XVI* (Springer, Berlin, 2007), pp. 421–428

R. Scheichl, Robust coarsening in multiscale PDEs, in *Domain Decomposition Methods in Science and Engineering XX*. Lecture Notes in Computational Science and Engineering, vol. 91 (Springer, Berlin, 2013), pp. 51–62

N. Spillane, V. Dolean, P. Hauret, F. Nataf, C. Pechstein, R. Scheichl, Abstract robust coarse spaces for systems of PDEs via generalized eigenproblems in the overlaps. Numer. Math. **126**(4), 741–770 (2014)

A. Toselli, O. Widlund, *Domain Decomposition Methods—Algorithms and Theory*. Springer Series in Computational Mathematics, vol. 34 (Springer, Berlin, 2005)

Optimized Schwarz Methods for Domain Decompositions with Parabolic Interfaces

Martin J. Gander and Yingxiang Xu

1 Introduction

Optimizing parameters involved in the transmission conditions of subdomain iterations leads to the well-known optimized Schwarz methods, see Gander (2006, 2008)
and references therein, where for analysis usually a model problem is considered
on $\mathbb{R}^2$, decomposed into two half planes with a straight interface. In applications
the interface is however seldom straight, which creates a gap between theory and
applications. After early steps in Gander (2011), several research efforts have been
devoted to close this gap: for a general curved interface, transmission conditions
involving the local interface curvature using micro-local analysis were derived in
Barucq et al. (2014), but they are not optimal. When the curved interface is simple,
for example a circle, it was shown in Gander and Xu (2014, 2017) that the curvature
enters the transmission parameters and the corresponding estimates of the convergence factors, and that optimized transmission parameters can be well approximated
using parameters from straight interface analysis, provided the curvature is included
through a proper scaling. For cylindrical interfaces, see Gigante et al. (2013). This
analysis can however not show if any other geometric characteristics enter the
optimized transmission parameters for a general curved interface, apart from the
curvature. We examine here the situation of a parabolically shaped interface, and
show that in addition to the interface curvature, other information of the interface

M.J. Gander
Section de Mathématiques, Université de Genève, 2-4 rue du Lièvre, CP 64, CH-1211 Genève,
Switzerland
e-mail: Martin.Gander@unige.ch

Y. Xu (✉)
School of Mathematics and Statistics, Northeast Normal University, Changchun 130024, China
e-mail: yxxu@nenu.edu.cn

© Springer International Publishing AG 2017

C.-O. Lee et al. (eds.), *Domain Decomposition Methods in Science
and Engineering XXIII*, Lecture Notes in Computational Science
and Engineering 116, DOI 10.1007/978-3-319-52389-7_33

will also enter the optimized transmission parameters. In applications with curved interfaces, optimized transmission parameters from the straight interface analysis are often used locally without any theoretical explanation and lead to fairly good performance, see for example Gander (2006). We will also compare our new results with this approach.

2   Schwarz Methods with Parabolic Interfaces

We consider the model problem

$$
\begin{aligned}
(\Delta - \eta)u &= f, \text{ in } \Omega, \\
u &= 0, \text{ on } \partial\Omega,
\end{aligned}
\tag{1}
$$

where $\eta > 0$ is a model parameter, $\Omega = \{(x, y)|x = \frac{1}{2}(\tau^2 - \sigma^2), y = \sigma\tau, \sigma \in (0, 1), \tau \in (0, 1)\}$. Using the so-called parabolic coordinates

$$
y = \sigma\tau, \qquad x = \frac{1}{2}(\tau^2 - \sigma^2),
\tag{2}
$$

we have $\Omega = \{(x(\sigma, \tau), y(\sigma, \tau))|0 < \sigma < 1, 0 < \tau < 1\}$. We introduce the decomposition $\Omega = \Omega_1 \cup \Omega_2$ with $\Omega_1 = \{(x(\sigma, \tau), y(\sigma, \tau))|0 < \sigma < \sigma_0 + L, 0 < \tau < 1\}$ and $\Omega_2 = \{(x(\sigma, \tau), y(\sigma, \tau))|\sigma_0 < \sigma < 1, 0 < \tau < 1\}$ where σ_0 is a constant satisfying $0 < \sigma_0 < 1$ and $L \geq 0$ is a constant that describes the overlap. If $L = 0$, there is no overlap. The curves $\Gamma_1 = \{(x(\sigma, \tau), y(\sigma, \tau))|\sigma = \sigma_0 + L, 0 < \tau < 1\}$ and $\Gamma_2 = \{(x(\sigma, \tau), y(\sigma, \tau))|\sigma = \sigma_0, 0 < \tau < 1\}$ are the artificial interfaces, see Fig. 1.

A general parallel Schwarz algorithm is then given by

$$
\begin{aligned}
(\Delta - \eta)u_i^n &= f & \text{in } \Omega_i, \\
u_i^n &= 0 & \text{on } \partial\Omega_i \backslash \Gamma_i, \\
\mathcal{B}_i(u_i^n) &= \mathcal{B}_i(u_j^{n-1}) & \text{on } \Gamma_i, 1 \leq i \neq j \leq 2,
\end{aligned}
\tag{3}
$$

where $\mathcal{B}_i, i = 1, 2$, are transmission conditions to be chosen. It is well known that for fast convergence, the transmission operators $\mathcal{B}_i, i = 1, 2$ should be chosen as $\partial_{n_i} + \mathcal{S}_i$, with $\mathcal{S}_i$ local differential operators along the interfaces approximating the Dirichlet to Neumann operators (Gander, 2006, 2008).

The Schwarz method (3) is usually analyzed with Fourier techniques, but in the case of parabolic interfaces this is not possible. Noting that the transform (2) is a conformal map with scale factor $H = \sqrt{\sigma^2 + \tau^2}$, the model problem (1) becomes

$$
\begin{aligned}
(\tfrac{1}{\sigma^2 + \tau^2}\Delta_{\sigma\tau} - \eta)u(\sigma, \tau) &= f(\sigma, \tau), \text{ in } \Omega, \\
u(\sigma, \tau) &= 0, \qquad \text{on } \partial\Omega.
\end{aligned}
\tag{4}
$$

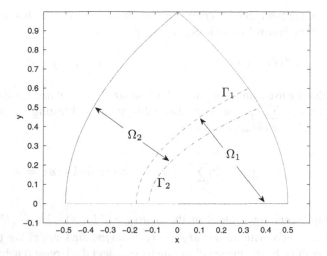

Fig. 1 Domain decomposition with parabolic interfaces

Choosing the transmission operators $\mathcal{B}_i$, $i = 1, 2$ as $\mathcal{B}_i = \partial_\sigma + \mathcal{S}_i$, we then obtain the Schwarz method (3) as

$$
\begin{aligned}
(\tfrac{1}{\sigma^2+\tau^2}\Delta_{\sigma\tau} - \eta)u_i^n(\sigma, \tau) &= f(\sigma, \tau) && \text{in } \Omega_i, \\
u_i^n(\sigma, \tau) &= 0 && \text{on } \partial\Omega_i \backslash \Gamma_i, \\
(\partial_\sigma + \mathcal{S}_i)(u_i^n) &= (\partial_\sigma + \mathcal{S}_i)(u_j^{n-1}) && \text{on } \Gamma_i, 1 \le i \ne j \le 2.
\end{aligned}
\tag{5}
$$

3 Optimized Local Transmission Conditions

We now determine the optimized local operators $\mathcal{S}_i$, $i = 1, 2$. Since the Fourier transform can not be used, we apply the technique of separation of variables, which has been employed successfully in analyzing optimized Schwarz methods for model problems with variable reaction term in Gander and Xu (2016). To this end, we assume that the function $u(\sigma, \tau)$ is separable, $u(\sigma, \tau) = \phi(\sigma)\psi(\tau)$, or equivalently, $u_i^n(\sigma, \tau) = \phi_i^n(\sigma)\psi(\tau)$, $i = 1, 2$. Inserting this ansatz into the first equation of (5) with homogeneous right hand side $f = 0$ gives

$$
-(\phi_i^n(\sigma))''\psi(\tau) - \phi_i^n(\sigma)\psi''(\tau) + (\sigma^2 + \tau^2)\eta\psi_i^n(\sigma)\psi(\tau) = 0, \quad i = 1, 2.
$$

Separating terms, we see that there must exist a positive constant α such that

$$
-\frac{(\phi_i^n(\sigma))''}{\phi_i^n(\sigma)} + \sigma^2\eta = \frac{\psi''(\tau)}{\psi(\tau)} - \tau^2\eta = -\alpha, \quad i = 1, 2.
$$

Together with the homogeneous boundary conditions, we obtain that α must be an eigenvalue of the Sturm-Liouville eigenvalue problem

$$\psi''(\tau) + (\alpha - \tau^2\eta)\psi(\tau) = 0, \quad \psi(0) = \psi(1) = 0. \tag{6}$$

Assuming that we use a uniform grid with mesh size $h = 1/N$ in the τ-direction, we then have $\psi(\tau) = \sum_{j=1}^{N} \psi_j \sin j\pi\tau$. Using this ansatz and testing (6) with $\sin k\pi\tau$ for $k = 1, \cdots, N$, we obtain for each k

$$(\alpha - k^2\pi^2)\psi_k - 2\eta \sum_{j=1}^{N} \psi_j \int_0^1 \tau^2 \sin j\pi\tau \sin k\pi\tau d\tau = 0.$$

Hence α represents eigenvalues of the matrix $\pi^2 \mathrm{diag}(1^2, 2^2, \cdots, N^2) + 2\eta M$, where M is a matrix with entries $M_{jk} = \int_0^1 \tau^2 \sin j\pi\tau \sin k\pi\tau d\tau$. We then denote the k-th eigenvalue by α_k, the smallest one by $\alpha_{\min}$ and the largest one by $\alpha_{\max}$.

For each eigenvalue α_k, $k = 1, \cdots, N$, we then need to consider

$$-(\phi_1^n(\sigma))'' + (\alpha_k + \sigma^2\eta)\phi_1^n(\sigma) = 0, \quad \phi_1^n(0) = 0,$$
$$-(\phi_2^n(\sigma))'' + (\alpha_k + \sigma^2\eta)\phi_2^n(\sigma) = 0, \quad \phi_2^n(1) = 0,$$

whose basic solutions are known in closed form,

$$\phi_{in}(\sigma; \alpha, \eta) = \frac{M(-\frac{1}{4}\frac{\alpha}{\sqrt{\eta}}, \frac{1}{4}, \sqrt{\eta}\sigma^2)}{\sqrt{\sigma}},$$
$$\phi_{de}(\sigma; \alpha, \eta) = \frac{W(-\frac{1}{4}\frac{\alpha}{\sqrt{\eta}}, \frac{1}{4}, \sqrt{\eta})}{M(-\frac{1}{4}\frac{\alpha}{\sqrt{\eta}}, \frac{1}{4}, \sqrt{\eta})} \frac{M(-\frac{1}{4}\frac{\alpha}{\sqrt{\eta}}, \frac{1}{4}, \sqrt{\eta}\sigma^2)}{\sqrt{\sigma}} + \frac{W(-\frac{1}{4}\frac{\alpha}{\sqrt{\eta}}, \frac{1}{4}, \sqrt{\eta}\sigma^2)}{\sqrt{\sigma}},$$

where W and M are Whittaker functions. Note that $\phi_{in}(\sigma; \alpha, \eta)$ increases monotonically in σ with $\phi_{in}(0; \alpha, \eta) = 0$ and $\phi_{de}(\sigma; \alpha, \eta)$ decreases monotonically in σ with $\phi_{de}(1; \alpha, \eta) = 0$.

Using the separation assumption $u_i(\sigma, \tau) = \phi_i(\sigma)\psi(\tau)$ also in the transmission conditions in (5) gives

$$(\partial_\sigma + \mathcal{S}_1)\phi_1^n(\sigma_0 + L)\psi(\tau) = (\partial_\sigma + \mathcal{S}_1)\phi_2^{n-1}(\sigma_0 + L)\psi(\tau),$$
$$(\partial_\sigma + \mathcal{S}_2)\phi_2^n(\sigma_0)\psi(\tau) = (\partial_\sigma + \mathcal{S}_2)\phi_1^{n-1}(\sigma_0)\psi(\tau).$$

Inserting $\psi(\tau) = \sum_{j=1}^{N} \psi_j \sin j\pi\tau$ and testing these equations by $\sin k\pi\tau$ we obtain for each $k = 1, 2, \cdots, N$

$$(\partial_\sigma + \mu_1(k))\phi_1^n(\sigma_0 + L) = (\partial_\sigma + \mu_1(k))\phi_2^{n-1}(\sigma_0 + L),$$
$$(\partial_\sigma + \mu_2(k))\phi_2^n(\sigma_0) = (\partial_\sigma + \mu_2(k))\phi_1^{n-1}(\sigma_0),$$

where $\mu_i(k)$, $i = 1, 2$ are the Fourier symbols of the operators $\mathcal{S}_i$.

Similar to the technique used in Gander and Xu (2016) (see also Gander 2006), we then obtain the convergence factor of algorithm (5),

$$\rho(L, \mu_1(k), \mu_2(k)) := \frac{(\partial_\sigma + \mu_1(k))\phi_{de}(\sigma_0 + L)}{(\partial_\sigma + \mu_1(k))\phi_{in}(\sigma_0 + L)} \frac{(\partial_\sigma + \mu_2(k))\phi_{in}(\sigma_0)}{(\partial_\sigma + \mu_2(k))\phi_{de}(\sigma_0)}. \tag{7}$$

As local approximations of the Dirichlet to Neumann operators, we consider

$$\mu_1^{app}(k) = p_1 + q_1\alpha_k, \quad \mu_2^{app}(k) = -p_2 - q_2\alpha_k,$$

which correspond to the local operators along the interfaces Γ_1 and Γ_2,

$$\mathcal{S}_1 = p_1 - q_1\partial_{\tau\tau} + q_1\tau^2\eta, \quad \mathcal{S}_2 = -p_2 + q_2\partial_{\tau\tau} - q_2\tau^2\eta.$$

Inserting $\mu_i^{app}(k)$, $i = 1, 2$ into (7) leads to the convergence factor

$$\rho_{opt}(\alpha_k, L, p_1, p_2, q_1, q_2) := \frac{(\partial_\sigma + p_1 + q_1\alpha_k)\phi_{de}(\sigma_0 + L)}{(\partial_\sigma + p_1 + q_1\alpha_k)\phi_{in}(\sigma_0 + L)} \frac{(\partial_\sigma - p_2 - q_2\alpha_k)\phi_{in}(\sigma_0)}{(\partial_\sigma - p_2 - q_2\alpha_k)\phi_{de}(\sigma_0)}. \tag{8}$$

The best choice for the free parameters p_i, q_i, $i = 1, 2$, minimizes the convergence factor, i.e. it is solution of the min-max problem

$$\min_{p_i > 0, q_i \geq 0, i=1,2} \max_{\alpha \in [\alpha_{min}, \alpha_{max}]} |\rho_{opt}(\alpha, L, p_1, p_2, q_1, q_2)|. \tag{9}$$

Using the theory of ordinary differential equations, one can prove

Lemma 1

(a) *For any fixed* $\alpha, \eta > 0$, $\phi_{in}(\sigma; \alpha, \eta)$ *is monotonically increasing in* σ *for* $\sigma > 0$. *For any fixed* $\sigma, \eta > 0$, $\frac{\partial_\sigma \phi_{in}(\sigma; \alpha, \eta)}{\phi_{in}(\sigma; \alpha, \eta)}$ *is monotonically increasing in* α *for* $\alpha > 0$.

(b) *For any fixed* $\alpha, \eta > 0$, $\phi_{de}(\sigma; \alpha, \eta)$ *is monotonically decreasing in* σ *for* $\sigma \in (0, 1)$. *For any fixed* $\sigma, \eta > 0$, $-\frac{\partial_\sigma \phi_{de}(\sigma; \alpha, \eta)}{\phi_{de}(\sigma; \alpha, \eta)}$ *is monotonically increasing in* α *for* $\alpha > 0$.

Let $G(\sigma, \alpha, \eta) := \frac{\partial_\sigma \phi_{in}(\sigma; \alpha, \eta)}{\phi_{in}(\sigma; \alpha, \eta)} - \frac{\partial_\sigma \phi_{de}(\sigma; \alpha, \eta)}{\phi_{de}(\sigma; \alpha, \eta)}$ and $G_{min} := G(\sigma_0; \alpha_{min}, \eta)$.

Theorem 1 *For the O00 (optimized of order 0) method, let* $p_1 = p_2 = p > 0$ *and* $q_1 = q_2 = 0$. *Then for small overlap,* $L > 0$, *the parameter* $p^* = 2^{-1} G_{min}^{\frac{2}{3}} L^{-\frac{1}{3}}$ *solves asymptotically the min-max problem (9) and*

$$\max_{\alpha \in [\alpha_{min}, \alpha_{max}]} |\rho_{opt}(\alpha, L, p^*, p^*, 0, 0)| = 1 - 4G_{min}^{\frac{1}{3}} L^{\frac{1}{3}} + O(L^{\frac{2}{3}}). \tag{10}$$

Proof Using Lemma 1, the results can be proved by the techniques used to prove Theorem 3.8 and Theorem 3.9 in Gander and Xu (2014).

Table 1 Optimized transmission parameters and the corresponding convergence factor estimate

	Type	Constraint	Optimized parameters	$\max |\rho_{opt}|$
$L > 0$	OO2	$p_1 = p_2 > 0$ $q_1 = q_2 > 0$	$p_1^* = p_2^* = 2^{-\frac{7}{5}} G_{\min}^{\frac{2}{5}} L^{-\frac{1}{5}}$ $q_1^* = q_2^* = 2^{\frac{1}{5}} G_{\min}^{-\frac{2}{5}} L^{\frac{3}{5}}$	$1 - 2^{\frac{12}{5}} G_{\min}^{\frac{1}{5}} L^{\frac{1}{5}} + O(L^{\frac{2}{5}})$
	O2s	$p_1 > 0, p_2 > 0$ $q_1 = q_2 = 0$	$p_1^* = 2^{-\frac{8}{5}} G_{\min}^{\frac{4}{5}} L^{-\frac{1}{5}}$ $p_2^* = 2^{-\frac{2}{5}} G_{\min}^{\frac{2}{5}} L^{-\frac{3}{5}}$	$1 - 2^{\frac{8}{5}} G_{\min}^{\frac{1}{5}} L^{\frac{1}{5}} + O(L^{\frac{2}{5}})$
$L = 0$	OOO	$p_1 = p_2 > 0$ $q_1 = q_2 = 0$	$p_1^* = p_2^* = 2^{-\frac{1}{2}} G_{\min}^{\frac{1}{2}} \alpha_{\max}^{\frac{1}{4}}$	$1 - 2^{\frac{3}{2}} G_{\min}^{\frac{1}{2}} \alpha_{\max}^{-\frac{1}{4}} + O(\alpha_{\max}^{-\frac{1}{2}})$
	OO2	$p_1 = p_2 > 0$ $q_1 = q_2 > 0$	$p_1^* = p_2^* = 2^{-\frac{5}{4}} G_{\min}^{\frac{3}{4}} \alpha_{\max}^{\frac{1}{8}}$ $q_1^* = q_2^* = 2^{-\frac{1}{4}} G_{\min}^{-\frac{1}{4}} \alpha_{\max}^{-\frac{3}{8}}$	$1 - 2^{\frac{9}{4}} G_{\min}^{\frac{1}{4}} \alpha_{\max}^{-\frac{1}{8}} + O(\alpha_{\max}^{-\frac{1}{4}})$
	O2s	$p_1 > 0, p_2 > 0$ $q_1 = q_2 = 0$	$p_1^* = 2^{-\frac{5}{4}} G_{\min}^{\frac{3}{4}} \alpha_{\max}^{\frac{1}{8}}$ $p_2^* = 2^{\frac{1}{4}} G_{\min}^{\frac{1}{4}} \alpha_{\max}^{\frac{3}{8}}$	$1 - 2^{\frac{5}{4}} G_{\min}^{\frac{1}{4}} \alpha_{\max}^{-\frac{1}{8}} + O(\alpha_{\max}^{-\frac{1}{4}})$

Similar results can also be proved for the OO2 (optimized of order 2) method and the O2s (optimized two-sided Robin) method for overlapping, and non-overlapping domain decompositions. The corresponding results are summarized in Table 1.

4 Geometric Characteristics Entering the Optimization

In Sect. 3 we obtained the optimized transmission conditions in the parabolic coordinates (σ, τ), where the interface is a line. In a real application, one would however compute in the standard Cartesian coordinates where the interface is a parabola in our model problem, and we study now how the optimized parameter of OOO looks in the standard Cartesian coordinates to see how geometric characteristics enter the optimization of the transmission parameters. Without loss of generality, we consider only the interface Γ_1, where the optimized transmission condition is

$$(\partial_\sigma + p^*) u_1^n (\sigma_0 + L, \tau) = (\partial_\sigma + p^*) u_2^{n-1} (\sigma_0 + L, \tau). \tag{11}$$

A direct calculation gives $\partial_{n_1} = \frac{1}{\sqrt{\sigma^2 + \tau^2}} \partial_\sigma$, and dividing both sides of (11) by $\sqrt{\sigma^2 + \tau^2}$ we get

$$\left(\partial_{n_1} + \frac{1}{\sqrt{\sigma^2 + \tau^2}} p^*\right) u_1^n (x, y) = \left(\partial_{n_1} + \frac{1}{\sqrt{\sigma^2 + \tau^2}} p^*\right) u_2^{n-1} (x, y), \text{ on } \Gamma_1. \tag{12}$$

A further direct calculation shows that $\sigma^2 + \tau^2 = \sqrt{x^2 + y^2} - x + \frac{y^2}{\sqrt{x^2+y^2}-x}$, and hence in Cartesian coordinates the optimized transmission parameter is given by

$(\sqrt{x^2 + y^2} - x + \frac{y^2}{\sqrt{x^2+y^2}-x})p^*$, i.e. it varies along the interface, instead of being a constant. To see how the interface curvature enters this optimized transmission condition, we compute the curvature of the interface Γ_1 and obtain $\kappa = \frac{\sigma}{(\sigma^2+\tau^2)^{\frac{3}{2}}} = \frac{\sigma}{H^3}$ with $\sigma = \sigma_0 + L$. Hence the optimized parameter in Cartesian coordinates is given by $(\frac{\sigma_0+L}{\kappa})^{-\frac{1}{3}}p^*$. Note that the constant $\sigma_0 + L$ describes the position of the parabolically shaped interface. Therefore, in addition to the interface curvature, other geometric characteristics (here the constant $\sigma_0 + L$) can enter as well the optimized transmission parameters.

5 Numerical Experiments

To show that our predicted transmission parameter from Theorem 1 is indeed asymptotically optimal, we first consider the model problem (1) in the parabolic coordinates (σ, τ), i.e. the OO0 variant of the Schwarz algorithm (5), with $\sigma_0 = 0.5$ and $S_i = p^*$, $i = 1, 2$. We discretize (5) using FreeFem++, and start with a random initial guess on the interfaces, simulating directly the error equations, i.e. $f = 0$. The number of iterations required to reach an error reduction of $1e - 6$ is shown in the first row of Table 2. A log-log plot of these results on the left in Fig. 2 shows good agreement with the estimate in Theorem 1. To show how our prediction p^* approximates the numerically optimal Robin parameter, we vary the Robin parameter p from 3 to 18 with 76 equidistant samples and record the corresponding number of iterations required by the Schwarz method with $N = 160$. The results are shown on the right in Fig. 2, and we see that our prediction p^* is very close to the numerically optimal Robin parameter.

We next solve the model problem (1) in Cartesian coordinates using Freefem++ like one would in a real application. We choose again the interface parameter $\sigma_0 = 0.5$, and use the transmission condition (12) on Γ_1 and a corresponding one on Γ_2. In this situation the overlap is the local distance between the interfaces Γ_1 and Γ_2. In Table 2 in the second row we show the number of iterations required by the

Table 2 Iteration numbers of the OO0 Schwarz method with overlap $1/N$ discretized in parabolic coordinates (first row), compared to discretization in Cartesian coordinates taking all geometric information into account (second row), and using the optimized parameter from the straight interface analysis (Gander, 2006) either locally scaled by the interface curvature (third row) or with $k_{\min} = \pi/c$, where c is the interface length (last row)

Coordinates	N	20	40	80	160	320
Parabolic	#iter(OO0)	8	11	13	17	23
Cartesian	#iter(OO0)	8	12	14	19	24
	#iter(OO0-Scaled)	10	12	16	22	28
	#iter(OO0-Straight)	10	13	16	22	28

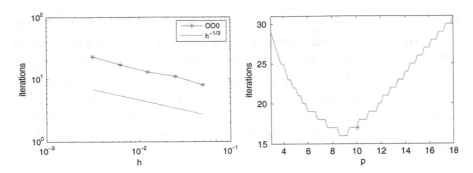

Fig. 2 *Left*: Log-log plot of the number of iterations from the first row in Table 2. *Right*: Number of iterations required by the OO0 Schwarz method in parabolic coordinates compared to other values of the Robin parameter p; the *red star* indicates our prediction p^*

optimized Schwarz method to reach an error reduction of $1e - 6$. Comparing with the first row, we see that our prediction of the optimized Robin parameter taking into account all geometric characteristics performs basically as when computing in the parabolic coordinates. In the third and last row of Table 2, we show the results obtained with the strategy suggested in Gander and Xu (2017), i.e. to use the optimized transmission parameter from the straight interface analysis (Gander, 2006), either scaled locally by the interface curvature, or choosing $k_{\min} = \pi/c$ with c the length of the interface.[1] These last two approaches also reach the same asymptotic convergence order and are comparable, but more iterations are needed than for our new approach which takes more geometric features into account.

6 Conclusion

To get a better understanding on the influence of geometry on optimized transmission conditions, we studied a model problem using a domain decomposition with parabolically shaped interfaces. Using separation of variables, we showed that the optimized parameter in Cartesian coordinates varies along the interface, and not only the interface curvature comes in, but also further geometric characteristics of the interface appear. We then showed numerically that indeed taking all these geometric characteristics into account the new optimized parameter outperforms the strategy of using only the local curvature or interface length to scale appropriately an optimized parameter from a straight interface analysis.

---

[1] The length of the interface $\sigma = \sigma_0$ is easy to calculate to be $\frac{\sigma_0^2}{2}\texttt{arcsinh}(\frac{1}{\sigma_0}) + \frac{1}{2}\sqrt{\sigma_0^2 + 1}$.

Acknowledgements The author "Y. Xu" was partly supported by NSFC-11671074, 11471047, CPSF-2012M520657 and the Science and Technology Development Planning of Jilin Province 20140520058JH.

References

H. Barucq, M.J. Gander, Y. Xu, On the influence of curvature on transmission conditions, in *Domain Decomposition Methods in Science and Engineering XXI*. Lecture Notes in Computational Science and Engineering (Springer, Berlin, 2014), pp. 323–331

M.J. Gander, Optimized Schwarz methods. SIAM J. Numer. Anal. **44**(2), 699–731 (2006)

M.J. Gander, Schwarz methods over the course of time. Electron. Trans. Numer. Anal. **31**(5), 228–255 (2008)

M.J. Gander, On the influence of geometry on optimized Schwarz methods. SeMA J. **53**(1), 71–78 (2011)

M.J. Gander, Y. Xu, Optimized Schwarz methods for circular domain decompositions with overlap. SIAM J. Numer. Anal. **52**(4), 1981–2004 (2014)

M.J. Gander, Y. Xu, Optimized schwarz methods for model problems with continuously variable coefficients. SIAM J. Sci. Comput. **38**(5), A2964–A2986 (2016)

M.J. Gander, Y. Xu, Optimized Schwarz methods with nonoverlapping circular domain decompositions. Math. Comput. **86**(304),637–660 (2017)

G. Gigante, M. Pozzoli, C. Vergara, Optimized Schwarz methods for the diffusion-reaction problem with cylindrical interfaces. SIAM J. Numer. Anal. **51**(6), 3402–3430 (2013)

A Mortar Domain Decomposition Method for Quasilinear Problems

Matthias A.F. Gsell and Olaf Steinbach

1 Introduction

As model problem for a quasilinear partial differential equation we consider the Richards equation, see, e.g., Berninger (2008),

$$n \frac{\partial \theta(p)}{\partial t} - \nabla \cdot \left(\frac{K}{\mu} k(\theta(p)) \nabla (p - d) \right) = f$$

to find the unknown pressure p. This equation results from the principle of mass balance and by using several laws from hydrology. The quantity $n(\mathbf{x})$ prescribes the porosity of the soil, $K(\mathbf{x})$ is the permeability of the soil, μ is just the constant viscosity of water, and $d(\mathbf{x}) := d(x_1, \ldots, x_d) = \varrho \, g \, x_d$ with the constant water density ϱ and with the gravitational constant g. The nonlinear parameter function θ describes the saturation of the soil in dependency of the pressure p. k is the relative permeability of the soil which depends on the saturation. There are several models available which describe the shape of θ and k. In this work we use the model of Brooks and Corey (1964) where the saturation is given as

$$\theta(p) := \begin{cases} \left(\dfrac{p}{p_b} \right)^{-\lambda} (\theta_{\max} - \theta_{\min}) + \theta_{\min} & \text{for } p \leq p_b, \\[2mm] \theta_{\max} & \text{for } p > p_b. \end{cases}$$

Here, $\theta_{\min}$ and $\theta_{\max}$ are the minimal and maximal saturation level, $p_b < 0$ is the so called bubbling pressure, and $\lambda > 0$ is the pore size distribution factor. The relative

M.A.F. Gsell (✉) • O. Steinbach
Institut für Numerische Mathematik, TU Graz, Steyrergasse 30, 8010 Graz, Austria
e-mail: gsell@tugraz.at; o.steinbach@tugraz.at

© Springer International Publishing AG 2017

C.-O. Lee et al. (eds.), *Domain Decomposition Methods in Science and Engineering XXIII*, Lecture Notes in Computational Science and Engineering 116, DOI 10.1007/978-3-319-52389-7_34

permeability is given as

$$k(\theta) := \left(\frac{\theta - \theta_{min}}{\theta_{max} - \theta_{min}} \right)^{3+\frac{2}{\lambda}}.$$

Hence we conclude

$$k(\theta(p)) = \begin{cases} \left(\dfrac{p}{p_b} \right)^{-3\lambda-2} & \text{for } p \leq p_b, \\ 1 & \text{for } p > p_b. \end{cases}$$

The considerations made so far are valid for a single soil type only, see Fig. 1. In the case of several layers of different soil types we have to consider parameter functions θ and k which depend explicitly on $\mathbf{x}$, see Fig. 2 where we have a decomposition of Ω into N non–overlapping subdomains Ω_i representing a soil layer each with local parameter functions θ_i and k_i. Hence we define global parameter functions as

$$\theta(\mathbf{x}, p(\mathbf{x}, t)) = \theta_i(p(\mathbf{x}, t)), \quad k(\mathbf{x}, \theta(\mathbf{x}, p(\mathbf{x}, t))) = k_i(\theta_i(p(\mathbf{x}, t))), \quad \mathbf{x} \in \Omega_i.$$

In what follows we will apply an implicit–explicit time discretization scheme and local Kirchhoff transformations to end up with a domain decomposition variational formulation of local linear elliptic partial differential equations, but with nonlinear transmission conditions. For the discretization we then use a mortar finite element approach.

Fig. 1 Single soil type

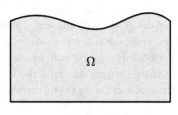

Fig. 2 Several soil layers

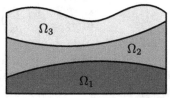

2 Variational Formulation

Let $\Omega \subset \mathbb{R}^d$, $d = 2, 3$, be a bounded Lipschitz domain with boundary $\partial\Omega$ which is decomposed into two mutually disjoint parts Γ_D and Γ_N where boundary conditions of Dirichlet and Neumann type are given, respectively. We assume meas $\Gamma_D > 0$, and let $\mathbf{n}$ be the outer unit normal. For $T > 0$ we consider the initial boundary value problem to find $p : \Omega \times (0, T) \to \mathbb{R}$ such that

$$n \frac{\partial \theta(p)}{\partial t} - \nabla \cdot \left(\frac{K}{\mu} k(\theta(p)) \nabla (p - d) \right) = f \qquad \text{in } \Omega \times (0, T), \tag{1a}$$

$$p = p_D \qquad \text{on } \Gamma_D \times (0, T), \tag{1b}$$

$$\frac{K}{\mu} k(\theta(p)) \nabla (p - d) \cdot \mathbf{n} = p_N \qquad \text{on } \Gamma_N \times (0, T), \tag{1c}$$

$$p = p_0 \qquad \text{at } \Omega \times \{0\} \tag{1d}$$

is satisfied.

For $M \in \mathbb{N}$ let $0 = t_0 < t_1 < \ldots < t_M = T$ be a decomposition of the time interval $(0, T)$. For an implicit time discretization we use a backward Euler method to approximate the time derivative,

$$\frac{\partial}{\partial t} \theta(\mathbf{x}, p(\mathbf{x}, t)) \Big|_{t=t_m} \approx \frac{\theta(p_m) - \theta(p_{m-1})}{\tau_m}, \quad \tau_m := t_m - t_{m-1}, \ p_m(\mathbf{x}) \approx p(t_m, \mathbf{x}).$$

After time discretization, the variational formulation of (1a) is to find, for all time steps $1 \le m \le M$, $p_m \in H^1(\Omega)$, $p_{m|\Gamma_D} = p_D(t_m)$, such that

$$\int_\Omega \frac{n}{\tau_m} \theta(p_m) v \, d\mathbf{x} + \int_\Omega \frac{K}{\mu} k(\theta(p_m)) \nabla (p_m - d) \cdot \nabla v \, d\mathbf{x} = \langle \widehat{F}, v \rangle_\Omega$$

is satisfied for all $v \in V := H^1_{0, \Gamma_D}(\Omega)$, where

$$\langle \widehat{F}, v \rangle_\Omega := \int_\Omega \left(f(t_m) + \frac{n}{\tau_m} \theta(p_{m-1}) \right) v \, d\mathbf{x} + \int_{\Gamma_N} p_N(t_m) \, v \, ds_\mathbf{x}.$$

For the remaining nonlinear term we apply an explicit discretization step,

$$k(\theta(p_m)) \nabla (p_m - d) \approx k(\theta(p_m)) \nabla p_m - k(\theta(p_{m-1})) \nabla d$$

where we keep the nonlinearity within the first term. Hence we end up with a variational formulation to find $p_m \in H^1(\Omega)$, $p_{m|\Gamma_D} = p_D(t_m)$, such that

$$\int_\Omega \frac{n}{\tau} \theta(p_m) v \, d\mathbf{x} + \int_\Omega \frac{K}{\mu} k(\theta(p_m)) \nabla p_m \cdot \nabla v \, d\mathbf{x} = \langle F, v \rangle_\Omega \tag{2}$$

is satisfied for all $v \in V$, where

$$\langle F, v \rangle_\Omega := \langle \widehat{F}, v \rangle_\Omega + \int_\Omega \frac{K}{\mu} k(\theta(p_{m-1})) \nabla d \cdot \nabla v \, \mathrm{dx}.$$

Theorem 1 *Assume $n, K \in L_\infty^+(\Omega) = \{u \in L_\infty(\Omega) \mid ess\,inf_{x \in \Omega}\, u > 0\}$, $\tau, \mu \in \mathbb{R}_+$. Let $\theta_i = \theta_{|\Omega_i} \in C^{0,1}(\mathbb{R})$ be monotonically increasing, and we assume $k_i = k_{|\Omega_i} \in C^{0,1}(\mathbb{R}) \cap L_\infty(\mathbb{R})$ and $k(s) \geq c > 0$ for all $s \in \mathbb{R}$. Then there exists a unique solution of the variational problem (2).*

To handle the nonlinear term in the variational formulation (2) we will apply the Kirchhoff transformation (Alt and Luckhaus, 1983; Berninger et al., 2015; Schreiber (2009)) locally within the subdomains Ω_i. Since this results in nonlinear Dirichlet transmission conditions, we will use a primal–hybrid formulation (Boffi et al., 2013; Raviart and Thomas, 1977) to split the global problem (2) into local ones with suitable transmission conditions.

In what follows we will skip the dependence on the time step, and we consider one time step only.

Let $\overline{\Omega} = \cup_{i=1}^N \overline{\Omega}_i$ be a nonoverlapping domain decomposition which resolves the different soil layers, see Fig. 3. When defining the primal space

$$X := \left\{ p \in L^2(\Omega) \middle| p_{|\Omega_i} \in H^1(\Omega_i) \right\},$$

the Lagrange multiplier space

$$M := \left\{ \mu \in \prod_{i=1}^N H^{-1/2}(\partial \Omega_i) \middle| \exists\, \mathbf{q} \in H_{0,\Gamma_N}(\mathrm{div}, \Omega) : \mathbf{q} \cdot \mathbf{n}_i = \mu \text{ on } \partial \Omega_i \right\},$$

and the bilinear form

$$b(p, v) := -\sum_{i=1}^N \langle p_{|\Omega_i}, v \rangle_{\partial \Omega_i},$$

Fig. 3 Decomposition

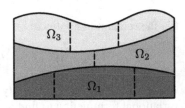

we obtain a variational problem to find $(p, \lambda) \in X \times M$ such that

$$\sum_{i=1}^{N} \left(\int_{\Omega_i} \frac{n}{\tau} \theta(p) v \, d\mathbf{x} + \int_{\Omega_i} \frac{K}{\mu} k(\theta(p)) \nabla p \cdot \nabla v \, d\mathbf{x} \right) + b(v, \lambda) = \langle F, v \rangle_{\Omega},$$

$$b(p, v) = -\langle p_D, v \rangle_{\partial \Omega}$$

is satisfied for all $(v, \nu) \in X \times M$. Now we are in the position to apply local Kirchhoff transformations to shift the remaining nonlinearities from the subdomains Ω_i to the local boundaries $\partial \Omega_i$. We therefore introduce the generalized pressure $u \in X$ as $u_{|\Omega_i} := \kappa_i(p_{|\Omega_i})$ which satisfies, see Marcus and Mizel (1979),

$$\nabla u_{|\Omega_i} = k_i(\theta_i(p_{|\Omega_i})) \nabla p_{|\Omega_i}.$$

The mapping κ_i is a superposition operator induced by $\kappa_i : \mathbb{R} \to \mathbb{R}$ which is defined as

$$\kappa_i(r) = \int_0^r k_i(\theta_i(s)) \, ds.$$

It can be shown that the nonlinear operators $\kappa_i : H^1(\Omega_i) \to H^1(\Omega_i)$ are continuous and bounded. If there exist positive constants $c_i > 0$ such that $k_i(s) \geq c_i$ for all $s \in \mathbb{R}$, i.e. κ_i being monotone, then the inverse operators κ_i^{-1} exist and are again continuous and bounded. Using these local nonlinear operators, we can define

$$\iota_i := \theta_i \circ \kappa_i^{-1}, \quad c(u, v) := -\sum_{i=1}^{N} \langle \kappa_i^{-1}(u_{|\Omega_i}), v \rangle_{\partial \Omega_i},$$

and we finally obtain a variational problem to find $(u, \lambda) \in X \times M$, such that

$$\sum_{i=1}^{N} \left(\int_{\Omega_i} \frac{n}{\tau} \iota(u) v \, d\mathbf{x} + \int_{\Omega_i} \frac{K}{\mu} \nabla u \cdot \nabla v \, d\mathbf{x} \right) + b(v, \lambda) = \langle F, v \rangle_{\Omega}, \tag{3}$$

$$c(u, v) = -\langle p_D, v \rangle_{\partial \Omega}$$

is satisfied for all $(v, \nu) \in X \times M$. The variational problem (3) is by construction equivalent to (2), and hence we conclude unique solvability of (3).

3 Mortar Finite Element Discretization

For the discretization of the variational problem (3) we use the mortar finite element method, see Wohlmuth (2001). Let $\mathcal{T}_{h,i}$ be a local triangulation of the subdomain Ω_i, $i = 1, \ldots, N$, see Fig. 4. Note that the local triangulations do not have to coincide at

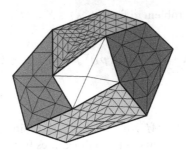

Fig. 4 Triangulation

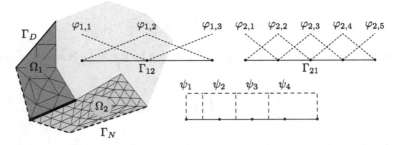

Fig. 5 Construction of ansatz space for Lagrange multiplier in $\mathbb{R}^2$

neighbouring interfaces. With $\Gamma_{D,i} := \Gamma_D \cap \partial\Omega_i$ we define for each subdomain Ω_i the space

$$H^1_\star(\Omega_i) := \begin{cases} H^1(\Omega_i) & \text{if meas } \Gamma_{D,i} = 0, \\ H^1_{0,\Gamma_{D,i}}(\Omega_i) & \text{else.} \end{cases}$$

We define the local finite element ansatz spaces $X_{h,i} := \mathcal{S}^1(\mathcal{T}_{h,i}) \cap H^1_\star(\Omega_i)$ as the space of all piecewise linear and continuous functions in Ω_i. The global ansatz space is then defined as $X_h := \prod_{i=1}^N X_{h,i}$. To define a discrete ansatz space for the Lagrange multiplier $\lambda \in M$ we consider each interface Γ_{ij} with $\Gamma_{ij} := \partial\Omega_i \cap \partial\Omega_j$, $i \neq j$, separately. For a nonempty interface Γ_{ij} we have two neighbouring subdomains and their triangulations $\mathcal{T}_{h,i}$ and $\mathcal{T}_{h,j}$. In view of a better approximation property, we choose the finer triangulation and denote its index by m_{ij}. The mesh $\mathcal{I}_{h,ij}$ of the interface Γ_{ij} is induced by $\mathcal{T}_{h,m_{ij}}$, that is $\mathcal{I}_{h,ij} = \mathcal{T}_{h,m_{ij}}|_{\Gamma_{ij}}$. By $\mathcal{I}'_{h,ij}$ we denote a modified dual mesh, i.e. we define $M_{h,ij} := \mathcal{S}^0(\mathcal{I}'_{h,ij})$ to be the space of all piecewise constant functions on the dual mesh, see Fig. 5. The global ansatz space is then defined as the product space $M_h := \prod_{\Gamma_{ij}} M_{h,ij}$. By construction, $u_h \in X_h$ satisfies $u_h = 0$ on Γ_D, and the discrete Lagrange multiplier $\lambda_h \in M_h$ are just defined on the interfaces within Ω. If we assume, that there exists a discrete extension $u_{h,D}$, satisfying the inhomogeneous Dirichlet boundary conditions, we obtain the following discrete nonlinear variational problem to find $(u_h, \lambda_h) \in X_h \times M_h$ such that $\widetilde{u}_h := u_h + u_{h,D}$ satisfies

$$\sum_{i=1}^{N} \left(\int_{\Omega_i} \frac{n}{\tau} \iota(\widetilde{u}_h) v_h \, d\mathbf{x} + \int_{\Omega_i} \frac{K}{\mu} \nabla \widetilde{u}_h \cdot \nabla v_h \, d\mathbf{x} \right) + b(v_h, \lambda_h) = \langle F, v_h \rangle_{\Omega},$$

$$c(u_h, v_h) = 0$$

for all $(v_h, \nu_h) \in X_h \times M_h$. Since $M_{h,ij} \subset L_2(\Gamma_{ij})$, we can rewrite

$$b(v_h, \lambda_h) := -\sum_{\Gamma_{ij}} (v_{h|\Omega_i} - v_{h|\Omega_j}, \lambda_h)_{\Gamma_{ij}}$$

as well as

$$c(v_h, \lambda_h) := b(\kappa^{-1}(v_h), \lambda_h) = -\sum_{\Gamma_{ij}} (\kappa_i^{-1}(v_{h|\Omega_i}) - \kappa_j^{-1}(v_{h|\Omega_j}), \lambda_h)_{\Gamma_{ij}}.$$

Since the discrete variational problem is still nonlinear, we apply Newton's method and obtain the linearized problem: For $\widetilde{w}_h := w_h + u_{h,D}$, $w_h \in X_h$, find $(u_h, \lambda_h) \in X_h \times M_h$, such that

$$\sum_{i=1}^{N} \left(\int_{\Omega_i} \frac{n}{\tau} \iota'(\widetilde{w}_h) u_h v_h \, d\mathbf{x} + \int_{\Omega_i} \frac{K}{\mu} \nabla u_h \cdot \nabla v_h \, d\mathbf{x} \right) + b(v_h, \lambda_h) = \langle \widetilde{F}, v_h \rangle_{\Omega},$$

$$c'(\widetilde{w}_h, u_h, v_h) = \langle \widetilde{G}, v_h \rangle_S \tag{4}$$

is satisfied for all $(v_h, \nu_h) \in X_h \times M_h$. The linear forms of the discrete and linearized variational problem (4) are

$$\langle \widetilde{F}, v_h \rangle_{\Omega} = \langle F, v_h \rangle_{\Omega} + \langle \overline{F}, v_h \rangle_{\Omega}, \quad \langle \widetilde{G}, v_h \rangle_S := c'(\widetilde{w}_h, w_h, v_h) - c(\widetilde{w}_h, v_h)$$

with $c'(\widetilde{w}_h, u_h, v_h) := b((\kappa^{-1})'(\widetilde{w}_h) u_h, v_h)$ and

$$\langle \overline{F}, v_h \rangle_{\Omega} := \sum_{i=1}^{N} \left(\int_{\Omega_i} \frac{n}{\tau} (\iota'(\widetilde{w}_h) \widetilde{w}_h - \iota(\widetilde{w}_h)) v_h \, d\mathbf{x} - \int_{\Omega_i} \frac{K}{\mu} \nabla u_{h,D} \cdot \nabla v_h \, d\mathbf{x} \right).$$

The stability and error analysis of the mixed formulation (4) follows from related stability conditions of the underlying bilinear forms and appropriate finite element methods, see Gsell (2016).

4   Numerical Example

As an example we consider the domain $\Omega = (0, 1) \times (0, 2) \subset \mathbb{R}^2$, see Fig. 6, with Dirichlet conditions on $\Gamma_D := (0, 1) \times \{2\}$, while on the remaining boundary Γ_N we have Neumann boundary conditions. The four layers behave like sand, sandy loam, loam and sand, see Gsell (2016). We assume that there are no sources or sinks within Ω, i.e. $f \equiv 0$. On Γ_D we prescribe a pressure which increases in time, that is

$$
p_D(\mathbf{x}, t) := \begin{cases} -0.5 \, (10 - t) & t < 10, \\ 0.0 & t \geq 10. \end{cases}
$$

On Γ_N we prescribe the no–outflow–condition $p_N(\mathbf{x}, t) \equiv 0$. Since we approximate the solution of the transformed variational problem (3), we have to consider the Dirichlet datum u_D for the generalized pressure which is given as $u_D(\mathbf{x}, t) = \kappa_i(p_D(\mathbf{x}, t))$ for $\mathbf{x} \in \Gamma_{D,i}$. The Neumann datum remains unchanged. The following snapshots show contour lines of the pressure p, which can be computed by the application of the inverse transformation, that is $p_{|\Omega_i} = \kappa_i^{-1}(u_{|\Omega_i})$. Due to the choice of the data, the problem evolutes to a pure diffusion equation. That is why the snapshots were taken at $t = 0, 250, 500, 1000, 2000, 4000, 8000, 10,000$.

Fig. 6 Triangulation

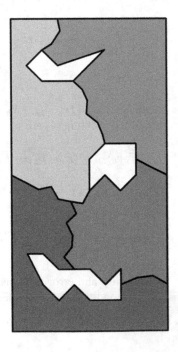

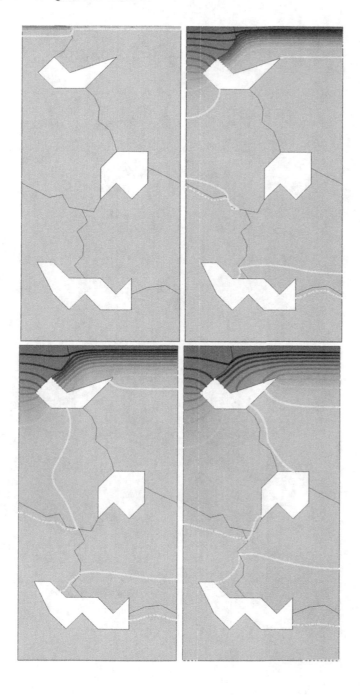

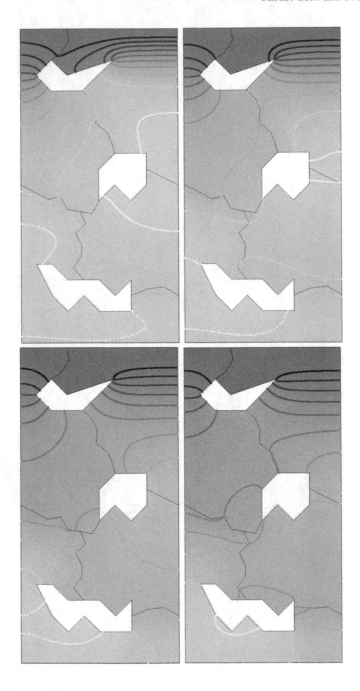

Acknowledgements This work was supported by the Austrian Science Fund (FWF) within the International Research Training Group IGDK 1754.

References

H.W. Alt, S. Luckhaus, Quasilinear elliptic–parabolic differential equations. Math. Z. **183**, 311–341 (1983)

H. Berninger, Domain decomposition methods for elliptic problems with jumping nonlinearities and application to the Richards equation, Ph.D. thesis, Freie Universität Berlin, 2008

H. Berninger, R. Kornhuber, O. Sander, A multidomain discretization of the Richards equation in layered soil. Comput. Geosci. **19**(1), 213–232 (2015)

D. Boffi, F. Brezzi, M. Fortin, *Mixed Finite Element Methods and Applications.* Springer Series in Computational Mathematics, vol. 44 (Springer, Heidelberg, 2013)

R.H. Brooks, A.T. Corey, *Hydraulic Properties of Porous Media* (Colorado State University, Fort Collins, 1964)

M.A.F. Gsell, Mortar domain decomposition methods for quasilinear problems and applications, Ph.D. thesis, TU Graz, in preparation, 2016

M. Marcus, V.J. Mizel, Every superposition operator mapping one Sobolev space into another is continuous. J. Funct. Anal. **33**(2), 217–229 (1979)

P.-A. Raviart, J.M. Thomas, Primal hybrid finite element methods for 2nd order elliptic equations. Math. Comput. **31**(138), 391–413 (1977)

S. Schreiber, *Nichtüberlappende Gebietszerlegungsmethoden für lineare und quasilineare (monotone und nichtmonotone) Probleme* (Universität Kassel, 2009)

B.I. Wohlmuth, *Discretization Methods and Iterative Solvers Based on Domain Decomposition.* Lecture Notes in Computational Science and Engineering, vol. 17 (Springer, Berlin, 2001)

Deflated Krylov Iterations in Domain Decomposition Methods

Y.L. Gurieva, V.P. Ilin, and D.V. Perevozkin

1 Introduction

The goal of this research is an investigation of some advanced versions of algebraic approaches to parallel domain decomposition algorithms for solving sparse large systems of linear algebraic equation (SLAEs) with nonsymmetric sparse matrices arising from some approximation of the multi-dimension boundary value problems (BVPs) in complicated computational domains on non-structured grids.

Algebraic domain decomposition methods (DDMs) are the main tool to provide high performance computing when solving very large SLAEs, which are the bottlenecks of contemporary interdisciplinary tasks. There are many publications on this topic, see Toselli and Widlund (2005), Dolean et al. (2015), Dubois et al. (2012) and Gurieva and Il'in (2015) and literature cited there, for example. They present a manifold of mathematical and technological contradictory problems. On the one hand, high convergence rate of iterative processes leads to high computational complexity of algorithms. On the other hand, performance of applied program packages depends on the data structures used and code adaptation to a particular parallel architecture.

We describe some essential aspects of the algorithms implemented on the basis of the multi-preconditioned semi-conjugate residual method and the coarse grid

Y.L. Gurieva • D.V. Perevozkin (✉)
Institute of Computational Mathematics and Mathematical Geophysics SB RAS, Novosibirsk, Russia
e-mail: foxillys@gmail.com

V.P. Ilin
Institute of Computational Mathematics and Mathematical Geophysics SB RAS, Novosibirsk, Russia

Novosibirsk State University, Novosibirsk, Russia

© Springer International Publishing AG 2017
C.-O. Lee et al. (eds.), *Domain Decomposition Methods in Science and Engineering XXIII*, Lecture Notes in Computational Science and Engineering 116, DOI 10.1007/978-3-319-52389-7_35

correction procedure with basic functions of different orders. In some sense, the proposed approaches present a further development of the ideas considered in papers by Saad (2003) and Bridson and Greif (2006).

This paper is organized as follows. Section 2 contains the formulation of the problems to be solved. Section 3 is devoted to the parallel structure of algorithms. Section 4 deals with demonstration of the numerical results. In conclusion, the results obtained are described.

2   Statement of the Problem

Let us have a boundary value problem

$$Lu = f(\mathbf{r}), \quad \mathbf{r} \in \Omega, \quad lu|_\Gamma = g(\mathbf{r}), \tag{1}$$

in a computational open domain Ω with a boundary Γ and a closure $\bar{\Omega} = \Omega \bigcup \Gamma$, where L and l are some linear differential operators. We suppose that (1) has a unique solution $u(\mathbf{r})$ which is smooth enough.

Let us decompose Ω into P subdomains (with or without overlapping):

$$\begin{aligned}
\Omega &= \bigcup_{q=1}^{P} \Omega_q, \quad \bar{\Omega}_q = \Omega_q \bigcup \Gamma_q, \\
\Gamma_q &= \bigcup_{q' \in \omega_q} \Gamma_{q,q'}, \quad \Gamma_{q,q'} = \Gamma_q \bigcap \bar{\Omega}_{q'}, \quad q' \neq q.
\end{aligned} \tag{2}$$

Here Γ_q is the boundary of Ω_q which is composed from the segments $\Gamma_{q,q'}, q' \in \omega_q$, and $\omega_q = \{q_1, \ldots, q_{M_q}\}$ is a set of M_q contacting, or conjuncted, subdomains. We can denote also by $\Omega_0 = R^d/\Omega$ the external subdomain:

$$\bar{\Omega}_0 = \Omega_0 \bigcup \Gamma, \quad \Gamma_{q,0} = \Gamma_q \bigcap \bar{\Omega}_0 = \Gamma_q \bigcap \Gamma, \quad \Gamma_q = \Gamma_q^i \bigcup \Gamma_{q,0}, \tag{3}$$

where $\Gamma_q^i = \bigcup_{q' \neq 0} \Gamma_{q,q'}$ and $\Gamma_{q,0} = \Gamma_q^e$ mean internal and external parts of the boundary of Ω_q. We define also an overlapping $\Delta_{q,q'} = \Omega_q \bigcap \Omega_{q'}$ of the neighbouring subdomains. If $\Gamma_{q,q'} = \Gamma_{q'.q}$ and $\Delta_{q,q'} = 0$ then overlapping of Ω_q and $\Omega_{q'}$ is empty.

The idea of DDM includes the definition of sets of boundary value problems for all subdomains which should be equivalent to the original problem (1):

$$\begin{aligned}
Lu_q(\mathbf{r}) &= f_q, \quad \mathbf{r} \in \Omega_q, \quad l_{q,q'}(u_q)\big|_{\Gamma_{q,q'}} = g_{q,q'} \equiv l_{q'.q}(u_{q'})\big|_{\Gamma_{q'.q}}, \\
q' &\in \omega_q, \quad l_{q,0} u_q|_{\Gamma_{q,0}} = g_{q,0}, \quad q = 1, \ldots, P.
\end{aligned} \tag{4}$$

At each segment of the internal boundaries of subdomains, the interface conditions in the form of the Robin boundary condition are imposed:

$$\alpha_q u_q + \beta_q \frac{\partial u_q}{\partial \mathbf{n}_q}\Big|_{\Gamma_{q,q'}} = \alpha_{q'} u_q + \beta_{q'} \frac{\partial u_{q'}}{\partial \mathbf{n}_{q'}}\Big|_{\Gamma_{q',q}},$$
$$|\alpha_q| + |\beta_q| > 0, \quad \alpha_q \cdot \beta_q \geq 0. \tag{5}$$

Here $\alpha_{q'} = \alpha_q, \beta_{q'} = \beta_q$ and $\mathbf{n}_q$ means the outer normal to the boundary segment $\Gamma_{q,q'}$ of the subdomain Ω_q.

We consider the iterative additive Schwarz method which can be interpreted as a sequential recomputation of the boundary condition:

$$L u_q^n = f_q, \quad l_{q,q'} u_q^n|_{\Gamma_{q,q'}} = l_{q',q} u_{q'}^{n-1}|_{\Gamma_{q',q}}. \tag{6}$$

In order to solve the considered problem numerically we need to perform its discretization. We introduce the grid computational domain Ω^h which consists of a set of the numbered nodes Q_l, $l = 1, \ldots, N$, where N is the total number of mesh points. Then we divide Ω^h into P grid subdomains Ω_q^h

$$\bar{\Omega}^h = \bigcup_{q=1}^{P} \bar{\Omega}_q^h, \quad \bar{\Omega}^h = \Omega_h \bigcup \Gamma^h, \quad \bar{\Omega}_q^h = \Omega_q^h \bigcup \Gamma_q^h, \tag{7}$$

In the case of a non-overlapping decomposition, for $q' \neq q''$ we have $\Omega_{q'}^h \bigcap \Omega_{q''}^h = \emptyset$, and $\Gamma_{q',q''}^h = \bar{\Omega}_{q'}^h \bigcap \bar{\Omega}_{q''}^h$ is the common boundary (a grid separator) between the contacting subdomains $\Omega_{q'}^h, \Omega_{q''}^h$.

After an approximation of the original continuous problem (1) on the non-structured grid Ω^h, one can obtain a SLAE

$$Au \equiv \sum_{l' \in \bar{\omega}_l} a_{l,l'} u_{l'} = f, \quad A = \{a_{l,l'}\} \in \mathcal{R}^{N,N}, \quad u = \{u_l\}, \quad f = \{f_l\} \in \mathcal{R}^N, \tag{8}$$

where the matrix A is supposed to be invertible and nonsymmetric in general. We consider the nodal grid equations only, i.e. each vector component u_l or f_l corresponds to some mesh point $Q_l \in \Omega^h$. Here $\bar{\omega}_l$ is the stencil of the grid point Q_l, and $N_{\omega_l} \ll N$ is the corresponding number of the neighbouring nodes. Also, we denote by N_q and $N_{q,q'}$ the numbers of the grid nodes in the grid subdomain Ω_q^h and the boundary segment $\Gamma_{q,q'}^h$ respectively.

3 Deflated DDM in Krylov Subspaces

From here after, we consider a decomposition of the grid computational domain without mesh separators. This means that the continuous internal boundaries $\Gamma_{q,q'}$ for $q \neq 0$ do not contain mesh points, and $\Gamma_{q,q'}^h \neq \Gamma_{q',q}^h$.

If we denote by $\hat{u}_q, \hat{f}_q \in \mathcal{R}^{N_q}$, $q = 1, \ldots, P$ the subvectors corresponding to a subdomain Ω_q, the system (8) can be written in the following block form

$$A_{q,q}\hat{u}_q = f_q - \sum_{r \in \omega_q} A_{q,r}\hat{u}_r \equiv \hat{f}_q, \quad A_{q,r} \in \mathcal{R}^{N_q, N_r}, \quad q = 1, \ldots, P. \tag{9}$$

The additive Schwarz method is then described by the following formula:

$$\begin{aligned} B_{q,q}\hat{u}_q^n &\equiv (A_{q,q} + C_{q,q})\hat{u}_q^n = \\ &= f_q + C_{q,q}\hat{u}_q^{n-1} - \sum_{r \in \omega_q} A_{q,r}\hat{u}_r^{n-1}, \quad n = 1, 2, \ldots \end{aligned} \tag{10}$$

Here we suppose that the preconditioning matrices $B_{q,q}$ are nonsingular ones and hence for $n \to \infty$ the iterative process (10) converges to a unique solution $u = \{\hat{u}_q\}$ of SLAE (8). The matrix $C_{q,q}$ in (10) is responsible for the interface condition between the subdomains and has nonzero entries for the near-boundary nodes of Ω_q only.

In the case of a decomposition without overlapping, the global solution vector is the direct sum of its subvectors, i.e. $u = \hat{u}_1 \oplus \ldots \oplus \hat{u}_P$. In general, the formulae of the iterative method within the Schwarz approach can differ from that above, and we use RAS (Restricted Additive Schwarz, see Dolean et al. 2015; Toselli and Widlund 2005) for a definition of the iterative process. Here we have to construct the grid domain decomposition in two steps. Firstly, we define a decomposition into some non-intersected subdomains, see (7). Let us denote by Γ_q^0 the grid boundary of Ω_q^h and define an extended subdomain $\Omega_q^1 = \Omega_q^h \bigcup \Gamma_q^0 = \bar{\Omega}_q^h$. At the second step we extend each subdomain layer-by-layer and define a set of the embedded subdomains:

$$\begin{aligned} \Gamma_q &\equiv \Gamma_q^0 = \{l' \in \omega_l, \; l \in \Omega_q, \; l' \notin \Omega_q, \; \Omega_q^1 = \bar{\Omega}_q^0 = \Omega_q \bigcup \Gamma_q^0\}, \\ \Gamma_q^t &= \{l' \in \omega_l, \; l \in \Omega_q^{t-1}, \; l' \in \Omega_q^{t-1}, \; \Omega_q^t = \bar{\Omega}_q^{t-1} = \Omega_q^{t-1}\Gamma_q^{t-1}\}, \\ t &= 1, \ldots, \Delta_q. \end{aligned} \tag{11}$$

Here Δ_q is a measure parameter of the extension of the subdomain $\Omega_q^{\Delta_q}$. The RAS iterative process can be described as $u_{RAS}^n = \{u_l^n, \; l \in \Omega_q^0\}$.

The conventional additive Schwarz (AS) method can be rewritten in more general form as

$$B_n(u^n - u^{n-1}) = f - Au^{n-1} \equiv r^{n-1}, \quad n = 1, 2, \ldots, \tag{12}$$

where the preconditioning matrix $B_n = \text{block-diag}\{B_{q,q}^n\}$ may be chosen differently at each iteration.

To solve SLAE (1), we apply a preconditioned iterative process in the Krylov subspaces instead of (12). In particular, we use multi-preconditioned semi-conjugate residual (MPSCR) method (Gurieva and Il'in, 2015), which is the unification of the ideas presented in Bridson and Greif (2006), Il'in and Itskovich (2007), Eisenstat et al. (1983) and Yuan et al. (2004). Let us have some rectangular matrices and vectors of iterative parameters

$$P_n = (p_1^n \ldots p_{m_n}^n) = \{p_k^n\} \in \mathcal{R}^{N,m_n}, \quad \bar{\alpha}_n = (\alpha_{n,1} \ldots \alpha_{n,m_n})^T = \{\alpha_k^n\} \in \mathcal{R}^{m_n}.$$

Then MPSCR iterations are defined by the recursions for $n = 0, 1, \ldots$:

$$r^0 = f - Au^0, u^{n+1} = u^n + P_n \bar{\alpha}_n, \quad r^{n+1} = r^n - AP_n \bar{\alpha}_n. \tag{13}$$

Let us suppose that at each n-th iteration we have m_n different nonsingular matrix preconditioners $B_n^{(k)}, k = 1, \ldots, m_n$. In this case the initial search vectors are chosen as $p_k^0 = (B_0^{(k)})^{-1} r^0$. Let these vectors be linearly independent and let the matrices P_n in (13) have full ranks m_n. Then under the orthogonality conditions

$$(Ap_k^n, Ap_{k'}^{n'}) = \rho_{n,k} \delta_{k,k'}, \quad \rho_{n,k} = (Ap_k^n, Ap_k^n), \tag{14}$$

where $\delta_{n,n'}$ is the Kronecker symbol, the formulas (13), with the coefficients

$$\alpha_k^n = (r^n, A(B_n^{(k)})^{-1} r^n)/\rho_{n,k}, \quad k = 1, \ldots, m_n, \tag{15}$$

provide the minimal norm $\|r^n\|$ of the residual in the block Krylov subspaces $\text{Span}\{AP_1, \ldots, AP_n\}$. The matrices $P_i, i = 1, \ldots, n+1$, are defined as

$$P_{n+1} = Q_{n+1} - \sum_{k=0}^{n} \sum_{l=0}^{m_k} \beta_{k,l}^n p_l^k, \quad Q_{n+1} = \{q_k^{n+1} = (B_{n+1}^{(k)})^{-1} r^{n+1}\},$$

$$\beta_{k,l}^n = (Ap_l^k, A(B_n^{(k)})^{-1} r^n)\rho_{n,l}, \quad k = 1, \ldots, m_n. \tag{16}$$

We apply MPSCR method with two types of preconditioners ($B_n^{(s)}$ and $B_n^{(c)}$) at each iteration. The first one corresponds to the block Jacobi–Schwarz preconditioner from (10) and (12), and the second one is responsible for a coarse grid correction, or aggregation, or deflation approach (Dolean et al., 2015; Toselli and Widlund, 2005). This procedure is based on the low rank approximation of the original matrix A (Gurieva and Il'in, 2015):

$$(B_n^{(c)})^{-1} \equiv \tilde{A}_n = W_n \hat{A}_n^{-1} W_n^T, \quad \hat{A}_n = W_n^T A W_n \in \mathcal{R}^{N_n^{(c)}, N_n^{(c)}},$$

$$W_n = (w_1 \ldots w_{N_n^{(c)}}) \in \mathcal{R}^{N, N_n^{(c)}}, \quad N_n^{(c)} \ll N. \tag{17}$$

Here W_n are some full rank rectangular matrices whose columns consist of the entries presenting the values of the finite basis functions $w_q(\mathbf{r})$ defined at some coarse grid with the number of the macro-nodes $N_n^{(c)} \ll N$ (this number can have different value at different iterations). This macrogrid can be independent of the domain decomposition, but we use $N_n^{(c)} = P$ and $w_q(\mathbf{r})$ with the entries equal one in Ω_q and the zero entries in other subdomains.

One disadvantage of SCR is the long recursions and high memory requirements to compute the search vectors p_k^n. More lightweight approach is in an application of the BiCGStab (Saad, 2003) with a deflation to improve the residual at the first iteration only. Having initial guess u^{-1}, we compute

$$u^0 = u^{-1} + (B_0^{(c)})^{-1} r^{-1}, \quad r^{-1} = f - Au^{-1},$$
$$r^0 = f - Au^0, \quad p^0 = r^0 - (B_0^{(c)})^{-1} r^0,$$
(18)

where $B_0^{(c)}$ is defined by (17). This trick provides the orthogonality properties $W_0^T r^0 = 0$, $W_0^T A p^0 = 0$. The next iterations are implemented by the corresponding steps of the conventional BiCGStab method.

4 Numerical Experiments

Consider solving a model Dirichlet boundary value problem for 2D and 3D diffusion-convection equation with constant coefficients p, q, r:

$$\frac{\partial^2 u}{\partial x^2} + \frac{\partial^2 u}{\partial y^2} + \frac{\partial^2 u}{\partial z^2} + p \frac{\partial u}{\partial x} + q \frac{\partial u}{\partial y} + r \frac{\partial u}{\partial z} = f(x, y, z),$$
$$(x, y, z) \in \Omega, \ u|_\Gamma = g(x, y, z), \ \Omega = [0, 1]^3.$$
(19)

Problem (19) is discretized by the monotone exponential finite volume scheme (Il'in, 2003) on a square (cubic) mesh with $N = N_x^d$ degrees of freedom, for different values of N_x. The stopping criterion for external iterations was $||r^n|| \leq \varepsilon^e = 10^{-7}$. All the experiments were carried out on the hybrid cluster (NKS-30T, 2017) where every MPI process was run on Intel Xeon E5450 processor.

The implementation of DDM was made via the hybrid programming with two levels of a parallelization. At the upper level, the iterative Krylov process over P subdomains has been organized on the basis of MPI approach which forms one MPI-process for every subdomain and provides data communications. The auxiliary SLAEs in subdomains were solved by PARDISO from Intel MKL which uses multithreading, thus giving one more level of parallelism.

Table 1 presents the results for the 2D problem (19) solved by the deflated BiCGStab-DDM method at the upper level of the iterative process with the Dirichlet interface condition. Acceleration of the method was done only before the iterations by the procedure (18). The boundary conditions and the right hand side were chosen in accordance with the known exact solution $u(x, y) = 3xy^2 - x^3$. The experiments were made on the square macro-grid of P^2 equal subdomains, with the number of $(N/P)^2$ mesh points in each subdomain. Here the number of iterations are given for the grids with the numbers of their points $N = 64^2, 128^2, 256^2$. Each four columns stand for the case without deflation, the case with the piece-wise constant, the linear and the quadratic basis functions w_k taken for the deflation matrices $W_0 \in \mathcal{R}^{N,P}$, respectively. Zero initial guess and overlapping parameter $\Delta = 0, 1, 2, 3$ were taken.

As we can see from these results, an application of the coarse grid correction gives the considerable improvement of the BiCGStab method, for different values of

Table 1 The numbers of iterations for BiCGStab method (2D problem) for different grids, macrogrids and basis functions in the deflation matrix, $\Delta = 0, 1, 2, 3$, $p = q = 4$

N	Δ	p^2											
		2^2				4^2				8^2			
64^2	0	19	21	23	17	27	27	25	19	38	34	33	26
	1	12	12	12	10	18	16	15	13	21	20	19	14
	2	9	10	9	8	13	13	11	11	17	16	14	11
	3	8	8	8	7	10	12	9	9	13	13	12	10
128^2	0	27	29	31	22	43	41	36	26	51	46	44	38
	1	16	18	18	14	24	22	21	17	30	27	25	16
	2	13	14	13	12	19	18	17	14	23	21	21	15
	3	11	12	11	10	15	15	14	12	19	18	16	11
256^2	0	42	35	46	35	65	52	45	33	98	73	65	32
	1	22	24	22	19	32	30	30	22	43	39	38	31
	2	17	20	19	14	26	25	22	18	34	31	30	24
	3	15	18	17	13	22	21	20	15	28	25	25	20

Table 2 The number of iterations and run times for SCR method with coarse grid corrections at every 5-th iteration and for block algorithm MPSCR, $p = q = r = 0$, $\Delta = 0$

N	Method	P				
		4	8	16	32	64
32^3	SCR	520.34	590.27	590.23	660.30	700.42
	MPSCR	450.48	540.34	540.32	620.38	670.48
64^3	SCR	664.81	822.71	1011.96	1021.72	1052.07
	MPSCR	595.35	703.18	852.39	982.32	1092.66
128^3	SCR	114,217.2	13,272.5	13,333.1	15,122.3	15,020.6
	MPSCR	101,226.3	11,179.1	13,443.2	15,632.8	15,930.7

coefficients p, q for the single usage of the acceleration before the first iteration only. Moreover, the efficiency of the deflation procedure increases when the smoothness of the basis functions grows. Another way to decrease the number of iterations is to use small subdomain overlapping, $\Delta = 1, 2, 3$. However, for big Δ values, the solution of BVPs in the subdomains becomes too expensive, and so we have the optimal parameters $\Delta \approx 4$, in the sense of the run time. These effects are especially valuable for the big numbers of subdomains and the degrees of freedom of the SLAE.

The second set of experiments is devoted to application of the SCR method with two preconditioners $B_n^{(s)}$ and $B_n^{(c)}$, the latter one formed using piecewise constant basis functions. Here we solved 3D Laplace equation ($p = q = r = f = 0$) in (19) with the exact solution $u = x^2 + y^2 + z^2$ and the initial guess $u^0 = 0$. Also, the domain decomposition was carried out without overlapping of the subdomains, with the Dirichlet interface conditions. In each cell of Table 2 we present the number of iterations and the run time for the grids $N = 32^3, 64^3, 128^3$, and for the number of subdomains (it is equal to the number of MPI-processes) $P = 4, 8, 16, 32, 64$. The results for the second set of experiments indicate that it may not be advantageous to employ coarse grid correction at every step of an iterative process, especially if low-order basis functions are used. This observation also correlates with the results obtained in the first set of experiments.

5   Conclusion

The numerical results presented demonstrate that multi-preconditioned DDM in the Krylov subspaces have reasonable efficiency. Our main goal is to investigate the scalability of parallel DDM with application of multi-preconditioned SCR iterative process and the coarse grid correction approach with different order of basis functions. Our numerical experiments with the proposed approaches have shown valuable improvement of the methods' behaviour for the test problems considered. However, further experimental investigations are needed to understand the properties of the algorithms and to arrive at a robust high-performance code and to define a niche of the approaches presented when used for some particular applied problems.

References

R. Bridson, C. Greif, A multipreconditioned conjugate gradient algorithm. SIAM J. Matrix Anal. Appl. **27**(4), 1056–1068 (2006)

V. Dolean, P. Jolivet, F. Nataf, *An Introduction to Domain Decomposition Methods: Algorithms, Theory, and Parallel Implementation*, vol. 144 (Society for Industrial and Applied Mathematics, Philadelphia, 2015)

O. Dubois, M.J. Gander, S. Loisel, A. St-Cyr, D.B. Szyld, The optimized Schwarz method with a coarse grid correction. SIAM J. Sci. Comput. **34**(1), A421–A458 (2012)

S.C. Eisenstat, H.C. Elman, M.H. Schultz, Variational iterative methods for nonsymmetric systems of linear equations. SIAM J. Numer. Anal. **20**(2), 345–357 (1983)

Y.L. Gurieva, V.P. Il'in, Parallel approaches and technologies of domain decomposition methods. J. Math. Sci. **207**(5), 724–735 (2015)

V.P. Il'in, On exponential finite volume approximations. Russ. J. Numer. Anal. Math. Model. **18**(6), 479–506 (2003)

V.P. Il'in, E.A. Itskovich, Semi-conjugate direction methods with dynamic preconditioning. Sibirskii Zhurnal Industrial'noi Matematiki **10**(4), 41–54 (2007)

NKS-30T, http://www2.sscc.ru/ENG/Resources.htm (2017)

Y. Saad, *Iterative Methods for Sparse Linear Systems*, 2nd edn. (Society for Industrial and Applied Mathematics, Philadelphia, PA, 2003)

A. Toselli, O. Widlund, *Domain Decomposition Methods—Algorithms and Theory*. Springer Series in Computational Mathematics, vol. 34 (Springer, Berlin, 2005)

J.Y. Yuan, G.H. Golub, R.J. Plemmons, W.A.G. Cecílio, Semi-conjugate direction methods for real positive definite systems. BIT Numer. Math. **44**(1), 189–207 (2004)

Parallel Overlapping Schwarz
with an Energy-Minimizing Coarse Space

Alexander Heinlein, Axel Klawonn, and Oliver Rheinbach

1 Introduction and Description of the Method

The GDSW preconditioner is a two-level overlapping Schwarz preconditioner introduced in Dohrmann et al. (2008a) with a proven condition number bound for the general case of John domains for scalar elliptic and linear elasticity model problems. It is algebraic in the sense that it can be constructed from the assembled system matrix. However, compared to FETI-DP (see Toselli and Widlund 2005) or BDDC methods, in GDSW the standard coarse space is relatively large, especially in three dimensions. In Dohrmann and Widlund (2010), a related hybrid preconditioner with a reduced coarse problem for three-dimensional elasticity was introduced. Here, the degrees of freedom (d.o.f.) corresponding to the faces are modified.

The GDSW preconditioner is a two-level additive overlapping Schwarz preconditioner with exact local solvers; cf. Toselli and Widlund (2005). It can be written as

$$M_{\mathrm{GDSW}}^{-1} = \Phi \left(\Phi^T A \Phi \right)^{-1} \Phi^T + \sum_{i=1}^{N} R_i^T \tilde{A}_i^{-1} R_i, \tag{1}$$

cf. Dohrmann et al. (2008b). The matrix Φ is the essential ingredient of the GDSW preconditioner. It is composed of coarse space functions which are discrete

A. Heinlein (✉) • A. Klawonn
Mathematisches Institut, Universität zu Köln, Weyertal 86-90, 50931 Köln, Germany
e-mail: alexander.heinlein@uni-koeln.de; axel.klawonn@uni-koeln.de

O. Rheinbach
Institut für Numerische Mathematik und Optimierung, Fakultät für Mathematik und Informatik, Technische Universität Bergakademie Freiberg, Akademiestr. 6, 09596 Freiberg, Germany
e-mail: oliver.rheinbach@math.tu-freiberg.de

© Springer International Publishing AG 2017
C.-O. Lee et al. (eds.), *Domain Decomposition Methods in Science and Engineering XXIII*, Lecture Notes in Computational Science and Engineering 116, DOI 10.1007/978-3-319-52389-7_36

harmonic extensions from the interface to the interior degrees of freedom of nonoverlapping subdomains. The values on the interface are restrictions of the nullspaces of the operator to the interface.

For $\Omega \subset \mathbb{R}^2$ being decomposed into John domains, the condition number of the GDSW preconditioner is bounded by

$$\kappa \left(M_{GDSW}^{-1} K \right) \le C \left(1 + \frac{H}{\delta} \right) \left(1 + \log \left(\frac{H}{h} \right) \right)^2, \tag{2}$$

cf. Dohrmann et al. (2008a) and Dohrmann et al. (2008b). Here, H is the size of a subdomain, h is the size of a finite element, and δ is the overlap.

Implementation Our parallel implementation of the GDSW preconditioner is based on Trilinos version 12.0; cf. Heroux et al. (2005). For the mesh partitioning, we use ParMETIS, cf. Karypis et al. (2011), the problems corresponding to the local level are solved using UMFPACK, cf. Davis and Duff (1997) (version 5.3.0), and the coarse level is solved using Mumps, cf. Amestoy et al. (2001) (version 4.10.0), in parallel mode. For the finite element implementation, we use the library LifeV; see Formaggia et al. (2016) (version 3.8.8).

On the JUQUEEN BG/Q supercomputer, we use the clang compiler 4.7.2 and ESSL 5.1 when compiling Trilinos and the GDSW preconditioner implementation. On the Cray XT6m at Universität Duisburg-Essen, we use the Intel compiler 11.1 and the Cray Scientific Library (libsci) 10.4.4.

2   Model Problems

We consider model problems in two and three dimensions, i.e. $\Omega = [0, 1]^2$ or $\Omega = [0, 1]^3$. The domain is decomposed either in a structured way, i.e., into squares or cubes, or in an unstructured way, using ParMETIS.

Laplacian in 2D The first model problem is: find $u \in H^1(\Omega)$

$$\begin{aligned} -\Delta u &= 1 && \text{in } \Omega, \\ u &= 0 && \text{on } \partial\Omega. \end{aligned} \tag{3}$$

Linear Elasticity in 2D and 3D The second model problem is: find $u \in (H^1(\Omega))^2$;

$$\begin{aligned} \operatorname{div} \sigma &= f && \text{in } \Omega, \\ \mathbf{u} &= 0 && \text{on } \partial\Omega_D = \partial\Omega \cap \{x = 0\} \end{aligned} \tag{4}$$

where $\sigma = 2\mu\varepsilon + \lambda \operatorname{trace}(\varepsilon)I$ is the stress and $\varepsilon = \frac{1}{2}(\nabla u + (\nabla u)^T)$ the strain. The Lamé parameters are $\lambda = 1/2.6$ and $\mu = 0.3/0.52$.

3 Numerical Results

We first show parallel scalability results in two and three dimensions. Finally, we show an application of the preconditioner within a block preconditioner in monolithic fluid-structure interaction. The model problems are discretized using piecewise quadratic (P2) finite elements. Our default Krylov method is GMRES and will be used also for the symmetric positive definite model problems. Our stopping criterion is the relative criterion $\left\| r^{(k)} \right\|_2 / \left\| r^{(0)} \right\|_2 \leq 10^{-7}$ with $r^{(0)}$ and $r^{(k)}$ being the initial and the k-th residual, respectively. In our experiments, each subdomain is assigned to one processor core.

Weak Scalability in 2D We use five different meshes with $H/h = 100$ and an increasing number of subdomains; see Tables 1 and 2. The results of weak scaling tests from 4 to 1024 processor cores for both model problems and an overlap $\delta = 1h$ or $\delta = 2h$ are presented in Figs. 1 and 2. The GDSW preconditioner

Table 1 Number of degrees of freedom of the total mesh, coarse and local space dimensions of the GDSW preconditioner for the weak scaling tests in Fig. 1

# Subdomains	4	16	64	256	1024
Total problem, P2 finite elements	160,801	641,601	2,563,201	10,246,401	40,972,801
Avg. first level, P2, overlap 1h	41,207.5	41,612.6	41,815.7	41,917.3	41,968.1
Avg. first level, P2, overlap 2h	42,020	42,837.8	43,248.7	43,454.7	43,557.8
Coarse level	5	33	161	705	2945
Avg. first level, P2, overlap 1h (ParMETIS)	41,581.5	41,841.9	42,101.8	42,225.7	42,263.1
Avg. first level, P2, overlap 2h (ParMETIS)	42,686.5	43,243.7	43,752.9	43,999.4	44,077.9
Coarse level (ParMETIS)	3	45	241	1129	4822

Table 2 Number of degrees of freedom of the total mesh, coarse and local space dimensions of the GDSW preconditioner for the weak scaling tests in Figs. 2 and 3

# Subdomains	4	16	64	256	1024
Total problem, P2	321,602	1,286,408	5,126,402	20,492,802	81,945,602
Avg. first level, P2, overlap 1h	82,415	83,225.2	83,631.3	83,834.6	83,936.3
Avg. first level, P2, overlap 2h	84,040	85,675.5	86,497.4	86,909.3	87,115.6
Coarse level	14	90	434	1890	7874
Coarse level, no rotations	10	66	322	1410	5890
Avg. first level, P2, overlap 1h (ParMETIS)	83,163	83,683.9	84,203.6	84,451.3	84,526.2
Avg. first level, P2, overlap 2h (ParMETIS)	85,373	86,487.4	87,505.8	87,998.7	88,155.9
Coarse level (ParMETIS)	9	120	633	2950	12,567
Coarse level, no rotations (ParMETIS)	6	90	482	2258	9644

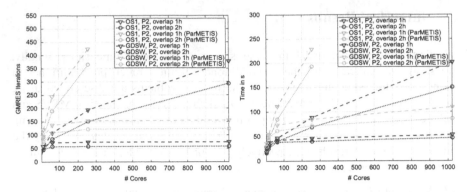

Fig. 1 Weak scaling for the Laplacian model problem in 2D, cf. (3), using P2 finite elements: number of iterations (*left*), runtimes (*right*). For the structured and the unstructured decomposition (ParMETIS), we have approximately 40,000 d.o.f. per subdomain

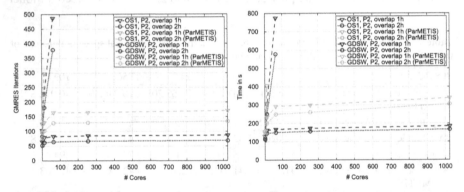

Fig. 2 Weak scaling for the linear elastic model problem in 2D, cf. (4), using P2 finite elements: number of iterations (*left*), runtimes (*right*). For the structured and the unstructured decomposition (ParMETIS), we have approximately 80,000 d.o.f. per subdomain

is numerically and parallel scalable, i.e., the number of iterations is bounded, both, for structured and unstructured decompositions, and the time to solution grows only slowly. The one-level preconditioner (OS1) does not scale numerically, and the number of iterations grows very fast. Indeed, for the unstructured decomposition, no convergence is obtained for OS1 within 500 iterations for more than 256 subdomains for the scalar problem and for more that 16 subdomains for elasticity. This is, of course, also due to the comparably small overlap. As a result of the better constant in (2), for the GDSW preconditioner, we observe better convergence for structured decompositions. Note that for the case of four subdomains the overlapping subdomains are significantly smaller.

A detailed analysis of different phases of the method is presented for linear elasticity in 2D in Fig. 3. We consider the standard full GDSW coarse space and the GDSW coarse space without rotations, i.e., the rotations are omitted from the

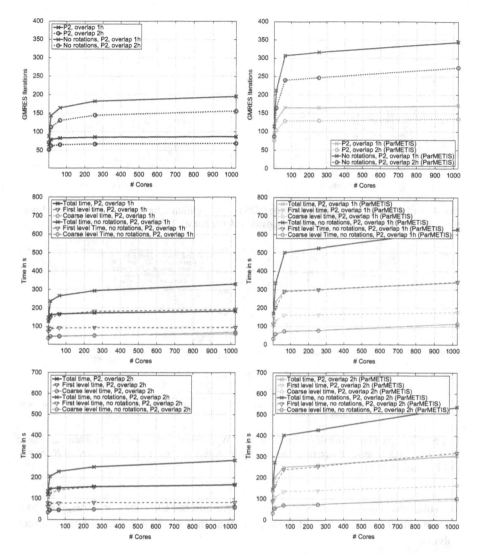

Fig. 3 Weak parallel scalability using the GDSW preconditioner for the model problem of linear elasticity in 2D, cf. (4): structured (*left*) and unstructured decomposition (*right*); number of iterations (*top*), timings for overlap $\delta = 1$ h (*middle*), and timings for overlap $\delta = 2$ h (*bottom*). For the structured and the unstructured decomposition (ParMETIS) we use a subdomain size of roughly 40,000 degrees of freedom

coarse space. This latter case is not covered by the bound (2), but the results indicate numerical and parallel scalability.

Strong Scalability in 2D Results for strong parallel scaling tests are shown in Fig. 4 for linear elasticity in 2D. We observe very good strong scalability for structured and unstructured domain decompositions. Note that the number of d.o.f.

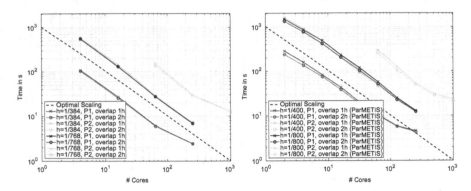

Fig. 4 Strong parallel scalability using the GDSW preconditioner for the model problem of linear elasticity in 2D, cf. (4): structured decomposition (*left*), ParMETIS decomposition (*right*)

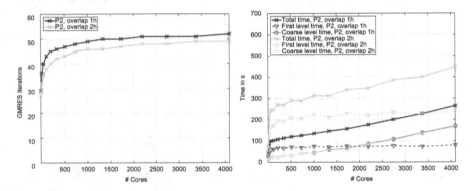

Fig. 5 Weak parallel scalability using the GDSW preconditioner for the problem of linear elasticity in 3D: number of iterations (*left*), timings (*right*). We use a subdomain size of $H/h = 6$ and P2 finite elements

per subdomain decreases when increasing the number of processor cores, and, to a certain extent, we thus benefit from an increasing speed of the local sparse direct solvers.

Weak Scalability for Linear Elasticity in 3D We present results of weak scalability runs for a linear elastic model problem in 3D from 8 to 4096 cores. We consider a structured decomposition of a cube and use the full GDSW coarse space in 3D. In Fig. 5, we present the number of iterations and the timings using P2 elements using an overlap δ of one or two elements. The number of iterations seems to be bounded by a constant number, whereas the solution times increases, i.e., the cost of the (parallel) sparse direct solver used for the coarse problem is noticeable in 3D.

Application in Fluid-Structure Interaction (FSI) We consider time-dependent monolithic FSI as in Balzani et al. (2015) but using a fully implicit scheme as in Deparis et al. (2015) and Heinlein et al. (2015). We apply a monolithic

Dirichlet-Neumann preconditioner applying the GDSW preconditioner for the structural block; see Balzani et al. (2015) and Heinlein et al. (2015) and the references therein. We use a pressure wave inflow condition for a tube using Mesh #1 from Heinlein et al. (2015). We consider a Neo-Hookean material for the tube; as opposed to Heinlein et al. (2015), we here use a fixed time step of 0.0005 s and show the runtimes during the simulation.

In Fig. 6, the runtimes of ten time steps using 128 cores of the Cray XT6m at Universität Duisburg-Essen are shown. We compare IFPACK, a one-level algebraic overlapping Schwarz preconditioner from Trilinos, our geometric one-level Schwarz preconditioner (OS1), the GDSW preconditioner without rotations (GDSW-nr), and the standard GDSW preconditioner for the structural block. We see that, although the computing times vary over the simulation time, the combination of the geometric overlap and a sufficiently large coarse space consistently reduces the runtime of the fully coupled monolithic FSI simulation by a factor of about two compared to the baseline given by IFPACK. Figure 7 shows the pressure and the deformation at $t = 0.007$ s where we have the largest computation time per timestep, cf. Fig. 6.

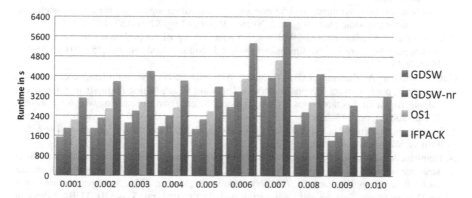

Fig. 6 Runtimes for the monolithic FSI simulation. For clarity, the runtimes of two subsequent time steps of size $\Delta t = 0.0005$ s are combined. The monolithic system has approximately 1.2 million d.o.f. We use a Neo-Hookean material. "OS1" is the one-level Schwarz preconditioner, "GDSW-nr" is the GDSW preconditioner without rotations, and "GDSW" is the GDSW preconditioner with full coarse space

Fig. 7 Pressure and deformation at time $t = 0.007$ s. The deformation is magnified by a factor of 10

Acknowledgements The authors acknowledge the use of the JUQUEEN BG/Q supercomputer (Stephan and Docter, 2015) at JSC Jülich, the use of the Cray XT6m at Universität Duisburg-Essen and the financial support by the German Science Foundation (DFG), project no. KL2094/3 and RH122/4.

References

P.R. Amestoy, I.S. Duff, J.-Y. L'Excellent, J. Koster, A fully asynchronous multifrontal solver using distributed dynamic scheduling. SIAM J. Matrix Anal. Appl. **23**(1), 15–41 (2001)

D. Balzani, S. Deparis, S. Fausten, D. Forti, A. Heinlein, A. Klawonn, A. Quarteroni, O. Rheinbach, J. Schröder, Numerical modeling of fluid-structure interaction in arteries with anisotropic polyconvex hyperelastic and anisotropic viscoelastic material models at finite strains. Int. J. Numer. Methods Biomed. Eng. (2015). ISSN 2040-7947. http://dx.doi.org/10.1002/cnm.2756.

T.A. Davis, I.S. Duff, An unsymmetric-pattern multifrontal method for sparse LU factorization. SIAM J. Matrix Anal. Appl. **18**(1), 140–158 (1997)

S. Deparis, D. Forti, G. Grandperrin, A. Quarteroni, FaCSI: a block parallel preconditioner for fluid-structure interaction in hemodynamics, Technical Report 13, MATHICSE, EPFL, Lausanne, 2015

C.R. Dohrmann, O.B. Widlund, Hybrid domain decomposition algorithms for compressible and almost incompressible elasticity. Int. J. Numer. Methods Eng. **82**(2), 157–183 (2010)

C.R. Dohrmann, A. Klawonn, O.B. Widlund, Domain decomposition for less regular subdomains: overlapping Schwarz in two dimensions. SIAM J. Numer. Anal. **46**(4), 2153–2168 (2008a). ISSN 0036-1429

C.R. Dohrmann, A. Klawonn, O.B. Widlund, A family of energy minimizing coarse spaces for overlapping Schwarz preconditioners, in *Domain Decomposition Methods in Science and Engineering XVII*. Lecture Notes in Computational Science and Engineering, vol. 60 (Springer, Berlin, 2008b), pp. 247–254

L. Formaggia, M. Fernandez, A. Gauthier, J.F. Gerbeau, C. Prud'homme, A. Veneziani, The LifeV Project. Web. http://www.lifev.org (2016)

A. Heinlein, A. Klawonn, O. Rheinbach, Parallel two-level overlapping Schwarz methods in fluid-structure interaction, in *Proceedings of the European Conference on Numerical Mathematics and Advanced Applications (ENUMATH)*, Ankara, September, 2015. Springer Lecture Notes on Computational Science and Engineering, vol. 112 (2016), pp. 521–530. TUBAF Preprint 15/2015: http://tu-freiberg.de/fakult1/forschung/preprints

M.A. Heroux, R.A. Bartlett, V.E. Howle, R.J. Hoekstra, J.J. Hu, T.G. Kolda, R.B. Lehoucq, K.R. Long, R.P. Pawlowski, E.T. Phipps, A.G. Salinger, H.K. Thornquist, R.S. Tuminaro, J.M. Willenbring, A. Williams, K.S. Stanley, An overview of the Trilinos project. ACM Trans. Math. Softw. **31**(3), 397–423 (2005)

G. Karypis, K. Schloegel, V. Kumar, ParMETIS - Parallel graph partitioning and sparse matrix ordering. Version 3.2, Technical Report, University of Minnesota, Department of Computer Science and Engineering, April 2011

M. Stephan, J. Docter, JUQUEEN: IBM Blue Gene/Q® Supercomputer System at the Jülich Supercomputing Centre. J. Large-Scale Res. Facil. **1**, A1 (2015). ISSN 2364-091X. doi:10.17815/jlsrf-1-18. http://dx.doi.org/10.17815/jlsrf-1-18

A. Toselli, O. Widlund, *Domain Decomposition Methods—Algorithms and Theory*. Springer Series in Computational Mathematics, vol. 34 (Springer, Berlin, 2005). ISBN 3-540-20696-5

Volume Locking Phenomena Arising in a Hybrid Symmetric Interior Penalty Method with Continuous Numerical Traces

Daisuke Koyama and Fumio Kikuchi

1 Introduction

When we compute numerical solutions of linear elasticity problems for nearly incompressible materials by using the P_1 conforming finite element method, we need to use sufficiently fine meshes in order to get numerical solutions with accuracy. This is referred to as *volume locking* (Babuška and Suri, 1992). It is well-known that discontinuous Galerkin (DG) methods are effective in eliminating locking (see, e.g., Hansbo and Larson 2002).

We investigate locking effects in a hybrid version of a symmetric interior penalty (SIP) method, which is one of DG methods, and is called the *HSIP* method in this paper. Unknowns in the HSIP method are approximations to the displacement of the elastic body and to the trace of the displacement on the skeleton. The latter is called the *numerical trace*. We consider two formulations of the HSIP method: the HSIP methods using discontinuous numerical traces (HSIP-D) and using continuous ones (HSIP-C). The degrees of freedom of the continuous numerical traces are less than those of the discontinuous ones. This gives the HSIP-C method an advantage over the HSIP-D method in practical computations. However, in Kikuchi (2015), it is numerically demonstrated that the HSIP-C method using P_1 elements for both the two unknowns causes volume locking phenomena. On the other hand, in Koyama and Kikuchi (2016), it is established that the HSIP-D is free from locking. In this paper, we mathematically prove that the HSIP-C method shows locking in the case

D. Koyama (✉)
The University of Electro-Communications, Chofugaoka 1-5-1, Chofu, Tokyo, Japan
e-mail: koyama@im.uec.ac.jp

F. Kikuchi
The University of Tokyo, Komaba 3-8-1, Meguro, Tokyo, Japan
e-mail: kikuchi@ms.u-tokyo.ac.jp

© Springer International Publishing AG 2017
C.-O. Lee et al. (eds.), *Domain Decomposition Methods in Science and Engineering XXIII*, Lecture Notes in Computational Science and Engineering 116, DOI 10.1007/978-3-319-52389-7_37

when P_1 elements are employed to approximate displacement and its trace on the skeleton.

We close this section with the introduction of several notations which will be used throughout this paper. For an arbitrary open subset Ω of $\mathbb{R}^2$, we denote by $L^2(\Omega)$ and by $H^s(\Omega)$ ($s > 0$) the usual space of real-valued square integrable functions on Ω and the real Sobolev space on Ω, respectively (see, e.g., Brenner and Scott 2008). We denote by $(\cdot, \cdot)_\Omega$ and by $\| \cdot \|_\Omega$ the inner product of $L^2(\Omega)$ and the associated norm, respectively. We equip $H^s(\Omega)$ with the usual norm denoted by $\| \cdot \|_{s,\Omega}$. We denote by $| \cdot |_{s,\Omega}$ the usual semi-norm of $H^s(\Omega)$. For the union Γ of arbitrary line segments in $\mathbb{R}^2$, we denote by $\langle \cdot, \cdot \rangle_\Gamma$ and by $| \cdot |_\Gamma$ the inner product of $L^2(\Gamma)$ and the associated norm, respectively. We use the same notations of the norm, the semi-norm, and the inner product for vector valued functions as well. In addition, C denotes a generic positive constant, and can be a different value at each of different places.

2   Linear Plane Strain Problem

For the two-dimensional displacement $\underline{u} = [u_1, u_2]^T$ of an elastic body, the strain tensor is given by $\underline{\underline{\varepsilon}}(\underline{u}) = \left[\frac{1}{2}\left(\partial u_i / \partial x_j + \partial u_j / \partial x_i\right)\right]_{1 \le i,j \le 2}$. We use an underline (resp. double underlines) to denote two dimensional vector (resp. 2×2 matrix) valued functions, operators, and their associated spaces. The isotropic linear elastic stress-strain relation is written by

$$\underline{\underline{\sigma}}(\underline{u}) = 2\mu\, \underline{\underline{\varepsilon}}(\underline{u}) + \lambda(\operatorname{div} \underline{u})\, \underline{\underline{\delta}},$$

where λ (> 0) and μ (> 0) are the Lamé parameters, and $\underline{\underline{\delta}}$ is the identity matrix. We consider the following linear plane strain problem:

$$\begin{cases} -\dfrac{\partial \sigma_{11}(\underline{u})}{\partial x_1} - \dfrac{\partial \sigma_{12}(\underline{u})}{\partial x_2} = f_1 \ \text{ in } \Omega, \\[2mm] -\dfrac{\partial \sigma_{21}(\underline{u})}{\partial x_1} - \dfrac{\partial \sigma_{22}(\underline{u})}{\partial x_2} = f_2 \ \text{ in } \Omega, \\[2mm] \hspace{3.5cm} \underline{u} = \underline{0} \ \text{ on } \partial\Omega, \end{cases} \tag{1}$$

where $\underline{\underline{\sigma}}(\underline{u}) = [\sigma_{ij}(\underline{u})]_{1 \le i,j \le 2}$, and $\underline{f} = [f_1, f_2]^T$ is a distributed external body force per unit in-plane area. We assume that Ω is a bounded polygonal domain of $\mathbb{R}^2$. In addition we fix $\mu > 0$.

3 The HSIP-D Method

Let $\mathcal{T}^h$ be a triangulation of Ω. We assume that $\mathcal{T}^h$ has no hanging nodes. The set of edges of $\mathcal{T}^h$ is denoted by $\mathcal{E}^h$. For each $K \in \mathcal{T}^h$, we define $\mathcal{E}^K := \{e \in \mathcal{E}^h \mid e \subset \partial K\}$. We define the skeleton Γ^h of $\mathcal{T}^h$ by $\Gamma^h := \bigcup_{e \in \mathcal{E}^h} \bar{e}$. The diameter of K is denoted by h_K, and the length of an edge $e \in \mathcal{E}^K$ by $|e|$. In addition, we set $h := \max_{K \in \mathcal{T}^h} h_K$. Assume that a family $\{\mathcal{T}^h\}_{h \in (0, \bar{h}]}$ of triangulations is regular.

The HSIP-D method seeks approximations to the solution $\underline{u}$ of (1) and to the trace of $\underline{u}$ on Γ^h by using functions belonging to

$$U^h := \prod_{K \in \mathcal{T}^h} P_k(K) \quad \text{and} \quad \widehat{U}^h := \prod_{e \in \mathcal{E}^h} P_k(e),$$

respectively, where P_k denotes the set of polynomial functions of order at most $k \geq 1$. So we consider their product space: $\underline{U}^h := U^h \times \widehat{U}^h \subset \underline{H}^1(\mathcal{T}^h) \times \underline{L}^2(\Gamma^h)$, where $H^s(\mathcal{T}^h) := \{v \in L^2(\Omega) \mid v|_K \in H^s(K) \; \forall K \in \mathcal{T}^h\}$ $(s > 0)$. We will denote the first and the second components of $\underline{v} \in \underline{H}^1(\mathcal{T}^h) \times \underline{L}^2(\Gamma^h)$ by $\underline{v}$ and $\hat{\underline{v}}$, i.e., $\underline{v} = \{\underline{v}, \hat{\underline{v}}\}$, unless specifically stated otherwise.

For each $K \in \mathcal{T}^h$ and for each $i = 1, 2$, we define local lifting operator R_i^K : $L^2(\partial K) \longrightarrow Q^K$ by $(R_i^K g, \varphi)_K = \langle g, \varphi n_i \rangle_{\partial K}$ for all $g \in L^2(\partial K)$ and for all $\varphi \in Q^K$, where $Q^K := P_{k-1}(K)$ and n_i is the ith component of the outward unit normal $\underline{n}$ on ∂K. We further define lifting operators $R_{\mathrm{div}}^K : \underline{L}^2(\partial K) \longrightarrow Q^K$ and $\underline{R}_{\underline{\varepsilon}}^K(\underline{g})$: $\underline{L}^2(\partial K) \longrightarrow \underline{Q}^K$ as follows (Kikuchi, 2015): $R_{\mathrm{div}}^K \underline{g} := \sum_{i=1}^2 R_i^K g_i$ and $\underline{R}_{\underline{\varepsilon}}^K(\underline{g}) := \left[\frac{1}{2}\left(R_i^K g_j + R_j^K g_i\right)\right]_{1 \leq i,j \leq 2}$ for $\underline{g} = [g_1, g_2]^T \in \underline{L}^2(\partial K)$.

We introduce the following three bilinear forms: for $\underline{u}, \underline{v} \in \underline{H}^2(\mathcal{T}^h) \times \underline{L}^2(\Gamma^h)$,

$$\tilde{a}_\eta^h(\underline{u}, \underline{v}) := 2\mu \sum_{K \in \mathcal{T}^h} \left[\left(\underline{\varepsilon}(\underline{u}), \underline{\varepsilon}(\underline{v})\right)_K + \left\langle \underline{\varepsilon}(\underline{u})\underline{n}, \hat{\underline{v}} - \underline{v} \right\rangle_{\partial K} \right.$$

$$\left. + \left\langle \hat{\underline{u}} - \underline{u}, \underline{\varepsilon}(\underline{v})\underline{n} \right\rangle_{\partial K} + \left(\underline{R}_{\underline{\varepsilon}}^K(\hat{\underline{u}} - \underline{u}), \underline{R}_{\underline{\varepsilon}}^K(\hat{\underline{v}} - \underline{v})\right)_K \right]$$

$$+ \eta \sum_{K \in \mathcal{T}^h} \sum_{e \in \mathcal{E}^K} \frac{1}{|e|} \langle \hat{\underline{u}} - \underline{u}, \hat{\underline{v}} - \underline{v} \rangle_e,$$

$$l^h(\underline{u}, \underline{v}) := \sum_{K \in \mathcal{T}^h} \left[(\mathrm{div}\,\underline{u}, \mathrm{div}\,\underline{v})_K + \langle (\mathrm{div}\,\underline{u})\underline{n}, \hat{\underline{v}} - \underline{v} \rangle_{\partial K} \right.$$

$$\left. + \langle \hat{\underline{u}} - \underline{u}, (\mathrm{div}\,\underline{v})\underline{n} \rangle_{\partial K} + \left(R_{\mathrm{div}}^K(\hat{\underline{u}} - \underline{u}), R_{\mathrm{div}}^K(\hat{\underline{v}} - \underline{v})\right)_K \right],$$

$$a_\eta^h(\underline{u}, \underline{v}) := \tilde{a}_\eta^h(\underline{u}, \underline{v}) + \lambda l^h(\underline{u}, \underline{v}), \tag{2}$$

where η is an interior penalty parameter ≥ 0, and $(\underline{\sigma}, \underline{\tau})_K := \sum_{i,j=1}^{2} \int_K \sigma_{ij}\tau_{ij} \, dx$ for $\underline{\sigma} = [\sigma_{ij}]_{1 \leq i,j \leq 2}$, $\underline{\tau} = [\tau_{ij}]_{1 \leq i,j \leq 2} \in \underline{L}^2(K)$.

We are now in a position to present a discrete problem, which provides the HSIP-D method: find $\underline{u}^h \in \underline{V}^h$ such that

$$a_\eta^h(\underline{u}^h, \underline{v}^h) = \left(\underline{f}, \underline{v}^h\right)_\Omega \quad \forall \underline{v}^h \in \underline{V}^h, \tag{3}$$

where $L_D^2(\Gamma^h) := \{\hat{v} \in L^2(\Gamma^h) \mid \hat{v} = 0 \text{ on } \partial\Omega\}$, $\widehat{V}^h := \widehat{U}^h \cap L_D^2(\Gamma^h)$, and $\underline{V}^h := \underline{U}^h \times \widehat{V}^h$.

Problem (3) has a unique solution for every $\underline{f} \in \underline{L}^2(\Omega)$ and for every $\eta > 0$ (see Koyama and Kikuchi 2016). Moreover the HSIP-D method is free from locking with respect to the solution set B_λ and the norm $\|\|\cdot\|\|_h$ in the sense of Babuška and Suri (1992) (see Koyama and Kikuchi 2016), where $B_\lambda := \{\underline{v} \in \underline{H}^2(\Omega) \cap \underline{H}_D^1(\Omega) \mid \|\underline{v}\|_{2,\Omega} + \lambda \|\operatorname{div}\underline{v}\|_{1,\Omega} \leq 1\}$, $H_D^1(\Omega) := \{v \in H^1(\Omega) \mid v = 0 \text{ on } \partial\Omega\}$, and

$$\|\|\underline{v}\|\|_h^2 := \sum_{K \in \mathcal{T}^h} \left[|\underline{v}|_{1,K}^2 + \sum_{e \in \mathcal{E}^K} \left(\frac{1}{|e|} |\hat{v} - \underline{v}|_e^2 + |e| \sum_{i,j=1}^{2} \left| \frac{\partial v_i}{\partial x_j} \right|_e^2 \right) \right].$$

We now introduce a semi-norm on $\underline{H}^1(\mathcal{T}^h) \times \underline{L}^2(\Gamma^h)$ as follows:

$$|\underline{v}|_h^2 := \sum_{K \in \mathcal{T}^h} \left(|\underline{v}|_{1,K}^2 + \sum_{e \in \mathcal{E}^K} \frac{1}{|e|} |\hat{v} - \underline{v}|_e^2 \right) \quad \forall \underline{v} \in \underline{H}^1(\mathcal{T}^h) \times \underline{L}^2(\Gamma^h).$$

This semi-norm can be a norm on $\underline{V}^h$ equivalent to $\|\|\cdot\|\|_h$, that is, there exists a positive constant C such that for all $h \in (0, \bar{h}]$ and for all $\underline{v}^h \in \underline{V}^h$,

$$C\|\|\underline{v}^h\|\|_h \leq |\underline{v}^h|_h \leq \|\|\underline{v}^h\|\|_h. \tag{4}$$

We define $\underline{\varepsilon}^h : \underline{U}^h \longrightarrow \underline{L}^2(\Omega)$ and $\mathbf{div}^h : \underline{U}^h \longrightarrow L^2(\Omega)$ as follows (Kikuchi, 2015): for every $\underline{v}^h \in \underline{U}^h$ and for every $K \in \mathcal{T}^h$,

$$\underline{\varepsilon}^h(\underline{v}^h)|_K := \underline{\varepsilon}(\underline{v}^h|_K) + \underline{\underline{R}}_\varepsilon^K(\hat{v}^h - \underline{v}^h),$$

$$\left(\mathbf{div}^h \, \underline{v}^h\right)|_K := \operatorname{div}(\underline{v}^h|_K) + R_{\operatorname{div}}^K(\hat{v}^h - \underline{v}^h). \tag{5}$$

For all $\underline{u}^h$, $\underline{v}^h \in \underline{U}^h$, we have

$$\tilde{a}_0^h(\underline{u}^h, \underline{v}^h) = 2\mu \left(\underline{\varepsilon}^h(\underline{u}^h), \underline{\varepsilon}^h(\underline{v}^h) \right)_\Omega, \tag{6}$$

$$l^h(\underline{u}^h, \underline{v}^h) = \left(\mathbf{div}^h \underline{u}^h, \mathbf{div}^h \underline{v}^h \right)_\Omega \quad (\text{see Kikuchi 2015}). \tag{7}$$

For all $\lambda > 0$, for all $\eta > 0$, for all $h \in (0, \bar{h}]$, and for all $\underline{v}^h \in \underline{V}^h$,

$$a_\eta^h(\underline{v}^h, \underline{v}^h) \geq \alpha \min\{1, \eta\} \|\|\underline{v}^h\|\|_h^2, \tag{8}$$

where α is a positive constant independent of λ, η, h, and $\underline{v}^h$ (see Koyama and Kikuchi 2016). Note that (8) holds for all $\eta > 0$ because bilinear form a_η^h includes the terms defined by lifting operators $\underline{R}_\varepsilon^K$ and R_{div}^K.

4 Volume Locking Phenomena in the HSIP-C Method

In this section, we fix η and assume that $k = 1$.

We introduce finite element spaces:

$$\begin{aligned}
U_c^h &:= U^h \cap H^1(\Omega), & V_c^h &:= U^h \cap H_D^1(\Omega), \\
\widehat{U}_c^h &:= \widehat{U}^h \cap C^0(\Gamma^h), & \widehat{V}_c^h &:= \widehat{U}_c^h \cap L_D^2(\Gamma^h), \\
\underline{U}_c^h &:= U^h \times \widehat{U}_c^h, & \underline{V}_c^h &:= U^h \times \widehat{V}_c^h.
\end{aligned}$$

Replacing $\underline{V}^h$ by $\underline{V}_c^h$ in (3), we can obtain the HSIP-C method.

We mathematically demonstrate that the HSIP-C method shows locking by following the method of proof due to Brenner and Scott (2008).

We can naturally identify $\widehat{U}_c^h$ with U_c^h, that is, there uniquely exists a linear operator $\mathcal{J}$ from $\widehat{U}_c^h$ onto U_c^h such that $\mathcal{J}\hat{v}_c^h = \hat{v}_c^h$ on ∂K for every $\hat{v}_c^h \in \widehat{U}_c^h$ and for every $K \in \mathcal{T}^h$.

Lemma 1 *There exists a positive constant C such that for all $h \in (0, \bar{h}]$, for all $\underline{v} \in \underline{H}^1(\Omega)$, and for all $\underline{v}^h \in \underline{U}_c^h$,*

$$|\underline{v} - \mathcal{J}\hat{v}^h|_{1,\Omega} \leq C|\underline{v} - \underline{v}^h|_h, \tag{9}$$

where $\underline{v} = \{\underline{v}, \underline{v}|_{\Gamma^h}\}$, and C is independent of h, $\underline{v}$, and $\underline{v}^h$.

Proof The usual scaling argument leads to that there exists a positive constant C such that for all $h \in (0, \bar{h}]$, for all $K \in \mathcal{T}^h$, and for all $v \in P_1(K)$,

$$\|v\|_{1,K} \leq C \left(\sum_{e \in \mathcal{E}^K} \frac{1}{|e|} |v|_e^2 \right)^{1/2}, \tag{10}$$

where C is independent of h, K, and v. For all $\underline{v} \in \underline{H}^1(\Omega)$ and for all $\underline{v}^h \in \underline{U}^h_c$,

$$|\underline{v} - \mathcal{J}\hat{\underline{v}}^h|^2_{1,\Omega} \leq 2 \sum_{K \in \mathcal{T}^h} \left(|\underline{v} - \underline{v}^h|^2_{1,K} + |\underline{v}^h - \mathcal{J}\hat{\underline{v}}^h|^2_{1,K} \right)$$

(by the triangle and the Schwarz inequalities)

$$\leq C \sum_{K \in \mathcal{T}^h} \left(|\underline{v} - \underline{v}^h|^2_{1,K} + \sum_{e \in \mathcal{E}K} \frac{1}{|e|} |\underline{v}^h - \hat{\underline{v}}^h|^2_e \right) \quad \text{(by (10))}.$$

This yields (9).   □

We now pose a hypothesis:

$$\{\underline{v}^h \in \underline{V}^h_c \mid \text{div}\, \underline{v}^h = 0\} = \{\underline{0}\}. \tag{L}$$

We understand from the following lemma that many triangulations satisfy (L) (cf. Brenner and Scott 2008, Exercise 11.x.14).

Lemma 2 *Let K_1 and K_2 be triangular elements whose vertices are $\{A, B, C\}$ and $\{B, C, D\}$, respectively. Let v^h_j ($j = 1, 2$) be continuous piecewise linear functions on $\overline{K_1 \cup K_2}$. Set $\underline{v}^h := [v^h_1, v^h_2]^T$. Assume that $\text{div}\, \underline{v}^h = 0$ and that $\underline{v}^h = \underline{0}$ on the sides AB and BD. If A, B, and D are not collinear, then $\underline{v}^h \equiv \underline{0}$ on $\overline{K_1 \cup K_2}$.*

We leave the proof to readers.

Lemma 3 *If (L) holds, then*

$$\text{Ker}(\mathbf{div}^h|_{\underline{V}^h_c}) = \{\{\underline{v}^h, \underline{0}\} \in \underline{V}^h_c \mid \underline{v}^h \in \underline{U}^h\}, \tag{11}$$

where $\mathbf{div}^h|_{\underline{V}^h_c}$ denotes the restriction of $\mathbf{div}^h$ to $\underline{V}^h_c$.

Proof We see from the Green formula that for every $\underline{v} \in \underline{P}_1(K)$,

$$\text{div}\, \underline{v} = R^K_{\text{div}}(\underline{v}) \quad \text{in } \mathbb{R}. \tag{12}$$

It follows from (5) and (12) that for all $\underline{v}^h \in \underline{U}^h$,

$$(\mathbf{div}^h\, \underline{v}^h)\big|_K = R^K_{\text{div}}(\hat{\underline{v}}^h) \quad \forall K \in \mathcal{T}^h. \tag{13}$$

This implies that $\mathbf{div}^h(\{\underline{v}^h, \underline{0}\}) = 0$ for every $\underline{v}^h \in \underline{U}^h$. Thus the right-hand side of (11) is included in $\text{Ker}(\mathbf{div}^h|_{\underline{V}^h_c})$.

Conversely, we suppose that $\underline{v}^h \in \underline{V}^h_c$ satisfies $\mathbf{div}^h\, \underline{v}^h = 0$. We find from (13) and (12) that for each $K \in \mathcal{T}^h$,

$$0 = (\mathbf{div}^h\, \underline{v}^h)\big|_K = R^K_{\text{div}}(\hat{\underline{v}}^h) = R^K_{\text{div}}\left((\mathcal{J}\hat{\underline{v}}^h)|_{\partial K}\right) = \text{div}\left((\mathcal{J}\hat{\underline{v}}^h)|_K\right),$$

and hence div $\left(\mathcal{J} \hat{\underline{v}}^h \right) = 0$ in Ω. Since $\mathcal{J} \hat{\underline{v}}^h \in \underline{V}_c^h$, it follows from hypothesis (L) that $\mathcal{J} \hat{\underline{v}}^h = \underline{0}$ in Ω. This implies that $\hat{\underline{v}}^h = \underline{0}$ on Γ^h. Thus $\underline{v}^h$ belongs to the right-hand side of (11). $\qquad\qquad\square$

We now define mapping $\mathbf{div}_1^h : \underline{V}_c^h / \mathrm{Ker}(\mathbf{div}^h \mid_{\underline{V}_c^h}) \longrightarrow L^2(\Omega)$ by

$$\mathbf{div}_1^h[\underline{v}^h] := \mathbf{div}^h \underline{v}^h \quad \forall \underline{v}^h \in \underline{V}_c^h,$$

where $[\underline{v}^h]$ is the set of equivalence class of $\underline{v}^h \in \underline{V}_c^h$. Since $\mathbf{div}_1^h$ is injective and $\underline{V}_c^h / \mathrm{Ker}(\mathbf{div}^h \mid_{\underline{V}_c^h})$ is finite dimensional, there exists a positive constant $C(h)$ such that for all $\underline{v}^h \in \underline{V}_c^h$,

$$\inf_{\underline{\chi}^h \in \underline{U}^h} \left\| \left\| \underline{v}^h + \{\underline{\chi}^h, \underline{0}\} \right\| \right\|_h \leq C(h) \left\| \mathbf{div}^h \underline{v}^h \right\|_\Omega . \tag{14}$$

Using (9) with $\underline{v} \equiv \underline{0}$ and (4), we get

$$\left| \mathcal{J} \hat{\underline{v}}^h \right|_{1,\Omega} \leq C \inf_{\underline{\chi}^h \in \underline{U}^h} \left\| \left\| \underline{v}^h + \{\underline{\chi}^h, \underline{0}\} \right\| \right\|_h \quad \forall \underline{v}^h \in \underline{V}_c^h. \tag{15}$$

Combining (14) and (15) gives us

$$\left| \mathcal{J} \hat{\underline{v}}^h \right|_{1,\Omega} \leq C(h) \left\| \mathbf{div}^h \underline{v}^h \right\|_\Omega \quad \forall \underline{v}^h \in \underline{V}_c^h. \tag{16}$$

Proposition 1 *Let $\underline{u} \in \underline{H}^2(\Omega) \cap \underline{H}_D^1(\Omega)$ satisfy*

$$\mathrm{div}\, \underline{u} = 0. \tag{17}$$

For each $\lambda > 0$, let $\underline{u}_\lambda^h \in \underline{V}_c^h$ satisfy

$$a_\eta^h(\underline{u}_\lambda^h, \underline{v}^h) = a_\eta^h(\underline{u}, \underline{v}^h) \quad \forall \underline{v}^h \in \underline{V}_c^h, \tag{18}$$

where $\underline{u} := \{\underline{u}, \underline{u}|_{\Gamma^h}\}$. Assume that (L) holds. Then we have

$$\left| \mathcal{J} \hat{\underline{u}}_\lambda^h \right|_{1,\Omega} \longrightarrow 0 \quad (\lambda \longrightarrow \infty). \tag{19}$$

Proof We first introduce the following trace inequality: for all $h \in (0, \bar{h}]$, for all $K \in \mathcal{T}^h$, for all $e \in \mathcal{E}^K$, and for all $v \in H^1(K)$,

$$|v|_e^2 \leq C \left(|e|^{-1} \|v\|_K^2 + |e||v|_{1,K}^2 \right), \tag{20}$$

where C is a positive constant independent of h, K, e, and v.

It follows from (18), (17), and (20) that we have

$$a_\eta^h(\underline{u}_\lambda^h, \underline{u}_\lambda^h) = a_\eta^h(\underline{u}, \underline{u}_\lambda^h)$$

$$= 2\mu \sum_{K \in \mathcal{T}^h} \left[\left(\underline{\underline{\varepsilon}}(\underline{u}), \underline{\underline{\varepsilon}}(\underline{u}_\lambda^h) \right)_K + \left\langle \underline{\underline{\varepsilon}}(\underline{u})\underline{n}, \hat{u}_\lambda^h - \underline{u}_\lambda^h \right\rangle_{\partial K} \right]$$

$$\leq C \|\underline{u}\|_{2,\Omega} \||\underline{u}_\lambda^h\||_h, \tag{21}$$

where C is a positive constant independent of h, λ, and $\underline{u}$. Using (8), we obtain

$$\||\underline{u}_\lambda^h\||_h \leq C\|\underline{u}\|_{2,\Omega}. \tag{22}$$

Combining (6), (7), (2), (21), and (22) leads us to

$$\|\mathbf{div}^h \underline{u}_\lambda^h\|_\Omega^2 \leq \lambda^{-1} C \|\underline{u}\|_{2,\Omega}^2 \longrightarrow 0 \quad (\lambda \longrightarrow \infty),$$

and thus, by (16), we get (19). □

Theorem 1 *Assume that (L) holds for every $h \in (0, \bar{h}]$. There exists a positive constant C independent of h such that*

$$\liminf_{\lambda \to \infty} \sup_{\underline{w} \in B_\lambda} \||\underline{w} - \underline{w}_\lambda^h\||_h \geq C \quad \forall h \in (0, \bar{h}], \tag{23}$$

where $\underline{w} := \{\underline{w}, \underline{w}|_{\Gamma^h}\}$ and $\underline{w}_\lambda^h \in \underline{V}_c^h$ is the solution of (18) after replacing $\underline{u}$ by $\underline{w}$.

Proof There exists a $\underline{u} \in \underline{H}^2(\Omega) \cap \underline{H}_D^1(\Omega)$ such that $\|\underline{u}\|_{2,\Omega} = 1$ and (17) holds (Brenner and Scott, 2008). Then $\underline{u} \in B_\lambda$ for all $\lambda > 0$. For every $h \in (0, \bar{h}]$ and for every $\lambda > 0$,

$$\sup_{\underline{w} \in B_\lambda} \||\underline{w} - \underline{w}_\lambda^h\||_h \geq \||\underline{u} - \underline{u}_\lambda^h\||_h \geq C \left| \underline{u} - \mathcal{J}\hat{\underline{u}}_\lambda^h \right|_{1,\Omega} \quad \text{(by (9))}$$

$$\geq C \left(|\underline{u}|_{1,\Omega} - \left| \mathcal{J}\hat{\underline{u}}_\lambda^h \right|_{1,\Omega} \right), \tag{24}$$

where C is independent of h and λ.

We can conclude from (19) and (24) that (23) holds. □

Remark 1 For a meaning of (23), see Brenner and Scott (2008). Using (23), we can also prove that the HSIP-C method with $k = 1$ shows locking of order h^{-1} with respect to the solution set B_λ and the norm $\||\cdot\||_h$ in the sense of Babuška and Suri (1992) (see Koyama and Kikuchi 2016).

References

I. Babuška, M. Suri, Locking effects in the finite element approximation of elasticity problems. Numer. Math. **62**(4), 439–463 (1992)

S.C. Brenner, L.R. Scott, *The Mathematical Theory of Finite Element Methods*, 3rd edn. Texts in Applied Mathematics, vol. 15 (Springer, New York, 2008)

P. Hansbo, M.G. Larson, Discontinuous galerkin methods for incompressible and nearly incompressible elasticity by nitsche's method. Comput. Methods Appl. Mech. Eng. **191**(17–18), 1895–1908 (2002)

F. Kikuchi, Finite element methods for nearly incompressible media. RIMS Kokyuroku (Kyoto University) **1971**, 28–46 (2015)

D. Koyama, F. Kikuchi, On volumetric locking in a hybrid symmetric interior penalty method for nearly incompressible linear elasticity. Submitted (2016)

Dual-Primal Domain Decomposition Methods for the Total Variation Minimization

Chang-Ock Lee and Changmin Nam

1 Introduction

Image denoising problem is one of classical problems in imaging science. In 1992, Rudin et al. (1992) proposed the following denoising model,

$$\min_{u \in BV(\Omega)} \left\{ \frac{\lambda}{2} \int_{\Omega} (u - f)^2 \, dx + \int_{\Omega} |\nabla u| \, dx \right\}, \tag{1}$$

where Ω is the domain of image and f is an observed image corrupted by noise. Here, the space of functions of bounded variation is defined as

$$BV(\Omega) = \left\{ u \in L^1(\Omega) : \sup_{\phi \in C_c^1(\Omega, R^2), \|\phi\|_\infty \le 1} \int_{\Omega} u(x) \, \text{div} \phi(x) \, dx < \infty \right\}.$$

This model has an anisotropic diffusion property so that the edge of the image is preserved.

Recently, as the number of CPUs and cores in a computer are increased, there have been attempts to solve this problem parally using the domain decomposition technique. For example, see Chang et al. (2015), Fornasier (2007), Fornasier et al. (2010), Fornasier and Schönlieb (2009), Hintermüller and Langer (2014), Lee et al. (2016) and Xu et al. (2010). Since the problem is nonsmooth and not separable, it is not easy to show the convergence of the domain decomposition algorithm. Tseng (2001) showed that if the function is separable, block Gauss-Seidel algorithm

C.-O. Lee • C. Nam (✉)
Department of Mathematical Sciences, KAIST, Daejeon 34141, South Korea
e-mail: colee@kaist.edu; ncm2200@kaist.ac.kr

© Springer International Publishing AG 2017

371

C.-O. Lee et al. (eds.), *Domain Decomposition Methods in Science and Engineering XXIII*, Lecture Notes in Computational Science and Engineering 116, DOI 10.1007/978-3-319-52389-7_38

converges to the minimizer, but (1) is not of this case. Fornasier et al. (2010) and Xu et al. (2010) used overlapping domain decomposition methods to overcome this difficulty. Also, Fornasier and Schönlieb (2009) proved the convergence of nonoverlaping domain decomposition method under certain assumptions.

The main point of the domain decomposition approach is that instead of solving one large problem, several small problems are solved in parallel to reduce the computing time. In Fornasier (2007), Fornasier pointed out that the subproblems should reproduce the original problem at smaller dimensions, but it is difficult to satisfy this requirement since the boundary conditions of local subdomain problems should be considered.

In this paper, we propose new domain decomposition techniques considering this requirement. First we decompose the domain of the dual form of (1), discovered by Chambolle (2004), into nonoverlapping rectangular subdomains. Then we change the local dual problems into the equivalent primal forms so that our methods use same algorithms to solve the original problem and local problems which can be solved in parallel.

2   Preliminaries

We assume that the image domain Ω consists of $N \times N$ discrete points, i.e.,

$$\Omega = [1, 2, \ldots, N] \times [1, 2, \ldots, N].$$

We define the function space V as a set of functions from Ω into $\mathbb{R}$ and V^* as a set of functions from Ω into $\mathbb{R}^2$ with the usual Euclidean inner product.

The operator $\nabla \colon V \to V^*$ is defined by

$$(\nabla u)_{ij}^1 = \begin{cases} u_{i+1,j} - u_{ij} & \text{for} \quad i = 1, \ldots, N-1, \\ 0 & \text{for} \quad i = N, \end{cases}$$

$$(\nabla u)_{ij}^2 = \begin{cases} u_{i,j+1} - u_{ij} & \text{for} \quad j = 1, \ldots, N-1, \\ 0 & \text{for} \quad j = N. \end{cases}$$

We define an operator div: $V^* \to V$ by $-\nabla^*$ (the adjoint of ∇).

For simplicity, we decompose the image domain Ω into two subsets Ω_1 and Ω_2 such that

$$\Omega_1 = [1, \ldots, N] \times [1, \ldots, N_1],$$
$$\Omega_2 = [1, \ldots, N] \times [N_1, \ldots, N].$$

Then the interface Γ is

$$\Gamma = [1, \ldots, N] \times [N_1].$$

For each subdomain, we define the local function spaces

$$V_1 = \{u \in V \mid \text{supp}(u) \subset \Omega_1\},$$
$$V_2 = \{u \in V \mid \text{supp}(u) \subset \Omega_2\},$$
$$V_1^* = \{\mathbf{p} \in V^* \mid \text{supp}(\mathbf{p}) \subset \Omega_1 \backslash \Gamma\},$$
$$V_2^* = \{\mathbf{p} \in V^* \mid \text{supp}(\mathbf{p}) \subset \Omega_2\}.$$

Note that $V = V_1 + V_2$, and $V^* = V_1^* \oplus V_2^*$.

We also define the local operators as the restriction of global operators ∇ and div to these spaces. More precisely, the operator $\nabla_{\Omega_1}: V_1 \to V_1^*$ is defined as

$$(\nabla_{\Omega_1} u)_{ij}^1 = \begin{cases} u_{i+1,j} - u_{ij} & \text{for } i = 1, \ldots, N-1, \\ 0 & \text{for } i = N, \end{cases}$$

$$(\nabla_{\Omega_1} u)_{ij}^2 = \begin{cases} u_{i,j+1} - u_{ij} & \text{for } j = 1, \ldots, N_1 - 1, \\ 0 & \text{for } j = N_1, \ldots, N. \end{cases}$$

We define $\nabla_{\Omega_2}: V_2 \to V_2^*$ with similar manner. We define $\text{div}_{\Omega_1}: V_1^* \to V_1$ by $-\nabla_{\Omega_1}^*$ and $\text{div}_{\Omega_2}: V_2^* \to V_2$ by $-\nabla_{\Omega_2}^*$.

3 Proposed Algorithms

We consider the following discrete version of (1),

$$\min_{u \in V} \left\{ \frac{\lambda}{2} \|u - f\|_V^2 + \sum_\Omega |\nabla u| \right\} \quad \text{for} f \in V. \tag{2}$$

Our result is based on the following two propositions which are summarized in Sect. 2 of Chambolle (2004).

Proposition 1 *The following two statements are equivalent.*

(i) $\displaystyle \bar{u} = \arg\min_{u \in V} \left\{ \frac{\lambda}{2} \|u - f\|_V^2 + \sum_\Omega |\nabla u| \right\}$

(ii) There exists $\mathbf{p} \in V^*$ *such that* $\begin{cases} f - \frac{1}{\lambda}\text{div}\mathbf{p} = \bar{u} \\ \mathbf{p} = \arg\min_{|\mathbf{p}| \le 1} \left\| \frac{1}{\lambda}\text{div}\mathbf{p} - f \right\|_V^2 \end{cases}$

Proposition 2 (Optimality Condition) *The following two statements are equivalent.*

$$\text{(i)} \quad \mathbf{p} = \arg\min_{|\mathbf{p}| \leq 1} \left\| \frac{1}{\lambda}\text{div}\mathbf{p} - f \right\|_V^2$$

$$\text{(ii)} \quad \begin{cases} -\nabla(\frac{1}{\lambda}\text{div}\mathbf{p} - f) + |\nabla(\frac{1}{\lambda}\text{div}\mathbf{p} - f)|\mathbf{p} = 0 \quad in \quad \Omega \\ |\mathbf{p}| \leq 1 \end{cases}$$

Now, we propose the block Gauss-Seidel algorithm for the primal problem (2).

Algorithm: Block Gauss-Seidel

Initialize $u_2^{(0)} := 0, f_2^{(0)} := 0$
For $n = 0, 1, \ldots$

$$(f_1^{(n+1)})_{ij} = (u_2^{(n)} - f_2^{(n)} + f)_{ij} \quad \text{for} \quad (i,j) \in \Omega_1$$

$$u_1^{(n+1)} = \arg\min_{u_1 \in V_1} \left\{ \frac{\lambda}{2}\|u_1 - f_1^{(n+1)}\|_{V_1}^2 + \sum_{\Omega_1 \backslash \Gamma} |\nabla_{\Omega_1} u_1| \right\}$$

$$(f_2^{(n+1)})_{ij} = (u_1^{(n+1)} - f_1^{(n+1)} + f)_{ij} \quad \text{for} \quad (i,j) \in \Omega_2$$

$$u_2^{(n+1)} = \arg\min_{u_2 \in V_2} \left\{ \frac{\lambda}{2}\|u_2 - f_2^{(n+1)}\|_{V_2}^2 + \sum_{\Omega_2} |\nabla_{\Omega_2} u_2| \right\}$$

$$u^{(n+1)} = f - f_1^{(n+1)} - f_2^{(n+1)} + u_1^{(n+1)} + u_2^{(n+1)}$$

end

Theorem 1 *The sequence $u^{(n)}$ of the block Gauss-Seidel algorithm converges to the minimizer of the problem (2).*

Proof By the Proposition 1, $u_1^{(n)}$, $u_2^{(n)}$, $f_1^{(n)}$, $f_2^{(n)}$, and $u^{(n)}$ are bounded sequences. Suppose that $u^{(\infty)}$ is the limit point of the sequence $u^{(n)}$. Then there exists a subsequence $u^{(n_k)}$ which converges to $u^{(\infty)}$. Now we claim that $u^{(\infty)}$ is the solution of (2).

By the Propositions 1 and 2, there exists $\mathbf{p}_1^{(n)} \in V_1^*$, $\mathbf{p}_2^{(n)} \in V_2^*$ for all $n \geq 1$ such that in $\Omega_1 \backslash \Gamma$,

$$\begin{cases} f_1^{(n)} - \frac{1}{\lambda}\text{div}_{\Omega_1}\mathbf{p}_1^{(n)} = u_1^{(n)}, \\ -\nabla_{\Omega_1}(\frac{1}{\lambda}\text{div}_{\Omega_1}\mathbf{p}_1^{(n)} - f_1^{(n)}) + |\nabla_{\Omega_1}(\frac{1}{\lambda}\text{div}_{\Omega_1}\mathbf{p}_1^{(n)} - f_1^{(n)})|\mathbf{p}_1^{(n)} = 0, \\ |\mathbf{p}_1^{(n)}| \leq 1, \end{cases}$$

and in Ω_2,

$$\begin{cases} f_2^{(n)} - \frac{1}{\lambda}\mathrm{div}_{\Omega_2}\mathbf{p}_2^{(n)} = u_2^{(n)}, \\ -\nabla_{\Omega_2}(\frac{1}{\lambda}\mathrm{div}_{\Omega_2}\mathbf{p}_2^{(n)} - f_2^{(n)}) + |\nabla_{\Omega_2}(\frac{1}{\lambda}\mathrm{div}_{\Omega_2}\mathbf{p}_2^{(n)} - f_2^{(n)})|\mathbf{p}_2^{(n)} = 0, \\ |\mathbf{p}_2^{(n)}| \le 1. \end{cases}$$

By refining the subsequences, we can assume that $f_1^{(n_{k_j})} \to f_1^{(\infty)}$, $f_2^{(n_{k_j})} \to f_2^{(\infty)}$, $p_1^{(n_{k_j})} \to p_1^{(\infty)}$, $p_2^{(n_{k_j})} \to p_2^{(\infty)}$, $p_2^{(n_{k_j}-1)} \to \tilde{p}_2^{(\infty)}$, $u_1^{(n_{k_j})} \to u_1^{(\infty)}$, and $u_2^{(n_{k_j})} \to u_2^{(\infty)}$. By the Proposition 2, the following monotone property holds for all $n \ge 1$;

$$\left\|\frac{1}{\lambda}\mathrm{div}(\mathbf{p}_1^{(n)} + \mathbf{p}_2^{(n)}) - f\right\| \ge \left\|\frac{1}{\lambda}\mathrm{div}(\mathbf{p}_1^{(n+1)} + \mathbf{p}_2^{(n)}) - f\right\|$$

$$\ge \left\|\frac{1}{\lambda}\mathrm{div}(\mathbf{p}_1^{(n+1)} + \mathbf{p}_2^{(n+1)}) - f\right\|$$

so that $\mathrm{div}(\mathbf{p}_1^{(\infty)} + \mathbf{p}_2^{(\infty)}) = \mathrm{div}(\mathbf{p}_1^{(\infty)} + \tilde{\mathbf{p}}_2^{(\infty)})$. As $j \to \infty$, in $\Omega_1 \backslash \Gamma$,

$$\begin{cases} f_1^\infty - \frac{1}{\lambda}\mathrm{div}_{\Omega_1}\mathbf{p}_1^{(\infty)} = u_1^{(\infty)}, \\ -\nabla_{\Omega_1}(\frac{1}{\lambda}\mathrm{div}_{\Omega_1}\mathbf{p}_1^{(\infty)} - f_1^{(\infty)})) + |\nabla_{\Omega_1}(\frac{1}{\lambda}\mathrm{div}_{\Omega_1}\mathbf{p}_1^{(\infty)} - f_1^{(\infty)})|\mathbf{p}_1^{(\infty)} = 0, \\ |\mathbf{p}_1^{(\infty)}| \le 1, \end{cases} \tag{3a}$$

and in Ω_2,

$$\begin{cases} f_2^{(\infty)} - \frac{1}{\lambda}\mathrm{div}_{\Omega_2}\mathbf{p}_2^{(\infty)} = u_2^{(\infty)}, \\ -\nabla_{\Omega_2}(\frac{1}{\lambda}\mathrm{div}_{\Omega_2}\mathbf{p}_2^{(\infty)} - f_2^{(\infty)}) + |\nabla_{\Omega_2}(\frac{1}{\lambda}\mathrm{div}_{\Omega_2}\mathbf{p}_2^{(\infty)} - f_2^{(\infty)})|\mathbf{p}_2^{(\infty)} = 0, \\ |\mathbf{p}_2^{(\infty)}| \le 1. \end{cases} \tag{3b}$$

Let $\mathbf{p}^{(\infty)} = \mathbf{p}_1^{(\infty)} + \mathbf{p}_2^{(\infty)}$. We claim that

(i) $f - \frac{1}{\lambda}\mathrm{div}\mathbf{p}^{(\infty)} = f - f_1^{(\infty)} - f_2^{(\infty)} + u_1^{(\infty)} + u_2^{(\infty)}$.

(ii) $-\nabla\left(\frac{1}{\lambda}\mathrm{div}\mathbf{p}^{(\infty)} - f\right) + \left|\nabla\left(\frac{1}{\lambda}\mathrm{div}\mathbf{p}^{(\infty)} - f\right)\right|\mathbf{p}^{(\infty)} = 0$.

(iii) $|\mathbf{p}^{(\infty)}| \le 1$.

The statement (i) is established by adding (3a) and (3b) and the statement (iii) is trivial. We have

$$\nabla_{\Omega_1}\left(\frac{1}{\lambda}\mathrm{div}_{\Omega_1}\mathbf{p}_1^{(\infty)} - f_1^{(\infty)}\right) = \nabla\left(\frac{1}{\lambda}\mathrm{div}_{\Omega_1}\mathbf{p}_1^{(\infty)} + \frac{1}{\lambda}\mathrm{div}_{\Omega_2}\tilde{\mathbf{p}}_2^{(\infty)} - f\right)$$

$$= \nabla\left(\frac{1}{\lambda}\mathrm{div}\mathbf{p}^{(\infty)} - f\right) \quad \text{in} \quad \Omega_1 \backslash \Gamma,$$

$$\nabla_{\Omega_2}\left(\frac{1}{\lambda}\mathrm{div}_{\Omega_2}\mathbf{p}_2^{(\infty)} - f_2^{(\infty)}\right) = \nabla\left(\frac{1}{\lambda}\mathrm{div}_{\Omega_1}\mathbf{p}_1^{(\infty)} + \frac{1}{\lambda}\mathrm{div}_{\Omega_2}\mathbf{p}_2^{(\infty)} - f\right)$$

$$= \nabla\left(\frac{1}{\lambda}\mathrm{div}\mathbf{p}^{(\infty)} - f\right) \quad \text{in} \quad \Omega_2,$$

which proves the statement (ii) and $u^{(\infty)}$ is the solution of (2). Since the solution of (2) is unique, the result follows. □

Next, we propose the relaxed block Jacobi algorithm as a parallel algorithm.

Algorithm: Relaxed Block Jacobi

Initialize $v_1^{(0)} := 0$, $v_2^{(0)} := 0$.
For $n = 0, 1, \ldots$

$$(f_1^{(n+1)})_{ij} = (-v_2^{(n)} + f)_{ij} \quad \text{for} \quad (i,j) \in \Omega_1$$

$$(f_2^{(n+1)})_{ij} = (-v_1^{(n)} + f)_{ij} \quad \text{for} \quad (i,j) \in \Omega_2$$

$$\tilde{u}_1^{(n+1)} = \arg\min_{u_1 \in V_1}\left\{\frac{\lambda}{2}\|u_1 - f_1^{(n+1)}\|^2 + \sum_{\Omega_1\backslash\Gamma}|\nabla_{\Omega_1}u_1|\right\}$$

$$\tilde{u}_2^{(n+1)} = \arg\min_{u_2 \in V_2}\left\{\frac{\lambda}{2}\|u_2 - f_2^{(n+1)}\|^2 + \sum_{\Omega_2}|\nabla_{\Omega_2}u_2|\right\}$$

$$v_1^{(n+1)} = \frac{v_1^{(n)} + f_1^{(n+1)} - \tilde{u}_1^{(n+1)}}{2}$$

$$v_2^{(n+1)} = \frac{v_2^{(n)} + f_2^{(n+1)} - \tilde{u}_2^{(n+1)}}{2}$$

$$u^{(n+1)} = f - v_1^{(n+1)} - v_2^{(n+1)}$$

end

Lemma 1 *In the relaxed block Jacobi algorithm, we have* $\|v_1^{(n+1)} - v_1^{(n)}\|_{V_1} \to 0$ *and* $\|v_2^{(n+1)} - v_2^{(n)}\|_{V_2} \to 0$ *as* $n \to \infty$.

Sketch of Proof By the Proposition 1, there exist $\tilde{\mathbf{p}}_1^{(n+1)} \in V_1^*$ and $\tilde{\mathbf{p}}_2^{(n+1)} \in V_2^*$ such that

$$\tilde{\mathbf{p}}_1^{(n+1)} = \arg\min_{p_1 \in V_1^*}\left\|\frac{1}{\lambda}\mathrm{div}_{\Omega_1}\mathbf{p}_1 + v_2^{(n)} - f\right\|_{V_1},$$

$$\tilde{\mathbf{p}}_2^{(n+1)} = \arg\min_{p_2 \in V_2^*}\left\|\frac{1}{\lambda}\mathrm{div}_{\Omega_2}\mathbf{p}_2 + v_1^{(n)} - f\right\|_{V_2}.$$

By the triangle inequality and minimization property, the result follows. □

With this lemma, one can easily prove the following theorem.

Theorem 2 *The sequence $u^{(n)}$ of the relaxed block Jacobi algorithm converges to the minimizer of the problem* (2).

4 Numerical Results

In this section, we compare our domain decomposition algorithms with the first order primal dual algorithm in Chambolle and Pock (2011). We used the following stop criterion to the relaxed block Jacobi algorithm and Algorithm 2 in Chambolle and Pock (2011) solving the full dimension problem (2):

$$\frac{\|u^{(n+1)} - u^{(n)}\|_V}{\|u^{(n+1)}\|_V} < 10^{-5}$$

with the parameters $\tau = 1/\sqrt{8}$, $\sigma = 1/\sqrt{8}$, $\gamma = 0.7\lambda$, which are used to run Algorithm 2 in Chambolle and Pock (2011). We choose the weight parameter λ in (1) as 7 empirically. For the local problems, we also used Algorithm 2 in Chambolle and Pock (2011) with the following stop criterion

$$\frac{\|u_i^{(n+1)} - u_i^{(n)}\|_V}{\|u_i^{(n+1)}\|_V} < 10^{-6}.$$

We tested two images of size 512×512 and 2048×3072, corrupted by additive zero mean Gaussian noise with variance 0.03. Table 1 shows the performance of the algorithm with the varying number of subdomains (Figs. 1 and 2).

Table 1 Results of the proposed algorithm

	Peppers 512×512			Boat 2048×3072		
		Virtual wall-clock			Virtual wall-clock	
Domain	Iter	time (s)	PSNR	Iter	time (s)	PSNR
1x1	1	3.59	27.39	1	115.48	28.79
2x2	54	6.69	27.39	39	324.12	28.79
4x4	66	2.26	27.39	52	153.13	28.79
8x8	81	1.44	27.39	63	24.83	28.79
16x16	96	1.12	27.39	75	10.28	28.79

The results for 1×1 domain are from Algorithm 2 in Chambolle and Pock (2011)

Fig. 1 (**a**) Original clean image of size 512×512, (**b**) Noisy image with Gaussian noise with zero mean and 0.03 variance (PSNR=15.66), (**c**) Denoised image with weight $\lambda = 7$ in (2)

Fig. 2 (**a**) Original clean image of size 2048×3072, (**b**) Noisy image with Gaussian noise with zero mean and 0.03 variance (PSNR=15.66), (**c**) Denoised image with weight $\lambda = 7$ in (2)

References

A. Chambolle, An algorithm for total variation minimization and applications. J. Math. Imaging Vision **20**, 89–97 (2004)

A. Chambolle, T. Pock, A first-order primal-dual algorithm for convex problems with applications to imaging. J. Math. Imaging Vision **40**, 120–145 (2011)

H. Chang, X.C. Tai, L.L. Wang, D. Yang, Convergence rate of overlapping domain decomposition methods for the Rudin-Osher-Fatemi model based on a dual formulation. SIAM J. Imag. Sci. **8**, 564–591 (2015)

M. Fornasier, Domain decomposition methods for linear inverse problems with sparsity constraints. Inverse Prob. **8**, 2505–2526 (2007)

M. Fornasier, C.B. Schönlieb, Subspace correction methods for total variation and l1-minimization. SIAM J. Numer. Anal. **47**, 3397–3428 (2009)

M. Fornasier, A. Langer, C.B. Schönlieb, A convergent overlapping domain decomposition method for total variation minimization. Numer. Math. **116**, 645–685 (2010)

M. Hintermüller, A. Langer, Non-overlapping domain decomposition methods for dual total variation based image denoising. SIAM J. Sci. Comput. (2014). doi:10.1007/s10915-014-9863-8

C.-O. Lee, J.H. Lee, H. Woo, S. Yun, Block decomposition methods for total variation by primal-dual stitching. J. Sci. Comput. **68**, 273–302 (2016)

L.I. Rudin, S. Osher, E. Fatemi, Nonlinear total variation based noise removal algorithms. Physics D **60**, 259–268 (1992)

P. Tseng, Convergence of a block coordinate descent method for nondifferentiable minimization. J. Optim. Theory Appl. **3**, 475–494 (2001)

J. Xu, X.C. Tai, L.L. Wang, A two-level domain decomposition method for image restoration. Inverse Prob. Imag. (2010). doi:10.3934/ipi.2010.4.523

A Parallel Two-Phase Flow Solver on Unstructured Mesh in 3D

Li Luo, Qian Zhang, Xiao-Ping Wang, and Xiao-Chuan Cai

The simulation of two-phase flow is important in many scientific and engineering processes, for instance, wetting, coating, painting, etc. There are many publications on phase field modelling of two-phase flows. Gao and Wang (2014) proposed a gradient stable semi-implicit finite difference scheme in 2D and 3D by using the convex splitting method for the Cahn-Hilliard equation and a projection method for the Navier-Stokes equations. Bao et al. (2012) presented a finite element method for phase field problems on 2D domains with rough boundary using unstructured meshes. The free interface problem is computationally very expensive especially in 3D; some parallelization strategies were adopted to accelerate the computation of certain two-phase flows. Shin et al. (2014) presented a parallel implementation of the Level Contour Reconstruction Method (LCRM) on structured meshes for simulating the splash of a drop onto a film of liquid, in which a weak scaling efficiency of 48% on 32,768 processors was reported.

In this paper, we present a new parallel finite element solver on unstructured 3D meshes and its implementation on a massively parallel computer. In order to construct a stable and efficient solver for the case of large density and viscosity ratio, we combine the stabilized schemes for the Cahn-Hilliard equation and projection-

L. Luo (✉)
Department of Mathematics, The Hong Kong University of Science and Technology, Kowloon, Hong Kong

Shenzhen Institutes of Advanced Technology, Shenzhen, People's Republic of China
e-mail: lluoac@ust.hk

Q. Zhang • X.-P. Wang
Department of Mathematics, The Hong Kong University of Science and Technology, Kowloon, Hong Kong

X.-C. Cai
Department of Computer Science, University of Colorado Boulder, Boulder, USA

© Springer International Publishing AG 2017

379

C.-O. Lee et al. (eds.), *Domain Decomposition Methods in Science and Engineering XXIII*, Lecture Notes in Computational Science and Engineering 116, DOI 10.1007/978-3-319-52389-7_39

type schemes for the Navier-Stokes equations to fully decouple the phase function, the velocity, and the pressure. The resulting decoupled systems are discretized by a piecewise linear finite element method in space and solved by a Krylov subspace method. Specifically, systems arising from implicit discretization of the Cahn-Hilliard equation and the velocity equation are solved by a restricted additive Schwarz preconditioned GMRES method, and the pressure Poisson system is solved by an algebraic multigrid preconditioned CG method. We show numerically that the proposed strategy works well for 3D problems with complex geometry and is highly scalable in terms of the number of iterations and the total computing time on a supercomputer with nearly 10,000 processors.

The paper is organized as follows. In Sect. 1, a phase field model is described. The fully decoupled scheme with a finite element discretization is also presented in this section. The domain decomposition techniques and scalable solvers are discussed in Sect. 2. In Sect. 3, we show two numerical experiments. Performance results of the parallel implementation are also reported. The paper is concluded in Sect. 4.

1   Mathematical Models and Discretization Schemes

Let Ω be a bounded domain in $\mathbb{R}^3$. The system of interest can be described by a coupled Cahn-Hilliard-Navier-Stokes equations, as follows:

$$\frac{\partial \phi}{\partial t} + \mathbf{u} \cdot \nabla \phi = \mathcal{L}_d \Delta \mu, \qquad \qquad \text{in} \quad \Omega, \qquad (1)$$

$$\mu = -\epsilon \Delta \phi - \frac{\phi}{\epsilon} + \frac{\phi^3}{\epsilon}, \qquad \qquad \text{in} \quad \Omega, \qquad (2)$$

$$Re\rho \left(\frac{\partial \mathbf{u}}{\partial t} + (\mathbf{u} \cdot \nabla)\mathbf{u} \right) = -\nabla p + \nabla \cdot (\eta D(\mathbf{u})) + \mathcal{B}\mu\nabla\phi, \qquad \text{in} \quad \Omega, \qquad (3)$$

$$\nabla \cdot \mathbf{u} = 0, \qquad \qquad \text{in} \quad \Omega. \qquad (4)$$

Here, a phase-field variable ϕ is introduced to describe the transition between the two homogeneous equilibrium phases $\phi_\pm = \pm 1$. μ is the chemical potential, ϵ is the ratio between interface thickness and characteristic length, and $\mu\nabla\phi$ is the capillary force. The mass density ρ and the dynamic viscosity η are interpolation functions of ϕ between fluid 1 and fluid 2, $\rho = \frac{1+\phi}{2} + \lambda_\rho \frac{1-\phi}{2}$, $\eta = \frac{1+\phi}{2} + \lambda_\eta \frac{1-\phi}{2}$, where $\lambda_\rho = \rho_2/\rho_1$ is the ratio of density between the two fluids and $\lambda_\eta = \eta_2/\eta_1$ is the ratio of viscosity. $\mathbf{u} = (u_x, u_y, u_z)$ where u_x, u_y, u_z are the velocity components along x, y, z directions, $D(\mathbf{u}) = \nabla\mathbf{u} + (\nabla\mathbf{u})^T$ is the rate of stress tensor, p is the pressure, $\mathcal{L}_d$ is the phenomenological mobility coefficient, Re is the Reynolds number and $\mathcal{B}$ measures the strength of the capillary force comparing to the Newtonian fluid stress (and $\mathcal{B}$ is inversely proportional to the capillary number). The motion of the contact line at solid boundaries Γ_w can be described by a relaxation boundary condition for the phase function and the generalized Navier boundary condition (GNBC) for velocity:

$$\frac{\partial \phi}{\partial t} + u_{\tau_1}\partial_{\tau_1}\phi + u_{\tau_2}\partial_{\tau_2}\phi = -\mathcal{V}_s L(\phi), \qquad \text{on} \quad \Gamma_w, \qquad (5)$$

$$(\mathcal{L}_s l_s)^{-1} u_{\tau_1} = \mathcal{B}L(\phi)\partial_{\tau_1}\phi/\eta - \mathbf{n}\cdot D(\mathbf{u})\cdot\boldsymbol{\tau}_1, \qquad \text{on} \quad \Gamma_w, \qquad (6)$$

$$(\mathcal{L}_s l_s)^{-1} u_{\tau_2} = \mathcal{B}L(\phi)\partial_{\tau_2}\phi/\eta - \mathbf{n}\cdot D(\mathbf{u})\cdot\boldsymbol{\tau}_2, \qquad \text{on} \quad \Gamma_w, \qquad (7)$$

where $\boldsymbol{\tau}_1$ and $\boldsymbol{\tau}_2$ are two unit tangent directions that are orthogonal to each other along the solid surface, $\boldsymbol{\tau}_1\cdot\boldsymbol{\tau}_2 = 0$. $\mathbf{n}$ is the unit outward normal direction of the solid surface. $\mathcal{V}_s$ is a phenomenological parameter. $L(\phi) = \epsilon\partial_n\phi + Q(\phi)$, $Q(\phi) = \partial\gamma_{wf}(\phi)/\partial\phi$ and $\gamma_{wf}(\phi) = -\frac{\sqrt{2}}{3}\cos\theta_s\sin(\frac{\pi}{2}\phi)$, θ_s is the static contact angle. $u_{\tau_1} = \mathbf{u}\cdot\boldsymbol{\tau}_1$ and $u_{\tau_2} = \mathbf{u}\cdot\boldsymbol{\tau}_2$. $\mathcal{L}_s$ is the slip length of liquid, $l_s = \frac{1+\phi}{2} + \lambda_{l_s}\frac{1-\phi}{2}$ is an interpolation between two different wall-fluid slip length, and $\lambda_{l_s} = l_{s2}/l_{s1}$ the ratio of slip length. In addition, the following impermeability conditions $u_n := \mathbf{u}\cdot\mathbf{n} = 0$, and $\partial_n\mu = 0$ are also imposed on the solid boundaries.

We present a semi-implicit finite element method for solving the above coupled systems on unstructured meshes in 3D. We apply a convex splitting of the free energy functional and treat the nonlinear term explicitly so that the resulting matrix does not change in time, and therefore can be pre-computed. In addition, we consider a pressure stabilized formulation (Guermond and Salgado, 2009) to decouple the Navier-Stokes equations into a convection-diffusion equation for velocity and a Poisson equation for pressure. Then, both of them can be easily approximated by the piecewise linear finite element methods.

Let Ω_h be a conforming mesh of Ω, and Γ_w^h is the solid boundary of Ω_h. In this paper, we only consider tetrahedral elements and P_1 functions. We define the following finite element spaces

$$W_h = \left\{w_h \in H^1(\Omega); \ w_h|_E \in P_1(E), \forall E \in \Omega_h\right\},$$

$$\mathbf{U}_h = \left\{\mathbf{u}_h \in \left[H^1(\Omega)\right]^3; \ \mathbf{u}_h\cdot\mathbf{n} = 0 \text{ on } \Gamma_w^h; \ \mathbf{u}_h|_E \in P_1(E)^3, \forall E \in \Omega_h\right\},$$

$$M_h = \left\{q_h \in W_h; \ \partial_n q_h = 0 \text{ on } \Gamma_w^h\right\}.$$

We denote by $(\cdot, \cdot)$ the $L^2(\Omega_h)$-inner product and by $\langle\cdot, \cdot\rangle_{\Gamma_w^h}$ the $L^2(\Gamma_w^h)$-inner product. Next, we introduce a time step $\delta t > 0$. The first-order temporal discretization in the weak form can be described in the following four steps:

Step 1: Solve the Cahn-Hilliard equation using a convex-splitting method: find $(\phi_h^{n+1}, \mu_h^{n+1}) \in W_h \times W_h$, such that for $\forall \ w_h \in W_h$,

$$\left(\frac{\phi_h^{n+1} - \phi_h^n}{\delta t}, w_h\right) + (\mathbf{u}_h^n\cdot\nabla\phi_h^n, w_h) = -\mathcal{L}_d(\nabla\mu_h^{n+1}, \nabla w_h), \qquad (8)$$

$$
\begin{aligned}
(\mu_h^{n+1}, w_h) &= \epsilon(\nabla\phi_h^{n+1}, \nabla w_h) + \frac{s}{\epsilon}(\phi_h^{n+1}, w_h) + \frac{1}{\epsilon}\left((\phi_h^n)^3 - (1+s)\phi_h^n, w_h\right) \\
&\quad + \left\langle \left(\frac{1}{\mathcal{V}_s}\left(\frac{\phi_h^{n+1} - \phi_h^n}{\delta t} + u_{\tau_1,h}^n \partial_{\tau_1}\phi_h^n + u_{\tau_2,h}^n \partial_{\tau_2}\phi_h^n\right) + Q\left(\phi_h^n\right)\right), w_h\right\rangle_{\Gamma_w}.
\end{aligned}
$$

$$(9)$$

Step 2: Update ρ_h^{n+1}, η_h^{n+1} and $l_{sh}^{n+1} \in W_h$:

$$
(\rho_h^{n+1}, \eta_h^{n+1}, l_{sh}^{n+1}) = \frac{1 + \phi_h^{n+1}}{2} + (\lambda_\rho, \lambda_\eta, \lambda_{l_s})\frac{1 - \phi_h^{n+1}}{2}. \tag{10}
$$

Step 3: Solve the velocity system of Navier-Stokes equations using a pressure stabilization scheme: find $\mathbf{u}_h^{n+1} \in U_h$, such that for $\forall\, \mathbf{v}_h \in U_h$,

$$
\begin{aligned}
Re\Bigg(\Bigg(&\frac{\frac{1}{2}(\rho_h^{n+1} + \rho_h^n)\mathbf{u}_h^{n+1} - \rho_h^n\mathbf{u}_h^n}{\delta t} + \rho_h^{n+1}(\mathbf{u}_h^n \cdot \nabla)\mathbf{u}_h^{n+1} + \frac{1}{2}\left(\nabla\cdot(\rho_h^{n+1}\mathbf{u}_h^n)\right)\mathbf{u}_h^{n+1}\Bigg), \mathbf{v}_h\Bigg) \\
= -&\left(\eta_h^{n+1}\left(\nabla\mathbf{u}_h^{n+1} + (\nabla\mathbf{u}_h^{n+1})^T\right), \nabla\mathbf{v}_h\right) + \mathcal{B}(\mu_h^{n+1}\nabla\phi_h^{n+1}, \mathbf{v}_h) - (2\nabla p_h^n - \nabla p_h^{n-1}, \mathbf{v}_h) \\
-&\left\langle\eta_h^{n+1}\left(\mathcal{L}_s l_{sh}^{n+1}\right)^{-1}u_{\tau_1,h}^{n+1}, v_{\tau_1,h}\right\rangle_{\Gamma_w} - \left\langle\eta_h^{n+1}\left(\mathcal{L}_s l_{sh}^{n+1}\right)^{-1}u_{\tau_2,h}^{n+1}, v_{\tau_2,h}\right\rangle_{\Gamma_w} \\
+&\, \mathcal{B}\left\langle\left(\epsilon\partial_n\phi_h^{n+1} + Q\left(\phi_h^{n+1}\right)\right)\partial_{\tau_1}\phi_h^{n+1}, v_{\tau_1,h}\right\rangle_{\Gamma_w} \\
+&\, \mathcal{B}\left\langle\left(\epsilon\partial_n\phi_h^{n+1} + Q\left(\phi_h^{n+1}\right)\right)\partial_{\tau_2}\phi_h^{n+1}, v_{\tau_2,h}\right\rangle_{\Gamma_w}.
\end{aligned}
$$

$$(11)$$

Step 4: Solve the pressure system of Navier-Stokes equations: find $p_h^{n+1} \in M_h$, such that for $\forall\, q_h \in M_h$,

$$
\left(\nabla(p_h^{n+1} - p_h^n), \nabla q_h\right) = -\frac{\bar{\rho}}{\delta t}Re(\nabla\cdot\mathbf{u}_h^{n+1}, q_h). \tag{12}
$$

In the above scheme, s is a stabilization parameter. $v_{n,h} = \mathbf{v}_h\cdot\mathbf{n}$, $v_{\tau_1,h} = \mathbf{v}_h\cdot\boldsymbol{\tau}_1$, $v_{\tau_2,h} = \mathbf{v}_h\cdot\boldsymbol{\tau}_2$, and $\bar{\rho} = \min(1, \lambda_\rho)$.

Remark 1 The time discretization scheme constructed above leads to a decoupled system for the phase function, the velocity, and the pressure. At each time step, we solve a convection-diffusion equation for $\mathbf{u}$, a system of convection-diffusion/elliptic equations for (ϕ, μ), and a Poisson equation for p. The matrices from the last two equations do not change in time, and can then be pre-computed for computational efficiency.

2   Scalable Solvers Based on Domain Decomposition and Algebraic Multigrid Techniques

In the scheme formulated in the previous section, there are three linear systems of equations to be solved at each time step. For the nonsymmetric problems in Step 1 and Step 3, we employ a restricted additive Schwarz preconditioned GMRES method to solve the linear systems of phase function and velocity. The choice of subdomain solver is critical to the Schwarz preconditioner. One of the popular choices is the incomplete LU (ILU) factorization. A large number of fill-ins levels helps in reducing iterations, but leads to an expensive solver in terms of the compute time and the memory usage. The impact of these factors will be discussed in numerical experiments. To solve the symmetric positive definite problem in Step 4, we employ an algebraic multigrid (AMG) preconditioned CG method. A scalable AMG solver BoomerAMG (Henson and Yang, 2002) is used as a preconditioner to effectively solve the pressure Poisson equation.

3   Numerical Experiments

In this section, we present some numerical experiments and analyze the parallel performance of the proposed algorithm. The algorithm is implemented using a finite element package libMesh (Kirk et al., 2006) for generating the stiffness matrices, and a parallel scientific computing library PETSc (Balay et al., 2016) for the preconditioned Krylov subspace solvers. The computational mesh is generated using Gmsh (Geuzaine and Remacle, 2009) and partitioned using MeTiS (Karypis and Kumar, 1995). Two numerical experiments will be presented including a droplet spreading over a rough surface and a two-phase flow in a bumpy channel.

We first consider a droplet spreading over a rough solid surface with parallel stripped texture. Along the y-axis the bottom surface is parametrized by a wave function $x = 0.025\sin(40y)$ with $y \in [-0.025\pi, 0.5\pi]$, and along the z-axis the function is translated from $z = 0$ to $z = 0.5\pi$. The height of the domain is 1.2. A spherical drop is initially located at $(0.35, 0.2375\pi, 0.25\pi)$ with radius 0.3. The initial speed is $(-1, 0, 0)$. A nonuniform mesh is generated such that near the bottom boundary the mesh is finer. The mesh has 3,055,992 elements and 535,509 vertices. The average mesh size near the bottom surface is $h = 5.64 \times 10^{-2}$ and the time step size is $\delta t = 2 \times 10^{-4}$. Other parameters used are as follows: $\lambda_\rho = 0.001$, $\lambda_\eta = 0.1$, $\lambda_{l_s} = 1$, $Re = 1000$, $\theta_s = 50°$, $\epsilon = 0.02$, $\mathcal{B} = 12$, $\mathcal{L}_d = 5 \times 10^{-4}$, $\mathcal{V}_s = 500$, $\mathcal{L}_s = 0.038$, and $s = 1.5$. The initial condition and the droplet spreading at $t = 0.4$ as well as a sample partition are shown in Fig. 1.

We next consider a flow of two immiscible fluids (red represents fluid 1 and blue represents fluid 2) in a bumpy channel is driven by the pressure gradient between

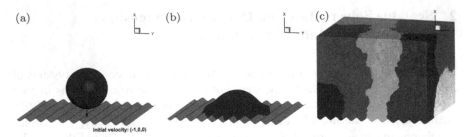

Fig. 1 (**a**) Initial condition, (**b**) the evolution of interface at $t = 0.4$, and (**c**) a sample partition into 16 subdomains for the droplet spreading case

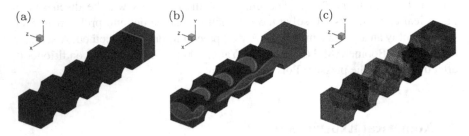

Fig. 2 (**a**) Initial condition, (**b**) the evolution of interface at $t = 0.28$, and (**c**) a sample partition into 8 subdomains for the bumpy channel flow case

the inflow boundary ($x = -0.5$, $p = 4000$) and the outflow boundary ($x = 0.5$, $p = 0$). The other boundaries are solid surfaces. The computational domain is $[-0.5, 0.5] \times [-0.075, 0.075] \times [-0.075, 0.075]$, and the radius of the cylinder bumps is 0.05. The mesh has 588,696 elements and 113,457 vertices. The average mesh size is $h = 9.15 \times 10^{-3}$ and the time step size is $\delta t = 10^{-4}$. Other parameters are as follows: $\lambda_\rho = 0.1$, $\lambda_\eta = 0.1$, $\lambda_{l_s} = 10$, $Re = 100$, $\theta_s = 120°$, $\epsilon = 0.005$, $\mathcal{B} = 12$, $\mathcal{L}_d = 5 \times 10^{-4}$, $\mathcal{V}_s = 200$, $\mathcal{L}_s = 0.0025$, and $s = 1.5$. The initial condition and the evolution of interface at $t = 0.28$ as well as a sample partition are shown in Fig. 2.

3.1 Parallel Performance

In this subsection, we focus on the bumpy channel flow case and report the parallel performance of the proposed solution algorithm. The scalability tests are performed on the Tianhe 2 supercomputer which ranks # 2 on the latest Top 500 list. Each node of Tianhe 2 has 24 processors and 64 GB memory. For the rest of the section, "np" denotes the number of processors, "GMRES" and "CG" denote the average number of GMRES and CG iterations per time step, respectively. "sp." represents

Table 1 The average number of iterations, compute time per time step, and speed up for solving Cahn-Hilliard system, the velocity system, and the pressure system

np	Subsolve	Cahn-Hilliard system #unknowns=102,540,706			Velocity system #unknowns=153,811,059			Pressure system #unknowns=51,270,353			
		GMRES	Time	sp.	GMRES	Time	sp.	Sweep	CG	Time	sp.
1920	ILU(1)	441.4	21.36	1	35	13.72	1	1	24.1	2.74	1
1920	ILU(2)	39.9	4.36	1	26.7	17.18	1	2	20.2	3.31	1
1920	ILU(3)	12.7	3.60	1	17.2	25.61	1	3	19.8	3.92	1
5760	ILU(1)	–	–	–	30	4.57	3.00	1	24.1	1.15	2.38
5760	ILU(2)	42.2	1.80	2.42	13.1	6.06	2.83	2	20.7	1.42	1.63
5760	ILU(3)	13.4	1.43	2.52	7	9.38	2.73	3	19.7	1.66	2.36
9600	ILU(1)	–	–	–	29.8	3.38	4.06	1	24.8	0.95	2.88
9600	ILU(2)	40.6	1.29	3.38	14.3	4.27	4.02	2	21	1.13	2.92
9600	ILU(3)	13.7	1.09	3.30	9.8	6.63	3.86	3	19.9	1.34	2.93

"–" means the case fails to converge

the speedup. All timings are reported in seconds. The restart value of GMRES is fixed at 50. 10^{-6} is used as the relative stopping condition for linear solvers.

The unstructured mesh has 301,412,352 elements and 51,270,353 vertices. We focus on how different levels of ILU fill-ins in the subdomain solver of Schwarz preconditioner affect the parallel efficiency. The overlapping size is fixed to 1. The number of processors increases from $np = 1920$ to 5760 to 9600. The results for different levels of ILU fill-ins at different np are summarized in the first 8 columns in Table 1. The results show that at least 2 levels of ILU fill-ins are needed for the Cahn-Hilliard system. Increasing the level of fill-ins helps reducing the number of GMRES iterations, this effect is more obvious for the Cahn-Hilliard system. However, higher level of fill-ins may cost more computation time. The table also suggests that ILU(3) is the best choice for the Cahn-Hilliard system and ILU(1) is the best choice for the velocity system. We have also considered the effect of varying the number of sweeps of the smoother in the AMG preconditioner for solving the pressure system. The last 4 columns in Table 1 shows that the number of CG iterations seems to be independent of np for all cases. However, increasing the number of sweeps does not improve the convergence of the linear solver much but requires more computational time, therefore one sweep of smoother is preferable for the multigrid method. Combining the above choices, we present the speedups and computational time for each system (marked as "total" including Step 1, 3, and 4 of the algorithm) starting from 1440 processors in Fig. 3. Excellent speedup is achieved when np is up to 2880 and the final speedup is 4.39 out of 6.67 on a fixed-size system which is reasonably good.

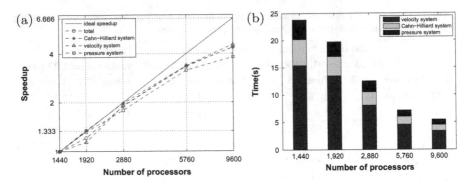

Fig. 3 Speedup (**a**) and distribution of total compute time (**b**) for the two-phase flow in a bumpy channel

4 Conclusions

In this paper we introduce a parallel finite element method on 3D unstructured meshes for the two-phase flow problem modelled by a phase-field model consisting of the coupled Cahn-Hilliard and Navier-Stokes equations. A restricted additive Schwarz preconditioned GMRES method is used to solve the systems arising from implicit discretization of the Cahn-Hilliard equation and the velocity equation, and an algebraic multigrid preconditioned CG method is used to solve the pressure Poisson system. Numerical experiments suggest that the overall algorithm scales well on unstructured meshes for problems with up to 150 millions unknowns and on machines with close to 10,000 processors.

References

S. Balay, S. Abhyankar, M.F. Adams, J. Brown, P. Brune, K. Buschelman, L. Dalcin, V. Eijkhout, W.D. Gropp, D. Kaushik, M.G. Knepley, L.C. McInnes, K.R., B.F. Smith, S. Zampini, H. Zhang, PETSc Users Manual, Technical Report, ANL-95/11 - Revision 3.7, Argonne National Laboratory, 2016

K. Bao, Y. Shi, S. Sun, X.-P. Wang, A finite element method for the numerical solution of the coupled Cahn-Hilliard and Navier-Stokes system for moving contact line problems. J. Comput. Phys. **231**, 8083–8099 (2012)

M. Gao, X.-P. Wang, An efficient scheme for a phase field model for the moving contact line problem with variable density and viscosity. J. Comput. Phys. **272**, 704–718 (2014)

C. Geuzaine, J.-F. Remacle, Gmsh: a three-dimensional finite element mesh generator with built-in pre- and post-processing facilities. Int. J. Numer. Methods Eng. **79**(11), 1309–1331 (2009)

J.-L. Guermond, A. Salgado, A splitting method for incompressible flows with variable density based on a pressure Poisson equation. J. Comput. Phys. **228**, 2834–2846 (2009)

V.E. Henson, U.M. Yang, BoomerAMG: a parallel algebraic multigrid solver and preconditioner. Appl. Numer. Math. **41**, 155–177 (2002)

G. Karypis, V. Kumar, METIS – Unstructured graph partitioning and sparse matrix ordering system, V2.0, Technical Report, 1995

B.S. Kirk, J.W. Peterson, R.H. Stogner, G.F. Carey, libMesh: A C++ library for parallel adaptive mesh refinement/coarsening simulations. Eng. Comput. **22**(3–4), 237–254 (2006)

S. Shin, J. Chergui, D. Juric, A solver for massively parallel direct numerical simulation of three-dimensional multiphase flows (2014), arXiv:1410.8568

Two New Enriched Multiscale Coarse Spaces for the Additive Average Schwarz Method

Leszek Marcinkowski and Talal Rahman

1 Introduction

We propose additive Schwarz methods with spectrally enriched coarse spaces for the standard finite element discretization of second order elliptic problems with highly varying and discontinuous coefficients. Such discontinuities may occur arbitrarily both inside and across subdomains. The convergence of the proposed methods depend linearly on the mesh parameter ratio H/h, and is independent of the distribution of the coefficient in the model problem when the coarse space is large enough. For similar work on domain decomposition methods addressing such problems, we refer to Galvis and Efendiev (2010); Spillane et al. (2014) and references therein.

The present method is an extension of a classical and an almost 20 years old additive Schwarz method, also known as the additive average Schwarz method, which was first proposed and analyzed in Bjørstad et al. (1997) for problems where the coefficients are constant in each subdomain, and later analyzed for varying coefficients in Dryja and Sarkis (2010). The condition number bound as shown in the last paper, depends quadratically on the mesh parameter ratio, and linearly on the contrast, that is the ratio between the maximum and the minimum value of the coefficient, in each subdomain boundary layer. Recently, the additive average Schwarz method has been extended to the case of Crouzeix–Raviart finite volume elements where, again, demonstrating that the method is robust with respect to

L. Marcinkowski (✉)
Faculty of Mathematics, University of Warsaw, Banacha 2, 02-097, Warszawa, Poland
e-mail: Leszek.Marcinkowski@mimuw.edu.pl

T. Rahman
Faculty of Engineering, Bergen University College, Inndalsveien 28, 5063, Bergen, Norway
e-mail: Talal.Rahman@hib.no

© Springer International Publishing AG 2017
C.-O. Lee et al. (eds.), *Domain Decomposition Methods in Science and Engineering XXIII*, Lecture Notes in Computational Science and Engineering 116, DOI 10.1007/978-3-319-52389-7_40

coefficients varying inside the subdomain but not along the subdomain boundary; cf. Loneland et al. (2016a,b). It is clear that, with standard coarse spaces it is hard to make an additive Schwarz method robust with respect to the contrast, unless some way of enrichment of the coarse spaces has been made.

Additive Schwarz methods for solving elliptic problems discretized by the finite element method have been studied extensively; see Toselli and Widlund (2005) for an overview. There are now several works on the additive average Schwarz method which exist in the literature, see e.g. Bjørstad et al. (1997); Dryja and Sarkis (2010). In the present work, borrowing some of the main ideas of Bjørstad and Krzyżanowski (2002); Chartier et al. (2003); Galvis and Efendiev (2010); Klawonn et al. (2015); Spillane et al. (2014), we propose to enrich the classical coarse space of the additive average Schwarz method by using a set of eigenfunctions of specially designed generalized eigenvalue problem in each subdomain. Those functions correspond to the eigenvalues that are larger than a given threshold. The analysis shows that the condition number bounds of the enriched method depend only on the threshold and the mesh parameter ratio. So, by enriching the coarse space, we are able to make the condition number to be independent of the contrast, thereby restore the bound which is known to be true for the case of piecewise constant coefficients.

The remainder of the paper is organized as follows: in Sect. 2, we introduce our model problem, and the finite element discrete formulation. Section 3 describes the classical Additive Average Schwarz method. In Sect. 4, we propose the two locally generalized eigenvalue problems in each subdomain, and show how we use their eigenfunctions to enrich the average coarse space of the method. In Sect. 5, we discuss the convergence of the method with the enrichment, and present some of the numerical results in Sect. 6.

2   Discrete Problem

In this paper we consider the following model elliptic partial differential equation:

$$- \nabla \cdot (\alpha(x) \nabla u) = f \quad \text{in} \quad \Omega, \qquad u = 0 \quad \text{on} \quad \partial \Omega, \tag{1}$$

where Ω is a polygonal domain in $\mathbb{R}^2$ and $f \in L^2(\Omega)$.

Let $\mathcal{T}_h$ be a quasi-uniform triangulation of Ω consisting of closed triangle elements such that $\bar{\Omega} = \bigcup_{K \in \mathcal{T}_h} K$. Let h_K be the diameter of K, and define $h = \max_{K \in \mathcal{T}_h} h_K$ as the largest diameter of the triangles $K \in \mathcal{T}_h$. We assume that there exists a nonoverlapping partitioning of Ω into open and connected Lipschitz polytopes $\{\Omega_i\}$, such that $\bar{\Omega} = \bigcup_{i=1}^N \bar{\Omega}_i$, which are aligned with the fine triangulation implying that an element of $\mathcal{T}_h$ can only be contained in one of the substructures Ω_i. Each subdomain then inherits a unique local triangulation $\mathcal{T}_h(\Omega_k)$ from $\mathcal{T}_h$. We also assume that the set of these subdomains form a coarse triangulation of the domain, which is shape regular in the sense of Brenner and Sung (1999). We define the sets of nodal points Ω_h, $\partial\Omega_h$, Ω_{ih} and $\partial\Omega_{ih}$ as the sets of vertices of the elements of $\mathcal{T}_h$ belonging to the regions Ω, $\partial\Omega$, Ω_i and $\partial\Omega_i$, respectively.

Let S_h be the standard continuous piecewise linear finite element space defined on the triangulation $\mathcal{T}_h$,

$$S_h = S_h(\Omega) := \{u \in C(\Omega) \cap H_0^1(\Omega) : v_{|K} \in P_1, \quad K \in \mathcal{T}_h\}.$$

The finite element approximation u_h of (1) is then defined as the solution to the following discrete problem: Find $u_h^* \in S_h$ such that

$$a(u_h^*, v) = (f, v), \qquad \forall v \in S_h, \tag{2}$$

where $a(u, v) = \sum_{K \in \mathcal{T}_h} \int_K \alpha \nabla u \nabla v \, dx$. Through scaling we can assume that $\alpha(x) \geq 1$. Also, since ∇u and ∇v are both piecewise constant on the elements of $\mathcal{T}_h$, $a(u, v)$ restricted to each element K can be written as $\int_K \alpha \nabla u \nabla v \, dx = (\nabla u)_{|K}(\nabla v)_{|K} \int_K \alpha(x) \, dx$, and hence we can assume that α is piecewise constant on each element of $\mathcal{T}_h$.

3 The Classical Additive Average Schwarz Method

In this section we introduce the Additive Average Schwarz method for the discrete problem (2).

We first introduce the average coarse space. For $u \in S_h(\Omega)$, we define the average operator $I_{av}u \in S_h(\Omega)$ as

$$I_{av}u := \begin{cases} u(x), & x \in \partial\Omega_{ih}, \\ \bar{u}_i, & x \in \Omega_{ih}, \end{cases} \quad i = 1, \ldots, N, \tag{3}$$

where

$$\bar{u}_i := \frac{1}{n_i} \sum_{x \in \partial\Omega_{i,h}} u(x). \tag{4}$$

Here, n_i is the number of nodal points on $\partial\Omega_i$, i.e., $\bar{u}_i$ is the discrete average of u over the boundary of the subdomain Ω_i.

The coarse space V_0 is defined as the image of the operator I_{av}, i.e.,

$$V_0 := Im(I_{av}). \tag{5}$$

Now, to introduce the local spaces, let $S_{h,k}$ be the restriction to $\overline{\Omega}_k$ of the function space S_h, i.e., $S_{h,k} = \{v \in C(\overline{\Omega}_k) : v_{|\tau} \in P_1, \tau \in \mathcal{T}_h(\Omega_k), v_{|\partial\Omega} = 0\}$, and the corresponding local subspace with zero boundary condition be $S_{h,k}^0 = S_{h,k} \cap H_0^1(\Omega_k)$.

Then we let the local spaces V_k to be equal to $S_{h,k}^0$. We decompose the finite element space S_h into $S_h(\Omega) = V_0 + \sum_{k=1}^N V_k$.

Note that this is a direct sum of the subspaces. However, only the local spaces are a-orthogonal to each other.

For $i = 0, \ldots, N$ we define projection like operators $T_i : S_h \to V_i$, as

$$a(T_i u, v) = a(u, v) \qquad \forall v \in V_i. \tag{6}$$

Now introducing $T := T_0 + \sum_{k=1}^N T_k$, we can replace the original problem by the equation

$$T u_h^* = g, \tag{7}$$

where $g = \sum_{i=0}^N g_i$ and $g_i = T_i u$. g_i is computed without knowing the solution u_h^* of (2):

$$a_i(g_i, v) = (f, v) \qquad \forall v \in V_i.$$

The bilinear form $a_i(\cdot, \cdot)$ is the restriction of $a(\cdot, \cdot)$ to Ω_i.

4   Eigenvalue Problems

In this section, we introduce the two generalized eigenvalue problems. We propose an extension of the coarse space by including some extensions of selected eigenfunctions of those problems in order to obtain better convergence properties of the method.

The layer corresponding to the subdomain Ω_k, consisting of elements of $\mathcal{T}_h(\Omega_k)$ touching the boundary $\partial \Omega_k$, is denoted by $\Omega_{k,\delta}$, cf. Fig. 1. For each subdomain and its layer, we define the maximum and the minimum values of the coefficient α as the following:

$$\begin{aligned}
\overline{\alpha}_{k,\delta} &:= \sup_{x \in \bar{\Omega}_{k,\delta}} \alpha(x), & \underline{\alpha}_{k,\delta} &:= \inf_{x \in \bar{\Omega}_{k,\delta}} \alpha(x), \\
\overline{\alpha}_k &:= \sup_{x \in \bar{\Omega}_k} \alpha(x), & \underline{\alpha}_k &:= \inf_{x \in \bar{\Omega}_k} \alpha(x).
\end{aligned} \tag{8}$$

The generalized eigenvalue problem is then defined as follows, with p as a superscript referring to the type of the problem: Find $(\lambda_j^{k,p}, \psi_j^{k,p}) \in \mathbb{R}_+ \times S_{h,k}^0$ such that

$$a_k(\psi_j^{k,p}, v) = \lambda_j^{k,p} b_k^{(p)}(\psi_j^{k,p}, v), \qquad \forall v \in S_{h,k}^0, \qquad p = 1, 2, \tag{9}$$

Fig. 1 The layer corresponding to the subdomain Ω_k, consisting of elements (*triangles*) of $\mathcal{T}_h(\Omega_k)$ touching the subdomain boundary $\partial\Omega_k$

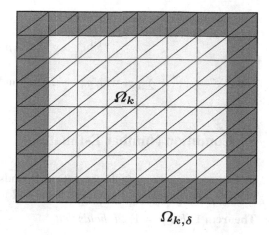

$$\Omega_{k,\delta}$$

where the bilinear forms are defined as

$$a_k(u, v) = a_{|\Omega_k}(u, v) = \int_{\Omega_k} \alpha \nabla u \nabla v \, dx, \tag{10}$$

$$b_k^{(1)}(u, v) = \underline{\alpha}_k (\nabla u, \nabla v)_{L^2(\Omega_k)}, \tag{11}$$

$$b_k^{(2)}(u, v) = \underline{\alpha}_{k,\delta} \int_{\Omega_{k,\delta}} \nabla u \nabla v \, dx + \int_{\Omega_k \setminus \Omega_{k,\delta}} \alpha \nabla u \nabla v \, dx, \tag{12}$$

with $\underline{\alpha}_k$ and $\underline{\alpha}_{k,\delta}$ being defined as in (8). Further, we extend $\psi_j^{k,p}$ to the rest of Ω by zero, and denote it by the same symbol; cf. also (13). We order the eigenvalues in the decreasing order as $\lambda_1^k \geq \lambda_2^k \geq \ldots \lambda_{M_k}^k$ where $M_k = dim(S_{h,k}^0)$. Then those bounds on the eigenvalues are true: $1 \leq \lambda_j^{k,p} \leq C_p$, where $C_1 = \frac{\overline{\alpha}_k}{\underline{\alpha}_k}$ and $C_2 = \frac{\overline{\alpha}_{k,\delta}}{\underline{\alpha}_{k,\delta}}$. Now define the local spectral component of the coarse space by

$$V_{k,0}^p = \mathrm{Span}(\psi_j^{k,p})_{j=1}^{n_k} \qquad k = 1, \ldots, N, \qquad p = 1, 2, \tag{13}$$

where $n_k \leq M_k = dim(S_{h,k}^0)$ is preset by the user or chosen adaptively for each subdomain. By adding this spectral component to the average coarse space, we propose a new and enriched coarse space defined as $V_0^{(p)} = V_0 + \sum_{k=1}^N V_{k,0}^p$, $p = 1, 2$. Accordingly, the new coarse operator $T_0^{(p)} : S_h \to V_0^p$ is defined as

$$a(T_0^{(p)} u, v) = a(u, v) \quad \forall v \in V_0^{(p)}, \qquad p = 1, 2. \tag{14}$$

With the local operators $T_k, k = 1, \ldots, N$ from the previous section, the new additive Schwarz operator $T^{(p)}$ becomes $T^{(p)} = T_0^{(p)} + \sum_{k=1}^N T_k$. The problem (2) is then replaced by the following ones:

$$T^{(p)} u_h^* = g^{(p)} \qquad p = 1, 2, \tag{15}$$

where $g^p = g_0^{(p)} + \sum_k g_k$ with $g_0^{(p)} = T_0^{(p)} u_h^*$ and $g_k = T_k u_h^*$ for $k = 1, \ldots, N$.

5 Condition Number Estimates

In this section, we provide theoretical bounds on the condition number of our method. The bounds are formulated in the following theorem.

Theorem 1 *For $p = 1, 2$ it holds that*

$$c \left(\min_k \frac{1}{\lambda_{n_k+1}^{k,p}} \right) \frac{h}{H} a(u, u) \leq a(T^{(p)} u, u) \leq C \, a(u, u), \qquad \forall u \in S_h,$$

where C, c are positive constants independent of the coefficient α, h and $H = \max_{k=1,\ldots,N} diam(\Omega_k)$.

The proof is based on the abstract framework for the additive Schwarz method, cf. e.g. Toselli and Widlund (2005).

Remark 1 In the original paper, cf. Bjørstad et al. (1997), where the authors assume that α is constant in each subdomain, the bound obtained for the Additive Average Schwarz method has the form: $\text{cond}(T) \leq C \frac{H}{h}$. For the multiscale problem, the bound as given in the paper (Dryja and Sarkis, 2010) has the following form: $\text{cond}(T) \leq C \max_k \frac{\bar{\alpha}_k}{\underline{\alpha}_k} \left(\frac{H}{h} \right)^2$.

Remark 2 If α is piecewise constant in each subdomain Ω_k, both eigenvalue problems become trivial, having only one eigenvalue which is equal to one. If the coefficient is constant in the boundary layers $\Omega_{k,\delta}$, although varying inside, in which case $\frac{\bar{\alpha}_{k,\delta}}{\underline{\alpha}_{k,\delta}} = 1$, the only eigenvalue of the second type of eigenvalue problem ($p = 2$) is also equal to one.

6 Numerical Experiments

For the numerical experiment we choose our model elliptic problem to be defined on a unit square, with homogeneous boundary condition and $f(x) = 2\pi^2 \sin(\pi x) sin(\pi y)$. For the coefficient α, we chose the following distribution, consisting of a background, channels crossing inside and stretching out of a subdomain, and inclusions along the boundary of a subdomain placed at the corners, where α takes different values. α_b, α_c, and α_i are the values of α respectively in the

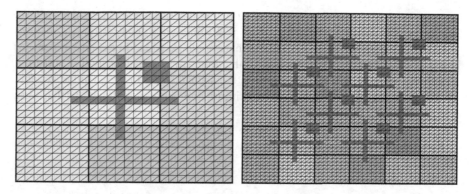

Fig. 2 Discretization and coarse partitioning of the unit square with different mesh sizes. The mesh size ratio $\frac{H}{h}$ are the same in this figure. Coefficient distribution includes both crossing channels and inclusions on the subdomain boundary

Table 1 Number of iterations and a condition number estimate (in parentheses) for each case, for the average Schwarz method, is shown

h	H					
	1/3	1/6	1/12	1/3	1/6	1/12
1/24	34 (5.73e1)			16 (1.46e1)		
1/48	56 (1.31e2)	49 (5.32e1)		28 (3.30e1)	25 (1.36e1)	
1/96	76 (2.80e2)	84 (1.20e2)	55 (5.35e1)	37 (7.04e1)	44 (3.03e1)	28 (1.36e1)

The left block of results correspond to the additive version, while the right block corresponds to the multiplicative version of the average Schwarz method. $\alpha_b = 1$, $\alpha_c = 1e4$, and $\alpha_i = 1e6$

Table 2 Number of iterations and a condition number estimate (in parentheses) for each case is shown

	None	2	4	6	8	10
Add	299 (2.72e6)	321 (7.98e5)	197 (1.36e4)	118 (7.10e3)	46 (4.48e1)	46 (4.44e1)
Mlt	159 (6.79e5)	163 (2.00e5)	99 (3.38e3)	59 (1.78e3)	23 (1.15e1)	23 (1.14e1)

The first line (Add) of results correspond to the additive version, while the second line (Mlt) corresponds to the multiplicative version of the method. $\alpha_b = 1$, $\alpha_c = 1e4$, and $\alpha_i = 1e6$. Each column corresponds to the number of eigenfunctions (preset) used in each subdomain for the test

background, in the channels, and in the inclusions. We have chosen one particular distribution of the coefficient for this paper, cf. Fig. 2.

The results are presented in Tables 1 and 2 using the average Schwarz method with the type 2 generalized eigenvalue problem. The tables show the number of iterations required to reduce the residual norm by 5e-6, and a condition number estimate (in parentheses), in each test case. Both the additive and the multiplicative version of the average method have been tried, the latter one converges twice as fast as the former one.

As seen from the first table, the proposed method is scalable and the condition number grow as the ratio $\frac{H}{h}$. For this table the eigenfunctions were chosen adaptively in each subdomain, those corresponding to the eigenvalues greater than 100. As we know it from the analysis that there is a minimum number of eigenfunctions (corresponding to the bad eigenvalues) that should be added in the enrichment for the method to be robust with respect to the contrast. For the distribution shown in Fig. 2, this number is eight as seen from the second table. In the adaptive version, cf. the same test case in Table 1, the maximum number of eigenfunctions that were used in this particular case was also eight.

Acknowledgements Leszek Marcinkowski was partially supported by Polish Scientific Project no 2016/21/B/ST1/00350.

References

P.E. Bjørstad, P. Krzyżanowski, Flexible 2-level Neumann-Neumann method for structural analysis problems, in *Proceedings of the 4th International Conference on Parallel Processing and Applied Mathematics, PPAM2001 Naleczow, Poland, September 9-12, 2001*. Lecture Notes in Computer Science, vol. 2328 (Springer, Berlin, 2002), pp. 387–394

P.E. Bjørstad, M. Dryja, E. Vainikko, Additive Schwarz methods without subdomain overlap and with new coarse spaces, in *Domain Decomposition Methods in Sciences and Engineering (Beijing, 1995)* (Wiley, Chichester, 1997), pp. 141–157

S.C. Brenner, L.-Y. Sung, Balancing domain decomposition for nonconforming plate elements. Numer. Math. **83**(1), 25–52 (1999)

T. Chartier, R.D. Falgout, V.E. Henson, J. Jones, T. Manteuffel, S. McCormick, J. Ruge, P.S. Vassilevski, Spectral AMGe (ρAMGe). SIAM J. Sci. Comput. **25**(1), 1–26 (2003). ISSN 1064-8275. doi:10.1137/S106482750139892X

M. Dryja, M. Sarkis, Additive Average Schwarz methods for discretization of elliptic problems with highly discontinuous coefficients. Comput. Methods Appl. Math. **10**(2), 164–176 (2010)

J. Galvis, Y. Efendiev, Domain decomposition preconditioners for multiscale flows in high contrast media: reduced dimension coarse spaces. Multiscale Model. Simul. **8**(5), 1621–1644 (2010). ISSN 1540-3459. doi:10.1137/100790112

A. Klawonn, P. Radtke, O. Rheinbach, FETI-DP methods with an adaptive coarse space. SIAM J. Numer. Anal. **53**(1), 297–320 (2015)

A. Loneland, L. Marcinkowski, T. Rahman, Additive average schwarz method for a crouzeix - raviart finite volume element discretization of elliptic problems, in Domain Decomposition Methods in Science and Engineering XXII, ed. by T. Dickopf, M.J. Gander, L. Halpern, R. Krause, L.F. Pavarino. Lecture Notes in Computational Science and Engineering, vol. 104 (Springer, 2016a), pp. 587–594

L.M. Loneland, T. Rahman, Additive average Schwarz method for a Crouzeix-Raviart finite volume element discretization of elliptic problems with heterogeneous coefficients. Numer. Math. **134**(1), 91-118 (2016b). doi:10.1007/s00211-015-0771

N. Spillane, V. Dolean, P. Hauret, F. Nataf, C. Pechstein, R. Scheichl, Abstract robust coarse spaces for systems of PDEs via generalized eigenproblems in the overlaps. Numer. Math. **126**(4), 741–770 (2014). ISSN 0029-599X. doi:10.1007/s00211-013-0576-y

A. Toselli, O. Widlund, *Domain Decomposition Methods—Algorithms and Theory*. Springer Series in Computational Mathematics, vol. 34 (Springer, Berlin, 2005). ISBN 3-540-20696-5

Relaxing the Roles of Corners in BDDC by Perturbed Formulation

Santiago Badia and Hieu Nguyen

1 Introduction

The Balancing Domain Decomposition by Constraints (BDDC) method was first introduced by Dohrmann (2003). Compared to its parent, the BDD method by Mandel (1993), one of the advances in BDDC method is the use of constraints to enforce equality of averages across faces, edges, or at individual dofs on substructure boundaries called corners. These constraints serve two purposes. First, they ensure that the coefficient matrix of the coarse problem is always invertible. Second, they induce a natural coarse space leading to fast convergence. While corner constraints do not have significant contribution in serving the second purpose, they are mainly responsible for the first one. In addition, in order to use positive definite sparse direct solvers, which are faster and more robust than their indefinite counterparts, the corners should be chosen so that the local matrix sub-assembled for all dofs in each substructure except corners is positive definite. Here we do not consider a change of basis, cf. Li and Widlund (2006), as it destroys good sparsity pattern of local matrices and is more complicated to implement.

S. Badia
Universitat Politecnica de Catalunya, Jordi Girona 1-3, Edifici C1, 08034 Barcelona, Spain
CIMNE - Centre Internacional de Mètodes Numèrics en Enginyeria, Parc Mediterrani de la
Tecnologia, UPC, Esteve Terradas 5, 08860 Castelldefels, Spain
e-mail: sbadia@cimne.upc.edu

H. Nguyen (✉)
Institute of Research and Development, Duy Tan University, 3 Quang Trung, Danang, Vietnam
CIMNE - Centre Internacional de Mètodes Numèrics en Enginyeria, Parc Mediterrani de la
Tecnologia, UPC, Esteve Terradas 5, 08860 Castelldefels, Spain
e-mail: nguyentrunghieu14@dtu.edu.vn

© Springer International Publishing AG 2017
C.-O. Lee et al. (eds.), *Domain Decomposition Methods in Science
and Engineering XXIII*, Lecture Notes in Computational Science
and Engineering 116, DOI 10.1007/978-3-319-52389-7_41

Different corner selection algorithms have been proposed by Dohrmann (2003), Lesoinne (2003), Klawonn and Widlund (2006) and Šístek et al. (2012) to guarantee such choices of corners. However, based on our experience, the implementation of this type of algorithms is an involved and time-consuming task, which does depend on the physical problem to be solved and also the type of FE formulation being used. Furthermore, the situation becomes far more complicated when subdomains are disconnected, or only connected by corners or edges. Unfortunately, the currently available parallel mesh partitioners, ParMETIS by Karypis et al. (1997) and PT-Scotch by Chevalier and Pellegrini (2008), cannot guarantee connected subdomains.

In this paper, we present a perturbed formulation of the BDDC method where the coarse coefficient matrix and the local stiffness matrices are guaranteed to be positive definite. For this new formulation, corner constraints are optional and should be selected only for the convergence purpose. Consequently, one can consider much smaller coarse problems, only involving faces and/or edges. This is particularly important when dealing with unstructured meshes and partitions generated by mesh partitioners, due to the proliferation of corners. Since the coarse problem is the bottleneck that can destroy scalability, these strategies are better suited for large scale simulations.

The presentation of this paper is concise, engineering-friendly and useful to quickly absorb the of essential ideas of the method for implementation. For a full mathematical treatment with complete analysis and additional numerical experiments, we refer the reader to Badia and Nguyen (2016).

2 BDDC Overview

Even though our results do apply for linear elasticity, our presentation, due to limited space, only features Poisson's equation: find $u(x) \in H_0^1(\Omega)$, for a given polygonal (polyhedral) domain $\Omega \subset \mathbb{R}^n$, $n = 2, 3$ and a source term $f(x) \in L^2(\Omega)$, such that

$$\underbrace{\int_\Omega \nabla u(x) \cdot \nabla v(x)\, dx}_{\equiv a(u,v)} = \underbrace{\int_\Omega f(x)v(x)\, dx}_{\equiv (f,v)}, \quad \text{for all } v(x) \in H_0^1(\Omega). \tag{1}$$

Let $\mathcal{T}_h$ be a shape-regular mesh of size h of Ω. Discretizing (1) using the space $V_h \subset H_0^1(\Omega)$ of linear piecewise polynomials defined on $\mathcal{T}_h$, we arrive at the following system of equations:

$$Au = f. \tag{2}$$

Let us also consider a nonoverlapping partition of Ω into subdomains, also known as substructures, $\bar{\Omega} = \cup_{i=1}^N \bar{\Omega}_j$ with the inter-subdomain interface $\Gamma = \cup_{i=1}^N \partial\Omega_j \backslash \partial\Omega$. We assume that the partition is quasi-uniform, and the subdomains are obtained by aggregation of elements in $\mathcal{T}_h$. We denote H_i, or generically H, the size of Ω_i.

Let $K^{(i)}$ be the stiffness matrix associated with substructure Ω_i. It should be noted that $K^{(i)}$ is symmetric positive semidefinite and is **singular** when Ω_i is a floating subdomain ($\partial\Omega_i \cap \partial\Omega = \emptyset$).

Denote by R_i the global to local mapping that restrict any vector u to its local counterpart u_i, i.e., $u_i = Ru$. It follows that

$$A = R^T KR, \quad \text{where } R = [R_1^T \ldots R_N^T]^T, \ K = \text{diag}(K^{(1)}, \ldots, K^{(N)}).$$

For simplicity, we assume that interior dofs are always ordered before interface dofs, namely

$$u = [u_I^T \ u_\Gamma^T]^T, \quad u_I = R_I u, \quad u_\Gamma = R_\Gamma u.$$

This leads to the following reordered block structures

$$A = \begin{bmatrix} A_{II} & A_{I\Gamma} \\ A_{\Gamma I} & A_{\Gamma\Gamma} \end{bmatrix}, \quad K = \begin{bmatrix} A_{II} & K_{I\Gamma} \\ K_{\Gamma I} & K_{\Gamma\Gamma} \end{bmatrix}, \quad \text{and} \quad K^{(i)} = \begin{bmatrix} A_{II}^{(i)} & A_{I\Gamma}^{(i)} \\ A_{\Gamma I}^{(i)} & K_{\Gamma\Gamma}^{(i)} \end{bmatrix}.$$

The BDDC preconditioner for solving the linear system (2) is completely defined by a weight matrix $W = \text{diag}(W^{(1)}, \ldots, W^{(N)})$ and a constraint matrix C. The matrix W forms a partition of unity, namely

$$R^T WR = \sum_{i=1}^{N} R_i^T W^{(i)} R_i = I.$$

We can now find the matrix of energy minimizing coarse basis functions Ψ and obtain the coefficient matrix of the coarse space K_c as follows

$$\underbrace{\begin{bmatrix} K & C^t \\ C & 0 \end{bmatrix}}_{K_{\text{BIG}}} \begin{bmatrix} \Psi \\ \Lambda \end{bmatrix} = \begin{bmatrix} 0 \\ R_c \end{bmatrix}, \quad K_c = \Psi^T K\Psi. \tag{3}$$

Finally, the BDDC preconditioner is formulated as

$$P_{\text{BDDC}} = P_1 + (I - P_1 A)P_2(I - AP_1), \tag{4}$$

$$P_1 = R_I^T A_{II}^{-1} R_I, \quad P_2 = R^T W(\Psi K_c^{-1} \Psi^T + P_3)WR, \tag{5}$$

where P_3 is defined by

$$\begin{bmatrix} K & C^t \\ C & 0 \end{bmatrix} \begin{bmatrix} P_3 v \\ \lambda \end{bmatrix} = \begin{bmatrix} v \\ 0 \end{bmatrix}, \quad \forall v. \tag{6}$$

For more details of the formulation and implementation of the BDDC method, we
refer the reader to Badia et al. (2014) and Dohrmann (2003, 2007).

3 Perturbed BDDC

Preconditioner Formulation Let $\widetilde{K} = \mathrm{diag}(\widetilde{K}^{(1)}, \ldots, \widetilde{K}^{(N)})$ be a perturbation of
K. Assume that $\widetilde{K}$ satisfies the following assumptions:

Assumption 1 *There exist two constants C_L and C_U which are independent of the
size of the domain (d), the size of the subdomains (H), and the number of the
subdomains (N) such that*

$$C_\mathrm{L}\, v^T K v \le v^T \widetilde{K} v \le C_\mathrm{U}\, v^T K v, \quad \text{for all } v \text{ of appropriate size.}$$

Assumption 2 *The matrix $\widetilde{K}^{(i)}$ is symmetric positive definite (s.p.d) for all i.*

Assumption 3 *There exists a constant C_ℓ which is independent of the size of the
domain (d), the size of the subdomains (H), and the number of the subdomains (N)
such that:*

$$C_\ell\, v_i^T K^{(i)} v_i \le v_i^T \widetilde{K}^{(i)} v_i, \quad \text{for all } v_i \text{ of appropriate size.}$$

Let $\widetilde{\Psi}, \widetilde{K}_c, \widetilde{P}_3$ be defined similarly to Ψ, K_c, P_3 as in (3) and (6), but with K
replaced by $\widetilde{K}$. Then the perturbed BDDC preconditioner is given as

$$\widetilde{P}_{\mathrm{BDDC}} = P_1 + (I - P_1 A)\widetilde{P}_2(I - A P_1),$$

$$\widetilde{P}_2 = R^T W (\widetilde{\Psi}\widetilde{K}_c^{-1}\widetilde{\Psi}^T + \widetilde{P}_3) W R,$$

Remark 1 If Assumption 2 holds, the matrix $\widetilde{K}$ is s.p.d. From (3), it follows that the
coarse matrix $\widetilde{K}_c$ is also s.p.d, thus is invertible. In addition, (3) and (6) can be solved
using positive definite sparse direct solvers when K is replaced by $\widetilde{K}$. Consequently,
corner constraints are not required in the perturbed formulation of BDDC.

Choices of Perturbation We present here two practical choices of perturbed local
stiffness matrices $\widetilde{K}^{(i)}$. The first one uses $M^{(i)}$, the mass matrix associated with
subdomain Ω_i:

$$\widetilde{K}^{(i)} = K^{(i)} + \frac{1}{d^2} M^{(i)}. \tag{7}$$

The second choice is to use

$$\widetilde{K}^{(i)} = K^{(i)} + \frac{H_i^{n-1}}{d^n} M_{\Gamma\Gamma}^{(i)}, \tag{8}$$

where $M_{\Gamma\Gamma}^{(i)}$ is the stiffness matrix associated with subdomain Ω_i assembled only for dofs on the interface. We call this choice Robin perturbation because the local Neumann problem in this case can be posed with Robin boundary condition $(H_j^{n-1}/D^n)\, u + \partial u/\partial n_i = 0$, where n_i is the outward normal vector of $\partial\Omega_i$.

It is not difficult to verify that the choices of $\widetilde{K}^{(i)}$ in (7) and (8) satisfy Assumptions 1, 2 and 3 with $C_\ell = C_{\mathrm{L}} = 1$ and $C_{\mathrm{U}} = 1 + C_\Omega$, where C_Ω depends only on the shape of Ω. Details can be found in Badia and Nguyen (2016).

4 Convergence Results

In this section, we present (without proofs) two main convergence results of the perturbed BDDC method. For detailed mathematical analysis, we refer the reader to Badia and Nguyen (2016).

Theorem 4 *There exists a positive constant C independent of h, H, N, C_{U}, C_{L} and C_ℓ such that*

$$\kappa(\widetilde{P}_{\mathrm{BDDC}}A) \leq C\, \frac{(C_{\mathrm{U}})^2}{C_{\mathrm{L}} \min\{C_\ell, C_{\mathrm{L}}\}} \left(1 + \ln\frac{H}{h}\right)^2 = \frac{\alpha_{\mathrm{M}}}{\alpha_{\mathrm{m}}},$$

where $\alpha_{\mathrm{m}} = C_{\mathrm{U}}^{-1}$ and α_{M} is consistently defined.

The proof of this theorem uses the fact that the spectrum of the preconditioned matrix of the whole system $\widetilde{P}_{\mathrm{BDDC}}A$ is the same as the spectrum of the preconditioned matrix of the Schur complement $\widetilde{B}_{\mathrm{BDDC}}S$ plus additional eigenvalues equals 1, cf. Dohrmann (2007) and Li and Widlund (2006). The estimates for eigenvalues in the spectrum of $\widetilde{B}_{\mathrm{BDDC}}S$ is documented in detail in Badia and Nguyen (2016).

Remark 2 Theorem 4 indicates that the perturbed BDDC method has the same polylogarithmic bound for the condition number as the standard one. The precondition number depends on the local problem size but not on the number of subdomains. In other word, the method is weakly scalable.

In order to be well-posed, the standard BDDC method need to have enough constraints to exclude all subdomain-wise constant functions for Poisson's equation and all rigid body modes for linear elasticity. This is no longer necessary for the perturbed BDDC method as its well-posedness is automatically guaranteed. However, the perturbed BDDC method still need to have sufficient constraints to achieve fast convergence.

The following theorem concerns the spectrum of the preconditioned system of the perturbed BDDC method when not all the subdomain-wise constant functions or the rigid body modes are excluded by selected constraints.

Theorem 5 *Assume that* $\ker(K_{\mathrm{BIG}}) \neq \emptyset$ *then the spectrum of the preconditioned system, counting multiplicities, can be decomposed as*

$$\sigma(\widetilde{P}_{\mathrm{BDDC}}A) = \mathcal{A}_1 \cup \mathcal{A}_2, \tag{9}$$

where $|\mathcal{A}_1| \leq \dim(\ker(K_{\mathrm{BIG}}))$, $\mathcal{A}_1 \subset [\alpha_m, \hat{\alpha}_M]$ *and* $\mathcal{A}_2 \subset [\alpha_m, \alpha_M]$. *Here, the constants* α_m *and* α_M *are defined in Theorem 4, and* $\hat{\alpha}_M > \alpha_M$.

Remark 3 When the seclected constraints fail to eliminate a small number of subdomain-wise constant functions or rigid body modes, namely $\ker(K_{\mathrm{BIG}}) \neq \emptyset$ and $\dim(\ker(K_{\mathrm{BIG}}))$ is small, Theorem 5 indicates that most of the eigenvalues of the preconditioned system can still be bounded by the usual bounds as in the case with sufficient constraints. Some of the remaining eigenvalues might be larger than the usual upper bound. However, they are isolated (the number of them is bounded from above by $\dim(\ker(K_{\mathrm{BIG}}))$. As large isolated eigenvalues can only delay the convergence of the CG method by few iterations, cf. Cavesson and Lindskog (1986), the perturbed BDDC method is still scalable.

5   Numerical Experiments

Both the standard and the perturbed BDDC preconditioners with different options of constraints will be used to solve (2) by the CG method. The number of CG iterations and the time (in second) to reduce the residual by at least a factor of 1e-6 will be reported.

In figures, legends C, E and F are used to indicate corner, edge and face constraints, respectively. The suffix 0 is for the standard BDDC formulation (no perturbation). The suffix CD is to emphasize that the corner selection algorithm by Šístek et al. (2012) and the standard BDDC formulation are used. If the legend is without a suffix, it represents a result with a perturbed BDDC formulation and that no corner selection algorithm is involved.

We present only results for perturbation by full mass matrices. For results using a Robin perturbation, we refer to Badia and Nguyen (2016). It is worth noting that the results of the two choices are very close.

We consider (1) with Ω being the unit cube and elasticity of a beam $[0\ 2] \times [0\ 0.5] \times [0\ 0.5]$. For the latter, (homogeneous) Dirichlet boundary condition is only imposed on one side of the beam (the plane $x = 0$).

We use uniform structured hexahedral meshes which are partitioned into $k \times k \times k$, $k = 3, \ldots, 11$ (Poisson's problem) and $4k \times k \times k$, $k = 2, \ldots, 11$ (elasticity) cubic subdomains. For weak scalability tests, when k increases (H decreases), we use smaller mesh size, h, to keep H/h constant.

From Figs. 1 and 2, we can conclude that the perturbed BDDC method, for all the considered choices of constraints, is weakly scalable, namely the numbers of iterations are almost constant when the number of subdomains increases. The

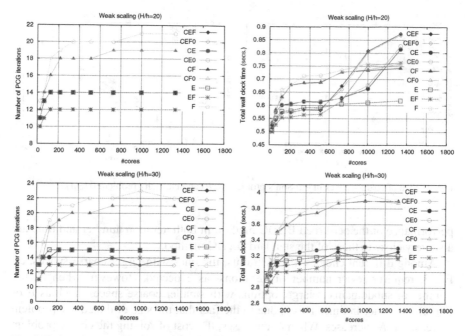

Fig. 1 Poisson's equation: Perturbation with full mass matrices

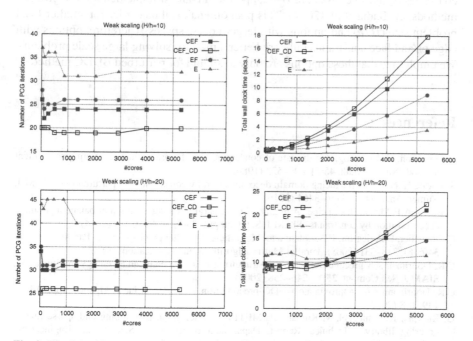

Fig. 2 Elasticity of a beam: Perturbation with full mass matrices

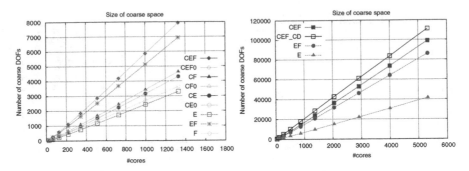

Fig. 3 Size of coarse spaces in Poisson's problem (*left*) and elasticity problem (*right*)

performance of the perturbed BDDC method in both iteration number and time are also very close to those of the standard BDDC method.

Among different choices of constraints, the ones with larger coarse spaces, cf. Fig. 3, requires fewer number of iterations, as expected. However, when N, the number of subdomains is large, options with smaller coarse spaces, such as E or F, perform better in time. This is due to the fact that the size of the coarse problem increases as N increases. When N increases, the cost of solving the coarse problem become more and more dominant and eventually dictates the time performance as coarse tasks and fine tasks are overlapped in advanced implementation of BDDC methods, cf. Badia et al. (2014). This phenomena exhibits earlier for smaller local problem size (H/h) and options with larger coarse spaces. Therefore, options with edge or/and face constraints only are better suited for solving large scale problems. We emphasize that these options are only available for perturbed BDDC method.

References

O. Axelsson, G. Lindskog, On the rate of convergence of the preconditioned conjugate gradient method. Numer. Math. **48**(5), 499–523 (1986)

S. Badia, H. Nguyen, Balancing domain decomposition by constraints and perturbation, SIAM J. Numer. Anal. **54**(6), 3436–3464 (2016)

S. Badia, A.F. Martín, J. Principe, A highly scalable parallel implementation of balancing domain decomposition by constraints. SIAM J. Sci. Comput. **36**(2), C190–C218 (2014)

C. Chevalier, F. Pellegrini, PT-Scotch: a tool for efficient parallel graph ordering. Parallel Comput. **34**(6–8), 318–331 (2008). Parallel Matrix Algorithms and Applications

C.R. Dohrmann, A preconditioner for substructuring based on constrained energy minimization. SIAM J. Sci. Comput. **25**(1), 246–258 (2003)

C.R. Dohrmann, An approximate BDDC preconditioner. Numer. Linear Algebra Appl. **14**(2), 149–168 (2007)

G. Karypis, K. Schloegel, V. Kumar, ParMETIS: parallel graph partitioning and sparse matrix ordering library, Technical Report, Department of Computer Science and Engineering, University of Minnesota, 1997

A. Klawonn, O.B. Widlund, Dual-primal FETI methods for linear elasticity. Commun. Pure Appl. Math. **59**(11), 1523–1572 (2006)

M. Lesoinne, A FETI-DP corner selection algorithm for three-dimensional problems, in *Domain Decomposition Methods in Science and Engineering XIV*, ed. by I. Herrera, D.E. Keyes, O.B. Widlund, R. Yates (National Autonomous University of Mexico (UNAM), Mexico City, Mexico, 2003), pp. 217–224

J. Li, O.B. Widlund, FETI-DP, BDDC, and block Cholesky methods. Int. J. Numer. Methods Eng. **66**(2), 250–271 (2006)

J. Mandel, Balancing domain decomposition. Commun. Numer. Methods Eng. **9**(3), 233–241 (1993)

J. Šístek, M. Čertíková, P. Burda, J. Novotný, Face-based selection of corners in 3D substructuring. Math. Comput. Simul. **82**(10), 1799–1811 (2012)

Simulation of Blood Flow in Patient-specific Cerebral Arteries with a Domain Decomposition Method

Wen-Shin Shiu, Zhengzheng Yan, Jia Liu, Rongliang Chen, Feng-Nan Hwang, and Xiao-Chuan Cai

1 Introduction

The high morbidity and mortality of stroke has caused a social and economic burden in contemporary society. The underlying mechanisms of stroke are not fully understood. Changes of cerebral hemodynamics might be one of the critical factors that cause stroke. There are several techniques to detect the hemodynamic alterations, one of which is through computer simulation by solving partial differential equations that describe the physics of the blood flow. For example, there are some numerical studies of blood flow through a total cavopulmonary connection (Bazilevs et al., 2009), the coronary (Taylor et al., 2013), cerebral aneurysms (Boussel et al., 2009; Cebral et al., 2005; Takizawa et al., 2011), and cerebrovascular arteries, which is the focus of this paper (Moore et al., 2005). In general, solving a fluid flow problem with complex geometry in 3D is difficult. In this work, we employ a Newton-Krylov-Schwarz (NKS) algorithm for solving large nonlinear systems arising from a fully implicit discretization of the incompressible Navier-Stokes equations using the Galerkin/least squares (GLS) finite element method. NKS has

W.-S. Shiu (✉) • Z. Yan • J. Liu • R. Chen
Shenzhen Institutes of Advanced Technology, Chinese Academy of Sciences, Shenzhen 518055, Guangdong, People's Republic of China
e-mail: whsin5@gmail.com; zz.yan@siat.ac.cn; jia.liu@siat.ac.cn; rl.chen@siat.ac.cn

F.-N. Hwang
Department of Mathematics, National Central University, Jhongli District, Taoyuan City 32001, Taiwan
e-mail: hwangf@math.ncu.edu.tw

X.-C. Cai
Department of Computer Science, University of Colorado, Boulder, CO 80309, USA
e-mail: cai@cs.colorado.edu

© Springer International Publishing AG 2017
C.-O. Lee et al. (eds.), *Domain Decomposition Methods in Science and Engineering XXIII*, Lecture Notes in Computational Science and Engineering 116, DOI 10.1007/978-3-319-52389-7_42

been applied for simple blood flow model problems previously (Hwang et al., 2010). In this work, we apply the algorithm to a patient-specific cerebrovascular problem that is more complicated, since the cerebrovascular artery has ischaemic stenosis, and the vessel wall is atherosclerotic. The rest of the paper is organized as follows. In the next section, we provide a description of the governing equations of blood flow in cerebral arteries, the finite element discretization, and the parallel NKS based solution algorithm. In Sect. 3, numerical results and parallel performance study are presented. Some concluding remarks are given in Sect. 4.

2 Blood Flow Model, Discretization, and Solution Algorithm

We assume that the blood flow is isothermal, incompressible, Newtonian and laminar, and modeled by the unsteady Navier-Stokes equations,

$$
\begin{cases}
\rho \left(\dfrac{\partial u}{\partial t} + u \cdot \nabla u \right) - \nabla \cdot \sigma = 0 & \text{in } \Omega \times (0, T), \\
\nabla \cdot u = 0 & \text{in } \Omega \times (0, T), \\
u = 0 & \text{on } \Gamma_{wall} \times (0, T), \\
u = g & \text{on } \Gamma_{in} \times (0, T), \\
\sigma \cdot n = 0 & \text{on } \Gamma_{out} \times (0, T), \\
u = u_0 & \text{in } \Omega \text{ at } t = 0,
\end{cases}
\tag{1}
$$

where $u = (u_1, u_2, u_3)^T$ is the velocity field, ρ is the fluid density, and σ is the Cauchy stress tensor defined as $\sigma = -pI + 2\mu D$, where p is the pressure, I is the identity tensor, μ is dynamic viscosity, and the deformation rate tensor $D = \frac{1}{2}[\nabla u + (\nabla u)^T]$. $\Omega \in R^3$ is the computational domain, with three boundaries Γ_{in}, Γ_{out} and Γ_{wall}; Γ_{in} is the surface of the inlet, Γ_{out} contains the surfaces of all outlets, and Γ_{wall} is the vessel wall. To close the flow system, some proper boundary conditions need to be imposed. We impose a uniform velocity, g, for the velocity on Γ_{in}; a stress-free boundary condition on Γ_{out}, and a no-slip boundary condition on Γ_{wall}.

To discretize (1), we employ a $P_1 - P_1$ GLS finite element method for the spatial domain, and an implicit first-order backward Euler scheme for the temporal domain (Wu and Cai, 2014). The GLS finite element takes the following form (Franca and Frey, 1992): Find $u_h^{(n+1)} \in V_h^g$ and $p_h^{(n+1)} \in P_h$, such that

$$
B(u_h^{(n+1)}, p_h^{(n+1)}; v, q) = 0, \quad \forall (v, q) \in V_h^0 \times P_h
$$

with

$$B(u, p; v, q) = \left(\frac{u - u^{(n)}}{\Delta t} + (\nabla u)u, v \right) + (\nu \nabla u, \nabla v) - (\nabla \cdot v, p)$$

$$+ \sum_{K \in \mathcal{T}^h} \left(\frac{u - u^{(n)}}{\Delta t} + (\nabla u)u + \nabla p, \tau_{GLS}((\nabla v)u - \nabla q) \right)_K$$

$$- (\nabla \cdot u, q) + (\nabla \cdot u, \delta_{GLS} \nabla \cdot v),$$

where V_h^0 and V_h^g are the weighting and trial velocity function spaces respectively. P_h is a linear finite element space for the pressure and used for both the weighting and trial pressure function spaces. $u^{(n)}$ is the velocity vector at the current time step, and u and p (we drop the superscript $(n + 1)$ here for simplicity) are unknown velocity and pressure at the next time step. ν is the kinematic viscosity. Δt is the time step size. Note that $\mathcal{T}^h = \{K\}$ is a tetrahedral mesh. We use the stabilization parameters τ_{GLS} and δ_{GLS} suggested in Franca and Frey (1992). The GLS formulation can be written as a nonlinear algebraic system

$$F(x) = 0, \tag{2}$$

where x is the vector of nodal values of the velocity and the pressure.

We apply NKS to solve (2). NKS is an inexact Newton method in which the Jacobian systems are solved by an one-level Schwarz preconditioned Krylov subspace method, briefly described as follows: Let $x^{(k)}$ be the current approximation of x, and $x^{(k+1)}$ the new approximation computed by the substeps:

Step 1: Solve the following preconditioned Jacobian system approximately by GMRES to find a Newton direction $s^{(k)}$,

$$J_k M_k^{-1} y = -F(x^{(k)}), \text{ with } s^{(k)} = M_k^{-1} y, \tag{3}$$

where J_k is the Jacobian of F evaluated at Newton step k, and M_k^{-1} is a right preconditioner.

Step 2: Obtain the new approximation with a linesearch method,

$$x^{(k+1)} = x^{(k)} + \lambda^{(k)} s^{(k)}, \tag{4}$$

where $\lambda^{(k)}$ is a step length parameter.

We define the additive Schwarz preconditioner in the matrix form as

$$M_k^{-1} = \sum_{i=1}^{N} (R_i^h)^T J_i^{-1} R_i^h,$$

where J_i^{-1} is the inverse of the subspace Jacobian $J_i = R_i^h J(R_i^h)^T$. We denote R_i^h as the global-to-local restriction operator and $(R_i^h)^T$ as the local-to-global prolongation operator. The multiplication of J_i^{-1} with a vector is solved by a direct solver such as sparse LU decomposition or an inexact solver such as ILU with some level of fill-ins.

3   A Case Study and Discussions

We consider a pair of patient-specific cerebrovascular geometries provided by the Beijing Tiantan Hospital, as shown in Fig. 1. The pair of cerebral arteries belongs to the same patient before and after the cerebral revascularization surgery respectively. In Fig. 1, the left artery has a stenosis in the middle, the right figure shows the same artery after the stenosis is surgically removed. Our numerical simulations provide a valuable tool to understand the change of the dynamics of the blood flow in the patient and the impact of the surgery. For convenience, let us denote the artery with a stenosis as "pre" and the repaired artery as "post". Table 1 lists the number of vertices, elements and unknowns of the finite element meshes that we generate for solving the flow problems.

The blood flow is characterized with density $\rho = 1.06\,g/cm^3$, and viscosity $\mu = 0.035\,g/(cm \cdot s)$. The inflow velocity profile is shown in Fig. 2. The time step size is $\Delta t = 10^{-2}\,s$. For the algorithm parameters, the overlapping size for the Schwarz preconditioner is set to be $\delta = 1$, and subdomain linear system is solved by ILU(1). The Jacobian system is solved inexactly by using an additive Schwarz preconditioned GMRES with relative stopping condition 10^{-4}. We define Newton convergence with a relative tolerance of 10^{-6} or an absolute tolerance of 10^{-10}. To observe the behavior of the blood flow in systolic and diastolic phases,

Fig. 1 3D tetrahedral meshes before and after the surgery. The narrowing cerebral artery with a local refinement at the stenosed segment (*left*) and the repaired cerebral artery (*right*)

Table 1 Mesh information for two cerebrovascular geometries

Mesh	# of vertices	# of elements	# of unknowns
Pre	441,475	2,208,337	1,765,900
Post	287,936	1,360,588	1,151,744

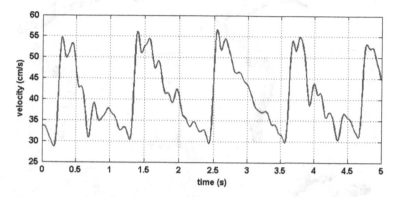

Fig. 2 Inflow velocity profile for 5 cardiac cycles discretized with 500 time steps

we respectively plot the numerical solutions at $t = 2.54$ s and $t = 3.2$ s. Figure 3 shows the relative pressure distributions, and Fig. 4 shows the streamlines whose color indicates the velocity magnitude.

We focus on the comparison between the "pre" and "post" cases. Figure 3 shows that the range of the relative pressure value of the "pre" case is more than double that of the "post" case at the systolic and diastolic phases. Moreover, as shown in the same figure, the relative pressure ratio between the anterior and posterior parts of the stenosed portion in the "pre" case is large, and the relative pressure value of the "post" case at the repaired portion has a smaller variation. From the streamline plots, the blood flow is more disordered in the "pre" case than in the "post" case during both the diastolic period and the systolic period. In addition, the maximum of the velocity occurs at the stenosed portion in the "pre" case, and the variation of the velocity distributions in the repaired portion is quite small. Similar to the pressure distribution, the range of velocity magnitude of the "pre" case is wider than the "post" case.

We use the "post" case to test the parallel performance, and the simulation is carried out for 10 time steps. Numerical results are summarized in Table 2. "np" is the number of processor cores. "NI" denotes the number of Newton iterations per time step, "LI" denotes the average number of GMRES iterations per Newton step, "T" represents the total compute time in seconds and "EFF" is the parallel efficiency. It is clear that for the iteration counts, the algorithm is not sensitive to the overlapping size δ. For fixed np, the number of average GMRES iterations decreases as the levels of fill-ins increases. The number of Newton iterations is

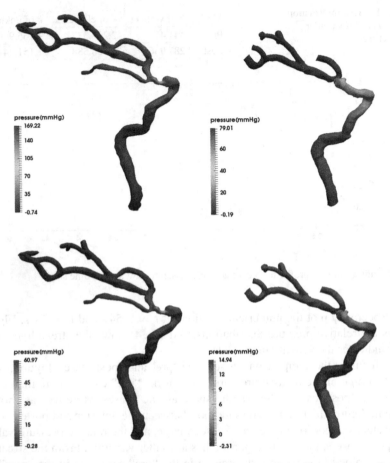

Fig. 3 Relative pressure distributions at $t = 2.54$ s (*top*) and $t = 3.2$ s (*bottom*) for pre (*left*) and post (*right*)

almost independent of the overlapping size for the Schwarz preconditioner and levels of fill-ins of subdomain solvers, and the average number of GMRES iterations increases slightly as the number of processor cores grows. Hence, we claim that NKS is quite robust for the test cases presented in this paper. For the best algorithmic parameter selection of ILU fill level 2, and small overlap of 0 or 1, about 70% relative efficiency is achieved in strong scaling between 32 and 128 processor cores.

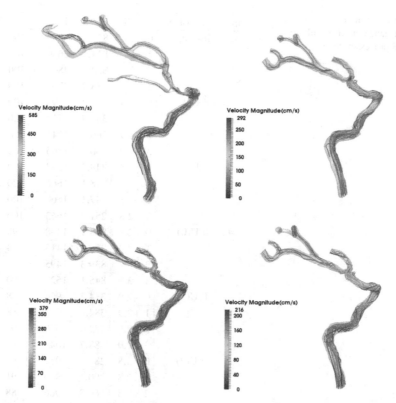

Fig. 4 Streamlines at $t = 2.54$ s (*top*) and $t = 3.2$ s (*bottom*) for pre (*left*) and post (*right*)

4 Concluding Remarks

We simulated blood flows in a pair of patient-specific cerebral arteries during 5 cardiac cycles by a fully implicit finite element discretization method and a Newton-Krylov-Schwarz algebraic solver. The simulations show clearly that the physics of the blood flow is more complicated before the surgery than after the surgery, and the stenosis causes a large variation of the pressure and velocity field. As to the NKS algorithm itself, we showed that the algorithm is robust with respect to the overlapping size for the Schwarz preconditioner and levels of fill-ins of subdomain solvers. A reasonably good scalability is observed with up to 128 processor cores.

Table 2 Parallel performance of NKS with up to 128 processor cores

np	Subsolver	δ	NI	LI	T	EFF (%)
32	ILU(0)	0	3	820.5	2860	100
		1	3	814.1	2650	100
		2	3	832.3	2805	100
		3	3	838.2	2761	100
	ILU(1)	0	2.9	351.8	1698	100
		1	2.9	351.9	1717	100
		2	2.9	360.7	1741	100
		3	2.9	366.5	1805	100
	ILU(2)	0	2.8	248.2	1563	100
		1	2.8	248.1	1666	100
		2	2.8	247.1	1600	100
		3	2.8	251.2	1663	100
64	ILU(0)	0	2.9	828.1	1438	99
		1	2.9	828.1	1413	94
		2	3	839.3	1495	94
		3	3	845.1	1527	90
	ILU(1)	0	2.9	384.2	966	88
		1	2.9	384.4	973	88
		2	2.9	372.0	970	90
		3	2.9	388.2	1042	87
	ILU(2)	0	2.8	289.5	931	84
		1	2.8	290.1	920	91
		2	2.8	266.3	906	88
		3	2.8	266.3	941	88
128	ILU(0)	0	3	842.9	845	85
		1	3	843.0	836	79
		2	3.6	876.5	1089	64
		3	3.9	914.0	1584	44
	ILU(1)	0	2.9	428.7	610	70
		1	2.9	428.2	617	70
		2	2.9	437.1	719	60
		3	2.9	443.1	932	48
	ILU(2)	0	2.8	324.8	570	69
		1	2.8	324.8	572	73
		2	2.8	300.9	583	69
		3	2.8	286.2	596	70

References

Y. Bazilevs, M.-C. Hsu, D.J. Benson, S. Sankaran, A.L. Marsden, Computational fluid–structure interaction: methods and application to a total cavopulmonary connection. Comput. Mech. **45**, 77–89 (2009)

L. Boussel, V. Rayz, A. Martin, G. Acevedo-Bolton, M.T. Lawton, R. Higashida, W.S. Smith, W.L. Young, D.S. Saloner, Phase-contrast MRI measurements in intracranial aneurysms in vivo of flow patterns, velocity fields, and wall shear stress: comparison with CFD. Magn. Reson. Med. **61**, 409–417 (2009)

J.R. Cebral, M.A. Castro, S. Appanaboyina, C.M. Putman, D. Millan, A.F. Frangi, Efficient pipeline for image–based patient–specific analysis of cerebral aneurysm hemodynamics: technique and sensitivity. IEEE Trans. Med. Imaging **24**, 457–467 (2005)

L.P. Franca, S.L. Frey, Stabilized finite element methods. II: the incompressible Navier–Stokes equations. Comput. Methods Appl. Mech. Eng. **99**, 209–233 (1992)

F.-N. Hwang, C.-Y. Wu, X.-C. Cai, Numerical simulation of three–dimensional blood flows using domain decomposition method on parallel computer. J. Chin. Soc. Mech. Eng. **31**, 199–208 (2010)

S.M. Moore, K.T. Moorhead, J.G. Chase, T. David, J. Fink, One–dimensional and three–dimensional model of cerebrovascular flow. Biochem. Eng. J. **127**, 440–449 (2005)

K. Takizawa, C. Moorman, S. Wright, J. Purdue, T. McPhail, P.P. Chen, J. Warren, T.E. Tezduyar, Patient–specific arterial fluid–structure interaction modeling of cerebral aneurysms. Int. J. Numer. Meth. Fluids **65**, 308–323 (2011)

C.A. Taylor, T.A. Fonte, J.K. Min, Computational fluid dynamics applied to cardiac computed tomography for noninvasive quantification of fracional flow reserve. J. Am. Coll. Cardiol. **61**, 2233–2241 (2013)

Y. Wu, X.-C. Cai, A fully implicit domain decomposition based ALE framework for three–dimensional fluid–structure interaction with application in blood flow computation. J. Comput. Phys. **258**, 524–537 (2014)

Editorial Policy

1. Volumes in the following three categories will be published in LNCSE:

i)   Research monographs
ii) Tutorials
iii) Conference proceedings

Those considering a book which might be suitable for the series are strongly advised to contact the publisher or the series editors at an early stage.

2. Categories i) and ii). Tutorials are lecture notes typically arising via summer schools or similar events, which are used to teach graduate students. These categories will be emphasized by Lecture Notes in Computational Science and Engineering. **Submissions by interdisciplinary teams of authors are encouraged.** The goal is to report new developments – quickly, informally, and in a way that will make them accessible to non-specialists. In the evaluation of submissions timeliness of the work is an important criterion. Texts should be well-rounded, well-written and reasonably self-contained. In most cases the work will contain results of others as well as those of the author(s). In each case the author(s) should provide sufficient motivation, examples, and applications. In this respect, Ph.D. theses will usually be deemed unsuitable for the Lecture Notes series. Proposals for volumes in these categories should be submitted either to one of the series editors or to Springer-Verlag, Heidelberg, and will be refereed. A provisional judgement on the acceptability of a project can be based on partial information about the work: a detailed outline describing the contents of each chapter, the estimated length, a bibliography, and one or two sample chapters – or a first draft. A final decision whether to accept will rest on an evaluation of the completed work which should include

–   at least 100 pages of text;
–   a table of contents;
–   an informative introduction perhaps with some historical remarks which should be accessible to readers unfamiliar with the topic treated;
–   a subject index.

3. Category iii). Conference proceedings will be considered for publication provided that they are both of exceptional interest and devoted to a single topic. One (or more) expert participants will act as the scientific editor(s) of the volume. They select the papers which are suitable for inclusion and have them individually refereed as for a journal. Papers not closely related to the central topic are to be excluded. Organizers should contact the Editor for CSE at Springer at the planning stage, see *Addresses* below.

In exceptional cases some other multi-author-volumes may be considered in this category.

4. Only works in English will be considered. For evaluation purposes, manuscripts may be submitted in print or electronic form, in the latter case, preferably as pdf- or zipped ps-files. Authors are requested to use the LaTeX style files available from Springer at http://www.springer.com/gp/authors-editors/book-authors-editors/manuscript-preparation/5636 (Click on LaTeX Template → monographs or contributed books).

For categories ii) and iii) we strongly recommend that all contributions in a volume be written in the same LaTeX version, preferably LaTeX2e. Electronic material can be included if appropriate. Please contact the publisher.

Careful preparation of the manuscripts will help keep production time short besides ensuring satisfactory appearance of the finished book in print and online.

5. The following terms and conditions hold. Categories i), ii) and iii):

Authors receive 50 free copies of their book. No royalty is paid.
Volume editors receive a total of 50 free copies of their volume to be shared with authors, but no royalties.

Authors and volume editors are entitled to a discount of 33.3 % on the price of Springer books purchased for their personal use, if ordering directly from Springer.

6. Springer secures the copyright for each volume.

Addresses:

Timothy J. Barth
NASA Ames Research Center
NAS Division
Moffett Field, CA 94035, USA
barth@nas.nasa.gov

Michael Griebel
Institut für Numerische Simulation
der Universität Bonn
Wegelerstr. 6
53115 Bonn, Germany
griebel@ins.uni-bonn.de

David E. Keyes
Mathematical and Computer Sciences
and Engineering
King Abdullah University of Science
and Technology
P.O. Box 55455
Jeddah 21534, Saudi Arabia
david.keyes@kaust.edu.sa

and

Department of Applied Physics
and Applied Mathematics
Columbia University
500 W. 120 th Street
New York, NY 10027, USA
kd2112@columbia.edu

Risto M. Nieminen
Department of Applied Physics
Aalto University School of Science
and Technology
00076 Aalto, Finland
risto.nieminen@aalto.fi

Dirk Roose
Department of Computer Science
Katholieke Universiteit Leuven
Celestijnenlaan 200A
3001 Leuven-Heverlee, Belgium
dirk.roose@cs.kuleuven.be

Tamar Schlick
Department of Chemistry
and Courant Institute
of Mathematical Sciences
New York University
251 Mercer Street
New York, NY 10012, USA
schlick@nyu.edu

Editor for Computational Science
and Engineering at Springer:
Martin Peters
Springer-Verlag
Mathematics Editorial IV
Tiergartenstrasse 17
69121 Heidelberg, Germany
martin.peters@springer.com

Lecture Notes
in Computational Science
and Engineering

24. T. Schlick, H.H. Gan (eds.), *Computational Methods for Macromolecules: Challenges and Applications.*

25. T.J. Barth, H. Deconinck (eds.), *Error Estimation and Adaptive Discretization Methods in Computational Fluid Dynamics.*

26. M. Griebel, M.A. Schweitzer (eds.), *Meshfree Methods for Partial Differential Equations.*

27. S. Müller, *Adaptive Multiscale Schemes for Conservation Laws.*

28. C. Carstensen, S. Funken, W. Hackbusch, R.H.W. Hoppe, P. Monk (eds.), *Computational Electromagnetics.*

29. M.A. Schweitzer, *A Parallel Multilevel Partition of Unity Method for Elliptic Partial Differential Equations.*

30. T. Biegler, O. Ghattas, M. Heinkenschloss, B. van Bloemen Waanders (eds.), *Large-Scale PDE-Constrained Optimization.*

31. M. Ainsworth, P. Davies, D. Duncan, P. Martin, B. Rynne (eds.), *Topics in Computational Wave Propagation.* Direct and Inverse Problems.

32. H. Emmerich, B. Nestler, M. Schreckenberg (eds.), *Interface and Transport Dynamics.* Computational Modelling.

33. H.P. Langtangen, A. Tveito (eds.), *Advanced Topics in Computational Partial Differential Equations.* Numerical Methods and Diffpack Programming.

34. V. John, *Large Eddy Simulation of Turbulent Incompressible Flows.* Analytical and Numerical Results for a Class of LES Models.

35. E. Bänsch (ed.), *Challenges in Scientific Computing - CISC 2002.*

36. B.N. Khoromskij, G. Wittum, *Numerical Solution of Elliptic Differential Equations by Reduction to the Interface.*

37. A. Iske, *Multiresolution Methods in Scattered Data Modelling.*

38. S.-I. Niculescu, K. Gu (eds.), *Advances in Time-Delay Systems.*

39. S. Attinger, P. Koumoutsakos (eds.), *Multiscale Modelling and Simulation.*

40. R. Kornhuber, R. Hoppe, J. Périaux, O. Pironneau, O. Wildlund, J. Xu (eds.), *Domain Decomposition Methods in Science and Engineering.*

41. T. Plewa, T. Linde, V.G. Weirs (eds.), *Adaptive Mesh Refinement – Theory and Applications.*

42. A. Schmidt, K.G. Siebert, *Design of Adaptive Finite Element Software.* The Finite Element Toolbox ALBERTA.

43. M. Griebel, M.A. Schweitzer (eds.), *Meshfree Methods for Partial Differential Equations II.*

44. B. Engquist, P. Lötstedt, O. Runborg (eds.), *Multiscale Methods in Science and Engineering.*

45. P. Benner, V. Mehrmann, D.C. Sorensen (eds.), *Dimension Reduction of Large-Scale Systems.*

46. D. Kressner, *Numerical Methods for General and Structured Eigenvalue Problems.*

47. A. Boriçi, A. Frommer, B. Joó, A. Kennedy, B. Pendleton (eds.), *QCD and Numerical Analysis III.*

48. F. Graziani (ed.), *Computational Methods in Transport.*

49. B. Leimkuhler, C. Chipot, R. Elber, A. Laaksonen, A. Mark, T. Schlick, C. Schütte, R. Skeel (eds.), *New Algorithms for Macromolecular Simulation.*

50. M. Bücker, G. Corliss, P. Hovland, U. Naumann, B. Norris (eds.), *Automatic Differentiation: Applications, Theory, and Implementations.*

51. A.M. Bruaset, A. Tveito (eds.), *Numerical Solution of Partial Differential Equations on Parallel Computers.*

52. K.H. Hoffmann, A. Meyer (eds.), *Parallel Algorithms and Cluster Computing.*

53. H.-J. Bungartz, M. Schäfer (eds.), *Fluid-Structure Interaction.*

54. J. Behrens, *Adaptive Atmospheric Modeling.*

55. O. Widlund, D. Keyes (eds.), *Domain Decomposition Methods in Science and Engineering XVI.*

56. S. Kassinos, C. Langer, G. Iaccarino, P. Moin (eds.), *Complex Effects in Large Eddy Simulations.*

57. M. Griebel, M.A Schweitzer (eds.), *Meshfree Methods for Partial Differential Equations III.*

58. A.N. Gorban, B. Kégl, D.C. Wunsch, A. Zinovyev (eds.), *Principal Manifolds for Data Visualization and Dimension Reduction.*

59. H. Ammari (ed.), *Modeling and Computations in Electromagnetics: A Volume Dedicated to Jean-Claude Nédélec.*

60. U. Langer, M. Discacciati, D. Keyes, O. Widlund, W. Zulehner (eds.), *Domain Decomposition Methods in Science and Engineering XVII.*

61. T. Mathew, *Domain Decomposition Methods for the Numerical Solution of Partial Differential Equations.*

62. F. Graziani (ed.), *Computational Methods in Transport: Verification and Validation.*

63. M. Bebendorf, *Hierarchical Matrices. A Means to Efficiently Solve Elliptic Boundary Value Problems.*

64. C.H. Bischof, H.M. Bücker, P. Hovland, U. Naumann, J. Utke (eds.), *Advances in Automatic Differentiation.*

65. M. Griebel, M.A. Schweitzer (eds.), *Meshfree Methods for Partial Differential Equations IV.*

66. B. Engquist, P. Lötstedt, O. Runborg (eds.), *Multiscale Modeling and Simulation in Science.*

67. I.H. Tuncer, Ü. Gülcat, D.R. Emerson, K. Matsuno (eds.), *Parallel Computational Fluid Dynamics 2007.*

68. S. Yip, T. Diaz de la Rubia (eds.), *Scientific Modeling and Simulations.*

69. A. Hegarty, N. Kopteva, E. O'Riordan, M. Stynes (eds.), *BAIL 2008 – Boundary and Interior Layers.*

70. M. Bercovier, M.J. Gander, R. Kornhuber, O. Widlund (eds.), *Domain Decomposition Methods in Science and Engineering XVIII.*

71. B. Koren, C. Vuik (eds.), *Advanced Computational Methods in Science and Engineering.*

72. M. Peters (ed.), *Computational Fluid Dynamics for Sport Simulation.*

73. H.-J. Bungartz, M. Mehl, M. Schäfer (eds.), *Fluid Structure Interaction II - Modelling, Simulation, Optimization.*

74. D. Tromeur-Dervout, G. Brenner, D.R. Emerson, J. Erhel (eds.), *Parallel Computational Fluid Dynamics 2008.*

75. A.N. Gorban, D. Roose (eds.), *Coping with Complexity: Model Reduction and Data Analysis.*

76. J.S. Hesthaven, E.M. Rønquist (eds.), *Spectral and High Order Methods for Partial Differential Equations.*

77. M. Holtz, *Sparse Grid Quadrature in High Dimensions with Applications in Finance and Insurance.*

78. Y. Huang, R. Kornhuber, O.Widlund, J. Xu (eds.), *Domain Decomposition Methods in Science and Engineering XIX.*

79. M. Griebel, M.A. Schweitzer (eds.), *Meshfree Methods for Partial Differential Equations V.*

80. P.H. Lauritzen, C. Jablonowski, M.A. Taylor, R.D. Nair (eds.), *Numerical Techniques for Global Atmospheric Models.*

81. C. Clavero, J.L. Gracia, F.J. Lisbona (eds.), *BAIL 2010 – Boundary and Interior Layers, Computational and Asymptotic Methods.*

82. B. Engquist, O. Runborg, Y.R. Tsai (eds.), *Numerical Analysis and Multiscale Computations.*

83. I.G. Graham, T.Y. Hou, O. Lakkis, R. Scheichl (eds.), *Numerical Analysis of Multiscale Problems.*

84. A. Logg, K.-A. Mardal, G. Wells (eds.), *Automated Solution of Differential Equations by the Finite Element Method.*

85. J. Blowey, M. Jensen (eds.), *Frontiers in Numerical Analysis - Durham 2010.*

86. O. Kolditz, U.-J. Gorke, H. Shao, W. Wang (eds.), *Thermo-Hydro-Mechanical-Chemical Processes in Fractured Porous Media - Benchmarks and Examples.*

87. S. Forth, P. Hovland, E. Phipps, J. Utke, A. Walther (eds.), *Recent Advances in Algorithmic Differentiation.*

88. J. Garcke, M. Griebel (eds.), *Sparse Grids and Applications.*

89. M. Griebel, M.A. Schweitzer (eds.), *Meshfree Methods for Partial Differential Equations VI.*

90. C. Pechstein, *Finite and Boundary Element Tearing and Interconnecting Solvers for Multiscale Problems.*

91. R. Bank, M. Holst, O. Widlund, J. Xu (eds.), *Domain Decomposition Methods in Science and Engineering XX.*

92. H. Bijl, D. Lucor, S. Mishra, C. Schwab (eds.), *Uncertainty Quantification in Computational Fluid Dynamics.*

93. M. Bader, H.-J. Bungartz, T. Weinzierl (eds.), *Advanced Computing.*

94. M. Ehrhardt, T. Koprucki (eds.), *Advanced Mathematical Models and Numerical Techniques for Multi-Band Effective Mass Approximations.*

95. M. Azaïez, H. El Fekih, J.S. Hesthaven (eds.), *Spectral and High Order Methods for Partial Differential Equations ICOSAHOM 2012.*

96. F. Graziani, M.P. Desjarlais, R. Redmer, S.B. Trickey (eds.), *Frontiers and Challenges in Warm Dense Matter.*

97. J. Garcke, D. Pflüger (eds.), *Sparse Grids and Applications – Munich 2012.*

98. J. Erhel, M. Gander, L. Halpern, G. Pichot, T. Sassi, O. Widlund (eds.), *Domain Decomposition Methods in Science and Engineering XXI.*

99. R. Abgrall, H. Beaugendre, P.M. Congedo, C. Dobrzynski, V. Perrier, M. Ricchiuto (eds.), *High Order Nonlinear Numerical Methods for Evolutionary PDEs - HONOM 2013.*

100. M. Griebel, M.A. Schweitzer (eds.), *Meshfree Methods for Partial Differential Equations VII.*

101. R. Hoppe (ed.), *Optimization with PDE Constraints - OPTPDE 2014.*

102. S. Dahlke, W. Dahmen, M. Griebel, W. Hackbusch, K. Ritter, R. Schneider, C. Schwab, H. Yserentant (eds.), *Extraction of Quantifiable Information from Complex Systems.*

103. A. Abdulle, S. Deparis, D. Kressner, F. Nobile, M. Picasso (eds.), *Numerical Mathematics and Advanced Applications - ENUMATH 2013.*

104. T. Dickopf, M.J. Gander, L. Halpern, R. Krause, L.F. Pavarino (eds.), *Domain Decomposition Methods in Science and Engineering XXII.*

105. M. Mehl, M. Bischoff, M. Schäfer (eds.), *Recent Trends in Computational Engineering - CE2014. Optimization, Uncertainty, Parallel Algorithms, Coupled and Complex Problems.*

106. R.M. Kirby, M. Berzins, J.S. Hesthaven (eds.), *Spectral and High Order Methods for Partial Differential Equations - ICOSAHOM'14.*

107. B. Jüttler, B. Simeon (eds.), *Isogeometric Analysis and Applications 2014.*

108. P. Knobloch (ed.), *Boundary and Interior Layers, Computational and Asymptotic Methods – BAIL 2014.*

109. J. Garcke, D. Pflüger (eds.), *Sparse Grids and Applications – Stuttgart 2014.*

110. H. P. Langtangen, *Finite Difference Computing with Exponential Decay Models.*

111. A. Tveito, G.T. Lines, *Computing Characterizations of Drugs for Ion Channels and Receptors Using Markov Models.*

112. B. Karazösen, M. Manguoğlu, M. Tezer-Sezgin, S. Göktepe, Ö. Uğur (eds.), *Numerical Mathematics and Advanced Applications - ENUMATH 2015.*

113. H.-J. Bungartz, P. Neumann, W.E. Nagel (eds.), *Software for Exascale Computing - SPPEXA 2013-2015.*

114. G.R. Barrenechea, F. Brezzi, A. Cangiani, E.H. Georgoulis (eds.), *Building Bridges: Connections and Challenges in Modern Approaches to Numerical Partial Differential Equations.*

115. M. Griebel, M.A. Schweitzer (eds.), *Meshfree Methods for Partial Differential Equations VIII.*

116. C.-O. Lee, X.-C. Cai, D. E. Keyes, H.H. Kim, A. Klawonn, E.-J. Park, O.B. Widlund (eds.), *Domain Decomposition Methods in Science and Engineering XXIII.*

For further information on these books please have a look at our mathematics catalogue at the following URL: www.springer.com/series/3527

Monographs in Computational Science and Engineering

1. J. Sundnes, G.T. Lines, X. Cai, B.F. Nielsen, K.-A. Mardal, A. Tveito, *Computing the Electrical Activity in the Heart.*

For further information on this book, please have a look at our mathematics catalogue at the following URL: www.springer.com/series/7417

Texts in Computational Science and Engineering

1. H. P. Langtangen, *Computational Partial Differential Equations.* Numerical Methods and Diffpack Programming. 2nd Edition

2. A. Quarteroni, F. Saleri, P. Gervasio, *Scientific Computing with MATLAB and Octave.* 4th Edition

3. H. P. Langtangen, *Python Scripting for Computational Science.* 3rd Edition

4. H. Gardner, G. Manduchi, *Design Patterns for e-Science.*

5. M. Griebel, S. Knapek, G. Zumbusch, *Numerical Simulation in Molecular Dynamics.*

6. H. P. Langtangen, *A Primer on Scientific Programming with Python.* 5th Edition

7. A. Tveito, H. P. Langtangen, B. F. Nielsen, X. Cai, *Elements of Scientific Computing.*

8. B. Gustafsson, *Fundamentals of Scientific Computing.*

9. M. Bader, *Space-Filling Curves.*

10. M. Larson, F. Bengzon, *The Finite Element Method: Theory, Implementation and Applications.*

11. W. Gander, M. Gander, F. Kwok, *Scientific Computing: An Introduction using Maple and MATLAB.*

12. P. Deuflhard, S. Röblitz, *A Guide to Numerical Modelling in Systems Biology.*

13. M. H. Holmes, *Introduction to Scientific Computing and Data Analysis.*

14. S. Linge, H. P. Langtangen, *Programming for Computations - A Gentle Introduction to Numerical Simulations with MATLAB/Octave.*

15. S. Linge, H. P. Langtangen, *Programming for Computations - A Gentle Introduction to Numerical Simulations with Python.*

For further information on these books please have a look at our mathematics catalogue at the following URL: www.springer.com/series/5151

Printed in the United States
By Bookmasters